Lecture Notes in Computer Science 3109

Commenced Publication in 1973
Founding and Former Series Editors:
Gerhard Goos, Juris Hartmanis, and Jan van Leeuwen

Springer
Berlin
Heidelberg
New York
Hong Kong
London
Milan
Paris
Tokyo

Suleyman Cenk Sahinalp   S. Muthukrishnan
Ugur Dogrusoz (Eds.)

Combinatorial Pattern Matching

15th Annual Symposium, CPM 2004
Istanbul, Turkey, July 5-7, 2004
Proceedings

Springer

Volume Editors

Suleyman Cenk Sahinalp
Simon Fraser University, School of Computing Science
Burnaby BC, V5A 1S6 Canada
E-mail: cenk@cs.sfu.ca

S. Muthukrishnan
Rutgers University, Department of Computer and Information Sciences
319 Core Bldg, 110 Frelinghuysen Rd, Piscataway, NJ 08854, USA
E-mail: muthu@cs.rutgers.edu

Ugur Dogrusoz
Bilkent University, Computer Engineering Department
06800 Ankara, Turkey
E-mail: ugur@cs.bilkent.edu.tr

Library of Congress Control Number: 2004108252

CR Subject Classification (1998): F.2.2, I.5.4, I.5.0, I.7.3, H.3.3, E.4, G.2.1, E.1

ISSN 0302-9743
ISBN 3-540-22341-X Springer-Verlag Berlin Heidelberg New York

Springer-Verlag is a part of Springer Science+Business Media

springeronline.com

© Springer-Verlag Berlin Heidelberg 2004
Printed in Germany

Typesetting: Camera-ready by author, data conversion by Olgun Computergrafik
Printed on acid-free paper   SPIN: 11014331   06/3142   5 4 3 2 1 0

Preface

The 15th Annual Symposium on Combinatorial Pattern Matching was held in Ciragan Palace Hotel, Istanbul, Turkey during July 5–7, 2004. CPM 2004 repeated the success of its predecessors; it even surpassed them in terms of the number of invited speakers, the number of submissions, and the number of papers accepted and presented at the conference.

In response to the call for papers, CPM 2004 received a record number of 79 high-quality submissions. Each submission was reviewed by at least three program committee members and the comments were returned to the authors. Following an extensive electronic discussion period, the Program Committee accepted 36 of the submissions to be presented at the conference. They constitute original research contributions in combinatorial pattern matching algorithms and data structures, molecular sequence analysis, phylogenetic tree construction, and RNA and protein structure analysis and prediction.

CPM 2004 had five invited speakers. In alphabetical order they were: Evan Eichler from the University of Washington, USA, Martin Farach-Colton from Rutgers University, USA, Paolo Ferragina from the University of Pisa, Italy, Piotr Indyk from MIT, USA, and Gene Myers from the University of California, Berkeley, USA.

It is impossible to organize such a successful program without the help of many individuals. We would like to express our appreciation to the authors of the submitted papers and to the program committee members and external referees, who provided timely and significant reviews.

July 2004

S.C. Sahinalp,
S. Muthukrishnan,
U. Dogrusoz

Organization

CPM 2004 was locally organized by Bilkent University, Ankara, Turkey. Within Bilkent University, the Center for Bioinformatics (BCBI) and the Computer Engineering Department cooperated.

Executive Committee

Organization Chair	Ugur Dogrusoz (Bilkent University)
Program Chairs	Suleyman Cenk Sahinalp
	(Simon Fraser University)
	S. Muthukrishnan
	(Rutgers Universtiy and AT&T)
Student Volunteer Chairs	Can Alkan (Simon Fraser University)
	Asli Ayaz (Bilkent University)
	Ozgun Babur (Bilkent University)
	Emek Demir (Bilkent University)
Social Events	Tasmanlar Tourism, Ankara, Turkey
Administrative Support	Gurkan Bebek (Case Western Reserve University)
	Emek Demir (Bilkent University)

Program Committee

Gerth Stølting Brodal	University of Aarhus, Denmark
Jeremy Buhler	Washington University, USA
Ugur Dogrusoz	Bilkent University, Turkey
Zvi Galil	Columbia University, USA
Ramesh Hariharan	Indian Institute of Science, India
Ming Li	University of Waterloo, Canada
Stefano Lonardi	University of California, Riverside, USA
Ian Munro	University of Waterloo, Canada
Craig Neville Manning	Google Inc., USA
S. Muthukrishnan	Rutgers University and AT&T, USA
Joseph Nadeau	Case Western Reserve University, USA
Meral Ozsoyoglu	Case Western Reserve University, USA
Ely Porat	Bar-Ilan University, Israel
Kunihiko Sadakane	Kyushu University, Japan
Suleyman Cenk Sahinalp	Simon Fraser University, Canada
Mona Singh	Princeton University, USA

Steering Committee

Alberto Apostolico	Purdue University, USA and University of Padua, Italy
Maxime Crochemore	University of Marne la Vallée, France
Zvi Galil	Columbia University, USA
Udi Manber	A9.com, USA

Sponsoring Organizations

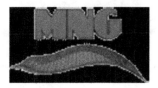 

External Referees

Can Alkan
Abdullah Arslan
Asli Ayaz
Kensuke BaBa
Ozgun Babur
Gary Benson
Petra Berenbrink
Bernard Chazelle
Tim Ting Chen
Richard Cole
Livio Colussi
Maxime Crochemore
Sergio De Agostino
Emek Demir
Zeynep Erson
Kimmo Fredriksson
Leszek Gasieniec
Erhan Giral
Danny Hermelin
Lucian Ilie
Tao Jiang
Haim Kaplan
Emre Karakoc
Carmel Kent
Ali Reza Khodabakhshi
Takuya Kida
Carl Kingsford
Moshe Koppel
Stefan Kurtz
Jesper Larsson

Moshe Lewenstein
Guohui Lin
Hao Lin
Bin Ma
Veli Makinen
Yishay Mansour
Giovanni Manzini
Gabriel Moruz
Milan Mosny
Giovanni Motta
Giulio Pavesi
Frederic Pio
Teresa Przytycka
Mathieu Raffinot
Francesco Rizzo
Igor B. Rogozin
Wojciech Rytter
Anoop Sarkar
Nira Shafrir
Dana Shapira
Tetsuo Shibuya
Dina Sokol
Masayuki Takeda
Andrey Utis
Anil Vullikanti
Oren Weimann
Kaizhong Zhang
Qiangfeng Zhang
Zefeng Zhang
Jie Zheng

Table of Contents

Sorting by Reversals in Subquadratic Time[*]

Eric Tannier[1] and Marie-France Sagot[1,2]

[1] INRIA Rhône-Alpes, Laboratoire de Biométrie et Biologie Évolutive
Université Claude Bernard, 69622 Villeurbanne cedex, France
{Eric.Tannier,Marie-France.Sagot}@inrialpes.fr
[2] King's College, London, UK

Abstract. The problem of sorting a signed permutation by reversals is inspired by genome rearrangements in computational molecular biology. Given two genomes represented as two signed permutations of the same elements (*e.g. orthologous genes*), the problem consists in finding a most parsimonious scenario of reversals that transforms one genome into the other. We propose a method for sorting a signed permutation by reversals in time $O(n\sqrt{n \log n})$. The best known algorithms run in time $O(n^2)$, the main obstacle to an improvement being a costly operation of detection of so-called "safe" reversals. We bypass this detection and, using the same data structure as a previous random approximation algorithm, we achieve the same subquadratic complexity for finding an *exact* optimal solution. This answers an open question by Ozery-Flato and Shamir whether a subquadratic complexity could ever be achieved for solving the problem.

1 Introduction

The problem of sorting a signed permutation by reversals is inspired by a problem of genome rearrangement in computational biology. Genome rearrangements such as reversals may change the order of the genes (or of other markers) in a genome, and also the direction of transcription. We identify the genes with the numbers $1, \ldots, n$, with a plus or minus sign to indicate such direction. Their order will be represented by a *signed permutation* of $\{\pm 1, \ldots, \pm n\}$, that is a permutation π of $\{\pm 1, \ldots, \pm n\}$ such that $\pi[-i] = -\pi[i]$, where $\pi[i]$ denotes the ith element in π. In the following, we indicate the sign of an element in a permutation only when it is minus.

The *reversal* of the interval $[i, j] \subseteq [1, n]$ $(i < j)$ is the signed permutation $\rho = 1, \ldots, i, -j, \ldots, -(i+1), j+1, \ldots, n$. Note that $\pi\rho$ is the permutation obtained from π by reversing the order and flipping the signs of the elements in the interval. If $\rho_1, \ldots, \rho_k$ is a sequence of reversals, we say that it sorts a permutation π if $\pi\rho_1 \cdots \rho_k = Id$ (Id is the all positive identity permutation $1, \ldots, n$). We denote by $d(\pi)$ the number of reversals in a minimum size sequence sorting π.

[*] Work supported by the French program bioinformatique Inter-EPST 2002 "Algorithms for Modelling and Inference Problems in Molecular Biology".

S.C. Sahinalp et al. (Eds.): CPM 2004, LNCS 3109, pp. 1–13, 2004.
© Springer-Verlag Berlin Heidelberg 2004

The problem of sorting by reversals has been the subject of an extensive literature. For a complete survey, see [2]. The first polynomial algorithm was given by Hannenhalli and Pevzner [4], and ran in $O(n^4)$. After many subsequent improvements, the currently fastest algorithms are those of Kaplan, Shamir and Tarjan [5] running in $O(n^2)$, a linear algorithm for computing $d(\pi)$ only (it does not give the sequence of reversals) by Bader, Moret and Yan [1], and an $O(n\sqrt{n\log n})$ random algorithm by Kaplan and Verbin [6] which gives most of the time an optimal sequence of reversals, but fails on some permutations with very high probability. A reversal ρ is said to be *safe* for a permutation π if $d(\pi\rho) = d(\pi)-1$. The bottleneck for all existing exact algorithms is the detection of safe reversals. Many techniques were invented to address this problem, but none has a better time complexity than linear, immediately implying a quadratic complexity for the whole method. In a recent paper [7], Ozery-Flato and Shamir compiled and compared the best algorithms, and wrote that: "A central question in the study of genome rearrangements is whether one can obtain a subquadratic algorithm for sorting by reversals".

In this paper, we give a positive answer to Ozery-Flato and Shamir's question. A good knowledge of the Hannenhalli-Pevzner theory and of the data structure used by Kaplan-Verbin is assumed.

In the next section, we briefly describe the usual tools (in particular the omnipresent breakpoint graph) for dealing with permutations and reversals. We mention some operations for sorting by reversals without giving any details (concerning, for instance, hurdle detection and clearing for which linear methods exist). The aim of this paper is to deal only with the most costly part of the method, that is sorting a permutation without hurdles. We introduce a new and elegant way of transforming the breakpoint graph of a permutation π by applying reversals either on π or on its *inverse permutation* π^{-1}. In section 3, we describe the method to optimally sort by reversals. With any classical data structure, the time complexity of this algorithm is $O(n^2)$, but even in this case it presents a special interest because it bypasses the detection of safe reversals which is considered as the most costly operation. Then in the last section, we indicate the data structure used to achieve subquadratic time complexity.

2   Mathematical Tools

To simplify exposition, we require that the first and last elements of a permutation remain unchanged. We therefore adopt the usual transformation which consists in adding the elements 0 and $n + 1$ to $\{1, \ldots, n\}$, with $\pi[0] = 0$, and $\pi[n+1] = n+1$. The obtained permutation is called an *augmented permutation*. In this paper, all permutations are augmented, and we omit to mention it from now on. The inverse permutation π^{-1} of π is the (signed) permutation such that $\pi\pi^{-1} = Id$.

2.1   Breakpoint Graphs for a Permutation and Its Inverse

The *breakpoint graph* $BG(\pi)$ of a permutation π is a graph with vertex set V defined as follows: for each integer i in $\{1, \ldots, n\}$, let i^a (the *arrival*) and i^d

(the *departure*) be two vertices in V; add to V the two vertices 0^d and $(n+1)^a$. Observe that all vertex labels are non negative numbers. For simplicity and to avoid having to use absolute values, we may later refer to vertex $(-i)^x$ (for $x = a$ or d). This will be the same as vertex i^x. The edge set E of $BG(\pi)$ is the union of two perfect matchings denoted by R, the *reality edges* and D, the *dream edges* (in the literature, reality and dream edges are sometimes called reality and desire edges, or, in a more prosaic way, black and gray edges):

- R contains the edges $(\pi[i])^d(\pi[i+1])^a$ for all $i \in \{0, \ldots, n\}$;
- D contains an edge for all $i \in \{0, \ldots, n\}$, from i^d if $\pi^{-1}[i]$ is positive, from i^a if $\pi^{-1}[i]$ is negative, to $(i+1)^a$ if $\pi^{-1}[i+1]$ is positive, and to $(i+1)^d$ if $\pi^{-1}[i+1]$ is negative.

The reality edges define the permutation π (what you have), and the dream edges define Id (what you want to have). Reality edges always go from a departure to an arrival, but in dreams, everything can happen. An example of a breakpoint graph for a permutation is given in Figure 1.

To avoid case checking, in the notation of an edge, the mention of departures and arrivals may be omitted. For instance, $i(i+1)$ is a dream edge, indicating nothing as concerns the signs of $\pi^{-1}[i]$ and $\pi^{-1}[i+1]$.

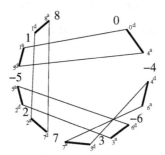

Fig. 1. The breakpoint graph of the permutation $0 - 4 - 6\ 3\ 7\ 2 - 5\ 1\ 8$. The bold edges are reality edges, and the thin ones are dream edges. The permutation should be read clockwise from 0 to $n+1$. This circular representation makes the cycles in the graph more visible. The edges that cross in the drawing correspond to crossing edges according to the definition.

A reality edge $(\pi[i])^d(\pi[i+1])^a$ is *oriented* if $\pi[i]$ and $\pi[i+1]$ have opposite signs, and *unoriented* otherwise. A dream edge $i(i+1)$ is *oriented* if $\pi^{-1}[i]$ and $\pi^{-1}[i+1]$ have opposite signs (that is, if the edge joins two departures or two arrivals), and *unoriented* otherwise. In the example of Figure 1, $(0,4)$, $(6,3)$, $(2,5)$ and $(5,1)$ are oriented reality edges, while $(3,4)$, $(6,7)$ are oriented dream edges.

To every dream edge $i(i+1)$, we associate the interval $[|\pi^{-1}[i]|, |\pi^{-1}[i+1]|]$ (or $[|\pi^{-1}[i+1]|, |\pi^{-1}[i]|]$ if $|\pi^{-1}[i]| > |\pi^{-1}[i+1]|$). Two dream edges are said to *cross* if their associated intervals intersect but one is not contained in the other. Only dream edges may cross in a breakpoint graph.

Dream and reality edges are trivially and uniquely decomposed into cycles (the sets of both types of edges are perfect matchings of the vertices). By the cycles of a permutation π, we mean the cycles of $R \cup D$ in $BG(\pi)$. We call the *size* of a cycle the number of dream edges it contains (it is half the usual length of a cycle). Two cycles are said to *cross* if two of their edges cross.

A *component* $\mathcal{C}$ of $BG(\pi)$ is an inclusionwise minimal subset of its cycles, such that no cycle of $\mathcal{C}$ crosses a cycle outside $\mathcal{C}$. A component is said to be *oriented* if it contains a cycle with an oriented edge, and *unoriented* otherwise. A *hurdle* is a special type of unoriented component. We do not define it more precisely, since we deal only with permutations without unoriented components, therefore without hurdles. See for example [5] for a complete description of what a hurdle is, and how to cope with hurdles when there are some. In the example of Figure 1, there is a single oriented component.

The following operation establishes the correspondence between dream and reality in $BG(\pi)$ and $BG(\pi^{-1})$. Let $(BG(\pi))^{-1}$ be the graph resulting from applying the following transformations to $BG(\pi)$:

- change each vertex label i^a into i^d and i^d into i^a whenever $\pi^{-1}[i]$ is negative;
- change each vertex label $(\pi[i])^a$ into i^a, and $(\pi[i])^d$ into i^d;
- change dream into reality and reality into dream.

The result of such a transformation applied to the example of Figure 1 is given in Figure 2.

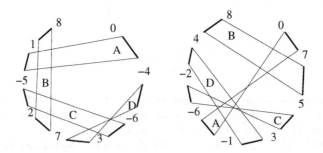

Fig. 2. A breakpoint graph and its inverse. The correspondence of the cycles in shown.

Lemma 1. $(BG(\pi))^{-1} = BG(\pi^{-1})$.

Proof. By definition, $(BG(\pi))^{-1}$ and $BG(\pi^{-1})$ have the same vertex set. There is a reality edge $(\pi[i])^d(\pi[i+1])^a$ in $BG(\pi)$ for all $i \in \{0, \ldots, n\}$. In $(BG(\pi))^{-1}$, it becomes a dream edge from i^d if $\pi[i]$ is positive or from i^a if $\pi[i]$ is negative, to $(i+1)^d$ if $\pi[i+1]$ is positive or to $(i+1)^a$ if $\pi[i+1]$ is negative. This corresponds exactly to the definition of a dream edge of $BG(\pi^{-1})$. Furthermore, there is a dream edge $i(i+1)$ in $BG(\pi)$ for all $i \in \{0, \ldots, n\}$. In $(BG(\pi))^{-1}$, it becomes a reality edge from $(\pi^{-1}[i])^d$ to $(\pi^{-1}[i+1])^a$.

Therefore, $(BG(\pi))^{-1}$ and $BG(\pi^{-1})$ have the same sets of dream and reality edges.   □

Observe that as a consequence, a cycle of π is also a cycle of π^{-1}.

2.2   Sorting Simple Permutations

A *simple permutation* is a permutation whose breakpoint graph contains only cycles of size 1 and 2 (called 1-cycles and 2-cycles). When dealing with simple permutations, we use "cycle" to mean 2-cycle, and "adjacency" to mean 1-cycle. Simple permutations are worth studying because of the following result.

Lemma 2. *[4] For any permutation π, there is a simple permutation π' such that $d(\pi) = d(\pi')$, and it is possible to deduce an optimal solution S for sorting π from any optimal solution S' for sorting π'. Transformations from π to π', and from S' to S are achievable in linear time.*

Let π be a simple permutation. The number of (2-)cycles is denoted by $c(\pi)$. By Lemma 1, $c(\pi) = c(\pi^{-1})$. The following is an easy but useful remark.

Lemma 3. *For any reversal ρ, $\rho = \rho^{-1}$, and if $\pi\rho_1 \cdots \rho_k = Id$, where $\rho_1, \ldots, \rho_k$ are reversals, then $\pi^{-1} = \rho_1 \cdots \rho_k$, $\pi^{-1}\rho_k \cdots \rho_1 = Id$, and $\rho_1 \cdots \rho_k\pi = Id$.*

Hannenhalli and Pevzner [4] proved the following fundamental result, which is the basis of the theory for sorting by reversals. We restrict ourselves to permutations without unoriented components, because the best algorithms all begin with a procedure to clear unoriented components in linear time, and we have nothing to add to this procedure. Therefore, we suppose unoriented components have already been cleared.

Lemma 4. *[4] If π is a simple permutation without unoriented components, $d(\pi) = c(\pi)$.*

This means that any optimal solution has to decrease the number of cycles by one at each step. The effect of the reversal of an interval $[i, j]$ on a breakpoint graph is to delete the two reality edges $\pi[i]\pi[i+1]$ and $\pi[j]\pi[j+1]$, and to replace them by two new reality edges $\pi[i](-\pi[j])$ and $(-\pi[i+1])\pi[j+1]$. The only way to decrease the number of cycles is to apply an inversion on the two reality edges of a unique cycle, and to replace them by two edges parallel to the two dream edges of the cycle, thus creating two adjacencies. In consequence, any reversal of an optimal sequence is associated with a cycle (the one it breaks). The set of reversals of any optimal sequence is therefore in bijection with the set of cycles of π. In the sequel, we use the same notation for reversals and cycles of π. For instance, we consider the permutation $\pi\rho$, for ρ a cycle of π. This means that if the two reality edges of ρ are $\pi[i]\pi[i+1]$ and $\pi[j]\pi[j+1]$, then the associated reversal ρ is the reversal of the interval $[i, j]$. We see in the next section the conditions for a reversal associated with a cycle to effectively create two adjacencies.

2.3   Contraction and Removal of Cycles

In what follows, permutations are always considered to be both without unoriented components and simple (the appropriate linear time tranformations are described in [3–5, 7]). Two edges are said to have the same orientation if they are both oriented or both unoriented.

Lemma 5. *In any cycle of a permutation, the two dream edges have the same orientation, and the two reality edges have the same orientation.*

Proof. Suppose two dream edges e, f of a same cycle had a different orientation, say e joins two departures, and f joins a departure and an arrival. The cycle would therefore have three departures and one arrival, which is impossible since reality edges always join one departure and one arrival. All the similar and dual cases are treated in the same way.   □

A cycle is said to be *contractible* if its dream edges are oriented. In the example of Figure 1, cycle D is contractible as both $(3, 4)$ and $(6, 7)$ are oriented dream edges.

A cycle is said to be *removable* if its reality edges are oriented. In the example of Figure 1, cycles A and C are removable because, respectively, $(0, 4)$ and $(5, 1)$, $(6, 3)$ and $(2, 5)$, are oriented reality edges. Cycle B is neither removable nor contractible as none of its edges is oriented.

It is straightforward from the definitions and from Lemma 1 that a cycle ρ is removable in π if and only if it is contractible in π^{-1}. Indeed, if a reality edge is oriented in $BG(\pi)$, it becomes an oriented dream edge in $BG(\pi)^{-1} = BG(\pi^{-1})$, and vice-versa. The following lemma is another important result in the theory of sorting by reversals.

Lemma 6. *[4] A cycle ρ of π is contractible if and only if $c(\pi\rho) = c(\pi) - 1$.*

Observe that this does not mean that $d(\pi\rho) = d(\pi) - 1$, because $\pi\rho$ may have unoriented components, and in this lies all the difficulty of sorting by reversals. From this and from Lemmas 1 and 3, we deduce immediately:

Lemma 7. *A cycle ρ of π is removable if and only if $c(\rho\pi) = c(\pi) - 1$.*

Let ρ be a cycle of π, with reality edges $(\pi[i])^d(\pi[i+1])^a$ and $(\pi[j])^d(\pi[j+1])^a$ (suppose $i < j$). Let $BG(\pi)/\rho$ be the graph obtained from $BG(\pi)$ by:

1. deleting the two reality edges of ρ, and replacing them by two edges parallel to the two dream edges of ρ;
2. changing the vertex label $(\pi[r])^a$ into $(\pi[r])^d$ and the vertex label $(\pi[r])^d$ into $(\pi[r])^a$ for every $r \in [i+1, j]$.

We call this operation the *contraction* of ρ in $BG(\pi)$.

Lemma 8. *[4] If a cycle ρ of a permutation π is contractible, then $BG(\pi)/\rho = BG(\pi\rho)$.*

It is possible to write a similar "dual" lemma for removability. Let $BG(\pi)\backslash\rho = (BG(\pi)^{-1}/\rho)^{-1}$. We call this operation the *removal* of ρ in $BG(\pi)$. It is clear that it consists in deleting the two dream edges of ρ, replacing them by two edges parallel to the two reality edges of ρ (thus obtaining two adjacencies), and changing the labels r of the vertices for every $r \in [|\pi[i+1]|, |\pi[j]|]$.

Lemma 9. *If a cycle ρ of a permutation π is removable, then $BG(\pi) \backslash \rho = BG(\rho\pi)$.*

The proof is straightforward from the previous Lemma together with Lemma 1. Contracting a contractible cycle in a breakpoint graph corresponds therefore exactly to applying the corresponding reversal to the permutation. In a similar way, removing a removable cycle corresponds to applying the corresponding reversal to the inverse of the permutation. In other words, contracting a cycle is changing the reality to make it closer to your dreams, and removing a cycle is changing your dreams to make them closer to the reality.

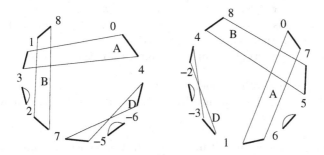

Fig. 3. Removal of a cycle in the permutation of Figure 2, and contraction of the same cycle in the inverse permutation.

The following facts about contraction and removal can easily be verified.

Lemma 10. *Let ρ, ρ' be two distinct cycles in π. If ρ is contractible, then ρ' is removable in π if and only if it is removable in $\pi\rho$. Conversely, if ρ is removable, ρ' is contractible in π if and only if it is contractible in $\rho\pi$. Moreover, if ρ is removable in π, then two edges distinct from the edges of ρ cross in π if and only if they cross in $\rho\pi$.*

Proof. Let $(\pi[r])^d(\pi[r+1])^a$ be a reality edge of ρ' in π. If r is in the interval of the reversal ρ, then $(\pi[r+1])^d(\pi[r])^a$ is a reality edge in $\pi\rho$, and both $\pi[r]$ and $\pi[r+1]$ changed sign. Otherwise, $(\pi[r])^d(\pi[r+1])^a$ is still a reality edge in $\pi\rho$, and $\pi[r]$ and $\pi[r+1]$ have same sign. In both cases, the edge preserves the same orientation. Consequently, ρ' is removable in π if and only if it is removable in $\pi\rho$. The dual statement is deducible from Lemmas 1 and 9.

Finally, we prove the last statement: let $i(i + 1)$ and $j(j + 1)$ (suppose $|\pi^{-1}[i]| < |\pi^{-1}[j]|$) be two dream edges of π. Let $i'(i' + 1)$ and $j'(j' + 1)$ be the same dream edges in $\rho\pi$, that is after the deletion of the cycle ρ in π (labels of the vertices may have changed). The edge $i(i+1)$ corresponds to the reality edge $\pi^{-1}[i]\pi^{-1}[i+1]$, and the edge $i'(i'+1)$ corresponds to the reality edge $(\rho\pi)^{-1}[i'](\rho\pi)^{-1}[i'+1] = \pi^{-1}\rho[i']\pi^{-1}\rho[i'+1]$. However, the numbers of the labels of the vertices do not change when contracting a cycle (see Lemma 8), therefore we have that $|\pi^{-1}\rho[i']| = |\pi^{-1}[i]|$ and $|\pi^{-1}\rho[i'+1]| = |\pi^{-1}[i+1]|$. The same applies with j. This means that $|\pi^{-1}[i]| < |\pi^{-1}[j]| < |\pi^{-1}[i+1]| < |\pi^{-1}[j+1]|$ if and only if $|(\rho\pi)^{-1}(i')| < |(\rho\pi)^{-1}(j')| < |(\rho\pi)^{-1}(i'+1)| < |(\rho\pi)^{-1}(j'+1)|$, which proves that the dream edges cross in $\rho\pi$ if and only if they cross in π. □

3   Constructing the Sequence of Reversals

We can sort each oriented component of π separately (again, see [4]). We therefore suppose without loss of generality that there is only one component in $BG(\pi)$. We call a *valid sequence* of a permutation π an ordering of a subset of its 2-cycles $\rho_1,\ldots,\rho_k$, such that for all $i \in \{1,\ldots,k\}$, ρ_i is a removable cycle of $\rho_{i-1}\ldots\rho_1\pi$. In other words, $\rho_1,\ldots,\rho_k$ is valid if $c(\rho_k\cdots\rho_1\pi) = c(\pi) - k$.

A valid sequence is said to be *maximal* if no cycle of $\rho_k\cdots\rho_1\pi$ is removable. It is *total* if $k = c(\pi)$, that is if $\rho_k\cdots\rho_1\pi = Id$.

The algorithm for sorting a simple permutation is the following:

1. compute a maximal valid sequence of π;
2. increase the size of the sequence by adding some cycles inside it while it is not total.

Observe that the above algorithm constructs a sequence of removable instead of contractible cycles. The latter would seem to be a more direct way of doing the same thing. It is probably possible to do so by directly contracting instead of removing cycles (that is, by sorting π instead of π^{-1}). However, we believe the proofs are much easier in this case because the structure of the breakpoint graph is much more stable after a removal than after a contraction.

The first step of the algorithm consists in simply detecting removable cycles and removing them while there exist removable cycles in the result. We now concentrate on the second step, which consists in, given a maximal valid sequence $\rho_1,\ldots,\rho_k$, adding some cycles to it.

We prove the following result:

Theorem 1. *If S is a maximal but not a total valid sequence of reversals for a permutation π, there is a nonempty sequence S' of reversals such that S may be split into two parts $S = S_1, S_2$, and S_1, S', S_2 is a maximal valid sequence of reversals for π.*

Proof. Let $S = \rho_1,\ldots,\rho_k$ be a maximal valid sequence of π. Let $\mathcal{C}$ be the set of cycles of the permutation $\rho_k\cdots\rho_1\pi$ (it is composed of all the cycles of π minus the cycles in S). The set $\mathcal{C}$ is a union of unoriented components (all reality edges are unoriented else there would remain a removable cycle, therefore all have same orientation making all dream edges unoriented). Since there is only one component in $BG(\pi)$, one cycle of S has to cross one cycle of $\mathcal{C}$. Choose such a cycle ρ_l in S, such that l is maximum for this property. Let $S_1 = \rho_1, \ldots, \rho_{l-1}$ and $S_2 = \rho_l, \ldots, \rho_k$. These will be the way of splitting S in two, as described in the theorem. Let $\pi_1 = \rho_{l-1} \cdots \rho_1 \pi$ (this is the permutation obtained by removing the cycles of S_1 in π).

We first prove some lemmas.

Lemma 11. *At least one cycle of each component of $\mathcal{C}$ crossed by ρ_l is removable in π_1.*

Proof. Since a removal does not change the orientation of the dream edges (Lemma 10), and $\mathcal{C}$ is unoriented in $\rho_k \cdots \rho_1 \pi$, no cycle of $\mathcal{C}$ is contractible in π or in π_1. Let ρ be a cycle of $\mathcal{C}$ intersecting ρ_l in π_1. Let $\pi_1[j]\pi_1[j']$ $(j < j')$ be a dream edge of ρ crossing ρ_l. Choose a reality edge $\pi_1[i]\pi_1[i+1]$ of ρ_l such that $j < i$ and $j' > i + 1$. Since $\pi_1[j]\pi_1[j']$ is unoriented, $\pi_1[j]$ and $\pi_1[j']$ have the same sign. Since ρ_l is removable in π_1, the reality edge $\pi_1[i]\pi_1[i + 1]$ is oriented and $\pi_1[i]$ and $\pi_1[i + 1]$ have opposite signs. Among the reality edges $(\pi_1[r])^d(\pi_1[r + 1])^a$ for $r \in [j, j']$, there is therefore a positive even number of oriented ones (there has to be an even number of changes of sign, and at least one between i and $i + 1$). Thus at least one removable cycle distinct from ρ_l has to cross ρ. By the maximality of l, this cycle is in $\mathcal{C}$. $\square$

Let $S' = \rho'_1, \ldots, \rho'_p$ be a valid sequence of cycles of $\mathcal{C}$ in π_1, such that $\mathcal{C} \setminus \{\rho'_1 \ldots \rho'_p\}$ contains no removable cycle in $\rho'_p \ldots \rho'_1 \pi_1$.

Lemma 12. *The cycle ρ_l is removable in $\rho'_p \cdots \rho'_1 \pi_1$.*

Proof. Let $\pi' = \rho'_{p-1} \cdots \rho'_1 \pi_1$. Let $\pi'[i]\pi'[i+1]$ and $\pi'[j]\pi'[j+1]$ be the two reality edges of ρ'_p in π'. Let M be the number chosen among $\pi'[i], \pi'[i+1], \pi'[j], \pi'[j+1]$ with highest absolute value, and m be the number with lowest absolute value. Since ρ'_p is removable in π', $\pi'[i]$ and $\pi'[i + 1]$ have opposite signs, and $\pi'[j]$ and $\pi'[j + 1]$ also have opposite signs. Recall that no cycle of $\mathcal{C}$ is contractible in π, so none is in π', from Lemma 10. Then the two dream edges $\pi'[i]\pi'[j + 1]$ and $\pi'[j]\pi'[i+1]$ of ρ'_p are unoriented. In consequence $\pi'[i]$ and $\pi'[j+1]$ have the same sign, and $\pi'[j]$ and $\pi'[i+1]$ also have the same sign. Note that $\pi'[j+1] = \pi'[i]+1$ and $\pi'[i + 1] = \pi'[j] + 1$, so M and m cannot be adjacent in the graph, and they have opposite signs. Now remove cycle ρ'_p from π'. Removing a cycle is deleting the dream edges and replacing them by two edges parallel to the reality edges, and changing the labels of the vertices of the graph whose absolute values are between $|m|$ and $|M|$, excluding m and M themselves (see Lemma 9). In the obtained graph, both M and m participate in two adjacencies, and they have opposite signs. In consequence, there is an odd number of changes of signs between M and m, therefore an odd number of oriented reality edges. Since oriented reality edges belong to a removable 2-cycle (they only go by pairs), at least one of these 2-cycles crosses ρ'_p in π'. Since there is no removable cycle of $\mathcal{C}$ in $\rho'_p \pi'$ (by hypothesis), and ρ_l is the only cycle outside $\mathcal{C}$ that may cross a cycle of $\mathcal{C}$, this removable cycle is ρ_l. $\square$

Lemma 13. *The sequence* S_1, S', S_2 *is a valid sequence of* π.

Proof. We already know that $\rho_1, \ldots, \rho_{l-1}, \rho'_1, \ldots, \rho'_p, \rho_l$ is a valid sequence, by Lemma 12. Let now $\pi_1 = \rho_l \cdots \rho_1 \pi$, and $\pi_2 = \rho_l \rho'_p \cdots \rho'_1 \rho_{l-1} \ldots \rho_1 \pi$. Let D_1 be the component of $BG(\pi_1)$ induced by the cycles $\rho_{l+1}, \ldots, \rho_k$ in π_1. Since each component of a breakpoint graph may be sorted separately, we can say that $\rho_{l+1}, \ldots, \rho_k$ is a total valid sequence for D_1. Let now D_2 be the component of $BG(\pi_1)$ induced by the cycles $\rho_{l+1}, \ldots, \rho_k$ in π_2. A cycle ρ_i is contractible in D_1 if and only if it is contractible in D_2, and two cycles cross in D_1 if and only if they cross in D_2 (they differ only by some removals, which do not change contractibility, nor "crossingness"). Then $\rho_{l+1}, \ldots, \rho_k$ is a total valid sequence for D_2. Finally, $\rho_1, \ldots, \rho_{l-1}, \rho'_1, \ldots, \rho'_p, \rho_l, \ldots, \rho_k$ is valid in π.   □

This concludes the proof of Theorem 1 because we have a valid sequence S_1, S', S_2, and S' is non empty (see Lemma 11).   □

In the example of Figure 1, with the labels of the cycles as indicated in Figure 2, the algorithm may work as follows: cycle C is removed, then cycle D. The sequence C, D is maximal since A and B are not removable in $DC\pi$. It is not total, so we must find the last cycle in the ordered sequence $\{C, D\}$ which crosses a cycle in $\{A, B\}$. This is cycle C which crosses B (C is thus ρ_l, $S_1 = \emptyset$ and $S_2 = \{C, D\}$). In $BG(\pi) \setminus \{C, D\}$, only A is removable in π. We therefore remove A. This makes B removable and we remove it. There are no more removable cycles in $BG(\pi) \setminus \{C, D\}$ (indeed, there are no more cycles), so A, B, C, D is a total valid sequence and $DCBA\pi = Id$.

Observe that in the case of this example, by working directly on π^{-1}, it is easy to find that D, C, B, A is a total valid sequence.

4   Complexity

With a classical data structure, applying a reversal and picking a contractible (or removable) cycle is achievable in linear time. As a consequence, any algorithm that has to apply a reversal at each step, even bypassing the search for a safe reversal, will run in $O(n^2)$. This is the case for our algorithm. In practice, even an $O(n^2)$ implementation should be an improvement on the existing algorithms. For small permutations, such quadratic implementation is even probably better than the use of another data structure. However, it is an important mathematical question whether sorting a signed permutation by reversals can be done in a theoretical subquadratic time complexity. We therefore use a more sophisticated implementation to make it run *exactly* in $O(n\sqrt{n \log n})$.

In 2003, Kaplan and Verbin [6] invented a clever data structure which allows to pick a contractible reversal and apply it in sublinear time. We use the same data structure, just adding some flags in order to be able to perform all the additional operations we need. Furthermore, we use the structure to represent the permutation π^{-1} instead of π. Indeed, we are looking for removable cycles in π, that is for contractible ones in π^{-1}. As in [6], we split the permutation into blocks of size $O(\sqrt{n \log n})$. We assign a flag to each block, turned "on" if the block should be read in the reverse order, changing the sign of the elements. We also add to each block a balanced binary tree (such as a splay tree for instance), where the elements of the blocks are stored in the nodes, sorted by increasing position of their successor in π^{-1}. This means that $\pi^{-1}[i]$ is before $\pi^{-1}[j]$ if $\pi(\pi^{-1}[i] + 1) < \pi(\pi^{-1}[j] + 1)$. Each node of the tree corresponds to a dream edge of π^{-1} (each 2-cycle is represented by two nodes). Each node keeps the information on the orientation of the associated dream edge (that is, on the contractibility and therefore also on the removability in π of the cycle in which the edge participates), and a "running total" storing the number of oriented edges in the subtree.

Up to this point, we have described exactly the data structure of Kaplan and Verbin. We now add some flags and numbers to the nodes of the trees in order to perform our own queries. Let $\mathcal{C}$ be a subset of the 2-cycles of π^{-1}. At the beginning, $\mathcal{C}$ is all the 2-cycles in $BG(\pi)$. To each node, a new flag is added, turned "on" if the corresponding edge is in a cycle of $\mathcal{C}$, and a "running total" stores the number of oriented edges that belong to $\mathcal{C}$ in the subtree. Figure 4 gives a representation of the data structure applied to the inverse permutation of Figure 2 (right side), split in two blocks. Observe that we do not need to know the number of all oriented edges, only of those which are in $\mathcal{C}$.

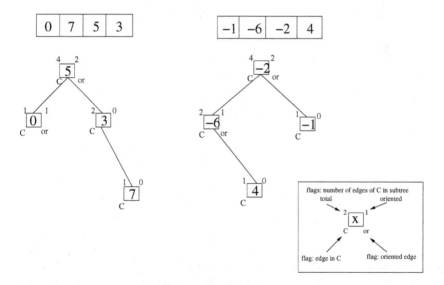

Fig. 4. The data structure for the permutation of Figure 2 (right side) at the beginning of the algorithm (all edges are in $\mathcal{C}$), if the permutation is split into two blocks.

In this way, the following queries are achievable in the same way, and same complexity, as in [6].

Lemma 14. [6] It is possible to know in constant time whether there is a removable cycle in π, and to pick one such cycle in time $O(\log n)$.

With the additional flags, and proceeding exactly in the same way, we have that:

Lemma 15. *It is possible to know in constant time whether there is a removable cycle of π in C, and if there is any, to pick one such cycle in time $O(\log n)$.*

Lemma 16. *It is possible to contract a cycle and maintain the data structure in time $O(\sqrt{n \log n})$.*

Proof. All we have to maintain as information besides what was done in [6] is whether an edge is in C. In order to do this, we just need to remove an edge from C when it is contracted and then to update the running totals. The number of elements of C is thus decreased by one in all the blocks containing the contracted cycle, and we calculate the difference between the number of elements in C and the former number of contractible cycles in C at each flipped block.   □

Figure 5 represents the data structure at a later step in the algorithm. As described before, the algorithm picks cycles C and D, removes them, and since there is no other removable cycle, replaces them. The result of this is shown in the figure.

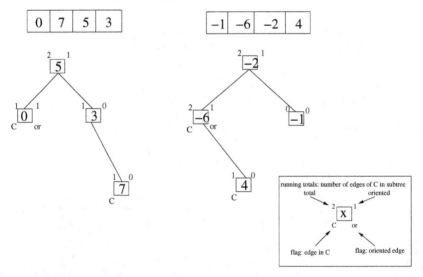

Fig. 5. The data structure for the permutation of Figure 2 (right side) at a later step of the algorithm, when cycles C and D have been removed, and then replaced (but remain removed from C).

Theorem 2. *There exists an algorithm which finds a total sorting sequence in time $O(n\sqrt{n \log n})$.*

Proof. We first find a maximal valid sequence $\rho_1, \ldots, \rho_k$. This takes $O(k\sqrt{n \log n})$ time. If the sequence is total, we are done. Otherwise, we are left with a set C of unremoved cycles.

We must then identify ρ_l in the sequence such that there is a removable cycle in $\mathcal{C}$ in $\rho_{l-1} \ldots \rho_1 \pi$, with l maximum for this property. This can be done by putting back (applying the reversals again) one by one the removed cycles $\rho_k, \ldots, \rho_l$, checking at each step (in constant time) if there is a removable cycle in $\mathcal{C}$, and stopping as soon as one is found. This is achievable in time $O((k - l)\sqrt{n \log n})$. The sequence $\rho_k, \ldots, \rho_l$ is a sequence of *safe* contractible reversals for π^{-1}, and is fixed once for ever. It will represent the last reversals of the final sequence. It is not modified nor looked at anymore in the algorithm, and this allows to maintain the complexity.

We then have to remove, while there are any, removable cycles in $\mathcal{C}$. This leads to the sequence $\rho_{k+1}, \ldots, \rho_{k'}$ which we insert after the sequence $\rho_1, \ldots, \rho_{l-1}$, while at the same time deleting it from $\mathcal{C}$ in the same way we have already seen. If $\rho_1, \ldots, \rho_{l-1}, \rho_{k+1}, \ldots, \rho_{k'}, \rho_l, \ldots, \rho_k$ is total, we are done. Otherwise, we start again replacing the removed cycles one by one in the reverse order but starting this from k' (the sequence $\rho_l, \ldots, \rho_k$ is not touched anymore). We do so until there is a removable cycle in the new $\mathcal{C}$, and start again while $\mathcal{C}$ is not empty.

Each cycle in the sequence is removed once, and replaced at most once in the whole procedure. The total complexity is then $O(n\sqrt{n \log n})$.   □

The remaining bottleneck of the new algorithm is now the complexity at each step of applying a reversal to a permutation and of keeping the set of contractible or removable cycles. This is what has to be improved in the future if one wishes to obtain a lower theoretical complexity for sorting a signed permutation by reversals.

References

1. D.A. Bader, B.M.E. Moret and M. Yan, "A linear-time algorithm for compting inversion distance between signed permutations with an experimental study", *Proceedings of the 7th Workshop on Algorithms and Data Structures* (2001), 365–376.
2. A. Bergeron, J. Mixtacki and Jens Stoye, "The Reversal Distance Problem", to appear in *Mathematics of Evolution and Phylogeny*, O. Gascuel, editeur, Oxford University Press.
3. P. Berman and S. Hannenhalli, "Fast sorting by reversals", *Proceedings of the 7th Symposium on Combinatorial Pattern Matching* (1996), Lecture Notes in Computer Science, vol. 1075, 168–185.
4. S. Hannenhalli and P. Pevzner, "Transforming cabbage into turnip (polynomial algorithm for sorting signed permutations by reversals", *Proceedings of the 27th ACM Symposium on Theory of Computing* (1995), 178–189.
5. H. Kaplan, R. Shamir and R.E. Tarjan, "Faster and simpler algorithm for sorting signed permutations by reversals", *SIAM Journal on Computing* 29 (1999), 880–892.
6. H. Kaplan and E. Verbin "Efficient data structures and a new randomized approach for sorting signed permutations by reversals", *Proceedings of the 14th Symposium on Combinatorial Pattern Matching* (2003), Lecture Notes in Computer Science, vol. 2676, 170–185.
7. M. Ozery-Flato and R. Shamir, "Two notes on genome rearrangement", *Journal of Bioinformatics and Computational Biology*, 1 (2003), 71–94.

Computational Problems in Perfect Phylogeny Haplotyping: Xor-Genotypes and Tag SNPs

Tamar Barzuza[1], Jacques S. Beckmann[2,3], Ron Shamir[4], and Itsik Pe'er[5]

[1] Dept. of Computer Science and Applied Mathematics, Weizmann Institute of Science, Rehovot 76100, Israel
tamar.barzuza@weizmann.ac.il
[2] Dept. of Molecular Genetics, Weizmann Institute of Science, Rehovot 76100, Israel
jacqui.beckmann@weizmann.ac.il
[3] Département de Génétique Médicale Ch-1011 Lausanne. Switzerland
jacques.beckmann@hospvd.ch
[4] School of Computer Science, Tel- Aviv University, Tel Aviv 69978, Israel
rshamir@post.tau.ac.il
[5] Medical and Population Genetics Group, Broad Institute, Cambridge MA 02142 US
peer@broad.mit.edu

Abstract. The perfect phylogeny model for haplotype evolution has been successfully applied to haplotype resolution from genotype data. In this study we explore the application of the perfect phylogeny model to other problems in the design and analysis of genetic studies. We consider a novel type of data, xor-genotypes, which distinguish heterozygote from homozygote sites but do not identify the homozygote alleles. We show how to resolve xor-genotypes under perfect phylogeny model, and study the degrees of freedom in such resolutions. Interestingly, given xor-genotypes that produce a single possible resolution, we show that the full genotype of at most three individuals suffice in order to determine all haplotypes across the phylogeny. Our experiments with xor-genotyping data indicate that the approach requires a number of individuals only slightly larger than full genotyping, at a potentially reduced typing cost. We also consider selection of minimum-cost sets of tag SNPs, i.e., polymorphisms whose alleles suffice to recover the haplotype diversity. We show that this problem lends itself to divide-and-conquer linear-time solution. Finally, we study genotype tags, i.e., genotype calls that suffice to recover the alleles of all other SNPs. Since most genetic studies are genotype-based, such tags are more relevant in such studies than the haplotype tags. We show that under the perfect phylogeny model a SNP subset of haplotype tags, as it is usually defined, tags the haplotypes by genotype calls as well.

Keywords: SNPs, Haplotypes, Perfect Phylogeny, Tag SNPs, Graph Realization.

1 Introduction

1.1 Background

Genetic information in nature is usually stored as a linear sequence, written in a molecular DNA alphabet of four letters (nucleotides), A, C, G and T. Higher organisms

S.C. Sahinalp et al. (Eds.): CPM 2004, LNCS 3109, pp. 14–31, 2004.

are *diploid*, i.e., have two near-identical copies of their genetic material arranged in paired molecules called *chromosomes*, one originating from each parent. Such chromosomes are *homologous*, that is, contain essentially the same genes in altered variants. Changes between variants comprise mainly of *Single Nucleotide Polymorphisms* (SNPs), i.e., sequence sites where one of two letters may appear [1]. These SNPs are numerous and it is estimated that any two homologous human chromosomes sampled at random from the population differ on average once in every thousand letters, accounting thus for a few million such differences along the entire genome. The variants of a SNP are called *alleles*. An individual is said to be *homozygous for a SNP* if both homologous chromosomes bear the same allele for this SNP and *heterozygous* otherwise. The sequence of alleles along a chromosome is called a *haplotype*. At first approximation a chromosome can be considered as a patchwork of haplotypes along its length. A *genotype* along homologous chromosomes lists the conflated (unordered pair of) alleles for each SNP (see Fig. 1).

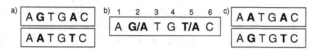

Fig. 1. An example of 6 SNPs along two homologous chromosomes of an individual. (a) This individual's haplotypes. (b) This individual's genotype. Here the Xor-genotype (set of heterozygous SNPs) would be {2,5}. (c) Another potential haplotype pair giving rise to the same genotype.

Both genotype and haplotype data are used in genetic studies. Haplotypes are often more informative [6]. Unfortunately, current experimental methods for haplotype determination are technically complicated and cost prohibitive. In contrast, a variety of current technologies offer practical tools for genotype determination [26]. Given genotype data, the haplotypes must be inferred computationally, in a process called *resolving, phasing* or *haplotyping* [7,8,9,10,11,27]. A single genotype may be resolved by different, equally-plausible haplotype pairs (see Fig. 1), but the joint inference of a set of genotypes may favor one haplotype pair over the others for each individual. Such inference is usually based on a model for the data. Informally, most models rely on the observed phenomenon that over relatively short genomic regions, different human genotypes tend to share the same small set of haplotypes [2,3].

1.2   The Perfect Phylogeny Model

During sexual reproduction, only one homologous copy of each chromosome is transmitted to the offspring. Moreover, that copy has alternating segments from the two homologous chromosomes of the parent, due to a segmental exchange process called (*meiotic*) *recombination*. Studies have shown that recombination occurs mainly in narrow regions called *hotspots*. The genomic segments between hotspots are called *blocks*. They show essentially no recombination [4] and their haplotype diversity is limited [2,3]. Within blocks, haplotypes evolve by mutations, i.e., replacement of one nucleotide by another at particular sites (other mutation types are not discussed here). Since mutations are relatively rare [5], it is often assumed, as we do here, that at most one mutation occurs in each site. The *perfect phylogeny model* for haplotype block evolution assumes that all haplotypes in the population have a common ancestor, no recombination and no recurrent mutation.

The *Perfect Phylogeny Haplotyping* problem (PPH) seeks to infer haplotypes that satisfy the perfect phylogeny model (we defer formal definitions to Section 1.5). PPH was first introduced by Gusfield [9], who presented an almost linear solution by reducing PPH to the classical *Graph Realization* problem. Simpler, direct solutions were later given [10,11], which take $O(nm^2)$ for n haplotypes and m SNPs.

1.3 Informative SNPs

Many medical genetics studies first determine the haplotypes for a set of individuals and then use these results to efficiently type a larger population. Having identified the restricted set of possible haplotypes for a region, the identity of a subset of the SNPs in the region may suffice to determine the complete haplotype of an individual. Such SNPs are called *tag SNPs*, and typing them alone would lose no information on the haplotypes. More generally, we may be interested only in a subset S of all SNPs (e.g., coding and regulatory SNPs only) but can use all available SNPs to tag them. In this setting we call S the set of *interesting SNPs,* and seek a smallest possible *informative SNP set*, i.e., is a subset of all SNPs that captures all the information on S (see Fig. 2). Hence, the identification of few informative SNPs may lead to substantial saving in typing costs. For this reason, the computational problems of finding a minimal tag (or informative) set have been studied [2,13,18].

```
                              SNPs
                    1 2 3 4  5 6  7 8  9 10 11
         ┌─────────────────────────────────┐
    4    │ A T T T A T C C T T T │
    3    │ C A T A G T A C T T T │
    2    │ A A T T G A C C A A G │
    1    │ C T A T G T A T A T G │
         └─────────────────────────────────┘
```

Haplotypes (labels 1 2 3 4 on left)

Fig. 2. Tag SNPs and informative SNPs. The set {1,2} is a tag SNP set. If {9,10,11} is the interesting SNP set, then the interesting set distinguishes the haplotypes 1, 2 and {3,4}, but does not distinguish between haplotypes 3 and 4. Therefore {1,2} and {6,8} are both informative SNPs sets but {4,5} and {2,3} are not. Notice that the same genotype **A/C T/A** is obtained for the tag SNP set {1,2} from the two pairs of haplotypes {1,2} and {3,4}.

Finding the minimum set of tag SNPs within an unconstrained block is NP-hard [12]. When the perfect phylogeny model is assumed, in the special case of a single interesting SNP, a minimal set of informative SNPs was shown to be detectable in $O(nm)$ time, for n haplotypes and m SNPs [13].

1.4 Contribution of This Work

We study here several problems arising under the perfect phylogeny model during genetic analysis of a region, along the process from haplotype determination toward their utilization in a genetic study. Our analysis focuses on a single block.

Some experimental methods such as DHPLC [14] can determine whether an individual is homozygous or heterozygous for each SNP, but cannot distinguish between the two homozygous sites. Typing SNPs in such manner will provide, for each individual, a list of the heterozygous sites, which we refer to as the individual's *xor-genotype*. Xor-genotypes are less informative than regular ("full") genotypes; but their generation may be less costly. Therefore, it is of interest to infer the haplotypes based primarily on xor-genotypes instead of full genotypes. In Section 2 we introduce the *Xor Perfect Phylogeny Haplotyping* problem (XPPH), study the limitations of using only xor-genotypes, and the additional genotype information required. Section 2.2 presents an efficient solution to XPPH based on the graph realization problem [15]. We implemented our solution and evaluated the XPPH strategy in Section 2.3. Our tests show that the method compares favorably with standard genotyping.

Section 3 studies informative SNPs under the perfect phylogeny model. We generalize the minimum informative set (and tag set) problems by introducing a cost function for SNPs, and seek minimum cost sets. The cost is motivated by the facts that typing some SNPs may be technically harder (e.g., those in repetitive or high GC regions), and that some SNPs are more attractive for direct typing (e.g., protein-coding SNPs, due to prior assumptions on their functionality). In section 3.2 we find minimal cost informative SNP sets in $O(m)$ time for any number of interesting SNPs, when the perfect phylogeny tree is given. This generalizes the result of [13]. Section 3.3 discusses a practical variant of the tag SNPs set, i.e., the phasing tag SNPs set: As we usually have only genotypic (conflated) information on the SNPs, a practical goal would be to find a set of SNPs that give the same information as tag SNPs, but instead of knowing their haplotype we only know their genotype. We prove that the set of informative SNPs is guaranteed to have this quality, and that this is guaranteed only under the perfect phylogeny model.

We conclude with a discussion in Section 4. Throughout the manuscript, many proofs are omitted, due to lack of space.

1.5 Preliminaries

We denote the two alleles for each SNP by 0 and 1. A haplotype is represented by a binary vector. A set of haplotypes is represented by a binary matrix H, where each row is a haplotype vector and each column is the vector of SNP alleles. We denote the allele of haplotype i for SNP j by H_{ij} or by h_j for the haplotype $h=H_i$. A *genotype* is the conflation (mixing) of two haplotypes. For example, the pair of haplotypes 00100 and 10001 gives rise to the genotype $\{0,1\}\{0,0\}\{0,1\}\{0,0\}\{0,1\}$.

The perfect phylogeny model is formalized as follows: Let $H_{n \times m}$ be a binary matrix of n distinct haplotypes over m SNPs. A *perfect phylogeny* for H is a pair (T,f) where $T=(V,E)$ is a tree with $\{1,...,n\} \subseteq V$ and $f:\{1,...,m\} \rightarrow E$ is an assignment of SNPs to edges such that (1) every edge of T is labeled at least once and (2) for any two rows k, l, $H_{kj} \neq H_{lj}$ iff the edge $f(j)$ lies on the unique path from node k to node l in T. The analysis of this model is heavily based on a fundamental property (cf. [21]):

Theorem 1: There is a perfect phylogeny for H iff H does not contain a 4×2 submatrix in which all four rows are different. Such a submatrix is called a *four-gamete*.

2 Xor Haplotyping

In this section we formally define the problem of Xor Perfect Phylogeny Haplotyping, provide an algorithm for the problem and discuss how to handle ambiguity in the data. We then show how to obtain the actual haplotypes efficiently using a small amount of additional full genotypes.

2.1 Problem Definition

Definition: A *xor-genotype* of a haplotype pair $\{h,h'\}$ is the set of their heterozygote SNPs, i.e., $\{s|h_s{\neq}h'_s\}$ (see Fig. 1). A set of haplotypes H *explains* a set of xor-genotypes X if each member of X is a xor-genotype of a pair of haplotypes in H.

Problem 1: Xor Perfect Phylogeny Haplotyping (XPPH)

Input: A set $X=\{X_1,...,X_n\}$ of xor-genotypes over SNPs $S=\{s_1,...,s_m\}$, such that $X_1\cup...\cup X_n =S$.

Goal: Find a haplotype matrix H with a perfect phylogeny (T,f), such that H explains X, or determine that no solution exists. (See Fig. 3)

Hereafter, we shall omit the term "or determine that no solution exists" from problem statements for brevity. This requirement is part of the goal of all the algorithms in this study.

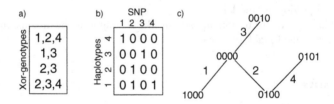

Fig. 3. An example of a solution to XPPH. (a) X -The four xor-genotypes. (b) H - The inferred haplotypes. (c) (T,f) - A perfect phylogeny for H. Notice that H explains X by taking the haplotype pairs $\{1,4\},\{3,4\},\{2,3\}$ and $\{1,3\}$. Note that T includes a haplotype (0000) that is not in H.

2.2 An Algorithm for XPPH

2.2.1 Inference up to Bit Flipping

A first step in understanding XPPH is the observation that the solution is never unique. Rather, flipping the alleles of SNPs in a solution yields yet another solution, as we now explain.

Definition: Two haplotype matrices $H_{n\times m}$ and $H'_{n\times m}$ are *equivalent up to bit flipping* (denoted $H\leftrightarrow H'$) if for any two rows k, l, $H_{kj}{\neq}H_{lj} \Leftrightarrow H'_{kj}{\neq}H'_{lj}$. $H\leftrightarrow H'$ iff one can be obtained from the other by exchanging the roles of 1 and 0 for some columns. Notice that $\leftrightarrow$ is a set-theoretic equivalence relation.

Observation 1: If $H \leftrightarrow H'$ then X can be explained by H iff it can be explained by H'.

Observation 1 implies that XPPH can only be solved up to bit flipping based on the data given by X.

In some cases, however, there may be several alternative sets of haplotypes that explain X and are not $\leftrightarrow$-equivalent. In that case, we will not be able to determine which of those sets is the one that really gave rise to X. Our only achievable goal is therefore to identify when the solution obtained is guaranteed to be the correct one. We will next show that this is guaranteed by the uniqueness of the solution. The analysis relies on the following property of perfect phylogenies:

Key property: Let (T,f) be a perfect phylogeny for H. If $H_{ij}=0$ then for all k, $H_{kj}=0$ iff nodes i and k are in the same component of $T\backslash f(j)$.

Definition: (T,f) is a *perfect phylogeny for X* if (T,f) is a perfect phylogeny for some H that explains X.

Proposition 1: When X has a unique perfect phylogeny then the haplotype matrix that explains it is unique up to bit flipping (i.e., up to $\leftrightarrow$-equivalence).

Proof: It suffices to prove that if (T,f) is a perfect phylogeny for H then there is no H' such that (T,f) is a perfect phylogeny for H' and $\neg(H \leftrightarrow H')$. First we observe that there is a unique correspondence between the nodes of T and the haplotypes in H. This correspondence is obtained as follows. We first identify the haplotype $h \in H$ of an arbitrary leaf v. This is done by observing the SNPs that correspond to the edge incident on v. h is the only haplotype that is distinct from all others in these SNPs. The haplotypes of all other nodes are now readily determined by the key property. This generates the uniuqe correspondence. The actual haplotypes are set by fixing arbitrarily the haplotype vector of one node and setting all others according to the key property. $\square$

Proposition 1 motivates a new formulation of Problem 1':

Problem 1': XPPH:

Input: A set $X=\{X_1,...,X_n\}$ of xor-genotypes over SNPs $S=\{s_1,...,s_m\}$, such that $X_1 \cup ... \cup X_n = S$.

Goal: Find a unique perfect phylogeny (T,f) for X, or determine that there are multiple perfect phylogenies for X.

Proposition 1 implies that a solution to Problem 1' (if unique) is guaranteed to be a perfect phylogeny for the correct set of haplotypes, i.e., the haplotypes that actually gave rise to X.

Gusfield's work [9] leads to an interesting and useful connection between xor-genotypes and paths along the perfect phylogeny tree, as follows.

Definition: We say that a pair (T,f) *realizes* $X_i \subseteq S$ if X_i is the union of edge labels that constitute a path in T. (T,f) is said to *realize a collection X of subsets* if each $X_i \in X$ is realized by (T,f).

Proposition 2: (T,f) is a perfect phylogeny for X iff X is realized by (T,f).

The following formulation for XPPH is finally obtained:

Problem 1″: XPPH:

Input: A set $X=\{X_1,...,X_n\}$ of xor-genotypes over SNPs $S=\{s_1,...,s_m\}$, such that $X_1\cup...\cup X_n =S$.

Goal: Find the unique realization (T,f) for X, or determine that there are multiple realizations for X.

Intuitively, the larger the number of typed SNPs, the greater the chances to have a unique realization. Occasionally, however, a dataset X may have multiple realizations even with many SNPS.. This is the case of the data including xor-equivalent SNPs:

Definition: We say that $s,s'\in S$ are *xor-equivalent w.r.t.* X and write $s\approx^X s'$ if for all i: $s\in X_i \Leftrightarrow s'\in X_i$.

Fortunately, xor-equivalent SNPs may be redundant. This is the case of the data including haplotype-equivalent SNPs:

Definition: We say that $s,s'\in S$ are *haplotype-equivalent w.r.t.* H and write $s\approx^H s'$ if for all i,j: $H_{is}\neq H_{js}\Leftrightarrow H_{is'}\neq H_{js'}$. Note that $\approx^H$ and $\approx^X$ are equivalence relations.

Observation 3: Haplotype-equivalence implies xor-equivalence but not vice versa. (See Fig 4).

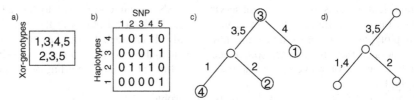

Fig. 4. Xor-equivalence and haplotype-equivalence. (a) X – The xor-genotypes. (b) H – The haplotypes matrix. Haplotypes 1 and 4 form the first xor-genotype, and haplotypes 2 and 3 form the second. The pairs of xor-equivalent SNPs are $\{1, 4\}$ and $\{3, 5\}$, while only 3 and 5 are haplotype-equivalent. (c) (T,f) – A realization for X that is a perfect phylogeny for H. (d) (T',f') – Another realization for X that is not a perfect phylogeny for H.

We next show that haplotype-equivalent SNPs are redundant.

Notation: Denote by $S^H\subseteq S$ the set that is obtained by taking one representative from each haplotype-equivalence class. Denote by H^H the haplotype matrix that is obtained by restricting H to S^H.

Observation 4: (1) To obtain a perfect phylogeny (T,f) for H, one can obtain a perfect phylogeny (T,f') for H^H and then set $f(s)=f'(s^H)$ for every $s\in S$ that is haplotype-equivalent to s^H. (2) (T,f') is a unique perfect phylogeny for H^H since S^H contains no haplotype-equivalent SNPs.

Observation 4 implies that haplotype-equivalent SNPs are redundant, hence may be merged to label a single edge in (T,f) (See Fig 4c); and by doing so, we discard the degrees of freedom that are due to haplotype-equivalent SNPs.

However, identifying haplotype-equivalent SNPs is not trivial when we only have xor-genotype information, which as Observation 3 implies may not suffice. In other words, the closest we can get to merging haplotype-equivalent SNPs is merging the xor-equivalent SNPs, which by Observation 3 may lead to information loss (See Fig 4d).

Definition: Denote by $S^X \subseteq S$ the set that is obtained by taking one representative from each xor-equivalence class. Denote by X^X the xor-genotypes that are obtained by restricting X to S^X. X^X is called the *canonic version of X*.

We show next that when the canonic version of X has a unique realization, then there was no information loss in merging xor-equivalent SNPs, since xor-equivalence implies haplotype-equivalence in this particular case.

Theorem 2: Let (T,f') be a unique realization for X^X. Extent the mapping f' to S by setting $f(s)=f'(s^X)$ for every s that is xor-equivalent to s^X. Then (T,f) is a perfect phylogeny for the correct haplotype matrix that gave rise to X.

Proof: By Proposition 2, (T,f') is a unique perfect phylogeny for X^X, and by Proposition 1 it is a perfect phylogeny for the correct haplotype matrix on S^X. We will next show that in the special case where (T,f') is unique, xor-equivalence implies haplotype-equivalence for the data set X. Then, by Observation 4, (T,f) is a perfect phylogeny for the correct haplotype matrix that gave rise to X. Suppose to the contrary that SNPs $s_1,s_2 \in S$ are xor-equivalent but not haplotype equivalent. Consider the unique perfect phylogeny (T^S,f^S) of H^H. Since s_1 and s_2 are not haplotype-equivalent they label distinct edges, e_1 and e_2 respectively, in T^S. Notice that $f^{-1}(e_1) \cup f^{-1}(e_2)$ are xor-equivalent. Let (T^S_1,f^S_1) be obtained from (T^S,f^S) by contracting e_1 (identifying e_1's nodes), and by taking $f^S_1(s)=e_2$ for $s \in f^{-1}(e_1)$. (T^S_2,f^S_2) is similarly obtained from (T^S,f^S) by contracting e_2. Then both (T^S_1,f^S_1) and (T^S_2,f^S_2) realize X^X, and $(T^S_1,f^S_1) \neq (T^S_2,f^S_2)$; in contradiction to the uniqueness of (T,f'). □

The formulation of Problem 1″ leads to a connection between XPPH and the graph realization problem:

Problem 2: The Graph Realization Problem (GR)

Input: A collection $P=\{P_j\}$ of subsets, $P_1,...,P_n \subseteq S$.

Goal: Find a pair (T,f) that realizes P.

Observation 2: Problem 1″ is now exactly the graph realization problem (when restricting the solution to GR to be unique).

The graph realization problem was first defined in matroid theory by Tutte [16], who proposed an algorithm of $O(mn^2)$ time, where $|P|=m$ and $|S|=n$. Gavril and Tamari [17] subsequently solved it in time $O(m^2n)$. Later, Bixby and Wagner [15] presented an $O(\alpha(m,n)mn)$ time algorithm, ($\alpha(m,n)$ is the inverse Ackermann function, $\alpha(m,n) \leq 4$ for all practical values of m,n). All three algorithms required linear space. These algorithms determine the existence of a graph realization and also the uniqueness of such a solution, hence they can be applied to solve XPPH.

The above discussion implies that the following procedure solves XPPH: Let M be the incidence matrix of X and S, i.e., $M_{ij}=1$ iff $s_j \in X_i$. Find S^X and X^X. (This can be done by a radix-sort of the columns of M in $O(nm)$ bitwise operations.) Then solve the graph realization problem on X^X. If the solution is unique it implies a perfect phylogeny for X.

In case that the xor-genotypes data cannot be augmented and there are several solutions to the GR problem, we may wish to choose one of them as a perfect phylogeny for X. Additional considerations may help in the choice [9]. We have developed a method for efficiently representing all the solutions for the graph realization problem by extending the algorithm in [17]. This representation is intuitive and implementation is straightforward. Details are omitted in this abstract.

2.2.2 Assigning Actual Haplotypes

In the previous section we concluded that even when XPPH has a single solution, the assignment of haplotypes to the tree nodes can be done only up to bit flipping. In order to obtain a concrete assignment, the input data must be augmented by additional genotyping of a selected set of individuals. We will prove that it suffices to fully genotype at most three individuals, and show how to select them. First, we explain how the additional genotype data are used to resolve the haplotypes. Denote by G_i the genotype of individual i (whose xor-genotype is X_i). Hereafter, we consider only those individuals with $X_i \neq \varnothing$.

Problem 3: Haplotyping on the Tree

Input: (a) A collection of non-empty xor-genotypes X; (b) a perfect phylogeny (T,f) for X, which is unique up to haplotype-equivalent SNPs; and (c) complete genotypes of the individuals $\{i_1,\ldots,i_p\}$.

Goal: Infer the haplotypes of all the individuals.

Haplotyping across the tree is based on the above key property, which determines the alleles of a SNP j for all haplotypes, based on its allele in some particular node. More specifically, all those alleles are determined given a genotype G_i, homozygote for SNP j, whose haplotypes correspond to identifiable nodes in T. Consequently, G_i resolves the bit-flip degree of freedom for each SNP $s \in S \backslash X_i$. Hence:

Proposition 3: The haplotypes can be completely inferred by $G_1,\ldots,G_p$ iff $X_1 \cap \ldots \cap X_p = \varnothing$.

The proposition brings about a criterion by which individuals should be selected for full genotyping. It motivates the following set-selection problem:

Problem 4: Minimum Tree Intersection (MTI)

Input: A collection of sets $X=\{X_1,\ldots,X_n\}$ and a perfect phylogeny (T,f) for X.

Goal: Find a minimum subset of X whose intersection is empty.

Note that the prefect phylogeny condition here is crucial: Without the condition that each X_i is a path in the tree, the problem is equivalent to the NP-hard set-cover problem.

Theorem 3: If $X_1 \cap \ldots \cap X_n = \varnothing$ then there is a minimum tree intersection set of size at most 3.

Proof: Consider the path X_1, and w.l.o.g. label the SNPs according to their order along that path as $(1,\ldots,k)$. For each i, the set $X_1 \cap X_i$ defines an interval in that order. If $X_1 \cap X_i = \varnothing$ for some i then $\{X_1, X_i\}$ are a solution. Otherwise all intervals overlap X_1. Denote these intervals by $[l_j, r_j]$ for $j=2,\ldots,n$. Take the interval that ends first and the interval that begins last, i.e., $L = \text{argmin}_j(r_j)$ and $R = \text{argmax}_j(l_j)$. Since $X_1 \cap \ldots \cap X_n = \varnothing$ then $[l_2, r_2] \cap \ldots \cap [l_n, r_n] = \varnothing$, hence it follows that $[l_R, r_R] \cap [l_L, r_L] = \varnothing$. We get $(X_1 \cap X_L \cap X_R) = \varnothing$. $\square$

In case no SNP is present in all X_i-s, the above proof provides an algorithm for finding three individuals whose full genotypes solve MTI. A slight modification allows finding two individuals instead of three when possible. The time complexity is $O(nm)$. Let $Y = X_1 \cap \ldots \cap X_n$.

Corollary 1: There are at most three individuals whose genotypes can resolve all the haplotypes on the SNP set $S \backslash Y$, and they can be found in $O(nm)$ time.

In case $Y \neq \varnothing$, the SNPs in Y can be inferred up to bit flipping.

2.3 Experimental Results

We implemented Gavril and Tamari's algorithm for Graph Realization [17]. Although it is not the asymptotically fastest algorithm available, it is simpler to implement and modify than [15]. Moreover, as the block size is usually bounded in practice by $m<30$, the quadratic dependence of the algorithm on m is not a handicap. Our implementation, GREAL, was written in C++, and is available at http://www.cs.tau. ac.il/ ~rshamir/greal. Another implementation due to Chung and Gusfield has recently been announced [19].

We used a standard population genetics simulator due to Hudson [22] to generate data samples under the perfect phylogeny model. In each run we generated $c=2400$ chromosomes with a prescribed number of SNPs, preserving the default values for all other simulation parameters. An important parameter in the experiments was the *minor allele frequency cutoff*, denoted by α: For a given value of α, we only used SNPs whose less frequent allele occurred in $\geq \alpha c$ chromosomes. The resulting haplotypes were randomly paired to generate xor-genotypes of individuals.

How many individuals are required to get a single solution?
We evaluated this measure by randomly adding individuals one by one and reapplying GR till the solution is unambiguous. The results (Fig. 5) show that for $\alpha \geq 0.03$, the number of individuals required to obtain a single solution is roughly an α-dependent constant, irrespective of the number of SNPs, and is practically bounded by 70. When rare alleles ($\alpha=0.01$) are present, the behavior is less predictable and the variance is very large. However, comprehensive sampling of the haplotypes is usually not achieved when rare alleles are present; fortunately, performance is satisfactory above the accepted α cutoff of 0.05.

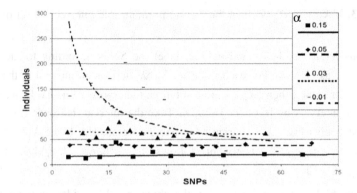

Fig. 5. Conditions for uniqueness of the solution. The plots show the number of xor-genotypes (y-axis) needed for obtaining a single solution for a given number of SNPs (x-axis). Different lines (or least squares curves) correspond to different thresholds on the minor allele frequency cutoff α. Note that the interpolated curve for α=0.01 is an extremely rough estimate.

XPPH vs. PPH

Since xor-genotypes contain less information, they may have a potential economic advantage over full genotypes. However, the number of individuals required for obtaining the haplotypes is larger. We compared the number of individuals needed by XPPH and by PPH. Chung and Gusfield [23] evaluated experimentally the number of individuals required for obtaining a unique solution to PPH. We computed the same statistic for XPPH (Fig. 6a). For 50 SNPs, 50 xor-genotypes guarantee ~90% chance of uniqueness, and increasing the number of individuals has only a minor effect. Essentially the same results hold for 100 SNPs. In comparison to [23], the chances for a unique XPPH solution with > 50 xor-genotypes is only a few percent lower than for PPH data with the same number of full genotypes.

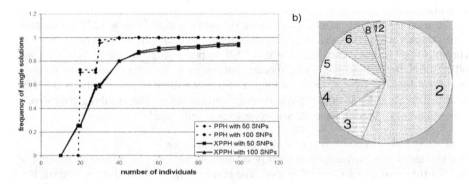

Fig. 6. The chance for a unique solution. (a) The frequency of a unique solution (y-axis) versus the number of individuals tested (x-axis). XPPH statistics are based on 5000 runs for 50 or 100 SNPs after filtering with α=0.05. PPH statistics from [23] are plotted for comparison. (b) The distribution of the number of non-unique solutions in deep coverage studies. Statistics were collected for 35 configurations of the number of SNPs (100-2000) and the number of individuals, which was at least 10 times the number of SNP equivalence classes. (α=0.05).

How high is the multiplicity of non-unique solutions?
We further focused on outlier ambiguous datasets, i.e., those that are ambiguous despite the examination of many individuals. For such datasets, the number of possible solutions is of much practical interest: If this number is limited, each solution may be tested separately. Indeed, the results (Fig. 6b) show that in this situation, when the solution is not unique, there are only a handful of solutions, usually only 2. Note that we assume equivalence of $\approx^H$ and $\approx^X$ for outlier datasets, which we confirmed for the datasets used here.

3 Informative SNPs

3.1 Problem Definition

In this section we study informative SNPs under the perfect phylogeny model. We begin by introducing some terminology, concordant with [13].

Definition: Let $H=\{H_1,\ldots,H_n\}$ be a set of haplotypes over a SNP set $S=\{s_1,\ldots,s_m\}$. Let $S''\subseteq S$ be a given subset of interesting SNPs. The set $S'\subseteq S\backslash S''$ is *informative* on H w.r.t. S'' if for each $1\le k,l\le n$, whenever there is a SNP $s''\in S''$ for which $H_{ks''}\neq H_{ls''}$, there is a SNP $s'\in S'$ for which $H_{ks'}\neq H_{ls'}$.

Note that that assumption that the informative and interesting SNPs are disjoint is made without loss of generality, since we can duplicate interesting SNPs as part of the candidates for the informative set. We generalize the Minimum Informative SNPs problem [13] by introducing a cost function, as follows:

Problem 5: Minimum-Cost Informative SNPs (MCIS):
Input: (a) A set of haplotypes $H=\{H_1,\ldots,H_n\}$ over a SNP set $S=\{s_1,\ldots,s_m\}$ along with a perfect phylogeny (T,f) for H.
(b) A set of interesting SNPs $S''\subseteq S$.
(c) A cost function $C:S\rightarrow R^+$.

Goal: Find a set $S'\subseteq S\backslash S''$ of minimum total cost that is informative w.r.t. S''.

(T,f) may already be known if H was found by solving XPPH. Alternatively, it can be computed in $O(mn)$ time from haplotypes [21].

A common task which is related to picking an informative SNP set is to describe all of the haplotype variation in the region [20]. Formally, we seek a *tag set* $S'\subseteq S$ s.t. for each $1\le l,k\le n$, there is $t\in S'$ for which $H_{kt}\neq H_{lt}$. In order to find tag SNPs of minimum cost, one could duplicate the SNP set S and define one of the copies as interesting. A solution to MCIS on the duplicated instance is a tag SNP set of minimal cost. Hence we shall focus on the more general MCIS problem.

3.2 An Algorithm for MCIS

3.2.1 Problem Decomposition
Recall that if $T=(V,E)$ is a perfect phylogeny for $H_{n\times m}$ then $\{1,\ldots,n\}\subseteq V$, i.e., the haplotypes of H label nodes in the perfect phylogeny. If a node of T is labeled by a haplo-

type from H we say it is *observed*. Otherwise we say it is *ancestral*. Ancestral nodes represent haplotypes that have been intermediate stages in evolution but did not survive to the present, or were not collected in the sample. It is easy to see that the leaves of T are always observed. The observed internal nodes in T can be used for a decomposition of T as follows:

Definition: An *ancestral component* is a subtree of T in which all the internal nodes are ancestral and all the leaves are observed.

Since the leaves of T are observed, T can be represented as a union of edge-disjoint ancestral components, where each union step merges two components by identifying copies of the same observed node. Two different components can share at most one observed node, but do not share ancestral node. Partitioning T into ancestral components is straightforward. We now show that in order to find informative SNPs we can divide the tree into ancestral components and find informative SNPs for each single component separately. The subproblem on a component is defined as follows: Denote an instance of MCIS by the input tuple $I=(H,S,C,T,f,S'')$. Let $T_1,...,T_p$ be T's ancestral components where $T_i=(V_i,E_i)$. Denote by $S_i \subseteq S$ the SNPs that label E_i. The input tuple for T_i is $I_i=(H_i,S_i,C_i,T_i,f_i,S_i'')$ where the sets and functions are the restriction of the original sets and functions to S_i,

Theorem 4: Suppose for every i, $IS(I_i)$ solves I_i. Then $IS(I)=IS(I_1)\cup...\cup IS(I_p)$ solves I.

Proof: We shall show that $IS(I)$ is informative w.r.t. S'' iff $IS(I_i)$ is informative w.r.t. S_i'' for all i; The theorem then will follow by the additivity of the cost function. If haplotypes k,l belong to the same observed component T_i, and there is a SNP s such that $H_{ks} \neq H_{ls}$, then by the key property it must be that $s \in S_i$. Therefore, the informativeness of $IS(I)$ implies the informativeness of $IS(I_i)$ for all i. For the opposite direction, suppose there are $t \in S''$ and $1 \leq l, k \leq n$ such that $H_{kt} \neq H_{lt}$. Let T_i be the subtree which contains the edge with label t (i.e., $t \in S_i$). Then by the key property, there are l',k' in T_i such that $H_{k't} \neq H_{l't}$, where l',k' are the observed nodes of T_i that are on the path from k to l in T. But then there is $s' \in IS(I_i) \subseteq IS(I)$ such that $H_{k's'} \neq H_{l's'}$. Hence, by the key property, $H_{ks'} \neq H_{ls'}$. $\square$

3.2.2 Solving MCIS on an Ancestral Component

In this section we solve MCIS restricted to a single ancestral component. We first reformulate it in terms of the tree edges, and then show how to solve it. We introduce the following notations: Edges labeled by interesting SNPs are called *target edges*. The set of target edges is $\tau=\{e|f^{-1}(e)\cap S''\neq\varnothing\}$. It specifies the interesting information in terms of tree edges. An edge is *allowed* if it is labeled by some non-interesting SNP. The set of allowed edges is $\alpha=\{e|f^{-1}(e)\cap(S\backslash S'')\neq\varnothing\}$. These are the edge-analogs of potentially informative SNPs. Edges in $\tau\backslash\alpha$ are called *forbidden*. Forbidden edges cannot be used as informative, but edges in $\tau\cap\alpha$ can.

We now expand the definition of the cost function to edges: The *cost of an edge e*, denoted $C(e)$, is the minimum cost of a non-interesting SNP that labels e. For $e \in \tau \backslash \alpha$ define $C(e)=\infty$. This allows us to provide an equivalent formulation for MCIS:

Problem 6: Minimum Cost Separating Set (MCSS)

Input: The same input as for MCIS.

Goal: Find $E' \subseteq E$ of minimum cost, such that in $G=(V,E\backslash E')$ there are no two observed nodes that are connected by a path containing a target edge.

Proposition 4: MCIS and MCSS are equivalent.

Proof: It suffices to show that an informative set for H w.r.t. S'' separates those observed nodes that are connected by a path containing edges from τ, and vice versa. Observed nodes of T, v_1 and v_2, have corresponding haplotypes of H, H_k and H_l, and vice versa. But then by the key property $H_{ks} \neq H_{ls}$ iff s labels an edge on the path from v_1 to v_2. $\square$

We are now ready to outline a dynamic programming algorithm for MCSS. W.l.o.g. assume $|V|>2$. Take some internal node $r \in V$ and root T at r. For $v \in V$ denote by $T_v=(V_v,E_v)$ the subtree of T that is rooted at v. For a solution $S_v \subseteq E_v$ of the induced sub instance $I(T_v)$, denote by R_v the connected component which contains v in $G_v=(V_v,E_v\backslash S_v)$. The algorithm will scan T from the leaves up and at each node v form an optimal solution for the subtree T_v based on the optimal solutions for the subtrees of its children. When combining such children solutions, we have to take into consideration the possibility that the combination will generate new paths between observed haplotypes, with or without target edges on them. To do this, we distinguish three types of solutions: S_v is called *empty* if there are no observed haplotypes in R_v. It is called *connected* if some observed haplotypes in R_v are connected to v via target edges. S_v is called *disconnected* otherwise, i.e., if there are observed haplotypes in R_v but there is no path connecting an observed haplotype to v via target edges. Let N_v, P_v and A_v denote the respective best empty, connected, or disconnected solutions. We define recursive formulae for their costs as follows:

- For a leaf node $v \in V$ we initialize: $C(N_v)=\infty$, $C(P_v)= \infty$, $C(A_v)=0$.
- For an internal node $v \in V$ with children $\{u_1,...,u_{k(v)}\}$ we write:

$$Tear(i)=\min\left\{C\left(N_{u_i}\right),C\left(P_{u_i}\right)+C\left(v,u_i\right),C\left(A_{u_i}\right)+C\left(v,u_i\right)\right\} \tag{1}$$

$$C\left(N_v\right)=\sum_{i=1}^{k(v)}Tear(i) \tag{2}$$

$$C\left(P_v\right)=\min\left\{\min_j\left\{C\left(P_{u_j}\right)+C\left(N_v\right)-Tear(j)\right\},\min_{j|(v,u_j)\in\tau}\left\{C\left(A_{u_j}\right)+C\left(N_v\right)-Tear(j)\right\}\right\} \tag{3}$$

If $\left\{i\left|\left(v,u_i\right)\notin \tau\right.\right\}=\phi$ then $C\left(A_v\right)=\infty$

Otherwise $C\left(A_v\right)=C\left(A_{u_j}\right)+\sum\limits_{\left(v,u_i\right)\notin\tau,i\neq j}\min\left\{C\left(A_{u_i}\right)Tear(i)\right\}+\sum\limits_{\left(v,u_i\right)\notin\tau}Tear(i)$ (4)

where $j=\underset{i\left|\left(v,u_i\right)\notin\tau\right.}{\arg\min}\left(C\left(A_{u_i}\right)-Tear(i)\right)$

The auxiliary value $Tear(i)$ measures the cost of an empty solution for the subtree including the edge (v,u_i) and the subtree of u_i. In computing $C(P_v)$ we have to either pick the cheapest of two alternatives: (a) all the subtrees are empty except one which is connected (first term in (3)), (b) all the subtrees are empty except one that is disconnected but incident on v via a target edge (second term). In computing $C(A_v)$ we find the best disconnected subtree, and allow the remaining subtrees to be either disconnected or empty. These formulae are implemented in a dynamic program as follows: (1) Visit V in postorder, computing $C(N_v)$, $C(P_v)$ and $C(A_v)$ for each $v \in V$. Obtain the minimal cost by $\min\{C(N_r),C(P_r),C(A_r)\}$. (2) Compute N_v, P_v and A_v by following the traceback pointers to get all those (v,u_i) edges that were chosen by the minimal cost while taking $C\left(P_{u_i}\right)+C\left(v,u_i\right)$ or $C\left(A_{u_i}\right)+C\left(v,u_i\right)$. The time complexity of this algorithm is $O(|S|)$.

3.3 Tag SNPs from Genotypes

Up until now we have followed the standard assumption in the computational literature [13,24,25] that tag SNPs need to reconstruct the full binary haplotypes from binary haplotypes of the tag set. As experiments that provide haplotypes are expensive, most studies seek to obtain experimentally only genotypes. For such data, the problem of finding tag SNPs should be reformulated to reflect the fact that the input is *genotypes*, rather than haplotypes: Recall that standard genotyping has three possible calls per site: $\{0,0\}$, $\{1,1\}$ and $\{0,1\}$, where the first two are homozygous and the latter is heterozygote. (The calls are often abbreviated to 0,1, and 2 respectively, and the genotype is represented as a vector over $\{0,1,2\}$.) The following question arises: Find a subset of SNPs given whose genotype calls one can completely identify the pair of haplotypes of an individual. We call such subset *phasing tag SNPs*.

Formally, let H be a set of haplotypes over a set S of SNPs, and consider genotypes formed from haplotype pairs in H. Denote by $g(k,l)_S$ the genotype formed from H_k and H_l on the SNP set S. We say that $\{i_1,i_2\}$ and $\{j_1j_2\}$ are *distinct with respect to S* if there is $s \in S$ such that $g(i_1,i_2)_s \neq g(j_1j_2)_s$.

Definition: $S' \subseteq S$ is a set of *phasing tag SNPs* if every two haplotype pairs from H are distinct with respect to S'. Hence, from the genotype calls of an individual for the set S', one can uniquely determine the exact sequence of the complete set S for each of its two haplotypes.

In general, the definitions of phasing tag SNPs and tag SNPs differ (see Fig. 2). The former is stronger:

Observation 5: If $S' \subseteq S$ are phasing tag SNPs then they are also tag SNPs.

Proof: All homozygous genotype-call vectors are distinct w.r.t. S': for all $i \neq j$, $g(i,i)_{S'} \neq g(j,j)_{S'}$. $\square$

We now show that, surprisingly, under the perfect phylogeny model, tag SNPs and phasing tag SNPs are equivalent. This identifies the commonly used definition with the more theoretically sound one, and therefore justifies the application of the current body of theory on tag SNPs to genotype data.

Theorem 5: Suppose that the haplotypes in H satisfy the perfect phylogeny model on S. A set $S' \subseteq S$ is a tag SNPs set if and only if S' is a phasing tag SNPs set.

Proof: It suffices to prove the "only if" direction. Suppose to the contrary that S' are tag SNPs but not phasing tag SNPs. Let $G_i = \{H_1, H_2\}$ and $G_j = \{H_3, H_4\}$ be distinct haplotype pairs with the same genotype call vector for S', i.e., $g(1,2)_{S'} = g(3,4)_{S'}$. Since S' is a tag SNP set, it distinguishes H_1 and H_3, so there must be $s_1 \in S'$ such that G_i and G_j are heterozygous to s_1, and H_1 and H_3 have different alleles for s_1. Similarly there must be $s_2 \in S'$ such that G_i and G_j are heterozygous to s_2, and H_1 and H_4 have different alleles for s_2. Therefore G_i and G_j are oppositely phased on s_1 and s_2. Since H_1, H_2, H_3, and H_4 are distinct, they violate the 4 gamete rule on s_1, s_2, in contradiction to Theorem 1. $\square$

4 Discussion

We studied here several questions arising in haplotype inference under the perfect phylogeny model. We introduced the model of xor-genotypes, and showed results that lay the computational foundation for the use of such data: (i) Inference of the sample haplotypes (up to negation) by adapting graph realization algorithms. (ii) Only two or three additional full genotypes are needed to completely resolve the haplotypes.

Simulations with genetic data show that xor genotypes are nearly as informative as full genotypes. Hence, genotyping methods that distinguish only between heterozygotes and homozygotes could potentially be applied to large scale genetic studies. Xor-genotypes may have economical advantage over the complete genotypes common today, since the information in a xor-genotype is only a fraction of the information given by a complete genotype. The feasibility and economic benefit of xor-genotype data cannot be appreciated by currently available technologies, but this work lays the foundation for evaluating the cost-effectiveness of technologies for obtaining such data.

The second part of the manuscript studied choosing a subset of the SNPs that fully describes the sample haplotypes. We provided efficient solutions to several optimization problems arising in this topic: We generalized previous results by finding optimal informative SNP set for any interesting set, and more generally, showed how to handle differential costs of SNPs. Finally, we have shown how to find tag SNPs for genotype data, which generalize the definition of tag SNPs to a more practical aspect.

Acknowledgements

We thank Orna Man for helpful ideas and practical knowledge. We thank Gadi Kimmel for productive discussions. We thank the reviewers for their helpful comments. I. P. was supported by an Eshkol postdoctoral fellowship from the Ministry of Science, Israel. R. S. was supported in part by the Israel Science Foundation (grant 309/02).

References

1. Sachidanandam R, et al. (International SNP Map Working Group). A map of human genome sequence variation containing 1.42 million single nucleotide polymorphisms. *Nature*, 2001; 409(6822): 928-33.
2. Patil N, et al. Blocks of Limited Haplotype Diversity Revealed by High Resolution Scanning of Human Chromosome 21. *Science*, 2001; 294(5547): 1719-23
3. Daly MJ, Rioux JD, Schaffner SF, Hudson TJ and Lander ES. High resolution haplotype structure in the human genome. *Nature Genetics*, 2001; 29(2): 229-32.
4. Jeffreys AJ, Kauppi L and Neumann R. Intensely punctate meiotic recombination in the class II region of the major histocompatibility complex. *Nature Genetics*, 2001; 29(2): 109-11.
5. Nachman MW and Crowell SL. Estimate of the mutation rate per nucleotide in humans. *Genetics*, 2000; 156(1): 297-304.
6. Gabriel SB, et al. The structure of haplotype blocks in human genome. *Science*, 2002; 296(5576): 2225-9.
7. Clark AG. Inference of haplotypes from PCR-amplified samples of diploid populations. *Molecular Biology and Evolution*, 1990; 7(2): 111-22
8. Excoffier L and Slatkin M. Maximum-likelihood estimation of molecular haplotype frequencies in a diploid population. *Molecular Biology and Evolution*, 1995; 12(5): 921-7.
9. Gusfield D. Haplotyping as Perfect Phylogeny: Conceptual Framework and Efficient Solutions. *Proceedings of the Sixth Annual International Conference on Computational Biology* 2002 (*RECOMB* '02): 166-75.
10. Bafna V, Gusfield D, Lancia G, and Yooseph S. Haplotyping as Perfect Phylogeny: A direct approach. Technical Report U.C. Davis CSE-2002-21, 2002.
11. Eskin E, Halperin E and Karp RM. Efficient reconstruction of haplotype structure via perfect phylogeny. To appear in the *Journal of Bioinformatics and Computational Biology* (*JBCB*), 2003.
12. Garey MR and Johnson DS Computers and Intractability, p. 222 Freeman, New York, 1979.
13. Bafna V, Halldórsson BV, Schwartz R, Clark AG and Istrail S. Haplotypes and informative SNP selection algorithms: don't block out information. *Proceedings of the Seventh Annual International Conference on Computational Biology* 2003 (*RECOMB* '03): 19-27.
14. Xiao W and Oefner PJ, Denaturing high-performance liquid chromatography: A review. *Human Mutation*, 2001; 17(6): 439-74.
15. Bixby RE and Wagner DK. An almost linear-time algorithm for graph realization, *Mathematics of Operations Research*, 1988; 13(1): 99-123.
16. Tutte WT. An Algorithm for determining whether a given binary matroid is graphic. *Proceedings of American Mathematical Society*, 1960; 11: 905-917.
17. Gavril F and Tamari R. An algorithm for constructing edge-trees from hypergraphs, *Networks* 1983; 13:377-388.

18. Zhang K, Deng M, Chen T, Waterman MS and Sun F. (2002) A dynamic programming algorithm for haplotype block partitioning. *Proceedings of the National Academy of Sciences*, 99(11): 7335-9.
19. Chung RH and Gusfield D. Perfect Phylogeny Haplotyper: Haplotype Inferral Using a Tree Model. *Bioinformatics*, 2002; 19(6): 780-781.
20. Johnson GC, et al. Haplotype tagging for the identification of common disease genes. *Nature Genetics*. 2001; 29(2): 233-7.
21. Gusfield D. Algorithms on Strings, Trees, and Sequences - Computer Science and Computational Biology. Cambridge University Press 1997.
22. Hudson RR. Generating samples under a Wright-Fisher neutral model of genetic variation. *Bioinformatics*, 2002; 18(2): 337-38.
23. Chung RH and Gusfield D. Empirical Exploration of Perfect Phylogeny Haplotyping and Haplotypers. *Proceedings of the nineth International Computing and Combinatorics Conference* 2003 (*COCOON* '03): 5-19.
24. Sebastiani P, Lazarus R, Weiss ST, Kunkel LM, Kohane IS and Ramoni MF. Minimal haplotype tagging. *Proceedings of the National Academy of Sciences of the USA*, 2003; 100(17): 9900-5.
25. Chapman JM, Cooper JD, Todd JA and Clayton DG. Detecting disease associations due to linkage disequilibrium using haplotype tags: a class of tests and the determinants of statistical power. *Human Heredity*, 2003; 56(1-3): 18-31.
26. Kwok PY. Genetic association by whole-genome analysis. *Science*, 2001; 294(5547): 1669-70.
27. Pe'er I and Beckmann JS. Resolution of haplotypes and haplotype frequencies from SNP genotypes of pooled samples. *Proceedings of the Seventh Annual International Conference on Computational Biology* 2003 (*RECOMB* '03): 237-246.

Sorting by Length-Weighted Reversals: Dealing with Signs and Circularity

Firas Swidan[1], Michael A. Bender[2,*], Dongdong Ge[2], Simai He[2],
Haodong Hu[2], and Ron Y. Pinter[1,**]

[1] Department of Computer Science,
Technion – Israel Institute of Technology, Haifa 32000, Israel
{firas,pinter}@cs.technion.ac.il
[2] Department of Computer Science,
SUNY Stony Brook, Stony Brook, NY 11794-4400, USA
{bender,dge,simaihe,huhd}@cs.sunysb.edu

Abstract. We consider the problem of sorting linear and circular permutations and 0/1 sequences by reversals in a length-sensitive cost model. We extend the results on sorting by length-weighted reversals in two directions: we consider the signed case for linear sequences and also the signed and unsigned cases for circular sequences. We give lower and upper bounds as well as guaranteed approximation ratios for these three cases. The main result in this paper is an optimal polynomial-time algorithm for sorting circular 0/1 sequences when the cost function is additive.

1 Introduction

Sorting by reversal (SBR) is of key importance in the study of genome rearrangement. To date, most of the algorithmic work on this problem applies to linear sequences (permutations or strings), and assumes the naïve unit-cost model. This model corresponds to a simple evolutionary model in which inversion mutations of any length are considered equally likely [1]; however, the mechanics of genome-rearrangement events (both inversions and other less ubiquitous transformations) suggest that the probabilities of reversals are dependent on fragment length.

Recently, a length-sensitive model was introduced [2,3], in which the cost of a reversal is a function of the length of the reversed sequence and the total cost is the sum of the costs of the individual reversals. Several nontrivial lower and upper bounds have been derived for the problem of SBR under this cost model. However, these results pertain only to the unsigned version of the problem, where the direction of the individual elements does not matter (as opposed to the signed version); in the study of genome rearrangement, the direction of the

* Supported in part by NSF Grants EIA-0112849, CCR-0208670, HRL Laboratories, and Sandia National Laboratories.
** Supported in part by the Bar-Nir Bergreen Software Technology Center of Excellence.

S.C. Sahinalp et al. (Eds.): CPM 2004, LNCS 3109, pp. 32–46, 2004.

elements is often important, since each element represents a whole gene (or a portion thereof) whose direction has significance. Furthermore, many interesting genomes are *circular*; these include bacterial, mitochondrial, and chloroplast genomes, which all play an important role in studies of comparative genomics.

In this paper we extend the recent work for the length-sensitive model on linear sequences in which the cost of reversing a subsequence of length ℓ is $f(\ell) = \ell^\alpha$ (where $\alpha \geq 0$). We consider both permutations as well as $0/1$ sequences representing the extreme cases of no repetitions at all among the elements on one hand, and as many repetitions as makes sense on the other hand. The elements are arranged in both linear and circular fashion, and we consider both the signed and unsigned cases.

Note that in general circularity offers more opportunities for lowering the optimal cost to sort a given sequence by reversals, and at the same time circularity presents challenges to providing efficient solutions. A non-unit cost model exacerbates the problems even farther. A sequence that exemplifies this distinction is $10^{n/2-1}1^{n/2-1}0$. One can sort the sequence into $0^{n/2}1^{n/2}$ with 2 reversals. Under a length-sensitive model with $\alpha = 1$ (i.e., the reversal cost is identical to the length of the sequence) the overall cost is $2n - 2$. In the circular case, where we adjoin the two ends of the sequence to form a circular arrangement, 1 reversal operation is enough to sort the sequence; its cost in the length sensitive model is 2. Thus, whereas the ratio between the costs of the two optimal solutions in the unit-cost model is 2, in the length-sensitive model it is $\Theta(n)$. Consequently, the treatment of circular sequences requires more care than the linear case.

Notice that the following relationships between the costs of the four sorting cases hold:

$$\text{unsigned circular} \leq \text{unsigned linear} \leq \text{signed linear} . \tag{1}$$

$$\text{unsigned circular} \leq \text{signed circular} \leq \text{signed linear} . \tag{2}$$

A summary of our results appears in Tables 1 and 2; note that the lower and upper bounds for the linear, unsigned case were extended verbatim from [3], whereas the approximation ratios vary somewhat among the different cases.

Table 1. Lower and upper bounds for SBR of signed or unsigned and linear or circular $0/1$ sequences and permutations.

α Value	Lower Bounds	Upper Bounds	
		Permutations	$0/1$'s
$0 \leq \alpha < 1$	$\Omega(n)$	$O(n \lg n)$	$\Theta(n)$
$\alpha = 1$	$\Omega(n \lg n)$	$O(n \lg^2 n)$	$\Theta(n \lg n)$
$1 < \alpha < 2$	$\Omega(n^\alpha)$	$\Theta(n^\alpha)$	$\Theta(n^\alpha)$
$\alpha \geq 2$	$\Omega(n^2)$	$\Theta(n^2)$	$\Theta(n^2)$

Considerable research has been reported on the algorithmics of SBR (mostly signed) for linear sequences in the unit-cost model, see e.g. [4, 5]. Recently, several papers have addressed the issues of circularity and duplication, most notably

Table 2. Approximation ratios for SBR of signed linear as well as signed and unsigned circular 0/1 sequences and permutations.

α Value	Signed Linear Permutations	0/1's	Unsigned Circular Permutations	0/1's	Signed Circular Permutations	0/1's
$0 \leq \alpha < 1$		$O(1)$		$O(1)$		$O(1)$
$\alpha = 1$	$O(\lg n)$	3	$O(\lg n)$	1	$O(\lg n)$	3
$1 < \alpha < 2$	$O(\lg n)$	$O(1)$				
$\alpha \geq 2$	$O(1)$	$O(1)$	$O(1)$	$O(1)$	$O(1)$	$O(1)$

Chen and Skiena [6] who deal with fixed-length reversals to sort both linear and circular permutations, Hartman [7] who deals with sorting by transpositions (which are much less common than reversals) on a circle, and Christie and Irving [8] who look at sorting (essentially) 0/1 linear sequences by both reversals and transpositions. None of these results, however, pertain to the length-sensitive model.

The rest of this paper is organized as follows. In Sect. 2 we provide algorithms for sorting 0/1 sequences, in Sect. 3 we deal with sorting circular permutations, and in Sect. 4 we show how to handle signed permutations in our length-sensitive model. Finally, in Sect. 5 we give sorting bounds for permutations and 0/1 sequences when inputs are signed or unsigned and circular or linear.

2   Exact and Approximation Algorithms for Sorting 0/1 Sequences

In this section we give exact and approximation algorithms for sorting linear signed as well as circular singed and unsigned 0/1 sequences. We first introduce approximation algorithms when $0 \leq \alpha < 1$. We then give an exact algorithm to sort unsigned circular 0/1 sequences when $\alpha = 1$. Finally, we introduce a reduction showing that the signed case can be approximated using the unsigned case when $\alpha \geq 1$.

2.1   Approximation Algorithms for $0 \leq \alpha < 1$

We now give $O(1)$ approximation algorithms for sorting linear signed as well as circular signed and unsigned sequences. The results are proved using potential function arguments.

Sorting Unsigned Circular Sequences. Given a circular sequence S, denote the length of the 0 and 1 blocks contained in S by $z_1, \ldots, z_k$ and $w_1, \ldots, w_k$ respectively. Let $Z = \max_{1 \leq i \leq k}\{z_i\}$ and $W = \max_{1 \leq i \leq k}\{w_i\}$. We define the potential function $P(S)$ as follows:

$$P(S) = \sum_{i=1}^{k}(z_i^\alpha + w_i^\alpha) - Z^\alpha - W^\alpha .$$

Lemma 1. *A reversal ρ of length r acting on a circular sequence S increases the value of the potential function $P(S)$ by at most $4r^\alpha$, that is, $P(S \cdot \rho) - P(S) \leq 4r^\alpha$.*

Proof. The proof is by case analysis: since a reversal can cut two blocks at most (the blocks at either edge of the reversal), a reversal increases the value of the term $\sum_{i=1}^{k}(z_i^\alpha + w_i^\alpha)$ by at most $2r^\alpha$. In addition, by similar reasoning, the value of either Z^α or W^α can decrease by at most r^α. □

Denote $S' = S \cdot \rho$. Notice that $S = S' \cdot \rho^{-1}$. Therefore, the reversal ρ decreases the potential value by the same amount that ρ^{-1} increases it. Thus, by applying Lemma 1 on the inverse reversal, we get a lower bound on the decrease of the potential value.

Corollary 1. *A reversal ρ of length r acting on a circular sequence S decreases the value of the potential function $P(S)$ by at most $4r^\alpha$, that is, $P(S \cdot \rho) - P(S) \geq -4r^\alpha$.*

Because the cost of a reversal of length r is r^α, and the value of $P(\cdot)$ equals 0 for a sorted sequence, we obtain the following lower bound.

Lemma 2. *The function $V(S) = \frac{1}{4}P(S)$ is a lower bound for sorting an unsigned circular sequence S.*

Proof. The proof is by induction on the number of reversals in an optimal solution.

Let m denote the number of reversals in an optimal sorting series. We prove that if a sorting solution uses exactly m reversals, then its cost is at least $V(S)$.

Base case: when $m = 0$, the sequence is already sorted. Induction step: suppose for all $m \leq k$ the claim holds. Consider a 0/1 sequence S having an optimal sorting series of length $m = k + 1$. Denote the first reversal by ρ and let r be its length. The reversal ρ changes the sequence S to S', which can be optimally sorted by k reversals. Applying the induction assumption on S', we get that $V(S')$ is a lower bound for sorting S'. By Corollary 1, $P(S') + 4r^\alpha \geq P(S)$. By the definition of $V(\cdot)$ we get: $V(S') + r^\alpha \geq V(S)$. Therefore:

$$\text{opt}(S) = \text{opt}(S') + r^\alpha \geq V(S') + r^\alpha \geq V(S) .$$

as needed. □

Lemma 2 motivates sorting a circular 0/1 sequence by fixing its two maximal blocks Z and W, and linearly sorting the two subsequences between them. The algorithm circularImprovedDC realizes this approach. Given a circular sequence S and two block indices i and j, consider the subsequence of blocks between i and j, counter clockwise, excluding the blocks i and j. If the subsequence starts with a 1, rename the 0's and 1's and define $S(i,j)$ to be the resulting subsequence. Define $S(j,i)$ similarly. In the following we use the algorithm improvedDC, introduced in [3], for sorting linear 0/1 sequences.

Theorem 1. *The algorithm circularImprovedDC has an approximation ratio of $O(1)$.*

Algorithm 1 circularImprovedDC(S)

1: Let $z_{i_0} = Z$ and $w_{j_0} = W$
2: Define $S_1 = S(i_0, j_0)$ and $S_2 = S(j_0, i_0)$
3: $s_1 \leftarrow$ improvedDC(S_1)
4: $s_2 \leftarrow$ improvedDC(S_2)
5: Output $s_1 + s_2$

Sorting Linear and Circular Signed Sequences. Consider a signed 0/1 sequence (linear or circular). Define a block in the sequence to be a contiguous segment of 0's (or 1's) having the same sign. Notice that there are four kinds of blocks in a signed 0/1 sequence.

Represent the 0/1 sequence as a sequence of blocks $b_1, \ldots, b_m$. Consider the potential function $V_1(S) = \frac{1}{2} \sum_{i=1}^{m} b_i^{\alpha}$ for linear sequences S and define $V_2(T)$ as in Lemma 2 for circular sequences T.

Lemma 3. *The potentials $V_1(S)$ and $V_2(T)$ are lower bounds on the cost of sorting linear and circular signed sequences respectively.*

Given a signed sequence S, let unsign(S) represent the sequence without the signs. The algorithm signedImprovedDC sorts signed linear sequences based on improvedDC.

Algorithm 2 signedImprovedDC(S)

1: $U \leftarrow$ unsign(S)
2: $u \leftarrow$ impovedDC(U)
3: Mimic the reversals used to sort U on S. Denote the resulting sequence by S'
4: Reverse elements of S' with a negative sign. Let s be the cost of this step
5: Output $s + u$

To sort circular signed sequences we modify the algorithm circularImprovedDC in a similar way. We refer to the modified algorithm as signedCircularImproved DC.

Theorem 2. *The algorithms signedImprovedDC and signedCircularImproved DC are $O(1)$ approximation algorithms.*

2.2 Optimal Algorithm for $\alpha = 1$

In this section we give a polynomial-time algorithm for sorting circular 0/1 sequences with additive cost functions ($\alpha = 1$). The sorting algorithm is based on dynamic programming. We give enough properties to constrain the set of candidate solutions, so that the optimal solution can be found in polynomial time. This approach was used in [3] to sort linear 0/1 sequences.

We first describe the three properties, "useless", "cutting" and "complex." The first two properties can be proved using the same techniques as in [3]. The main contribution of this part is in proving the third property, which is the most critical in establishing the optimality of the algorithm. Its proof is substantially different from the linear case, reflecting the essential divergence caused by the circularity. This is explained in detail below after we review some preliminaries.

Consider the two extreme blocks in a reversal. If the values of these blocks are identical, that is both equal 0 or 1, we call the reversal *useless*; if one of these two blocks is contained in a bigger block, we call the reversal *cutting*. If the reversal affects more than two blocks, we call the reversal *complex*. We call a reversal that is not complex *simple*. We can show that there are no useless and cutting reversals in a circular optimal solution. As mentioned earlier, the proof follows the same lines as in [3]. Roughly speaking, the same techniques apply because these proofs are "local," involving changes of reversals where the changes affect a single block of 0s or 1s (even though the reversal is longer). This explanation is elaborated below.

In contrast, the proof of the third property for linear sequences introduced in [3] depends heavily on the linearity and does not apply to circular sequences. In the proof by contradiction, one considers the last complex reversal. The changes that are introduced affect the two extreme blocks that the reversal contained. In the case of linear sequences, these blocks are far apart and one can subsequently make changes independently to the reversals involving these blocks. This property is not true for circular sequences. Specifically, reversals that involve one extreme block may also affect the other, because of "wrap-around".

In the following, all the sequences are circular unless mentioned otherwise.

Given a reversal series $\rho_1, \ldots, \rho_m$ acting on a 0/1 sequence $S = s_1, \ldots, s_q$, where $s_i \in \{0, 1\}$, denote the number of reversals in which element s_i participates by $N(s_i)$. Call $N(s_i)$ the *reversal count* of s_i.

When a subsequence $s_i, \ldots, s_j$ of S is never cut by a reversal series $\rho_1, \ldots, \rho_m$, we denote the number of reversals in which the subsequence takes part by $N(s_i, \ldots, s_j)$.

We show that for additive cost functions no optimal reversal series contains useless or cutting reversals, and there exists an optimal reversal series containing no complex reversals. Equation (3), relating the reversal counts to the reversal series cost is useful for the proofs.

$$\sum_{i=1}^{m} |\rho_i| = \sum_{j=1}^{q} N(s_j) \ . \tag{3}$$

The proofs of *useless* and *cutting*, appearing in [3], hold for the circular case.

Lemma 4. *A reversal series containing a useless reversal cannot be optimal.*

Lemma 5. *A reversal series containing a cutting reversal cannot be optimal.*

The following definitions are needed to prove that an optimal reversal series containing no complex reversals exists. Given a reversal ρ affecting a circular

0/1 sequence S, notice that we can cut S, making it linear, without cutting the reversal. Such a linearization is done in a counter clockwise direction. This linearization is implicitly used throughout the section.

We represent a 0/1 sequence $1^{w_1}, \ldots, 0^{w_{2\ell}}$ by the lengths of each block, that is we let $w = w_1, \ldots, w_{2\ell}$ denote a 0/1 sequence. We refer to w as a weighted sequence. A subsequence of w, $sg = w_i, \ldots, w_j$ or for short (i, j), is called a segment of w. Let $sg_k = w_{i_k}, \ldots, w_{j_k}$ for $k \in \{1, 2\}$ be two segments of w. We say that sg_1 is a sub-segment of sg_2, or contained in sg_2, if $i_2 \le i_1 < j_1 \le j_2$.

We denote a reversal acting on a segment (i, j) of w by $\rho(i, j)$. Let $w = w_1, \ldots, w_{2\ell}$ be a weighted sequence, and let $\rho(i, j)$ be a reversal acting on it. If $i \equiv j + 1 \pmod 2$, or for short $i \equiv j + 1$, we say that ρ unifies w_i with w_{j+1}, and w_j with w_{i-1}. The reversal ρ is called a unifying reversal. In addition, we say that block v_{i-1} of $v = w \cdot \rho = v_1, \ldots, v_{2\ell-2}$ contains blocks w_{i-1} and w_j of w. Similarly we say that v_{j-1} of v contains blocks w_{j+1} and w_i of w.

A reversal series $\rho_1, \ldots, \rho_m$ acting on w separates a segment sg of w, if $\rho_1, \ldots, \rho_m$ unifies all the 0 and 1 blocks of sg. If $sg = w$ then $\rho_1, \ldots, \rho_m$ sorts w.

In the sorting process the essence of a reversal is not directly connected to the coordinates (i, j) that define the reversal, but is designated by the blocks that the reversal affects. This essence plays a crucial rule when we try to define commutative reversals.

Given a reversal ρ acting on w and a reversal η acting on $w \cdot \rho$, we say that η commutes with ρ if two reversals η' and ρ' exist such that $w \cdot \rho \cdot \eta = w \cdot \eta' \cdot \rho'$, where η' and ρ' affect the same blocks of w that η and ρ affected respectively (in the following we drop the primes).

One can verify that the following proposition holds.

Proposition 1. *Let w be a weighted sequence, ρ be a reversal acting on it, and η be a reversal acting on $w \cdot \rho$. The three following conditions are equivalent:*

1. *The reversal η commutes with ρ.*
2. *The blocks that η affects in $w \cdot \rho$ are contiguous in w.*
3. *The segments that ρ and η affect are either contained one in another or disjoint.*

Given a weighted sequence w and a minimum sorting series $\rho_1, \ldots, \rho_m$, define $w^j = w \cdot \rho_1 \cdots \rho_j$. Let ρ_j be the greatest index complex reversal and s the segment that ρ_j affects. Denote by $\bar{s}$ the rest of the sequence. Let ρ_k for $k > j$ be the smallest index reversal such that s is separated in w^k. Consider msg, the maximal segment that contains s and is separated by $\rho_j, \ldots, \rho_k$. If $msg \ne w$, the circularity of w is not used in the separation of msg. Therefore, we can consider msg as a linear 0/1 sequence, and ρ_j as a complex reversal used through the separation of a linear 0/1 sequence. Under these assumptions, [3] proved that msg can be separated optimally without the use of complex reversals. Therefore, without loss of generality, we assume that $msg = w$. That is, s is separated only when w is sorted. The following lemmas characterize the reversals performed after ρ_j.

Lemma 6. *Let w, j, ρ_j, w^j, s, and $\bar{s}$ be as above. Let ρ_k for $k > j$ be a reversal that does not commute with ρ_j. Then ρ_k affects at least one block that contains mixed elements from s and $\bar{s}$. In addition, all the blocks that ρ_k affects become part of a mixed block.*

Lemma 7. *Let w, j, ρ_j, w^j, s, and $\bar{s}$ be as in Lemma 6. Let ρ_k for $k > j$ be the smallest index reversal that commutes with ρ_j. Then ρ_k commutes with all the reversals ρ_q for $k > q > j$.*

Corollary 2. *Let w, j, ρ_j, w^j, s, and $\bar{s}$ be as in Lemma 7. We can rearrange the reversal series so that the reversals ρ_k for $k > j$ do not commute with ρ_j.*

The reversals ρ_k for $k > j$ are simple reversals and by Corollary 2 do not commute with ρ_j.

The following lemmas characterize the sorting process under simple reversals. They are used for proving that the reversal series cannot be optimal if ρ_j remains complex after performing the rearrangement described in Corollary 2, and for calculating the optimal cost in polynomial time.

Given a weighted sequence $w = w_1, \ldots, w_{2\ell}$ and a reversal series $\rho_1, \ldots, \rho_m$ containing no cutting reversals, define c_i to be the number of reversals in which w_i takes part, i.e. $c_i = N(w_i)$. Notice that c_i is well defined, since $\rho_1, \ldots, \rho_m$ contains no cutting reversals.

Lemma 8. *Let $w = w_1, \ldots, w_{2\ell}$ be a weighted sequence and let $\rho_1, \ldots, \rho_m$ be a sorting series having no useless, no cutting, and no complex reversals. Denote $w^k = w \cdot \rho_1 \cdots \rho_k$, and let $w^0 = w$. Then each block of w^k contains a block of w that takes part during $\rho_1, \ldots, \rho_k$ in zero reversals.*

Proof. By induction on k. Base case: the claim is trivial for $k = 0$. Induction step: suppose each block of w^k contains a block of w that takes part during $\rho_1, \ldots, \rho_k$ in zero reversals. We need to prove that each block of w^{k+1} contains a block of w that takes part during $\rho_1, \ldots, \rho_{k+1}$ in zero reversals. Since ρ_{k+1} is simple, we know that it acts on two successive blocks of w^k. Since ρ_{k+1} is not cutting, ρ_{k+1} must be of the form $\rho(i-1, i)$. Such a reversal unifies the block w_i^k with w_{i-2}^k, and w_{i-1}^k with w_{i+1}^k. The other blocks of w^k are not affected by ρ_{k+1}, therefore, they contain by the induction assumption a block of w that takes part during $\rho_1, \ldots, \rho_k$ in zero reversals. The same block takes part during $\rho_1, \ldots, \rho_k$ in zero reversals as well.

Since ρ_{k+1} affects w_i^k unifying it with w_{i-2}^k, and by the induction assumption w_{i-2}^k contains a block v of w that takes part during $\rho_1, \ldots, \rho_k$ in zero reversals, the unification of w_i^k with w_{i-2}^k in w^{k+1} contains v. Since v is not affected by ρ_{k+1}, v takes part during $\rho_1, \ldots, \rho_k$ in zero reversals, and thus fulfills the requirement.

A similar analysis applies to the unification of w_{i-1}^k with w_{i+1}^k in w^{k+1}.   □

Lemma 9. *Let $w = w_1, \ldots, w_{2\ell}$ be a weighted sequence and let $\rho_1, \ldots, \rho_m$ be a sorting series having no useless, no cutting, and no complex reversals. There exist indices i and j, where i corresponds to a block of 1's and j corresponds to a block of 0's, such that $c_i = c_j = 0$.*

Lemma 9 is helpful in proving that the outcome of Corollary 2 is a reversal series containing no complex reversals.

Lemma 10. *Let $w, j, \rho_j, w^j, s,$ and $\bar{s}$ be the outcome of Corollary 2, that is ρ_k for $k > j$ do not commute with ρ_j. If ρ_j remains a complex reversal, the reversal series $\rho_1, \ldots, \rho_m$ cannot be optimal.*

Proof. We show that a reversal series having a lower cost exists. Consider the sequence w^j. The reversals ρ_k for $k > j$ are simple. By Lemma 9 there exists indices i' and p', where i' and p' correspond to a block of 1's and 0's respectively, such that $c_{i'} = c_{p'} = 0$. Notice that each of the blocks i' and p' could be contained as a unit in w^{j-1}, or could correspond to two blocks, one in s and one in $\bar{s}$. In the latter case, pick the part of the block that is in $\bar{s}$. In the former pick the blocks themselves. Denote the indices of the picked blocks in w^{j-1} by i and p. We divide the analysis into three cases:

1. The indices i and p are contained in $\bar{s}$. In this case we get $c_p = c_i = 0$, and therefore i and p divide the circular sequence into two linear subsequences. We consider the subsequence between i and p that contains s, denote it v, and the restriction of $\rho_1, \ldots, \rho_m$ to v. This restriction separates v and contains a complex reversal followed by simple reversals that do not commute with it. As mentioned before, [3] proved that such a series cannot be optimal.

2. The indices i and p are contained in s. This imposes that $|i - p| = 1$, or else the reversals acting on the blocks between i and p commute with ρ_j. Suppose that i appears before p in s when scanning the circular sequence in the counter clockwise direction. The other case is handled by renaming the 0's and 1's. Performing the reversals ρ_k for $k > j$ restricted to s and to $\bar{s}$ separates $\bar{s}$, while s gets the form $0^+1^+0^+1^+$. The notation 0^+ corresponds to an arbitrary strictly positive weight of 0's. The whole sequence is of the

$$\overset{\bar{s}}{\overbrace{0^+}}\ \overset{s}{\overbrace{0^+1^+0^+1^+}}\ \overset{\bar{s}}{\overbrace{1^+}}$$

form $0^+\ 0^+1^+0^+1^+\ 1^+$. Notice that the orientation of s and $\bar{s}$ is opposed since ρ_j is not performed on s. To sort the sequence, perform a last reversal

$$\overset{\bar{s}}{\overbrace{0^+}}\ 0^+\ \overset{\text{last reversal}}{\overbrace{1^+0^+}}\ 1^+\ \overset{\bar{s}}{\overbrace{1^+}}$$

$0^+\ 0^+\ 1^+0^+\ 1^+\ 1^+$. The cost of the modified series is not greater than the cost of the original one, since the last reversal's length is not greater than ρ_j's length. However, by Lemma 6, the restriction of ρ_{j+1} to s or to $\bar{s}$ produces a reversal containing a single block. If we omit this reversal from the modified series the cost of the modified series becomes smaller than the original one. A contradiction to the optimality of the original series.

3. The index i is contained in s and p contained in $\bar{s}$ or vice versa. We analyse the former case. The latter is handled by renaming the 0's and 1's. Notice that after performing ρ_j, the indices i and p divide the circle into two independent components. The reversal ρ_m can affect only one of these components. Consider the following division of the sequence w^j (counter

clockwise): $0^+\ \cdots\ \cdots\ 1^+\ \cdots\ \cdots$, where $\bar{s}_1$ and s_1 are the segments

Table 3. Available reversal counts summary.

Segment	$\bar{s}_2$		i	s_1			$\bar{s}_2$		
Block	1^+	0^+	1^+	1^+	1^+	0^+	0^+	1^+	0^+
Available reversal count	2	2	1	1	1	1	1	1	0
Explanation: reversals skipped	ρ_j,ρ_m	ρ_j,ρ_m	ρ_j	ρ_j	ρ_j	ρ_j	ρ_m	ρ_m	-

of $\bar{s}$ and s respectively lying between p and i, while $\bar{s}_2$ and s_2 are the segments of $\bar{s}$ and s respectively lying between i and p. Suppose that ρ_m affects $\bar{s}_2$ and s_2. Then the reversals ρ_k for $m > k > j$ separate s_1 and $\bar{s}_1$. Notice that omitting ρ_j does not affect the separation state of these segments. On the other hand, performing ρ_k for $m > k > j$ on $\bar{s}_2$ and s_2 does not separate them. Modify the reversal series as follows: Perform ρ_k for $m > k > j$ on w^{j-1} restricted to each of $\bar{s}_1, s_1, \bar{s}_2$, and s_2 (do not perform ρ_j and ρ_m). The resulting sequence is of the form: $\overset{p}{0^+} \ \overset{\bar{s}_1}{0^+ 1^+} \ \overset{s_2}{1^+ 0^+ 1^+} \ \overset{i}{1^+} \ \overset{s_1}{1^+ 0^+} \ \overset{\bar{s}_2}{0^+ 1^+ 0^+}$.
Table 3 summarizes the yet available reversal counts for part of the segments. To sort the remaining blocks perform the following reversals (reversals indicated by brackets $[,]$):

$$\overset{s_2}{1^+ [0^+ 1^+} \ \overset{i}{1^+} \ \overset{s_1}{1^+] 0^+} \ \text{and} \ \overset{s_2}{1^+ [0^+} \ \overset{s_1}{0^+} \ \overset{\bar{s}_2}{0^+ 1^+] 0^+} \ .$$

The two reversals sort the sequence, and do not violate the available reversal counts. Thus the modified reversal series has a cost not greater than the original one. A contradiction is established similarly to previous cases.

Putting it all together we prove the claim.   □

Lemmas 4, 5, and 10 establish Theorem 3.

Theorem 3. *An optimal reversal series contains no useless and no cutting reversals. In addition, an optimal reversal series containing no complex reversals exists.*

Theorem 3 and Lemma 9 implies that finding the optimal sorting cost can be done by taking the minimum over the following set: for all pair of indices (i, j), where $i < j$ and $i - j \equiv 1$, sort the segments $sg_1 = (i, j)$ and $sg_2 = (j, i)$ under the restriction $c_i = c_j = 0$. The sum of the two sorting costs is a candidate for the optimal solution. The calculation of the sorting costs for all pairs (i, j) and (j, i) can be done by running zerOneSort on ww, where w is an arbitrary linearization of a circular $0/1$ sequence. The algorithm zerOneSort, introduced in [3], is an exact algorithm for sorting unsigned linear sequences when $\alpha = 1$. The algorithm circularZerOneSort implements this idea.

Theorem 4. *The algorithm circularZerOneSort sorts a circular $0/1$ sequence optimally with a time complexity in $O\left(n^3\right)$ and a space complexity in $O\left(n^2\right)$*

Algorithm 3 circularZerOneSort(w)

1: Run zerOneSort on ww. Let A be the dynamic programming matrix built by zerOneSort
2: $opt \leftarrow \infty$
3: **for** $i = 0$ to $|w| - 1$ in steps of 1 **do**
4:    **for** $j = i + 1$ to $|w| - 1$ in steps of 2 **do**
5:       $tmp \leftarrow A(i, j, c_i = c_j = 0) + A(j, i + |w|, c_j = c_i = 0)$
6:       **if** $tmp < opt$ **then**
7:          $opt \leftarrow tmp$
8:       **end if**
9:    **end for**
10: **end for**
11: Output opt

2.3 Approximation Algorithms for $\alpha \geq 1$

We now give a reduction showing that a signed circular or linear sorting algorithm can be derived from an unsigned circular or linear sorting algorithm
respectively, while retaining the asymptotic approximation ratio.

Let A be an algorithm for sorting unsigned circular $0/1$ sequences and T
be a signed circular sequence. We can sort T by running A on unsign(T) and
correct the signs of the resulting sequence elements. Lemma 11 shows that this
method approximates the optimal sorting cost. A similar result applies to the
singed linear case.

Lemma 11. *The sum of A(unsign(T)) and the cost of signs correction is less
than* opt$(T) + 2A$(unsign(T)).

By combining the optimality of circularZerOneSort (Sect. 2.2) and Lemma 11,
we obtain an approximation algorithm for signed circular sequences when $\alpha = 1$.
A similar result applies to signed linear sequences.

Corollary 3. *The algorithm circularZerOneSort followed by signs correction is
a 3-approximation algorithm for sorting signed circular sequences when $\alpha = 1$.*

In [3] its shown that bubble sort is optimal for sorting unsigned $0/1$ sequences
when $\alpha \geq 2$. By Lemma 11 we get an approximation algorithm for sorting signed
sequences when $\alpha \geq 2$.

Corollary 4. *Bubble sort followed by signs correction is a 3-approximation algorithm for sorting signed linear or circular sequences when $\alpha \geq 2$.*

3 Sorting Circular Permutations

In this section we present an approximation algorithm for sorting circular permutations when $\alpha = 1$. The algorithm is based on an analysis that there exists
a cut that changes the circular permutation into a linear permutation, such that

the sorting cost for the linear permutation is only a constant factor larger than the sorting cost for the circular permutation. The algorithm is therefore to try all possible cuts. The proof follows an accounting argument given in this section, and is heavily dependent on the additivity of the cost function. The main idea of the argument is that we can mimic the circular reversals that we cut with linear ones, such that the linear reversals go the other way around the circle, not crossing the cut. The accounting argument shows that the total cost increases by at most a constant factor.

Let Π be a circular permutation on n elements. We present Π as a sequence of integers. Denote the n linear permutations equivalent to the circular identity permutation by I_j for $1 \leq j \leq n$. In this section, unless mentioned otherwise, permutations are circular.

Associate an *index* with each space between two successive elements of Π. A reversal *affects* an index i, if the reversal affects the two elements neighboring i. Map each index i to the number of reversals r_i affecting the index. We refer to r_i as the *index reversal count*. Let i_0 be an index such that $r_{i_0} = \min_i \{r_i\}$.

In the following we establish a lower bound on the sorting cost using r_{i_0}. Denote the circular optimal sorting cost of a circular permutation Π by $\mathrm{cosc}(\Pi)$. The linear optimal sorting cost of a linear permutation $\hat{\Pi}$ is denoted by $\mathrm{losc}(\hat{\Pi})$.

Lemma 12. *Let Π be a circular permutation, $\rho_1, \ldots, \rho_m$ be an optimal reversal sequence, and i_0 be as above. The following holds:*

$$\mathrm{cosc}(\Pi) \geq n \cdot r_{i_0} \ . \tag{4}$$

Proof. By definition, a reversal affects an index if and only if it affects its two neighbors. Let i be an index with two neighbors π_j and π_{j+1}. We get that $2 \cdot r_i \leq N(\pi_j) + N(\pi_{j+1})$, where $N(\pi_j)$ is the number of reversals in which element π_j takes part. Notice that the set of neighbors of all odd or all even indices does not contain repetition. Thus, we get:

$$\sum_j N(\pi_j) \geq \max \{ \sum_{i \text{ odd}} 2 \cdot r_i, \sum_{i \text{ even}} 2 \cdot r_i \} \ .$$

For optimal sorting series, one can verify that $\sum_j N(\pi_j) = \mathrm{cosc}(\Pi)$. Hence, $\mathrm{cosc}(\Pi) \geq \max \{ \sum_{i \text{ odd}} 2 \cdot r_i, \sum_{i \text{ even}} 2 \cdot r_i \}$. From the definition of i_0 we get: $\mathrm{cosc}(\Pi) \geq n \cdot r_{i_0}$. $\qquad\square$

Lemma 12 enables us to mimic circular reversals by linear reversals, while keeping the sorting cost within a constant factor. Given an index i, define Π_i to be the linear permutation derived by cutting Π at index i. Define Π_{lin} to be Π_{i_0}.

Lemma 13. *Let Π, r_i, and i_0 be as in Lemma 12. Consider the linear permutation Π_{lin} derived by cutting Π at index i_0. Let ρ be a reversal affecting i_0. The reversal ρ can be mimicked by two reversals on Π_{lin} with a total cost less than $2n$.*

Proof. Represent Π and Π_{lin} as follows: $\overbrace{i_0}^{\text{index}} \overbrace{\cdots}^{y\text{ elements}} \overbrace{\cdots}^{n-y-x\text{ elements}} \overbrace{\cdots}^{x\text{ elements}}$,
where ρ affects the x and y elements. The reversal ρ can be mimicked on Π_{lin} by the following two reversals (the brackets $[,]$ indicate the reversal):

$$[\overbrace{\cdots}^{y\text{ elements}} \overbrace{\cdots}^{n-y-x\text{ elements}} \overbrace{\cdots}^{x\text{ elements}}] .$$

$$\overbrace{\cdots}^{x\text{ elements reversed}} [\overbrace{\cdots}^{n-y-x\text{ elements reversed}}] \overbrace{\cdots}^{y\text{ elements reversed}} .$$

The final result is equivalent to the circular permutation $\Pi \cdot \rho$ and the cost of the two reversals is bounded by $2n$. □

Since there are r_{i_0} reversals to mimic, the total cost of mimicking the circular sorting procedure by a linear one is bounded by $2 \cdot n \cdot r_{i_0}$. This yields a constant factor approximation.

Theorem 5. *The optimal circular sorting cost of Π can be approximated up to a constant factor by the optimal linear distance of Π_{lin} to one of the n linear permutations equivalent to the circular identity permutation.*

Proof. Define i_0 as in Lemma 13. By Lemma 13, we can mimic all reversals in an optimal circular solution by reversals in a linear solution, ending up with a linear permutation I_{j_0} equivalent to the circular identity permutation. The following bound holds for the optimal linear distance between Π_{lin} and I_{j_0}.

$$d\left(\Pi_{\text{lin}}, I_{j_0}\right) \leq \text{cosc}\left(\Pi\right) + 2 \cdot n \cdot r_{i_0} .$$

By Lemma 12 we get: $d\left(\Pi_{\text{lin}}, I_{j_0}\right) \leq 3\text{cosc}\left(\Pi\right)$. □

Theorem 5 enables approximating the optimal circular sorting cost using algorithms for sorting linear permutations: cut the circular permutation at each index, approximate the distance between the cut and all I_j, and return the minimum result. Sorting linear permutations is done by the algorithm reorderReversalSort, introduced in [3], which is a $O\left(\log n\right)$ approximation algorithm. The algorithm circulaReordeReversalSort is based on this idea.

Theorem 6. *The algorithm circulaReordeReversalSort has an approximation ratio of $O\left(\log n\right)$.*

4   Sorting Signed Permutations

The algorithms for sorting unsigned permutations are modified to sort signed permutations, while retaining their approximation ratios. This is done using the Bafna-Pevzner reduction [9].

Theorem 7. *The algorithm reorderReversalSort can be modified to sort signed permutations, while retaining the algorithm's approximation ratio.*

A similar result applies to the circular case.

Algorithm 4 $circulaReordeReversalSort(\Pi)$

1: $opt \leftarrow \infty$
2: **for all** indices i **do**
3:     **for all** $j \in \{1, 2, \ldots, |\Pi|\}$ **do**
4:         $sortingCost \leftarrow$ reorderReversalSort $\left(I_j^{-1} \cdot \Pi_i\right)$
5:         **if** $sortingCost < opt$ **then**
6:             $opt \leftarrow sortingCost$
7:         **end if**
8:     **end for**
9: **end for**
10: Output opt

---

5   Basic Sorting Bounds

In this section we give sorting bounds for linear signed as well as circular singed and unsigned permutations and 0/1 sequences The results are extensions of the linear unsigned results introduced in [3]. Equations (1) and (2) make extending the upper and lower bounds straightforward.

Upper Bounds

By (1) and (2), the singed linear case is an upper bound to all other cases. The cost of changing the signs of all elements is in $O(n)$. Thus, the unsigned linear cost summed to $O(n)$ is an upper bound. This fact combined with the upper bounds in [3] implies the results shown in Table 1.

Lower Bounds

A lower bound for the unsigned circular case is by (1) and (2) a lower bound for all other cases. The potential function arguments introduced in [3] can be readily modified to fit the unsigned circular case. Summary of the lower bound results, for both permutations and 0/1, appears in Table 1.

References

1. Nadeau, J.H., Taylor, B.A.: Lengths of chromosomal segments conserved since divergence of man and mouse. Proc. Natl. Acad. Sci. USA **81** (1984) 814–818
2. Pinter, R., Skiena, S.: Sorting with length-weighted reversals. In: Proc. 13th International Conference on Genome Informatics (GIW). (2002) 173–182
3. Bender, M., Ge, D., He, S., Hu, H., Pinter, R., Skiena, S., Swidan, F.: Improved bounds on sorting with length-weighted reversals. In: Proc. 15th ACM-SIAM Symposium on Discrete Algorithms (SODA). (2004) 912–921
4. Pevzner, P.A.: Computational Molecular Biology - an Algorithmic Approach. MIT Press, Cambridge MA (2000)
5. Bergeron, A.: A very elementary presentation of the hannenhalli-pevzner theory. In: Proc. 12th Symp. Combinatorial Pattern Matching (CPM). (2001) 106–117

6. Chen, T., Skiena, S.: Sorting with fixed-length reversals. Discrete Applied Mathematics **71** (1996) 269–295
7. Hartman, T.: A simpler 1.5-approximation algorithm for sorting by transpositions. In: Proc. 14th Symp. Combinatorial Pattern Matching (CPM). (2003) 156–169
8. Christie, D.A., Irving, R.W.: Sorting strings by reversals and by transpositions. SIAM J. on Discrete Math **14** (2001) 193–206
9. Bafna, V., Pevzner, P.: Genome rearrangements and sorting by reversals. SIAM J. Computing **25** (1996) 272–289

Optimizing Multiple Spaced Seeds for Homology Search

Jinbo Xu[1], Daniel G. Brown[1], Ming Li[1], and Bin Ma[2]

[1] School of Computer Science, University of Waterloo,
Waterloo, Ontario N2L 3G1, Canada
{j3xu,browndg,mli}@uwaterloo.ca
[2] Department of Computer Science, University of Western Ontario,
London, Ontario N6A 5B8, Canada
bma@csd.uwo.ca

Abstract. Optimized spaced seeds improve sensitivity and specificity in local homology search [1]. Recently, several authors [2–4] have shown that multiple seeds can have better sensitivity and specificity than single seeds. We describe a linear programming-based algorithm to optimize a set of seeds. Our algorithm offers a performance guarantee: the sensitivity of a chosen seed set is at least 70% of what can be achieved, in most reasonable models of homologous sequences. Our method achieves performance comparable to that of a greedy algorithm, but our work gives this area a mathematical foundation.

1 Introduction

Heuristic homology search programs, such as BLASTN [5], are used extensively in genome analysis. To achieve a fast runtime, BLASTN only builds alignments between regions sharing a common region of k letters, called a seed. It then verifies which seeds come from good alignments using a local method. The various BLAST programs have contributed much to scientific research since its birth, and their introductory paper is the most cited paper of the 1990s.

BLASTN does not identify its seeds optimally. PatternHunter [1], introduced the use of spaced seeds in local alignment. This significantly increases sensitivity and computational efficiency. Since this introduction, spaced seeds have been studied by many researchers. Keich *et al.* [6] proposed a dynamic programming algorithm to compute the exact hit probability of a seed on a random region under simple distribution. Buhler *et al.*[2] expanded upon this and built a heuristic algorithm to find optimal seeds for alignments when the model of homologous regions is a Markov chain. Choi *et al.* [7] also studied how to calculate the exact hitting probability of a spaced seed and proposed a new algorithm for identifying the optimal spaced seed. Brejova *et al.*[8] considered conserved regions determined by hidden Markov models (HMMs), particularly for coding regions. These regions have three-periodic structure, and variation in conservation level throughout the region. Later, they extended the idea of spaced seeds to vector seeds [4], which encapsulate more freedom in describing a single seed.

A single spaced seed cannot achieve the sensitivity of exhaustive, quadratic runtime programs such as the Smith-Waterman algorithm [9]. There is a tradeoff between sensitivity and the false positive rate that was described in both the paper for PatternHunter

S.C. Sahinalp et al. (Eds.): CPM 2004, LNCS 3109, pp. 47–58, 2004.

II [3] and the vector seeds work of Brejova *et al.* [4]. These papers show that the use of multiple seeds (or one vector seed, which is equivalent) can have greater sensitivity than a single seed, even at the same false positive rate.

Here, we show how to optimize the choice of multiple seeds for homology search. Optimizing a single seed is *NP*-hard [3], and the exponential-runtime algorithms used for choosing the optimized single seed are not appropriate to the multiple-seed case [4, 7]. Instead, we formulate the optimal multiple spaced seeds problem as a *maximum coverage* problem and round the solution to a linear program for that problem. Our algorithm works regardless of the probabilistic model of homologous regions, giving a set of seeds whose sensitivity is guaranteed to be within a factor of $1 - 1/e \approx .63$ of the best possible. In some cases, we can achieve a much better factor.

In practice, our algorithm returns seed sets whose sensitivity is comparable to the greedily chosen set [3, 4]. The primary advantage of our algorithm is that not only are our seed sets good, we also have a bound on the sensitivity of the best possible set.

2   Problem Formulation

2.1   Alignments and Seeds

We model ungapped pairwise nucleotide alignments, which we call homologous regions, by binary strings, where 1 represents a match, and 0 a mismatch. The length of a region R is $|R|$, and $R[i]$ is the ith bit of R. We model these regions probabilistically, using a random distribution on binary strings. A random region model specifies the distributions of its length and its position-by-position distributions as well.

Here, we consider the following two region models, introduced in Ma *et al.* [1] and Brejova *et al.* [8]:

1. *PH* model [1]. In this model, each bit of the region is set to one independently with probability 0.7, and the length of the region is a constant 64 bits.
2. *HMM* model [8]. This model represents coding regions. Here, each three bits of the region are produced as triplets, and the conservation level in one triplet depends on the previous triplet's conservation level. This models regions of high and low conservation. The mean region length is approximately 90.

With models for alignments, we turn to seeds. A spaced seed S is a binary string. Its length is $|S|$, while its weight is its number of ones. A 1 in the seed S corresponds to a position with a required match and a 0 means "don't care."

Definition 1. *A* hit *to a seed S in a region R is an offset i such that $R[i + k] \geq S[k]$ for all $1 \leq k \leq |S|$. R is* hit *by the seed S if there is a hit to S in R. Given a set of seeds, $A = \{S_1, S_2, \ldots S_k\}$, R is* hit *by A when R is hit by at least one seed in A.*

2.2   Optimization Problems

Given a probabilistic model P for regions and two integers W and M, the *optimal single spaced seed problem* is to find the spaced seed S^* with weight W and length no more than M that maximizes the probability of a hit to a region chosen from P. Similarly, for $k > 1$, the *optimal k-seeds problem* is to find a set A^* of k seeds, each with weight W and length no more than M, that maximizes the hit probability. These problems are known to be *NP*-hard.

We can also choose a seed from a collection of seeds. Let $A = \{S_1, S_2, \ldots, S_m\}$ be a collection of spaced seeds, and suppose we have a probabilistic model P for homologous regions. The *seed specific optimal k-seeds problem* is to find the k-element subset Q^* of A that maximizes the probability of a hit to P.

We will analyze this problem by generating many regions from the distribution of P, and then choosing a collection of seeds with high sensitivity to these instances of P. We will guarantee that our set of regions is representative by choosing enough regions. Given a collection of N such regions, $R = \{R_1, \ldots, R_N\}$, and a set A of possible seeds, the *seed-region specific optimal k-seeds problem* is to find a k-element subset Q^* of A that hits the most regions in R.

A reduction from the maximum coverage problem shows that the seed-region specific optimal k-seeds problem is itself *NP*-hard. The following sections describe how to formulate the seed-region specific optimal seeds problem as a maximum k-coverage problem and solve it by an LP based randomized algorithm.

2.3   Bounds on Number of Instances

We solve the seed-specific optimal k-seeds problem by first generating a large set of random regions, and then choosing a good seed set for them. Chernoff bounds tell us how large a set we need so the hit probability for the regions is close to the model hit probability. In particular, the following is due to Chernoff bounds:

Theorem 1. *Let A be a set of seeds, P a model for random regions, and $R_1, \ldots, R_N$ independent regions from distribution P. Assume that A has probability p of hitting regions from P, and that the fraction of the N seeds that are hit by A is p'. For any $0 < \delta < 1$, the following bounds on p' and p hold:*

1. $Pr[p' - p > \delta] \le e^{-2N\delta^2}$,
2. $Pr[p' - p < \delta] \le e^{-2N\delta^2}$.

Hence, if we choose large enough N, we can approximate the desired probability, p, with the sample probability, p'. Once we choose the random regions, we cast the problem as a maximum k-coverage problem. Suppose we have a collection $R = \{R_1, \ldots, R_N\}$ of regions, and a set $S = \{S_1, \ldots, S_m\}$ of seeds. Let H_i be all regions that seed S_i hits. Our selection problem is to find the set of k sets H_i whose union is as large as possible; this is the maximum coverage problem. The obvious greedy algorithm approximates this to a $1 - 1/e$ fraction [10], and Feige has shown that this bound cannot be improved for the general maximum coverage problem [11].

3   Choosing Seed Sets by Linear Programming

Our approach to optimizing the maximum coverage is through linear programming and randomized rounding. This approach improves upon the performance of the greedy algorithm in our special case of maximum coverage. The integer program is straightforward: binary decision variables x_i are 1 when seed S_i is selected, and y_j are 1 if region R_j is hit. Our problem is then modeled this way:

$$\max \frac{1}{N} \sum_{j=1}^{N} y_j \tag{1}$$

subject to:

$$y_j \le \sum_{i:j\in H_i} x_i, \text{ for } j = 1, \ldots, N, \tag{2}$$

$$\sum_{i=1}^{m} x_i = k \tag{3}$$

$$x_i \in \{0,1\}, \text{ for } i = 1, \ldots, m, \tag{4}$$

$$y_j \in \{0,1\}, \text{ for } j = 1, \ldots, N. \tag{5}$$

Constraint 2 guarantees that a region R_j is hit only if at least one seed that hits it is chosen, while constraint 3 permits only k seeds to be chosen.

3.1   A Simple LP-Rounding Bound

No polynomial-time algorithm is known to exist for integer programs. Instead, we relax the integrality constraints on x_i and y_j and solve the resulting linear program. Then, we use a random rounding technique to pick a set of seeds. Assume the optimal linear solution to the LP is $(\mathbf{x}^*, \mathbf{y}^*)$. We create a probability distribution on the m seeds that consists of the scaled x^* values, x_i^*/k, and choose k indices, $s_1, \ldots, s_k$, from this probability distribution. We may choose the same index multiple times. Our rounded solution to the LP, (x^+, y^+) has $x_i^+ = 1$ if we chose index i, and $y_j^+ = 1$ if we chose one of the seeds that hits region R_j. The set $\mathcal{S} = \{S_i\}$ of seeds whose indices we have chosen form our set of seeds. (This set may be smaller than k.)

The performance of this algorithm is as good as the obvious greedy algorithm.

Theorem 2. *The LP-rounding algorithm generates a solution with expected approximation ratio at least $1 - 1/e$ for the seed-region specific optimal multiple seeds problem, and can be derandomized to ensure this performance deterministically.*

To prove this theorem, we first note that the probability that a particular seed S_i is *not* chosen is simple to estimate:

$$\Pr[x_i^+ = 0] = (1 - \frac{x_i^*}{k})^k \le e^{-x_i^*}.$$

Now, let us consider a single region and estimate its probability of being missed.

$$\Pr[y_j^+ = 0] = \Pr[\bigcap_{k:j\in H_k} x_k^+ = 0],$$

since a region is missed if no seed that hits it was chosen. If one seed is not chosen, other seeds are more likely to be chosen (since one seed is chosen in each round), so

$$\Pr[\bigcap_{k:j\in H_k} x_k^+ = 0] \leq \prod_{k:j\in H_k} Pr[x_k^+ = 0] \leq \prod_{k:j\in H_k} e^{-x_k^*}.$$

Moving sums to products, we see that $\Pr[y_j^+ = 0] \leq \exp(-\sum_{k:j\in H_k} x_k^*)$. Since $y_j^* \leq \sum_{i:j\in H_i} x_i^*$, this probability is at most $\exp(-y_j^*)$, which for y between 0 and 1 is bounded above by $1 - (1 - 1/e)y_j^*$. Thus, the probability that $y_j^+ = 1$ is at least $(1 - 1/e)y_j^*$. The expected fraction of the regions that are hit by the chosen set of seeds is at least a $(1 - 1/e)$ fraction of the optimal objective function value for the linear program. Since the LP bound is an upper bound on the optimal integer programming solution, the expected approximation ratio of the algorithm is at least $1 - 1/e$. This algorithm can also be derandomized by standard techniques [12, 13].

3.2   A Better Bound for the Rounding Algorithm

In some cases, we can prove a better performance bound than $1 - 1/e$. If we have lower and upper bounds on the number of regions hit by every possible seed, the seed set chosen by the randomized LP algorithm may have provable expected performance better than what can be proved in general. For example, in the *PH* model, any seed with weight 11 and length no more than 18 hits at least 30% of the sample regions if the number of samples is big enough.

In this direction, we can prove the following theorem in the appendix:

Theorem 3. *If each seed in A hits at least θN regions and the optimal linear solution of the LP has objective function value l^* ($0 \leq l^* \leq 1$), then the rounding algorithm has expected approximation ratio at least $\frac{\theta k - l^*}{(k-1)l^*}(1 - e^{-k}) + (1 - \frac{\theta k - l^*}{(k-1)l^*})(1 - e^{-1})$ for the seed-region specific optimal multiple seeds problem.*

Table 1 give the provable approximation ratio for the region models, for all seeds with length at most 18 and weight 11. In the *PH* model, $\theta \approx 0.30$, while for the *HMM* model, $\theta \approx 0.73$. In addition, in computing these two tables, we replace l^* with 1.

In both cases, the approximation ratio is better than $1 - 1/e$ and improves as k grows. However, if the seed weight is larger than 11, then Theorem 3 cannot prove an approximation ratio larger than $1 - 1/e$. But, if the seed weight is 10, or each region of the *PH* model is longer than 64, the theoretical approximation ratio will be better.

Table 1. Approximation ratios for region models and seeds of length at most 18 and weight 11.

number of seeds	2	4	6	8	10	12	14	16
ratio for *PH* model	0.632	0.655	0.691	0.706	0.714	0.719	0.723	0.725
ratio for *HMM* model	0.739	0.856	0.879	0.886	0.890	0.892	0.893	0.894

4 Experimental Results

Here, we present experiments with our seed search algorithm. We show that our algorithm finds high quality seed sets, and we compare our algorithm and the simple greedy algorithm, documenting that both algorithms choose good seeds.

4.1 Probabilistic Model Results

We used our algorithms to find good seed sets for our two random region models, *PH* and *HMM*. We generated 5000 instances of each model, $R_M = \{R_1 \ldots R_N\}$. For any seed set A, let $p_R(A)$ be the fraction of these 5000 instances that were hit by the set.

If our algorithm finds a seed set $\hat{A}$, we can use the algorithms described by Keich *et al.*[6] or Brejova *et al.*[8] to compute the fraction $p_M(\hat{A})$ of regions from the model M hit by the seed. As an upper bound on p_M, we use either 1 or a bound based on the LP solution. Let A^* be the optimal set of seeds for the model, and let I^* and l^* be the IP and LP objective function optima, respectively. We know that $p_R(A^*) \leq I^* \leq l^*$, since the LP bound is an upper bound on the IP bound, and A^* may not perform better than the IP solution on R. Theorem 1 shows that with probability at least 0.99, $p_M(A^*) \leq p_R(A^*) + 0.0215$. Thus, an upper bound on the best possible seed performance, with probability 0.99, is $l^* + 0.0215$. To estimate the optimization ratio of our algorithm, we compare $p_M(\hat{A})$ to $l^* + 0.0215$.

Table 2. Estimated approximation ratio of the seed sets generated by our algorithm with PH model and HMM model.

	PH model		HMM model	
# of seeds	$W = 10,$ $M \leq 18$	$W = 11,$ $M \leq 18$	$W = 10,$ $M \leq 18$	$W = 11,$ $M \leq 18$
2	0.916	0.899	0.970	0.962
4	0.934	0.912	0.961	0.960
6	0.939	0.919	0.959	0.960
8	0.943	0.931	0.953	0.956
10	0.948	0.931	0.954	0.953
12	0.950	0.936	0.953	0.950
14	0.951	0.937	0.953	0.950
16	0.952	0.938	0.951	0.948

Table 2 shows the estimated approximation ratio of the seed sets generated by our algorithm for seeds of weight 10 or 11 and length at most 18. Our seed sets are at least 90% as good as the best possible (with probability 0.99). Solving one linear programming instance takes seven hours to terminate on a single 500MHZ CPU, which we note is affordable, given that one only computes these seed sets once.

4.2   Seed Sets for ESTs

Here, we test the sensitivity of our seed sets against real alignments generated by comparing human and mouse EST sequences. Out of a collection of 29715 mouse EST sequences and 4407 human EST sequences, we randomly chose three million pairs to test for sensitivity. We computed the optimal alignments between these sequences using the Smith-Waterman algorithm (with matches scoring +1, mismatches and gap extensions -1, and gap openings -5), and tested to see how many of them were hit by a seed set. We identified true positives as being hits to alignments scoring at least 16, while false positives were hits to alignments with lower score. Please refer to Li *et al.*[3] for the detailed description of this procedure.

Table 3. sensitivity and false positive values of chosen seed sets for PH model and HMM model. TP: hit rate of the pairs with alignment score at least 16. FP: the ratio between the number of hit pairs and the number of all the pairs.

	PH model				HMM model			
	$W = 11, M \leq 18$		$W = 10, M \leq 18$		$W = 11, M \leq 18$		$W = 10, M \leq 18$	
#seeds	TP	FP	TP	FP	TP	FP	TP	FP
1	0.341	0.167	0.661	0.485	0.363	0.171	0.647	0.490
2	0.529	0.310	0.835	0.677	0.539	0.317	0.869	0.703
3	0.648	0.411	0.907	0.771	0.675	0.419	0.932	0.793
4	0.715	0.483	0.948	0.825	0.751	0.506	0.958	0.836
5	0.763	0.535	0.964	0.854	0.801	0.568	0.973	0.865
6	0.810	0.595	0.973	0.874	0.840	0.619	0.983	0.888
7	0.838	0.639	0.986	0.895	0.859	0.651	0.989	0.900
8	0.883	0.678	0.988	0.903	0.882	0.687	0.991	0.909
9	0.888	0.700	0.993	0.916	0.902	0.712	0.993	0.917
10	0.898	0.718	0.994	0.922	0.914	0.734	0.995	0.924
11	0.919	0.747	0.995	0.926	0.930	0.758	0.996	0.930
12	0.931	0.763	0.996	0.931	0.938	0.771	0.997	0.934
13	0.937	0.777	0.996	0.935	0.941	0.781	0.997	0.937
14	0.947	0.792	0.997	0.940	0.952	0.799	0.998	0.941
15	0.952	0.804	0.998	0.943	0.955	0.808	0.998	0.943
16	0.954	0.810	0.998	0.944	0.959	0.818	0.998	0.946

We used the seed sets chosen in the previous experiment to test how well they did in this practical test, and Table 3 shows the results. Most notably, the seed sets all quickly converge to high sensitivity. It is worth noting that four well-chosen seeds of weight 11 will often have better sensitivity and specificity than one seed of weight 10. Figure 1 further compares the performance of four seeds with weight 11 chosen to optimize the *HMM* model and a single seed with weight 10, which have similar false positive rates. The set of four weight 11 seeds is much more sensitive on the pairs with an alignment score between 16 and 40 than the single *HMM* seed with weight 10. For a pair with score at least 40, any seed can easily detect it. Our findings show that using seeds with weight 11 can detect more subtly similar pairs than using seeds with weight 10 at the same false positive level.

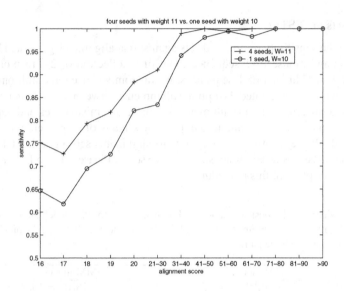

Fig. 1. Sensitivity comparison of four seeds with weight 11 chosen for the *HMM* model and a single seed with weight 10.

4.3   Curated Coding Alignments

We further tested our algorithm on two protein coding DNA alignment data sets, one of human/mouse alignments, and one of human/fruit fly alignments. Each data set is treated separately. Please refer to Brejova *et al.* [8] for a detailed description of these data. In the fly/human set, 972 alignments are used for training and 810 for testing, while in the mouse/human set, 2171 alignments are used to train and 1660 to test. These are available at
http://monod.uwaterloo.ca/supplements/03seeds. We used the training samples to search for the optimal seed sets with our LP algorithm (note that no probabilistic model is used here), and then tested our seed sets on the test samples.

Table 4 shows the sensitivity of the seed sets with weight 10 and weight 11. As shown in these two figures, four seeds with weight 11 are 2% more sensitive than a single seed with weight 10. However, there is no difference between the sensitivity of k ($k = 2, 3, 4$) seeds with weight 10 and $4k$ seeds with weight 11, which is because the alignments themselves are fairly high-scoring. A single optimal seed with weight 10 hits 86% of the human/fruit alignments and 91% of human/mouse alignments.

4.4   Comparison with Greedy Algorithm

We also implemented the greedy algorithms described in Li *et al.* [3] and Brejova *et al.* [4] and compared it with our LP-based randomized algorithm. Table 5 compares the sensitivity of the seed sets found by two algorithms on the two curated sets of coding alignments; the results are comparable. The advantage is that the LP based algorithm enables us to estimate a usually much better approximation ratio of the solutions, while the greedy algorithm only gives a very poor $(1 - 1/e)$ approximation ratio.

Table 4. Sensitivity of the seeds on coding region alignment data

	fruit fly vs. human		mouse vs. human	
# seeds	$W = 10, M \leq 18$	$W = 11, M \leq 18$	$W = 10, M \leq 18$	$W = 11, M \leq 18$
1	0.863	0.781	0.919	0.872
2	0.920	0.831	0.952	0.910
3	0.944	0.868	0.961	0.930
4	0.956	0.885	0.975	0.938
5	0.956	0.910	0.980	0.948
6	0.964	0.912	0.981	0.948
7	0.974	0.917	0.984	0.961
8	0.973	0.920	0.989	0.968
9	0.981	0.928	0.989	0.972
10	0.981	0.933	0.988	0.970
11	0.983	0.937	0.989	0.973
12	0.980	0.940	0.990	0.970
13	0.985	0.947	0.990	0.981
14	0.985	0.948	0.993	0.977
15	0.989	0.949	0.993	0.980
16	0.986	0.957	0.997	0.978

Table 5. Performance comparison of seeds generated by the greedy algorithm and the LP based randomized algorithm.

	fruit fly vs. human				mouse vs. human			
	$W = 10, M \leq 18$		$W = 11, M \leq 18$		$W = 10, M \leq 18$		$W = 11, M \leq 18$	
#seeds	Greedy	LP	Greedy	LP	Greedy	LP	Greedy	LP
1	0.863	0.863	0.781	0.781	0.919	0.919	0.872	0.872
2	0.920	0.920	0.831	0.831	0.952	0.952	0.902	0.910
3	0.944	0.944	0.860	0.868	0.970	0.961	0.922	0.930
4	0.952	0.956	0.886	0.885	0.977	0.975	0.941	0.938
6	0.963	0.964	0.911	0.912	0.984	0.981	0.957	0.948
8	0.970	0.973	0.925	0.920	0.991	0.989	0.964	0.968
10	0.978	0.981	0.938	0.933	0.993	0.988	0.970	0.970
12	0.985	0.980	0.944	0.940	0.994	0.990	0.973	0.970
14	0.989	0.985	0.948	0.948	0.995	0.993	0.977	0.977
16	0.989	0.986	0.951	0.957	0.996	0.997	0.981	0.978

5   Discussion

The experimental result with real data demonstrates that using more seeds with a bigger weight is better than using fewer seeds with a smaller weight in detecting the subtly similar sequence pairs. However, using more seeds with a bigger weight requires much more memory to index the sequence database.

As we know, for a general maximum k-coverage problem, the greedy algorithm and the LP-based algorithm have the same approximation ratio, $(1 - e^{-1})$. An interesting question is if the greedy algorithm can also guarantee a better theoretical bound for the optimal multiple seeds problem, given that the experimental results described in this paper indicate that the greedy algorithm has comparable performance to our LP-based algorithm.

Acknowledgements

The authors would like to thank Brona Brejova for her help and insightful comments. Our research is supported by the National Science and Engineering Research Council of Canada and the Human Frontier Science Program.

References

1. B. Ma, J. Tromp, and M. Li. Patternhunter: Faster and more sensitive homology search. *Bioinformatics*, 18:440–445, 2002.
2. Jeremy Buhler, Uri Keich, and Yanni Sun. Designing seeds for similarity search in genomic DNA. In *Proceedings of the Seventh Annual International Conference on Computational Molecular Biology (RECOMB03)*, pages 67–75, Berlin, Germany, April 2003.
3. Ming Li, Bin Ma, Derek Kisman, and John Tromp. Patternhunter II: Highly sensitive and fast homology search. *Journal of Bioinformatics and Computational Biology*, 2004. In Press.
4. Brona Brejova, Daniel Brown, and Tomas Vinar. Vector seeds: an extension to spaced seeds allows substantial improvements in sensitivity and specificity. In G. Benson and R. Page, editors, *Algorithms and Bioinformatics: 3rd International Workshop (WABI)*, volume 2812 of *Lecture Notes in Bioinformatics*, pages 39–54, Budapest, Hungary, September 2003. Springer.
5. S.F. Altschul, W. Gish, W. Miller, E.W. Myers, and D.J. Lipman. Basic local alignment search tool. *Journal of Molecular Biology*, 215:403–410, 1990.
6. Uri Keich, Ming Li, Bin Ma, and John Tromp. On spaced seeds for similarity search. *Discrete Applied Mathematics*, 2004. In Press.
7. Kwok Pui Choi and Louxin Zhang. Sensitivity analysis and efficient method for identifying optimal spaced seeds. *Journal of Computer and System Sciences*, 68:22–40, 2004.
8. Brona Brejova, Daniel G. Brown, and Tomas Vinar. Optimal Spaced Seeds for Hidden Markov Models, with Application to Homologous Coding Regions. In R. Baeza-Yates, E. Chavez, and M. Crochemore, editors, *Combinatorial Pattern Matching, 14th Annual Symposium (CPM)*, volume 2676 of *Lecture Notes in Computer Science*, pages 42–54, Morelia, Mexico, June 25–27 2003. Springer.
9. T.F. Smith and M.S. Waterman. Identification common molecular subsequences. *Journal of Molecular Biology*, 147:195–197, 1981.
10. Dorit S. Hochbaum. Approximating covering and packing problems:set cover, vertex cover, independent set, and related problems. In Dorit S. Hochbaum, editor, *Approximation Algorithms for NP-hard Problems*, chapter 3, pages 135–137. PWS, 1997.
11. Uriel Feige. A threshold of ln n for approximating set cover. *Journal of the ACM*, 45(4):634–652, 1998.
12. R. Motwani and P. Raghavan. *Randomized Algorithm*. Cambridge University Press, New York, 1995.
13. P. Raghavan. Probabilistic construction of deterministic algorithms: Approximating packing integer programs. *Journal of Computer and System Sciences*, 37:130–143, October 1988.

A Appendix: Proof of Theorem 3

We now prove the theorem from above. First, we give two lemmas.

Lemma 1. *Given a_1, a_2, b_1, $b_2 > 0$. If $\frac{a_1}{a_2} \geq \beta_1$, $\frac{b_1}{b_2} \geq \beta_2$, $\frac{a_2}{a_2+b_2} \geq \alpha > 0$, and $\beta_1 \geq \beta_2 > 0$, then $\frac{a_1+b_1}{a_2+b_2} \geq \alpha\beta_1 + (1-\alpha)\beta_2$.*

This lemma is a simple number fact, and trivial to verify.

Lemma 2. *Given z_1, z_2, ..., z_n, $b \geq z_i \geq 1$, $b > 1$, $\sum_{i=1}^{n} z_i = Z$. Then we have $\sum_{i=1}^{n}(1 - e^{-z_i}) \geq \frac{Z-n}{b-1}(1-e^{-b}) + (n - \frac{Z-n}{b-1})(1-e^{-1})$.*

Proof. If there exist z_i, z_j, $(1 \leq i < j \leq n)$ such that $1 < z_i, z_j < b$. Without loss of generality, assume $z_i \leq z_j$. For any $0 < \epsilon \leq \min(z_i - 1, b - z_j)$, we have

$$1 - e^{-(z_i-\epsilon)} + 1 - e^{-(z_j+\epsilon)} \leq 1 - e^{-z_i} + 1 - e^{-z_j}$$

since

$$(e^\epsilon - 1)(e^{-z_i} - e^{-z_j-\epsilon}) \geq 0$$

This indicates that if there are two z_i, z_j which are not equal to 1 or b, then we can always adjust their values to decrease the value of $\sum_{i=1}^{n}(1 - e^{-z_i})$ while keep $\sum_{i=1}^{n} z_i$ unchanged. Therefore, $\sum_{i=1}^{n}(1 - e^{-z_i})$ becomes minimum if and only if at most one z_i is not equal to 1 or b. If all z_i equal to 1 or b, then the number of z_i equal to b is at least $\frac{Z-n}{b-1}$. The lemma holds. If there is one k $(1 \leq k \leq n)$ such that $1 < z_k < b$, then by simple calculation we have

$$\sum_{i=1}^{n}(1 - e^{-z_i}) \geq \frac{Z-n}{b-1}(1-e^{-b}) + (n - \frac{Z-n}{b-1})(1-e^{-1})$$

$$+ \frac{z_k - 1}{b-1}(e^{-b} - e^{-1}) + e^{-1} - e^{-z_k}$$

$$> \frac{Z-n}{b-1}(1-e^{-b}) + (n - \frac{Z-n}{b-1})(1-e^{-1}).$$

Now, assume that each seed hits at least θN $(0 < \theta < 1)$ sample regions. Summing up the right hand side of (2), for any feasible solution of the linear program, we see that $\sum_{j=1}^{N} \sum_{i:j\in H_i} x_i \geq \theta(\sum_{i=1}^{m} x_i) = \theta kN$.

Let N_1 be the number of y variables with y^* less than one, and N_2 the number of y variables with y^* equal to one. Let l^* denote the objective function value of the optimal linear solution. Let Y_1^* be the sum of the y_j^* that are less than one, and X_1^* be the sum of the corresponding right hand sides of (2). These are equal, since we raise Y_1^* as much as possible. Similarly, let Y_2^* be the sum of the y_j^* which equal one (Obviously, $Y_2^* = N_2$), and let X_2^* the sum of the corresponding right hand sides of (2). Then we have the following equations:

$$X_1^* + X_2^* \geq \theta kN$$
$$X_2^* \leq kN_2$$
$$Y_1^* + Y_2^* = Nl^*$$
$$N_1 \geq X_1^*$$

Based on the above equations, simple calculations show:

$$\frac{X_2^* - N_2}{Nl^*} = \frac{X_2^* - Y_2^*}{Nl^*}$$

$$\geq \frac{\theta k N - Y_1^* - Y_2^*}{Nl^*}$$

$$\geq \frac{\theta k - l^*}{l^*} \tag{6}$$

Now we finish the proof of Theorem 3. According to the discussion in Section 3.1, we have the following two inequalities:

$$Pr[y_j^+ = 1] \geq (1 - 1/e)y_j^* \tag{7}$$

$$\Pr[y_j^+ = 1] \geq 1 - \exp(-\sum_{k:j\in H_k} x_k^*) \tag{8}$$

Based on Equation 7, we have

$$\frac{\sum_{j:y_j^*<1} E[y_j^+]}{\sum_{j:y_j^*<1} y_j^*} \geq 1 - e^{-1} \tag{9}$$

We need to deal with the case that y_j^* equals to 1. Based on Equation 8, we have

$$\frac{\sum_{j:y_j^*=1} E[y_j^+]}{\sum_{j:y_j^*=1} y_j^*} \geq \frac{1}{N_2} \sum_{j:y_j^*=1} (1 - e^{-\sum_{i:j\in H_i} x_i})$$

$$\geq \frac{1}{N_2}(\frac{X_2^* - N_2}{k-1}(1 - e^{-k}) + (N_2 - \frac{X_2^* - N_2}{k-1})(1 - e^{-1})) \tag{10}$$

$$\geq \frac{X_2^*/N_2 - 1}{k-1}(1 - e^{-k}) + (1 - \frac{X_2^*/N_2 - 1}{k-1})(1 - e^{-1}) \tag{11}$$

where (10) comes from Lemma 2.

Based on (9), (11), and Lemma 1, we have

$$\frac{\sum_{j=1}^N E[y_j^+]}{\sum_{j=1}^N y_j^*} = \frac{\sum_{j:y_j^*=1} E[y_j^+] + \sum_{j:y^*j<1} E[y_j^+]}{\sum_{j:y_j^*=1} y_j^* + \sum_{j:y^*j<1} y_j^*}$$

$$\geq (\frac{\frac{X_2^*}{N_2} - 1}{k-1}(1 - e^{-k}) + (1 - \frac{\frac{X_2^*}{N_2} - 1}{k-1})(1 - e^{-1}))\frac{Y_2^*}{Nl^*}$$

$$+(1 - e^{-1})(1 - \frac{Y_2^*}{Nl^*}) \tag{12}$$

$$\geq \frac{X_2^* - N_2}{(k-1)Nl^*}(1 - e^{-k}) + (1 - \frac{X_2^* - N_2}{(k-1)Nl^*})(1 - e^{-1})$$

$$\geq \frac{\theta k - l^*}{(k-1)l^*}(1 - e^{-k}) + (1 - \frac{\theta k - l^*}{(k-1)l^*})(1 - e^{-1}) \tag{13}$$

where (13) comes from (6).

This completes the proof of this theorem.

Approximate Labelled Subtree Homeomorphism

Ron Y. Pinter[1,*], Oleg Rokhlenko[1,*], Dekel Tsur[2], and Michal Ziv-Ukelson[1,**]

[1] Dept. of Computer Science, Technion – Israel Institute of Technology,
Haifa 32000, Israel
{pinter,olegro,michalz}@cs.technion.ac.il
[2] Caesarea Rothschild Institute of Computer Science, University of Haifa,
Haifa 31905, Israel
dekelts@cs.haifa.ac.il

Abstract. Given two undirected trees T and P, the Subtree Homeomorphism Problem is to find whether T has a subtree t that can be transformed into P by removing entire subtrees, as well as repeatedly removing a degree-2 node and adding the edge joining its two neighbors. In this paper we extend the Subtree Homeomorphism Problem to a new optimization problem by enriching the subtree-comparison with node-to-node similarity scores. The new problem, denoted $ALSH$ (Approximate Labelled Subtree Homeomorphism) is to compute the homeomorphic subtree of T which also maximizes the overall node-to-node resemblance. We describe an $O(m^2 n/\log m + mn \log n)$ algorithm for solving $ALSH$ on unordered, unrooted trees, where m and n are the number of vertices in P and T, respectively. We also give an $O(mn)$ algorithm for rooted ordered trees.

1 Introduction

The matching of labelled tree patterns, as opposed to linear sequences (or strings), has many important applications in areas such as bioinformatics, semistructured databases, and linguistics. Examples include the comparison among metabolic pathways, the study of alternative evolutionary trees (phylogenies), processing queries against databases and documents represented in e.g. *XML*, and many fundamental operations in the analysis of natural (and formal) languages. In all these scenarios, both the labels on the nodes as well as the structure of the underlying trees play a major role in determining the similarity between a pattern and the text in which it is to be found.

There are several, increasingly complex ways to model these kinds of problems. A starting point is the *subtree isomorphism problem* [15, 16, 23]: Given a pattern tree P and a text tree T, find a subtree of T which is isomorphic to P (i.e. find if some subtree of T, that is identical in structure to P, can be obtained by removing entire subtrees of T) or decide that there is no such tree.

---

* Research supported in part by the Bar-Nir Bergreen Software Technology Center of Excellence.
** Research supported in part by the Aly Kaufman Post Doctoral Fellowship and by the Bar-Nir Bergreen Software Technology Center of Excellence.

S.C. Sahinalp et al. (Eds.): CPM 2004, LNCS 3109, pp. 59–73, 2004.

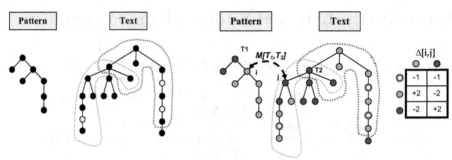

Fig. 1. Subtree homeomorphism. The white nodes in the text are degree-2 nodes that have been removed. Two subtrees in the text that are homeomorphic to the pattern are circled by the dashed lines.

Fig. 2. Approximate Labelled Subtree Homeomorphism. The node-label similarity scores are specified in table Δ, and $D = -1$. Two subtrees in the text that are homeomorphic to the pattern are circled by the dashed lines. The left homeomorphic subtree has an LSH score of 5, while the right one has an LSH score of 12.

The *subtree homeomorphism problem* [3, 20] is a variant of the former problem, where degree 2 nodes can be deleted from the text tree (see Figure 1). The *constrained tree inclusion* problem [26] is a variant of labelled subtree homeomorphism where label equality between pairs of aligned nodes in the compared subtrees is required. Note that all the tree matching problems mentioned so far have polynomial-time algorithms. The more advanced *tree inclusion problem* [14] is that of locating the smallest subtrees of T that include P, where a tree is included in another tree if it can be obtained from the latter by deleting nodes. Here, deleting a node v from a tree entails deleting all edges incident to v and inserting edges connecting the parent of v with the children of v. *Tree inclusion* on unordered trees is NP-complete [14]. Schlieder and Naumann [21] extended the tree inclusion problem to an *approximate tree embedding problem* in order to retrieve and rank search results using a tree-similarity measure whose semantics are tailored to *XML* data. The complexity of their algorithm, which is based in dynamic programming and processes the query tree in a bottom up fashion, is exponential.

This paper addresses queries on labelled trees. The trees could be either ordered or unordered, rooted or unrooted – depending on the application at hand. The main point is, however, that even though the trees are labelled, label equality is not required in a match. Instead, a node-to-node similarity measure is used to rank the resemblance or distance between pairs of aligned nodes. This extension is motivated by the aforementioned applications, where a more meaningful match can be found if – in addition to the topological structure similarity between subtrees – node-to-node resemblance is also taken into account. For example, in bioinformatics, the similarity between metabolic pathways [17, 22] is based both on the resemblance of the elements constituting the pathways (e.g. proteins) as well as on the likeness of their network structure (the former would reflect closeness between proteins, based on either a BLAST sequence comparison score or on functional homology, and the latter would be the topological equivalent of multi-source unrooted trees [19]). A query that searches for a small pathway fragment in a larger tree should take into account the topology similarities (in the form of subtree homeomorphism) as well as the pairwise similarity between proteins which make up the aligned subtree nodes. Another example is when designing a semantic query language for semistructured databases that are represented as *XML* documents [21]: here the node-to-node similarity score may reflect content or tag resemblance, and the topological difference allows flexibility in document structure matching [2]. Finally, in natural language processing, trees are often used to represent sentences, where nodes are labelled by words and sentential forms (or constituents); matching that takes into account synonyms and elisions is very useful in order to detect semantically close phrases.

Thus, in this paper we address the challenge of extending labelled subtree homeomorphism into a new optimization problem, by introducing node-to-node similarity measures that are combined with the topological distance between the pattern and text to produce a single, comprehensive score expressing how close they are to each other.

Let Δ denote a predefined node-to-node similarity score table and $\mathcal{D}$ denote a predefined (usually negative) score for deleting a node from a tree (Figure 2). A *mapping* $\mathcal{M}[T_1, T_2]$ from T_1 to T_2 is a partial one-to-one map from the nodes of T_1 to the nodes of T_2 that preserves the ancestor relations of the nodes. We define the following similarity measure for two homeomorphic trees.

Definition 1. *Consider two labelled trees T_1 and T_2, such that T_2 is homeomorphic to T_1, and let $\mathcal{M}[T_1, T_2]$ denote a node-to-node, homeomorphism-preserving mapping from T_1 to T_2. The* **Labelled Subtree Homeomorphism Similarity Score** *of $\mathcal{M}[T_1, T_2]$, denoted $LSH(\mathcal{M}[T_1, T_2])$, is*

$$LSH(\mathcal{M}[T_1, T_2]) = \mathcal{D} \times (|T_2| - |T_1|) + \sum_{(u,v) \in \mathcal{M}} \Delta[u, v]$$

Correspondingly,

Definition 2. *The* **Approximate Labelled Subtree Homeomorphism (ALSH) Problem** *is, given two undirected labelled trees P and T, and a scoring table that specifies the similarity scores between the label of any node appearing in T and the label of any node appearing in P, as well as a predefined node deletion penalty, to find a homeomorphism-preserving mapping $\mathcal{M}[P, t]$ from P to some subtree t of T, such that $LSH(\mathcal{M}[P, t])$ is maximal.*

In this paper we show how to compute optimal, bottom-up alignments between P and every homeomorphic subtree t of T, which maximize the LSH score between P and t. Our algorithms are based on the close relationship between subtree homeomorphism and weighted assignments in bipartite graphs. The ALSH problem is recursively translated into a collection of smaller ALSH problems, which are solved using weighted assignment algorithms (Figure 3). (Simpler, maximum bipartite perfect matching algorithms were applied in the algorithms for exact subtree morphisms [3, 9, 13, 23, 26].)

Our approach yields an $O(m^2n/\log m + mn\log n)$ algorithm for solving ALSH on unordered, unrooted trees, where m and n are the number of vertices in P and T, respectively. Note that the time complexity of the exact subtree isomorphism/homeomorphism algorithms [23, 26], which do not take into account the node-to-node scores, is $O(m^{1.5}n/\log m)$. Thus, the enrichment of the model with the node similarity information only increases the time complexity by half an order. We also give an $O(mn)$ algorithm for the problem on rooted ordered trees.

Also note that a related set of problems, where dynamic programming is combined with weighted bipartite matching, is that of finding the maximum agreement subtree and the maximum refinement subtree of a set of trees [12, 24]. Such algorithms utilize the special constraints imposed by the properties of evolutionary trees (internal nodes contain less information than leaves, similarity assumption allows for scaling, etc).

The rest of the paper is organized as follows. Section 2 includes weighted bipartite matching preliminaries. In Section 3 we describe the basic $O(m^2n + mn\log n)$ time ALSH algorithm for rooted unordered trees and show how to extend it to unrooted unordered trees without increasing the time complexity. These solutions are further improved in Section 4 to yield an $O(m^2n/\log m + mn\log n)$ solution for both rooted and unrooted unordered trees. In Section 5 we describe the $O(mn)$ time ALSH algorithm for rooted ordered trees.

Note that due to lack of space, most proofs are omitted. The proofs will be given in the journal paper.

2   Weighted Assignments and Min-Cost Max Flows

Definition 3. *Let $G = (V = X \cup Y, E)$ be a bipartite graph with edge weights $w(x, y)$ for all edges $(x, y) \in E$, where $x \in X$ and $y \in Y$. The Assignment Problem is to compute a matching M, i.e. a list of monogamic pairs $(x \in X, y \in Y)$, such that the size of M is maximum among all the matchings in G, and $\sum_{(x,y)\in M} w(x, y)$ is maximum among all the matchings with maximum size.*

Although researchers have developed several different algorithms for the assignment problem [1], many of these algorithms share common features. The successive shortest path algorithm for the minimum cost max flow problem appears to lie at the heart of many assignment algorithms [11]. The reduction from an assignment problem to a minimum cost max flow problem is as follows. Let s, t be two new vertices. Construct a graph G' with vertex set $V \cup \{s, t\}$, source s, sink t, and capacity-one edges: an edge (s, x) of cost zero for every $x \in X$, an edge (y, t) of cost zero for every $y \in Y$, and and edge (x, y) of cost $-w(x, y)$ for every $(x, y) \in E$. An integral flow f on G' defines a matching on G of size $|f|$ and weight $-cost(f)$ given by the set of edges $\{x, y\}$ such that (x, y) has flow one. Conversely, a matching M on G defines a flow of value $|M|$ and cost

$\sum_{(x,y)\in M} -w(x,y)$. This means that we can solve a matching problem on G by solving a flow problem on G'.

Edmonds and Karp's algorithm [5] finds a minimum cost maximum flow in G' in $O(EV \log V)$ time. In each stage of this algorithm, Dijkstra's algorithm [4] is used to compute an augmentation path for the existing flow f. Each run of Dijkstra's algorithm is guaranteed to increase the size of M by 1, and thus the algorithm requires at total of $O(V)$ phases to find a maximum score match. Fredman and Tarjan [7] developed a new heap implementation, called *Fibonacci heap*, that improves the running time of Edmond and Karp's algorithm to $O(VE + V^2 \log V)$. The latter bound is the best available strongly polynomial time bound (the running time is polynomially bounded in only V and E, independent of the edge costs) for solving the assignment problem.

Under the assumption that the input costs are integers in the range $[-C, \ldots, C]$, Gabow and Tarjan [8] used cost-scaling and blocking flow techniques to obtain an $O(V^{1/2}E \log(VC))$ time algorithm for the assignment problem. Two-phase algorithms with the same running time appeared in [10, 18]. The scaling algorithms are conditioned by the *similarity assumption* (i.e. the assumption that $C = O(n^k)$ for some constant k) and have the integrality of cost function restriction.

The following lemma, which was first stated and proven in [11], will serve as the basis for the proofs of Lemma 2 and Lemma 4.

Lemma 1 ([5, 11, 25]). *A flow f has minimum cost iff its residual graph R has no negative cost cycle.*

3   Dynamic Programming Algorithms for ALSH

3.1   The Basic Algorithm for Rooted Unordered Trees

We use $d(v)$ to denote the number of neighbors of a node v in an unrooted tree, and $c(v)$ to denote the number of children of a node v in a rooted tree.

Let $T^r = (V_T, E_T, r)$ be the text tree which is rooted in r, and $P^{r'} = (V_P, E_P, r')$ be the pattern tree which is rooted in r'. Let $p_u^{r'}$ denote a subtree of $P^{r'}$ which is rooted in node u of $P^{r'}$, and t_v^r denote a subtree of T^r which is rooted in node v of T^r.

We define $RScores[v \in V_T, u \in V_P]$ as follows.

Definition 4. *For each node $v \in V_T$ and each node $u \in V_P$, $RScores[v, u]$ is the maximal LSH similarity score between a subtree $p_u^{r'}$ of $P^{r'}$ and a corresponding homeomorphic subtree of t_v^r, if such exists. Otherwise, $RScores[v, u]$ is $-\infty$.*

The first algorithm, denoted Algorithm 1, computes the values $RScores[v, u]$ recursively, in a *postorder* traversal of T^r. First, $RScores[v, u]$ are computed for all leaf nodes of T^r and $P^{r'}$. Next, $RScores[v, u]$ are computed for each node $v \in V_T$ and $u \in V_P$, based on the values of the previously computed scores for all children of v and u as follows. Let u be a node of $P^{r'}$ with children

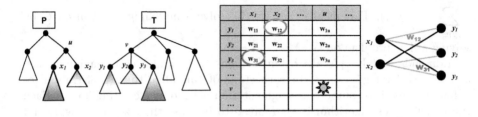

Fig. 3. The work done by the first algorithm during the computation of $RScores[v, u]$. The bipartite graph is constructed in order to compute the optimal weighted assignment between the children of u and those of v. w_{ij} marks the optimal LSH similarity score for aligning the subtrees rooted at $y_j \in T$ and $x_i \in P$.

$x_1, \ldots, x_{c(u)}$ and v be a node of T^r with children $y_1, \ldots, y_{c(v)}$. After computing $RScores[y_j, x_i]$ for $i = 1, \ldots, c(u)$ and $j = 1, \ldots, c(v)$, a bipartite graph G is constructed with bipartition X and Y, where X is the set of children of u, Y is the set of children of v, and each node in X is connected to each node in Y. An edge (x_i, y_j) in G is annotated with weight $RScores[y_j, x_i]$ (Figure 3).

$RScores[v, u]$ is then computed as the maximum between the following two terms:

1. The node-to-node similarity value $\Delta[v, u]$, plus the sum of the weights of the matched edges in the maximal assignment over G. Note that this term is only computed if $c(u) \leq c(v)$.
2. The weight $RScores[y_j, u]$ for the comparison of u and the best scoring child y_j of v, updated with the penalty for deleting v.

Upon completion, the optimal LSH similarity score, denoted *best_score*, where $best\_score = \max_{j=1}^{n} RScores[y_j, x_m]$, is found, and every node $y_j \in T$ with $RScores[y_j, x_m] = best\_score$ is reported as a root of a subtree of T which bears maximal similarity to P under the LSH measure.

Time Complexity Analysis

Theorem 1.

1. *Algorithm 1 computes the optimal ALSH solution for two rooted unordered trees in $O(m^2 n + mn \log n)$ time.*
2. *Under the similarity assumption, Algorithm 1 yields an $O(m^{1.5} n \log(nC))$ time complexity.*

Proof. The time complexity analysis of Algorithm 1 is based in the following observation.

Observation 1. $\sum_{u=1}^{m} c(u) = m - 1$, and $\sum_{v=1}^{n} c(v) = n - 1$.

Algorithm 1 computes a score once for each node pair $(v \in T, u \in P)$. The dominant part of this computation is spent in obtaining the weighted assignment between a bipartite graph with $c(v)+c(u)$ nodes and $c(v) \times c(u)$ edges. Therefore, based on Fredman and Tarjan's algorithm [7], each assignment computation takes $O(c(u)(c(u) \times c(v) + c(v) \log c(v))) = O(c(u)^2 \times c(v) + c(u) \times c(v) \log c(v))$ time.

Summing up the work spent on assignment computations over all node pairs, we get

$$O(\sum_{u=1}^{m} \sum_{v=1}^{n} (c(u)^2 c(v) + c(u)c(v) \log c(v)) \stackrel{\text{observation 1}}{=}$$

$$O(\sum_{u=1}^{m} c(u)^2 n + c(u)n \log n) \stackrel{\text{observation 1}}{=} O(m^2 n + mn \log n).$$

Under the (similarity) assumption that all scores assigned to the edges of G are integers in the range $[-C, \ldots, C]$, where $C = O(n^k)$ for some constant k, Algorithm 1 can be modified to run in time $O(m^{1.5} n \log(nC))$ by employing the algorithm of Gabow and Tarjan [8] for weighted bipartite matching with scaling. □

3.2   Extending the Algorithm to Unrooted Unordered Trees

Let $T = (V_T, E_T)$ and $P = (V_P, E_P)$ be two unrooted trees. The ALSH between P and T could be computed in a naive manner as follows. Select an arbitrary node r of T to get the rooted tree T^r. Next, for each $u \in P$ compute rooted ALSH between P^u and T^r. This method will yield an $O(m^3 n + m^2 n \log n)$ strongly-polynomial algorithm for ALSH on unrooted unordered trees, and an $O(m^{2.5} n \log(nC))$ time algorithm under the similarity assumption with integrality cost function restriction. In this section we show how to reduce these bounds back to $O(m^2 n + mn \log n)$ and $O(m^{1.5} n \log(nC))$ time, correspondingly, by utilizing the decremental properties of the weighted assignment algorithm.

The second algorithm, denoted Algorithm 2, starts by selecting a vertex r of T to be the designated root. T^r is then traversed in postorder, and each internal vertex $v \in T^r$ is compared with each vertex $u \in P$. Let $y_1, \ldots, y_{c(v)}$ be the children of $v \in T^r$, and let $x_1, \ldots, x_{d(u)}$ be the neighbors of $u \in P$. When computing the score for the comparison of u and v we need to take into account the fact that, since P is unrooted, each of the neighbors of u may eventually serve as a parent of u in a considered mapping of a subtree of P, which is rooted in u, with a homeomorphic subtree of t_v^r. Suppose that node x_i, which is a neighbor of u, serves as the parent of u in one of these subtree alignments. Since x_i is the chosen parent, the children set of u includes all its neighbors but x_i. Therefore, the computation of the similarity score for v versus u requires the consideration of the highest scoring weighted matching between the children of v, and all neighbors of u except x_i. Since each neighbor x_i of u, $i = 1, \ldots, d(u)$, is a potential parent of a subtree rooted in u, our algorithm will actually need to compute a series of weighted assignments for the graphs $G_i = (X_i \cup Y, E_i)$, $i = 1, \ldots, d(u)$, where Y consists of the children of v in T^r, while X_i consists of all neighbors of u except x_i. Therefore, we define $UScores[v \in V_T, u \in V_P, x_i \in neighbors(u)]$ as follows.

Definition 5. *For each vertex $v \in T^r$, each vertex $u \in P$, and every vertex $x_i \in neighbors(u)$, $UScores[v, u, x_i]$ is the maximal LSH similarity score between a subtree $p_u^{x_i}$ of P and a corresponding homeomorphic subtree of t_v^r, if one exists. Otherwise, $UScores[v, u, x_i]$ is set to $-\infty$.*

For the graphs G_i defined above, the weight of an edge $(x_j, y_{j'})$ in G_i is $UScores[y_{j'}, x_j, u]$.

The computation of $UScores[v, u, x_i]$ is carried out as follows: $UScores[v, u, x_i]$ is set to the maximum between the following two terms:

1. The node-to-node similarity value $\Delta[v, u]$, plus the sum of the weights of the matched edges in the maximal weighted assignment over G_i. Note that this term is computed only if $d(u) - 1 \leq c(v)$.
2. The weight $UScores[y_j, u, x_i]$ of the comparison of u and the best scoring child y_j of v, updated with the penalty for deleting v.

Since each node $u \in P$ is also a potential root for P in the sought alignment, an additional entry $UScores[v, u, \phi]$ is computed, which stores the maximal LSH similarity score between a subtree of P^u and a corresponding homeomorphic subtree of t_v^r, if one exists. $UScores[v, u, \phi]$ is computed from the weighted assignment for the graph $G = (X \cup Y, E)$, where X is the set of neighbors of u.

Upon completion, the maximal LSH similarity score $best\_score = \max_{v \in V_T, u \in V_P} UScores[v, u, \phi]$ is computed, and every node pair $(j \in T, i \in P)$ with $UScores[j, i, \phi] = best\_score$ is reported as a subtree t_j^r of T^r which bears maximal similarity to tree P^i under the LSH measure.

Time Complexity Analysis

The time complexity bottleneck of Algorithm 2 is based in the need to compute, for each comparison of a pair $u \in P$ and $v \in T$, weighted assignments for a series of bipartite graphs $G_i = (X_i \cup Y, E)$ for $i = 1, \ldots, d(u)$. This, in contrast to Algorithm 1, where only one weighted assignment is computed for each comparison of a pair $u \in P$ and $v \in T$. However, the next lemma shows that the decremental nature of weighted bipartite matching makes it possible to compute assignments for the full series of G_i graphs in the same time complexity as the assignment computation for the single graph G. In the following, let $k = |X|$ and $\ell = |Y|$. Note that $k - 1 \leq l$.

Lemma 2. *Let G be the bipartite graph constructed by Algorithm 2 for some pair of vertices $v \in V_T$ and $u \in V_P$. Given an optimal assignment of G, the optimal assignment of G_i can be computed via one run of a single-source shortest-path computation on a subgraph of G.*

Proof. The proof is based on the reduction of the assignment problem to min-cost max-flow, as described in Section 2. We will show that after having computed the

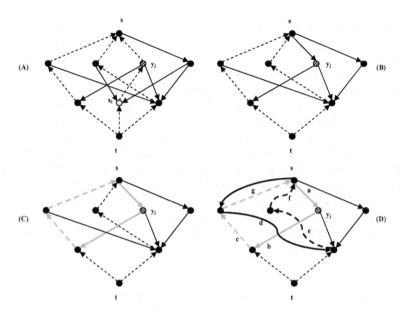

Fig. 4. The computation of the optimal assignment for $X_i \cup Y$. Note that, for the sake of graphical clarity, some of the edges of the graph are not shown in the figure. (A) The residual graph R for flow f. The dashed lines denote the backward edges and the solid lines denote the forward edges. The white circle denotes vertex x_i and the lined circle denotes its matched neighbor y_i. (B) The residual graph R' received from R by removing x_i and its adjacent edges, as well as reversing the flow on edge (s, y_i). (C) The augmentation path p in R' is marked by grey edges. (D) The negative cost cycle c is marked by curves. Edges [a,b,c,g] correspond to the "correction path" p, edges [g,d,e,f] correspond to the negative cost cycle c, and edges [a,b,c,d,e,f] mark the path p' with a cost which is less then that of p.

min-cost max-flow of G', which defines the optimal assignment for G, the optimal assignment for G_i can be obtained via one more augmenting path computation.

Let f denote the minimum cost maximum flow on G', defining the the optimal weighted assignment M, and let R denote the residual graph for f (Figure 4.A).

Since f is optimal, by Lemma 1, there are no negative cost cycles in R. Also, since $\ell \geq k$, $|M| = k$. Let y_i be the matched neighbor of x_i in M, i.e., $(x_i, y_i) \in M$. Let M' denote the matching $M \setminus (x_i, y_i)$, i.e. the matching obtained by removing the pair (x_i, y_i) from M. We say that y_i is "newly widowed" in M'. Clearly, $|M'| = k - 1$. Let f' be the flow defined by M'. Let G'' denote the subgraph of G' obtained by removing vertex x_i and all edges adjacent to x_i in G'. Let R' be the "decremented" residual subgraph of R, obtained by removing vertex x_i and all edges adjacent to x_i in R, as well as cancelling the flow on edge (s, y_i) in R. (Figure 4.B). Our objective is to compute a min-cost max-flow (of value $k - 1$) of G'', which will correspondingly define the optimal assignment of size $k - 1$ on G_i.

Clearly, the value of f' is $k - 1$. But is f' min-cost? By Lemma 1, if there are no negative cost cycles in R', then f' is a min-cost flow among all $k - 1$-valued flows in G''.

Otherwise, there are one or more negative cost cycles in R'. Each one of these negative cost cycles in R' must begin with edge (s, y_i), since this edge is the only edge of R' which did not exist in R, and R had no negative cost cycles.

Among all such negative cost cycles in R', all of which start with edge (s, y_i), we wish to find the one which contributes the most to decrementing the value of f'. Therefore, let p denote the **"correction path"** for R', defined as the the min-cost path in R' among all paths which originate in s, pass through the newly widowed vertex y_i and end back in s (Figure 4.C). Clearly, p can not start from some other edge than (s, y_i) since all $\ell - k$ vertices in Y which were unmatched in M are "out of the game" and will remain unmatched in the optimal assignment of size $|M| - 1$ on G_i. Otherwise, the assumed optimality of M would be contradicted.

We claim that:

1. If p has negative total cost, then M' is not optimal.
2. If M' is not optimal, then p is the optimal "correction path" for R'. That is, the flow obtained by pushing one unit of flow in R' from s through y_i and back to s via path p, is the minimal cost maximum value flow for network graph G'' and defines, respectively, the optimal assignment for G_i.

The above claims are proven as follows.

1. This is immediate from Lemma 1, since p is by definition a cycle in R'.
2. Let f'' be obtained from f' by correcting along p, and let R'' denote the residual graph for flow f''. We will prove that f'' is min-cost in G'', by proving that there are no negative cost cycles in R''. Suppose there was some negative cost cycle c in R''. Since R had no negative cost cycles, and the only new edge $(s, y_i) \in R'$ has been saturated by p in R'' (therefore all previous cycles are broken in R''), we know that c has at least one edge which is the reversal of an edge in p. Therefore, cycle c consists of edges in R' and one or more edges whose reversals are on p. Therefore, p and c share at least one vertex.

   Let $p \oplus c$ be the set of edges on p or on c except for those occurring on p and reversed on c. The cost of $p \oplus c$ is $cost(p) + cost(c)$, since the $\oplus$ operation has removed pairs of edges and their reversals (the score is negated in the reversal) in $p \cup c$. Since c is a negative cost cycle, $cost(p) + cost(c) < cost(p)$. Furthermore, $p \oplus c$ can be partitioned into a path p' originating in s, going through y_i and ending in s, plus a collection of cycles. (Note that none of these cycles includes edge (s, y_i), since this edge is already occupied by p'). Clearly, all the cycles must have nonnegative cost since they consist of edges only from R' which has no negative cost cycles by Lemma 1. Thus path p' is a "correction path" of cost less than p (Figure 4.D). This contradiction implies the lemma.

Note that the above proof assumes $k \leq \ell$, in which case $|M| = k$. However, the special case $k = \ell + 1$ can be easily handled by adding a dummy vertex to Y and connecting it by edges to all the vertices in X. The weight of these edges should be set to N for some very large number N.   □

Lemma 3.

1. *Computing the weighted assignments for the bipartite graphs $G_1, \ldots, G_k$ can be done in $O(k^2\ell + k\ell \log \ell)$ time.*
2. *Under the similarity assumption, computing the weighted assignments for the bipartite graphs $G_1, \ldots, G_k$ can be done in $O(k^{1.5}\ell \log(\ell C))$ time [12].*

Theorem 2.

1. *Algorithm 2 computes the optimal ALSH solution for two unrooted unordered trees in $O(m^2 n + mn \log n)$ time.*
2. *Under the similarity assumption, Algorithm 2 yields an $O(m^{1.5}n \log(nC))$ time complexity.*

Proof. We use the following simple observation:

Observation 2. The sum of vertex degrees in an unrooted tree P is $\sum_{u=1}^{m} d(u) = 2m - 2$.

Algorithm 2 computes a score for each node pair $(v \in T, u \in P)$. During this procedure, the algorithm computes weighted assignments of several bipartite graphs. Due to the decremental relationship between these graphs, it is shown in Lemma 3 that computing the maximal assignment for all graphs in the series can be done in the same time bound taken by computing a single assignment. This computation, which is the dominant part of the procedure's time, can be done in $O(d(u)^2 c(v) + d(u)c(v) \log c(v))$ time. Therefore, the total complexity is

$$O(\sum_{u=1}^{m} \sum_{v=1}^{n} (d(u)^2 c(v)) + d(u)c(v) \log c(v)) \overset{\text{observation 1}}{=}$$

$$O(\sum_{u=1}^{m} d(u)^2 n + d(u)n \log n) \overset{\text{observation 2}}{=} O(m^2 n + mn \log n).$$

Under the (similarity) assumption that all scores assigned to the edges of G are integers in the range $[-C, \ldots, C]$, where $C = O(n^k)$ for some constant k, the algorithm can be modified to run in time $O(m^{1.5}n \log(nC))$ by employing the algorithm of Gabow and Tarjan [8] for weighted bipartite matching with scaling and [12] for cavity matching.   □

Note that the first term of the sum $O(m^2 n + mn \log n)$ dominates the time complexity of Algorithm 2, since the second term only takes over when $m \leq \log n$. Therefore, in the next section we will show how to reduce the time complexity of the first, dominant term in the time complexity of Algorithm 2.

4   A More Efficient ALSH Algorithm for Unordered Trees

In this section we show how the dominant term in the time complexity of the algorithms described in the previous section can be reduced by a $\log m$ factor, assuming a constant-sized label alphabet. We will use the previous algorithm but solve the maximum matching problems more efficiently by employing the notion of clique partition of a bipartite graph [6, 23]. The modified algorithm, denoted Algorithm 3, is the same as Algorithm 2 with the exception that, in step 12, we solve the assignment problem differently. Let v denote some vertex in T^r whose children are $Y = y_1, \ldots, y_{c(v)}$ and let u denote a vertex in P whose neighbors are $X = x_1, \ldots, x_{d(u)}$. We now attempt to compute the Assignment of $G = (X \cup Y, E)$. We will partition the edges of G into complete bipartite graphs $C_1, C_2, \ldots, C_{clusters_u}$. Let $Key(x_i)$ be a vector containing the weights of the edges $(x_i, y_1), \ldots, (x_i, y_{c(v)})$. We do the partition in the following way: First, we sort the vertices of X in lexicographic order where the key of a vertex x is $Key(x)$. Afterwards, we split X into sets of equal keys $X^1, X^2, \ldots, X^{clusters_u}$ (i.e., for any two vertices $x, x' \in X_i$, any edge (x, y_j) has the same weight as edge (x', y_j) for all vertices $y_j \in Y$).

Now, for $1 \leq i \leq clusters_u$ we set C_i to be the subgraph induced by the vertices of X^i and all their neighbors in Y. We now follow the method of [6, 23] and build a network G^* whose vertices are $V^* = X \cup Y \cup \{c_1, \ldots, c_{clusters_u}, s, t\}$. The edges are $E^* = E_1 \cup E_2 \cup E_3$ where $E_1 = \{[s, x_i] : x_i \in X\} \cup \{[y_i, t] : y_i \in Y\}$, $E_2 = \{[x_i, c_j] : j \leq clusters_u, x_i \in C_j\}$, and $E_3 = \{[c_j, y_i] : j \leq clusters_u\}$. All edges have capacity 1. Edges from sets E_1 and E_2 are assigned a cost of zero. An edge of type E_3 from c_j to y_i is assigned a cost which is identical to the cost of edge (x, y_i) where $x \in X$ is any vertex belonging to the set C_j. The source is s and the sink is t. We find a min-cost max flow f^* in G^* using Fredman and Tarjan's algorithm, and construct from this flow the assignment in G.

Lemma 4. *A min-cost max flow in f^* corresponds directly to a min-cost max flow in f.*

We denote by $D(u)$ the number of distinct trees in the forest $P^u_{u_1}, \ldots, P^u_{u_k}$.

Lemma 5. *The assignment between $u \in P$ and $v \in T$ can be computed in $O(d(u) (D(u)c(v) + c(v) \log c(v)))$ time.*

Time Complexity Analysis

Algorithm 3 activates procedure $ComputeScoresForTextNode$ once for each node pair $(v \in T, u \in P)$. The dominant part of this procedure's work is spent in computing the weighted assignment between a bipartite graph with $c(v) + d(u)$ nodes and $c(v) \times d(u)$ edges.

Therefore, based on Lemma 5, the total work spent on assignment computations is

$$O(\sum_{u=1}^{m} \sum_{v=1}^{n} (d(u)(clusters_u \times c(v)) + c(v) \log c(v))) \overset{\text{observation 1}}{=}$$

$$O(\sum_{u=1}^{m} d(u) \times clusters_u \times n + d(u)n \log n) \overset{observation\ 2}{=}$$

$$O(n \sum_{u=1}^{m} d(u)\ D(u) + mn \log n).$$

We will now turn to bound the summation $O(\sum_{u=1}^{m} d(u)\ D(u))$. We start with next lemma, which sets an asymptotically tight bound on the number of distinct labelled rooted trees in a forest of n vertices, denoted $f(n)$.

Lemma 6. *Assuming constant label alphabet, $f(n) = O(n/\log n)$.*

Lemma 7. $\sum_{u=1}^{m} d(u)\ D(u) = O(m^2/\log m)$.

We have thus proved the following theorem.

Theorem 3. *Algorithm 3 computes the optimal ALSH solution for two unrooted unordered trees in $O(m^2 n/\log m + mn \log n)$ time.*

5   Solving ALSH on Ordered Rooted Trees

A simplified version of the first algorithm, denoted Algorithm 4, can be used to solve the ALSH problem on ordered rooted trees. The simplification is based on the fact that the assignment problems now turn into maximum weighted matching problems on ordered bipartite graphs, where no two edges are allowed to cross. (We refer the reader to [26] for a discussion of non-crossing plain bipartite matching in the context of exact ordered tree homeomorphism.) Also note that, in the context of our ALSH solutions, we actually apply a rectangular case of the perfect assignment problem on $G = (X \cup Y, E)$, i.e. $|X| \leq |Y|$ and all nodes in X must eventually be paired with some node in Y. Therefore, the assignment computation reduces to the *Approximate Weighted Episode Matching* optimization problem, as defined below.

Definition 6. *The* **Approximate Weighted Episode Matching Problem:** *Given pattern string X, a source string Y, and a character-to-character similarity table $\Delta[\Sigma_X, \Sigma_Y]$, find among all $|X|$-sized subsequences of Y the subsequence Q which is most similar to X, that is, the sum $\sum_{i=1}^{|X|} \Delta[Q_i, X_i]$ is maximized.*

Lemma 8. *The highest-scoring approximate episode occurrence of a pattern X of size k characters in a source string Y of size ℓ characters can be computed in $O(k \times \ell)$ time.*

Proof. This can be achieved by applying the classical dynamic programming string alignment algorithm to a ($k + 1$ rows by $\ell + 1$ columns) dynamic programming graph (see Figure 5). All horizontal edges in the graph, corresponding to

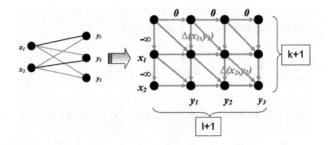

Fig. 5. Approximate Weighted Episode Matching Calculation.

character deletions from Y, are assigned a score of zero. All vertical edges in the graph, corresponding to character deletions from X, are assigned a score of $-\infty$. A diagonal edge leading into vertex (x_i, y_j) corresponds to the string-edit operation of substituting the ith character of X with the jth character of Y, and is therefore assigned the score $\Delta[i, j]$.   □

Time Complexity Analysis

Theorem 4. *ALSH of two rooted ordered trees can be computed in $O(mn)$ time.*

Proof. Algorithm 4 computes the assignment once for each node pair $(v \in T, u \in P)$. In this section we showed that, for rooted unordered trees, the dominant work of the assignment computation can be narrowed down to $O(c(v) \times c(u))$ time. Therefore, the total time complexity is:

$$\sum_{v=1}^{n} \sum_{u=1}^{m} O(c(v) \times c(u)) \overset{\text{observation 1}}{=} \sum_{v=1}^{n} O(m \times c(v)) \overset{\text{observation 1}}{=} O(m \times n). \quad □$$

Acknowledgements

We thank Seffi Naor and Gabriel Valiente for helpful discussions and comments.

References

1. R.K. Ahuja, T.L. Magnanti, and J.B. Orlin. *Network Flows: Theory, Algorithms and Applications*. Prentice-Hall, 1993.
2. D. Carmel, N. Efrati, G.M. Landau, Y.S. Maarek, and Y. Mass. An extension of the vector space model for querying xml documents via xml fragments. In *XML and Information Retrieval (Workshop) Tampere, Finland*, 2002.
3. M. J. Chung. $O(N^{2.5})$ time algorithms for the subgraph homeomorphism problem on trees. *J. Algorithms*, 8:106–112, 1987.
4. E. W. Dijkstra. A note on two problems in connection with graphs. *Numerische Mathematik*, 1:269–271, 1959.
5. J. Edmonds and R. M. Karp. Theoretical improvements in algorithmic efficiency for network flow problems. *J. of the Assoc. for Comput. Mach.*, 19(2):248–264, 1972.

6. T. Feder and R. Motwani. Clique partitions, graph compression and speeding-up algorithms. In *Proceedings 23rd Symposium on the Theory of Computing (STOC 91)*, pages 123–133, 1991.

7. M. L. Fredman and R. E. Tarjan. Fibonacci heaps and their uses in improved network optimization algorithms. *Journal of the Association for Computing Machinery*, 34(3):596–615, 1987.

8. H. N. Gabow and R. E. Tarjan. Faster scaling algorithms for network problems. *SIAM J. Comput.*, 18(5):1012–1036, 1989.

9. P. B. Gibbons, R. M. Karp, and D. Soroker. Subtree isomorphism is in random NC. *Discrete Applied Mathmatics*, 29:35–62, 1990.

10. A. V. Goldberg, S. A. Plotkin, and P.M Vidya. Sublinear-time parallel algorithms for matching and related problems. *J. Algorithms*, 14:180–213, 1993.

11. L. R. Ford Jr. and D. R. Fulkerson. *Flows in Networks*. Princeton, NJ: Princeton University Press, 1962.

12. M. Y. Kao, T. W. Lam, W. K. Sung, and H. F. Ting. Cavity matchings, label compressions, and unrooted evolutionary trees. *SIAM J.Comput.*, 30(2):602–624, 2000.

13. M. Karpinski and A. Lingas. Subtree isomorphism is NC reducible to bipartite perfect matching. *Information Processing Letters*, 30(1):27–32, 1989.

14. P. Kilpelainen and H. Mannila. Ordered and unordered tree inclusion. *SIAM J. Comput.*, 24(2):340–356, 1995.

15. D. W. Matula. An algorithm for subtree identification. *SIAM Rev.*, 10:273–274, 1968.

16. D. W. Matula. Subtree isomorphism in $O(n^{5/2})$. *Ann. Discrete Math.*, 2:91–106, 1978.

17. H. Ogata, W. Fujibuchi, S. Goto, and M. Kanehisa. A heuristic graph comparison algorithm and its application to detect functionally related enzyme clusters. *Nucleic Acids Research*, 28:4021–4028, 2000.

18. J. B. Orlin and R. K. Ahuja. New scaling algorithms for the assignment and minimum cycle mean problems. *Math. Prog.*, 54:541–561, 1992.

19. R. Y. Pinter, O. Rokhlenko, E. Yeger-Lotem, and M. Ziv-Ukelson. A new tool for the alignment of metabolic pathways. *Manuscript*, 2004.

20. S. W. Reyner. An analysis of a good algorithm for the subtree problems. *SIAM J. Comput.*, 6:730–732, 1977.

21. T. Schlieder and F. Naumann. *Approximate Tree Embedding for Querying XML Data*. ACM SIGIR Workshop on XML and IR, 2000.

22. F. Schreiber. Comparison of metabolic pathways using constraint graph drawing. In *Proceedings of the Asia-Pacific Bioinformatics Conference (APBC'03), Conferences in Research and Practice in Information Technology*, 19, pages 105–110, 2003.

23. R. Shamir and D. Tsur. Faster subtree isomorphism. *Journal of Algorithms*, 33(2):267–280, 1999.

24. M. A. Steel and T. Warnow. Kaikoura tree theorems: Computing the maximum agreement subtree. *Information Processing Letters*, 48(2):77–82, 1993.

25. R. E. Tarjan. *Data Structures and Network Algorithms*. SIAM, Philadelphia, 1982.

26. G. Valiente. Constrained tree inclusion. In *Proceedings of 14th Annual Symposium of Combinatorial Pattern Matching (CPM '03)*, volume 2676 of *Lecture Notes in Computer Science*, pages 361–371, 2003.

On the Average Sequence Complexity

Svante Janson[1], Stefano Lonardi[2], and Wojciech Szpankowski[3]

[1] Dept. of Mathematics – Uppsala University – Uppsala, Sweden
[2] Dept. of Computer Science – University of California – Riverside, CA
[3] Dept. of Computer Sciences – Purdue University – West Lafayette, IN

Abstract. In this paper we study the average behavior of the number of distinct substrings in a text of size n over an alphabet of cardinality k. This quantity is called the *complexity index* and it captures the "richness of the language" used in a sequence. For example, sequences with low complexity index contain a large number of repeated substrings and they eventually become periodic (e.g., tandem repeats in a DNA sequence). In order to identify unusually low- or high-complexity strings one needs to determine how far are the complexities of the strings under study from the average or maximum string complexity. While the maximum string complexity was studied quite extensively in the past, to the best of our knowledge there are no results concerning the average complexity. We first prove that for a sequence generated by a mixing model (which includes Markov sources) the average complexity is asymptotically equal to $n^2/2$ which coincides with the maximum string complexity. However, for memoryless source we establish a more precise result, namely the average string complexity is $n^2/2 - n \log_k n + (1 + (1-\gamma)/\ln k + \phi_k(\log_k n) + o(1))n$ where $\gamma \approx 0.577$ and $\phi_k(x)$ is a periodic function with a small amplitude for small alphabet size.

1 Introduction

In the last decades, several attempts have been made to capture mathematically the concept of "complexity" of a sequence. The notion is connected with quite deep mathematical properties, including the rather elusive concept of randomness of a string (see, e.g., [14, 19, 22]).

In this paper, we are interested in studying a measure of complexity of a sequence called the *complexity index*. The complexity index captures the "richness of the language" used in a sequence. Formally, the *complexity index* $c(x)$ of a string x is equal to the number of distinct substrings in x (see e.g. [20]). The measure is simple but quite intuitive. Sequences with low complexity index contain a large number of repeated substrings and they eventually become periodic. However, in order to classify low complexity sequences one needs to determine average and maximum string complexity. In this paper we concentrate on the average string complexity.

We assume that sequences are generated by some probabilistic source (e.g., Bernoulli, Markov, etc.). As a consequence, the number $c(x)$ of distinct substrings can be modeled by a random variable over a discrete domain. Given a

S.C. Sahinalp et al. (Eds.): CPM 2004, LNCS 3109, pp. 74–88, 2004.

source emitting strings of size n over an alphabet of cardinality k, we call this random variable $C_{n,k}$. The main objective of this study is to give a detailed characterization of the expectation of the random variable $C_{n,k}$.

A related notion is that of the *l-subword complexity* or *l-spectrum* $c^l(x)$ of a string x, which is the number of distinct l-mers in x, for $1 \leq l \leq |x|$. We define $C_{n,k}^l$ to be the random variable associated with the number of distinct words of size l in a random string of size n over an alphabet of cardinality k. Clearly, $C_{n,k} = \sum_{l=1}^{n} C_{n,k}^l$.

The idea of using the complexity index or the l-spectrum to characterize sequence statistically has a long history of applications in several fields, such as data compression, computational biology, data mining, computational linguistics, among others.

In dictionary-based data compression, the average length of the pointer is connected with the expected size of the dictionary which in turns depends on the number of distinct subwords (see, e.g., [6]). Low-complexity strings contain more repeated substrings and therefore one can expect them to be more compressible than strings with high complexity index. For example, in [14] bounds between subword complexity and Lempel-Ziv complexity are established.

In the analysis of biosequences, the problem of characterizing the "linguistic complexity" of DNA or proteins is quite old. In the early days of computational biology, it was almost routine to compute the number and/or the frequency of occurrences of distinct l-mers and draw conclusions about the string under study based on those counts (see [23, 7, 15, 12, 16], just to mention a few).

In these and several other application domains, the typical problem associated with the complexity index is to determining whether a particular sequence x has a *statistically significant* complexity index. An example of a significance score proposed in a recent paper by Troyanskaya et al. [31] in the context of the analysis of prokaryotic genomes, is the following

$$s(x) = c(x) - \max\{C_{n,k}\} = c(x) - \sum_{i=1}^{n} \min(k^i, n - i + 1).$$

Here, the authors compare the observed complexity $c(x)$ with the maximum possible complexity for a string of size n over an alphabet of cardinality k. Note however, that the score disregards both the distribution of $C_{n,k}$, and the probabilistic characteristics of the source.

A more statistically-sound approach would entail the following steps. First, select an appropriate model for the source that emitted x (Bernoulli, Markov, etc.). Then, measure the statistical significance as a function of the discrepancy between the observed complexity $c(x)$ and the model-based expectation.

This approach of standardizing the value of an observation with respect to the expectation and the standard deviation of the associated random variable is common practice in Statistics. The underlying assumption is that the random variable is normally distributed. The standardized z-score for the complexity index would be

$$z(x) = \frac{c(x) - \mathbf{E}(C_{n,k})}{\sqrt{\mathbf{Var}(C_{n,k})}}$$

for a given string x. Although we do not know under which conditions $C_{n,k}$ is distributed normally, such a score is nonetheless more sound that other *ad hoc* scores.

A similar situation takes place when describing the significance of other events in texts, like the number of occurrences, periodicities, etc. Although the normality of the corresponding random variables can be proved under specific conditions, there is a consensus that standardized z-scores should be always preferred over simpler scores (see, e.g., [25] and references therein).

Given an observation x of the source, we would like to compute the statistical significance $z(x)$ of its complexity index. As far as we know, however, the moments $\mathbf{E}(C_{n,k})$ and $\mathbf{Var}(C_{n,k})$ have never been characterized before. The goal of this paper is to study $\mathbf{E}(C_{n,k})$. The asymptotic analysis of the variance remains an open problem.

In order to proceed, we need to introduce some standard concepts and notation about strings. The set Σ denotes a nonempty *alphabet* of *symbols* and a *string* over Σ is an ordered sequence of symbols from the alphabet. In the rest of the paper, we assume that $|\Sigma| = k$. Given a string x, the number of symbols in x defines the *length* $|x|$ of x. Henceforth, we assume $|x| = n$.

The *concatenation* (or *product*) of two strings x and y is denoted by xy, and corresponds to the operation of appending y to the last symbol of x. Let us decompose a text x in uvw, i.e., $x = uvw$ where u, v and w are strings over Σ. Strings u, v and w are called *substrings*, or *words*, of x.

We write $x_{[i]}$, $1 \leq i \leq |x|$ to indicate the i-th symbol in x. We use $x_{[i,j]}$ shorthand for the substring $x_{[i]}x_{[i+1]} \ldots x_{[j]}$ where $1 \leq i \leq j \leq |x|$, with the convention that $x_{[i,i]} = x_{[i]}$. Substrings in the form $x_{[1,j]}$ corresponds to the *prefixes* of x, and substrings in the form $x_{[i,n]}$ to the *suffixes* of x.

Finally, we recall that the *subword complexity* function $c^l(x)$ of a string x is defined as the number of distinct substrings of x of length l. The quantity $c(x) = \sum_{l=1}^{n} c^l(x)$ is the *complexity index* of x. Observe first that $c^l(x)$ is upper bounded by $\min(k^l, n - l + 1)$ since there are precisely $n - l + 1$ words of length l, of which at most k^l can be distinct. Therefore

$$c(x) \leq \sum_{l=1}^{n} \min(k^l, n - l + 1).$$

Note that if we choose $n \leq k$, then $\min(k^l, n-l+1) = n-l+1$ for all $1 \leq l \leq n$. Therefore when $n \leq k$, the bound simplifies to $c(x) \leq n(n+1)/2$.

The value $c(x)$ is strongly connected with the structure of the *non-compact suffix trie* for x. A non-compact suffix trie of a string x is a digital search tree built from all the suffixes of x. The trie for a string of size n has $n + 1$ leaves, numbered 1 to $n + 1$, where leaf $n + 1$ correspond to an extra unique symbol $\$ \notin \Sigma$, and each edge is labeled with a symbol in Σ. No two edges outgoing from a node can have the same label. The trie has the property that for any leaf i, the concatenation of the labels on the path from the root the the leaf i spells out exactly the suffix of x that starts at position i, that is $x_{[i,n]}$. The substrings of x can be obtained by spelling out the words from the root to any internal

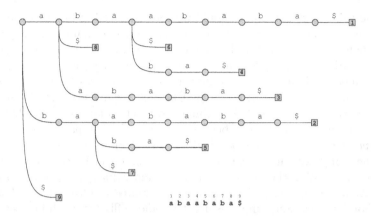

Fig. 1. The non-compact suffix trie T_x for $x = $ abaababa. There are 24 internal non-root nodes, therefore $c(x) = 24$.

node of the tree. In other words, each internal node (except the root) is in one-to-one correspondence with each distinct substring of x. As a consequence, *the complexity index $c(x)$ can be obtained by counting the non-root internal nodes in the non-compact suffix trie for x.* This would take, however, $O(n^2)$ time and space. The non-compact suffix trie for abaababa is illustrated in Figure 1.

A faster solution to compute $c(x)$ involves the use of the *suffix tree* $\bar{T}_x$ of x. The suffix tree can obtained by "path-compression" of the non-compact suffix trie, that is, by deleting internal nodes with only one child and coalescing consecutive edges in a single edge labeled by the concatenation of the symbols. If one deletes unary nodes only at the *bottom* of the non-compact suffix tries, the resulting tree is called *compact suffix trie*. The compact suffix trie and the suffix tree for abaababa are shown in Figure 2.

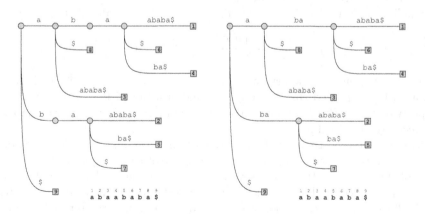

Fig. 2. The compact suffix trie (LEFT) and the suffix tree (RIGHT) for $x = $ abaababa.

In practice, suffix trees can be build without the need of building the suffix trie first. In fact, several $O(n \log |\Sigma|)$ constructions are available. The algorithm by McCreight [21] and the one by Chen and Seiferas [9] are variation of the Weiner's algorithm [34]. Note that these algorithms take only linear time for finite alphabets. All these constructions are *off-line* because they process the text right to left. An on-line algorithm by Ukkonen [32] achieves also linear time. Recently, Farach [11] proposed an optimal construction for large alphabets.

The unary nodes which have been removed in the compaction process are called *implicit nodes*. An edge labeled by a string of length $m+1$ has m implicit nodes. The complexity index $c(x)$ of a text-string x can be computed on the suffix tree by counting the number of implicit and explicit (non-root) nodes in the tree. As a consequence, the $c(x)$ can be computed in $O(n)$ time and space. The relation between non-compact suffix tries and suffix trees will be used later in paper to obtain the leading term of the complexity for a general probabilistic model.

Finally, we briefly describe some recent results regarding the maximum of $c^l(x)$. It is known that $c^l(x)$ is also strongly connected with the structure of the suffix tree $\bar{T}_x$. De Luca [10] proved that

$$\max\{c^l(x) : 1 \le l \le n\} = n - \max\{R, K\} + 1 = n - \max\{L, H\} + 1$$

where K is the length of the shortest suffix of x that occurs only once, H is the length of the shortest prefix of x that occurs only once, $R - 1$ is the height of the deepest branching node in the suffix tree $\bar{T}_x$ and $L - 1$ is the height of the deepest branching node in the suffix tree $\bar{T}_{x^R}$. De Luca also gave a closed form for $c(x)$

$$c(x) = 1 + \frac{(n+K)(n-K+1)}{2} - \sum_{j=2}^{|\Sigma|} \sum_{i=0}^{R} ig(j, i)$$

where $g(j, i)$ is the count of the words of length i which are branching nodes of the suffix tree with at least j children [10].

Shallit [26] derived a simpler bound for $c(x)$ for binary alphabets ($k = 2$)

$$c(x) \le \frac{(n-d+1)(n-d)}{2} + 2^{d+1} - 1 \sim \frac{n^2}{2}$$

where d is the unique integer such that $2^d + d - 1 \le n < 2^{d+1} + d$. More importantly, he proved that the upper bound is attained for all n by using a property of the *de Bruijn* graphs. An extension of this result to larger alphabets was recently described in [13].

Kása [17] studied the probability distribution of random variable associated with the complexity index for the family of words of length equal to the size of the alphabet ($n = k$). He proved several facts about the random variable $C_{k,k}$, and he also conjectured a property of the smallest value of the complexity after which all the frequencies are non-zero. The conjecture, proved later by Leve and Seebold [18], states that if one chooses $k = \frac{l(l+1)}{2} + 2 + i$ where $l \ge 2$ and $0 \le i \le l$ then

$\mathbf{P}\left(C_{k,k}=t\right)>0$ for all t such that $\dfrac{l(l^2-1)}{2}+3l+2+i(l+1)\le t\le \dfrac{k(k+1)}{2}$

In this paper we mostly deal with the average string complexity. First, for a general probabilistic model (e.g., Markov source) we prove that the average complexity is asymptotically equal to $n^2/2$ which coincides with the maximum string complexity. We shall strengthen this result for unbiased memoryless sources. In particular, we prove that

$$\mathbf{E}(C_{n,k})=\binom{n+1}{2}-n\log_k n+\left(\frac{1}{2}+\frac{1-\gamma}{\ln k}+\phi_k(\log_k n)\right)n+O(\sqrt{n\log n})$$

where $\gamma\approx 0.577$ and $\phi_k(x)$ is a continuous function with period 1 and small amplitude for small alphabet size (e.g., $|\phi_2(x)|<2\cdot 10^{-7}$). To prove this result we use the Stein–Chen method together with the Mellin transform approach.

2   Main Results

As a warm-up exercise, we studied the closed forms for $\mathbf{E}\left(C_{n,k}\right)$ and $\mathbf{Var}\left(C_{n,k}\right)$ for short strings (e.g., $n\le 5$) for a symmetric memoryless source. Some facts about $\mathbf{P}(C_{n,k}=t)$ are immediate. For example

$$\mathbf{P}\left(C_{n,k}<n\right)=0$$
$$\mathbf{P}\left(C_{n,k}=n\right)=k^{1-n}$$
$$\mathbf{P}\left(C_{n,k}>\frac{n(n+1)}{2}\right)=0\ \text{when}\ n\le k$$
$$\mathbf{P}\left(C_{n,k}>\sum_{i=1}^{n}\min(k^i,n-i+1)\right)=0\ \text{when}\ n>k$$

Following the lines by Kása [17], we were able to obtain closed form for the cases shown in Table 1 assuming a symmetric Bernoulli model for the source. Given the discrete distribution of $C_{n,k}$ one can easily compute the expectation and the variance of $C_{n,k}$.

Corollary 1. *The expectation and the variance of the random variable $C_{n,k}$ for $2\le n\le 5$ over any alphabet of size k, under a symmetric Bernoulli source is*

$$\mathbf{E}\left(C_{2,k}\right)=3-(1/k)$$
$$\mathbf{E}\left(C_{3,k}\right)=6-(3/k)$$
$$\mathbf{E}\left(C_{4,k}\right)=10-(6/k)+(1/k^2)-(1/k^3)$$
$$\mathbf{E}\left(C_{5,k}\right)=15-(10/k)+(4/k^2)-(6/k^3)+(2/k^4)$$

and

$$\mathbf{Var}\left(C_{2,k}\right)=(k-1)/k^2$$
$$\mathbf{Var}\left(C_{3,k}\right)=3(k-1)/k^2$$
$$\mathbf{Var}\left(C_{4,k}\right)=(k-1)(6k^4-5k^3+12k^2-k+1)/k^6$$
$$\mathbf{Var}\left(C_{5,k}\right)=2(k-1)(5k^6-10k^5+33k^4-28k^3+16k^2-10k+2)/k^8$$

Table 1. The probability distribution of $C_{n,k}$ for $2 \le n \le 5$ over any alphabet of size k, under a symmetric Bernoulli source.

$n \to$	2	3	4	5
$\mathbf{P}\,(C_{n,k} = 2)$	$1/k$	0	0	0
$\mathbf{P}\,(C_{n,k} = 3)$	$1 - 1/k$	$1/k^2$	0	0
$\mathbf{P}\,(C_{n,k} = 4)$	0	0	$1/k^3$	0
$\mathbf{P}\,(C_{n,k} = 5)$	0	$3(k-1)/k^2$	0	$1/k^4$
$\mathbf{P}\,(C_{n,k} = 6)$	0	$(k-1)(k-2)/k^2$	0	0
$\mathbf{P}\,(C_{n,k} = 7)$	0	0	$3(k-1)/k^3$	0
$\mathbf{P}\,(C_{n,k} = 8)$	0	0	$4(k-1)/k^3$	0
$\mathbf{P}\,(C_{n,k} = 9)$	0	0	$6(k-1)(k-2)/k^3$	$3(k-1)/k^4$
$\mathbf{P}\,(C_{n,k} = 10)$	0	0	$(k-1)(k-2)(k-3)/k^3$	0
$\mathbf{P}\,(C_{n,k} = 11)$	0	0	0	$10(k-1)/k^4$
$\mathbf{P}\,(C_{n,k} = 12)$	0	0	0	$2(k-1)(3k-5)/k^4$
$\mathbf{P}\,(C_{n,k} = 13)$	0	0	0	$19(k-1)(k-2)/k^4$
$\mathbf{P}\,(C_{n,k} = 14)$	0	0	0	$10(k-1)(k-2)(k-3)/k^4$
$\mathbf{P}\,(C_{n,k} = 15)$	0	0	0	$(k-1)(k-2)(k-3)(k-4)/k^4$
$\mathbf{P}\,(C_{n,k} \ge 16)$	0	0	0	0

As it turns out, obtaining a closed form for $\mathbf{E}\,(C_{n,k})$ and $\mathbf{Var}\,(C_{n,k})$ for any n, k is a very challenging problem. In practical applications, moreover, having the closed form only for small values of n would be of limited interest. It is certainly more valuable to study the behavior of the moments of $C_{n,k}$ asymptotically, that is, when n is very large.

The main result of our study is a characterization of $\mathbf{E}\,(C_{n,k})$ for large n. In our first result we show that for quite general sources the average complexity asymptotically coincides with the maximal string complexity. We consider mixing sources in which the probability of two events A and B defined on two substrings separated by g symbols is bounded as follows: $(1 - \psi(g))\mathbf{P}(A)\mathbf{P}(B) \le \mathbf{P}(AB) \le (1 + \psi(g))\mathbf{P}(A)\mathbf{P}(B)$ where the mixing coefficient $\psi(g) \to 0$ as $g \to \infty$ (cf. [28] for a detailed definition).

Theorem 1. *Let $C_{n,k}$ be the complexity index of a string of length n generated by a strongly mixing stationary source. Then, for large n,*

$$\mathbf{E}C_{n,k} = \binom{n+1}{2} - O(n \log n).$$

Hence $C_{n,k} = n^2/2 + O_p(n \log n)$, i.e. $(n^2/2 - C_{n,k})/n \log n$ is bounded in probability.

Proof. We start with a simple observation. For a given sequence x of size n, build a non-compact suffix trie and a compact suffix trie. Recall that in a compact trie we collapse all unary links at the *bottom* of the suffix trie. In other words, in a compact suffix trie a path from the root to an external node (containing a suffix) is the minimal prefix of any two suffixes that distinguishes them. Also recall that the string complexity of x is the number of non-root internal nodes in the non-compact trie. We shall argue below that the most contribution to the string complexity comes from the nodes that are in the non-compact trie but not in the compact trie. The upper bound follows immediately from

$$C_{n,k} \le n(n+1)/2.$$

To find a matching lower bound, we consider the compact and non-compact suffix tries. We know that a typical depth and height in a compact suffix trie is $O(\log n)$. More precisely let H_n be the height of a compact suffix tree. It was shown in [27] that (at least for $\psi(g)$ satisfying $\sum_{g \ge 0} \psi^2(g) < \infty$) $H_n/\ln n \to 2/h_1$ a.s., where h_1 is the Renyi's entropy (cf. [28, p. 157]). More precisely, the proof shows (for any $\psi(g) \to 0$) that for any $\epsilon > 0$

$$\mathbf{P}\left(H_n \le \frac{2}{h_1}(1+\epsilon)\log n\right) = 1 - O(\log n/n^\epsilon). \tag{1}$$

We claim that the main contribution to $C_{n,k}$ comes from strings that are in the non-compact trie but not in the compact trie. In fact, the i-th suffix string has $n - i$ internal nodes of which at least $n - i - H_n$ are not in the compact trie. These nodes all correspond to unique substrings of x, and thus

$$C_{n,k} \ge \sum_{i=1}^{n}(n - i - H_n) = \tfrac{1}{2}n(n+1) - nH_n.$$

By (1), for a suitable constant B and large n, $\mathbf{P}(H_n > B \log n) < n^{-1}$ and thus

$$\mathbf{E}\left(\tfrac{1}{2}n(n+1) - C_{n,k}\right) \le n\mathbf{E}H_n \le n\big(B \log n + n\mathbf{P}(H_n > B \log n)\big) = O(n \log n),$$

which completes the proof.

However, from theoretical point of view the most interesting case is when the string is generated by an unbiased source (such a source should have the largest complexity). In this case, we are able to characterize very precisely the average complexity.

Theorem 2. *Let $C_{n,k}$ be the complexity index of a string generated by an unbiased memoryless source. Then the average l-subword complexity is*

$$\mathbf{E}(C_{n,k}^l) = k^l(1 - e^{-nk^{-l}}) + O(l) + O(nlk^{-l}). \tag{2}$$

Furthermore, for large n the average complexity index becomes

$$\mathbf{E}(C_{n,k}) = \binom{n+1}{2} - n \log_k n + \left(\frac{1}{2} + \frac{1 - \gamma}{\ln k} + \phi_k(\log_k n)\right)n + O(\sqrt{n \log n})$$

where $\gamma \approx 0.577$ is Euler's constant and

$$\phi_k(x) = -\frac{1}{\ln k}\sum_{j \ne 0}\Gamma\left(-1 - \frac{2\pi i j}{\ln k}\right)e^{2\pi i j x}$$

is a continuous function with period 1. $|\phi_k(x)|$ is very small for small k: $|\phi_2(x)| < 2 \cdot 10^{-7}$, $|\phi_3(x)| < 5 \cdot 10^{-5}$, $|\phi_4(x)| < 3 \cdot 10^{-4}$.

Interestingly enough, the term $O(n)$ of the average complexity contains a fluctuating term $\phi_k(x)$. (Note that $\phi_2(\log_2 n)$ equals $-P_0(n)$ in [28, p. 359].) The formula

$$|\Gamma(-1-iy)| = |\Gamma(-iy)/(-1-iy)| = \left(y(1+y^2)\sinh(\pi y)/\pi\right)^{-1/2} \quad (3)$$

[28, p. 42] shows that the coefficients in ϕ_k are small and decrease geometrically. Numerical evaluations give the bounds for $k = 2, 3, 4$ stated in the theorem and also, for example, $|\phi_k(x)| < 0.01$ for $k \le 12$ and $|\phi_k(x)| < 0.1$ for $k \le 200$. Even for very large k, this term is not very large; we have, still using (3) (we omit the details), $|\phi_k(x)| < 0.5$ for $k \le 10^9$, and $|\phi_k(x)| < \ln\ln(k)/\pi$ for all k. The fluctuating phenomenon is quite common in asymptotics of discrete problems.

3 Proof of Theorem 2

In this section we prove Theorem 2, however, to simplify our notation we restrict ourselves to the binary case $k = 2$. Extension to $k > 2$ is straightforward.

Recall that $C^l_{n,2}$ is the number of distinct substrings of length l in our random binary string of size n, and let

$$A_l = \mathbf{E}(C^l_{n,2}).$$

Thus $C_{n,2} = \sum_{l=1}^n C^l_{n,2}$ and $\mathbf{E}(C_{n,2}) = \sum_{l=1}^n A_l$. Define

$$\delta_l = A_l + l - 1 - 2^l\left(1 - e^{-n2^{-l}}\right) \quad (4)$$

$$= A_l - (n+1-l) + 2^l\left(e^{-n2^{-l}} - 1 + n2^{-l}\right). \quad (5)$$

Then

$$\mathbf{E}(C_{n,2}) = \sum_{l=1}^n A_l = \sum_{l=1}^n \left((n+1-l) + \delta_l - 2^l\left(e^{-n2^{-l}} - 1 + n2^{-l}\right)\right)$$

$$= \binom{n+1}{2} - \sum_{l=1}^n 2^l\left(e^{-n2^{-l}} - 1 + n2^{-l}\right) + \sum_{l=1}^n \delta_l. \quad (6)$$

Below we will use several times the estimates (for $x > 0$) $0 < 1 - e^{-x} < \min(1, x)$ and $0 < e^{-x} - 1 + x < \min(x, x^2)$. In particular,

$$0 < 2^l\left(1 - e^{-n2^{-l}}\right) < \min(2^l, n), \quad (7)$$

$$0 < 2^l\left(e^{-n2^{-l}} - 1 + n2^{-l}\right) < n^2 2^{-l}. \quad (8)$$

We begin by estimating δ_l. First (for short strings, i.e., for small l), we use $1 \le C^l_{n,2} \le 2^l$, and thus $0 \le A_l \le 2^l$ and, using (4) and (7),

$$\delta_l = O(2^l). \quad (9)$$

For long substrings, i.e., for large l's, there is another simple estimate. Clearly, $0 \le C_{n,2}^l \le n - l + 1$. Observe that

$$\mathbf{E}(n - l + 1 - C_{n,2}^l) \le \mathbf{E}\left(\left|\{(i,j) : i < j \text{ and } x_{[i,i+l-1]} = x_{[j,j+l-1]}\}\right|\right)$$
$$= \binom{n-l+1}{2} 2^{-l} \le n^2 2^{-l}$$

i.e.

$$0 \le n - l + 1 - A_l \le n^2 2^{-l}.$$

Hence, using (5) and (8),

$$\delta_l = O(n^2 2^{-l}). \tag{10}$$

Note that (9) and (10) easily yield $\sum_{l=1}^n \delta_l = O(n)$, which by (6) yields $\mathbf{E}(C_{n,2})$ up to $O(n)$. In order to obtain our more precise result, we need a better estimate of δ_l when $2^l \approx n$.

Let x be a sequence generated by an i.i.d. source (i.e., $p(0) = p(1) = 1/2$) and let us define

$$\mathbf{P}(n, l) = \mathbf{P}(x_{[1,l]} \ne x_{[j,j+l-1]} \text{ for } j = 2, \ldots, n - l + 1)$$

By counting each repeated substring in x only the last time it occurs, it is easily seen (by shift invariance) that

$$A_l = \sum_{m=l}^{n} \mathbf{P}(m, l) \tag{11}$$

Now fix l and m, and let us define $I_j = \mathbf{1}[x_{[1,l]} = x_{[j,j+l-1]}]$ for $j = 2, \ldots, m-l+1$. Then

$$\mathbf{E}(I_j) = \mathbf{P}(I_j = 1) = \mathbf{P}[x_{[1,l]} = x_{[j,j+l-1]}] = 2^{-l} \text{ for every } j \ge 2.$$

In the next lemma we establish that I_i and I_j are uncorrelated when $i, j > l$.

Lemma 1. *If $i, j \ge l + 1$ and $i \ne j$ then $\mathbf{E}(I_i I_j) = \mathbf{P}(x_{[1,l]} = x_{[i,i+l-1]} = x_{[j,j+l-1]}) = 2^{-2l}$.*

Proof. Assume $i < j$. Scan the three substring left to right. Each bit in $x_{[i,i+l-1]}$ is either a fresh random bit or coincides with an earlier bit in $x_{[j,j+l-1]}$. In any case, it fixes one new bit each in $x_{[1,l]}$ and $x_{[j,j+l-1]}$, so the probability of success in this step is 2^{-2}.

Observe that, if $j \ge l+1$, to condition on I_j is the same as to change every bit in $x_{[j,j+l-1]}$ to the corresponding bit in $x_{[1,l]}$, leaving all other bits untouched. Let $x^{(j)}$ be the resulting string, and let $J_{ij} = \mathbf{1}[x_{[1,l]}^{(j)} = x_{[i,i+l-1]}^{(j)}]$. Clearly, $J_{ij} = I_i$ if $|i - j| \ge l$. Note also that Lemma 1 yields, when $i \ne j$ and $i, j > l$,

$$\mathbf{E}(J_{ij}) = \mathbf{E}(I_i \mid I_j = 1) = \frac{\mathbf{E}(I_i I_j)}{\mathbf{E}(I_j)} = 2^{-l}.$$

Let us now set, for fixed integers l and $m \geq 2l$,

$$W = \sum_{j=l+1}^{m-l+1} I_j.$$

Let $d_{TV}(X, Y)$ be the total variation distance between random variables X and Y. Then, with $\mathrm{Po}(\lambda)$ being the Poisson distribution with mean λ, by the Stein–Chen method (cf. [5, page 24]) we find, with $\lambda = \mathbf{E}W = (m - 2l + 1)2^{-l}$,

$$d_{TV}\left(W, \mathrm{Po}((m - 2l + 1)2^{-l})\right) \leq \frac{\min(1, \lambda)}{\lambda} \sum_j \mathbf{E}(I_j)\mathbf{E}\left| I_j + \sum_{i \neq j}(I_i - J_{ij}) \right|$$

$$\leq \frac{1}{\lambda} \sum_j \mathbf{E}(I_j)\left(\mathbf{E}(I_j) + \sum_{0 < |i-j| < l}(\mathbf{E}(I_i) + \mathbf{E}(J_{ij}))\right)$$

$$\leq \frac{1}{\lambda} \sum_{j=l+1}^{m-l-1} 2l \cdot 2 \cdot 2^{-2l}$$

$$= 4l2^{-l}.$$

In particular,

$$\left| \mathbf{P}(W = 0) - e^{-(m-2l+1)2^{-l}} \right| \leq d_{TV}\left(W, \mathrm{Po}((m - 2l + 1)2^{-l})\right) = O(l2^{-l}). \quad (12)$$

Moreover, by the first moment method

$$\mathbf{P}\left(\sum_{j=2}^{l} I_j \neq 0\right) \leq (l - 1)\mathbf{E}(I_j) = (l - 1)2^{-l}. \quad (13)$$

Observe that

$$\mathbf{P}(m, l) = \mathbf{P}\left(\sum_{j=2}^{l} I_j + W = 0\right).$$

Then by (13) and (12)

$$\mathbf{P}(m, l) = \mathbf{P}(W = 0) + O(l2^{-l}) = e^{-(m-2l+1)2^{-l}} + O(l2^{-l}) = e^{-(m-l)2^{-l}} + O(l2^{-l}). \quad (14)$$

We have assumed $m \geq 2l$. However, by the first moment method directly, the same estimate holds for $l \leq m < 2l$ too.

We thus have, by (11) and (14) and summing a geometric series,

$$A_l = \sum_{m=l}^{n} \mathbf{P}(m, l) = \frac{1 - e^{-(n+1-l)2^{-l}}}{1 - e^{-2^{-l}}} + O(nl2^{-l}).$$

Since

$$\frac{1 - e^{-(n+1-l)2^{-l}}}{1 - e^{-2^{-l}}} = \frac{1 - e^{-(n+1-l)2^{-l}}}{2^{-l}(1 + O(2^{-l}))}$$

$$= (2^l + O(1))(1 - e^{-(n+1-l)2^{-l}})$$

$$= 2^l(1 - e^{-(n+1-l)2^{-l}}) + O(n2^{-l})$$

$$= 2^l(1 - e^{-n2^{-l}}) + O(l) + O(n2^{-l}),$$

we find

$$A_l = 2^l(1 - e^{-n2^{-l}}) + O(l) + O(nl2^{-l})$$

which proves (2). Thus by (4)

$$\delta_l = O(l) + O(nl2^{-l}). \tag{15}$$

Using (9) for $1 \le l \le \log_2 \sqrt{n \ln n}$, (15) for $\log_2 \sqrt{n \ln n} < l \le 2 \log_2 n$, and (10) for $2 \log_2 n < l \le n$, we obtain

$$\sum_{l=1}^{n} \delta_l = O(n \log n)^{1/2}. \tag{16}$$

We turn to the first sum in (6). Let, for $x > 0$,

$$f(x) = \frac{e^{-x} - 1 + x1[x < 1]}{x}.$$

Then $|f(x)| < x$ for $0 < x < 1$ and $|f(x)| < 1/x$ for $x \ge 1$. Hence,

$$\sum_{l=1}^{n} 2^l\left(e^{-n2^{-l}} - 1 + n2^{-l}\right) = n \sum_{l=1}^{n} \left(f(n2^{-l}) + 1[2^l \le n]\right)$$

$$= n \sum_{l=1}^{n} f(n2^{-l}) + n\lfloor \log_2 n \rfloor$$

$$= n \sum_{l=-\infty}^{\infty} f(n2^{-l}) + O(1) + n\lfloor \log_2 n \rfloor \tag{17}$$

$$= n\psi(n) + n \log_2 n + O(1),$$

where, with $\langle x \rangle = x - \lfloor x \rfloor$, the fractional part of x,

$$\psi(x) = \sum_{l=-\infty}^{\infty} f(x2^{-l}) - \langle \log_2 x \rangle, \qquad x > 0.$$

(The series converges by the estimates above.) It is easily verified that ψ is bounded, continuous (also at powers of 2), and periodic in $\log_2 x$, i.e., $\psi(2x) = \psi(x)$. Hence $\psi(2^y)$ has period 1 and may be expanded in a Fourier series

$$\psi(2^y) = \sum_{\nu=-\infty}^{\infty} c_\nu e^{2\pi i \nu y} \tag{18}$$

where

$$c_\nu = \int_0^1 e^{-2\pi i \nu y} \psi(2^y)\, dy$$

$$= \int_0^1 e^{-2\pi i \nu y} \left(\sum_{l=-\infty}^{\infty} f(2^{y-l}) - y \right) dy$$

$$= \sum_{l=-\infty}^{\infty} \int_0^1 e^{-2\pi i \nu y} f(2^{y-l})\, dy - \int_0^1 e^{-2\pi i \nu y} y\, dy \qquad (19)$$

$$= \int_{-\infty}^{\infty} e^{-2\pi i \nu y} f(2^y)\, dy - \int_0^1 e^{-2\pi i \nu y} y\, dy.$$

Further, changing variables back to $(0, \infty)$,

$$\int_{-\infty}^{\infty} e^{-2\pi i \nu y} f(2^y)\, dy = \frac{1}{\ln 2} \int_0^{\infty} x^{-2\pi i \nu / \ln 2} f(x) \frac{dx}{x} = \frac{1}{\ln 2} \mathcal{M}[f(x); -2\pi i \nu / \ln 2], \qquad (20)$$

where $\mathcal{M}[f(x); s] = \int_0^{\infty} x^{s-1} f(x)\, dx$ is the Mellin transform of f at the point s (in our case $s = -2\pi i \nu / \ln 2$. (See [28, Chapter 9] for definition and basic properties.) Since $|f(x)| < \min(x, x^{-1})$, the Mellin transform $\mathcal{M}[f(x); z]$ is analytic in the strip $-1 < \Re z < 1$. In the smaller strip $0 < \Re z < 1$ we have, by [28, Table 9.1],

$$\mathcal{M}[f(x); z] = \mathcal{M}\left(\frac{e^{-x}-1}{x}; z\right) + \mathcal{M}\left(\mathbf{1}[x < 1]; z\right)$$

$$= \mathcal{M}(e^{-x} - 1; z - 1) + \mathcal{M}\left(\mathbf{1}[x < 1]; z\right) \qquad (21)$$

$$= \Gamma(z - 1) + \frac{1}{z}.$$

By analyticity, this extends to the strip $-1 < \Re z < 1$. In particular, taking the limit as $z \to 0$,

$$\mathcal{M}[f(x); 0] = \lim_{z \to 0} \left(\frac{\Gamma(1+z)}{z(z-1)} + \frac{1}{z} \right) = \lim_{z \to 0} \frac{\Gamma(1+z) - 1 + z}{z(z-1)} = -(\Gamma'(1) + 1) = \gamma - 1. \qquad (22)$$

Moreover, elementary integration yields

$$\int_0^1 e^{-2\pi i \nu y} y\, dy = \begin{cases} -1/2\pi i \nu, & \nu \neq 0, \\ 1/2, & \nu = 0. \end{cases} \qquad (23)$$

By (19)–(23),

$$c_\nu = \frac{1}{\ln 2} \Gamma\left(-1 - \frac{2\pi i \nu}{\ln 2}\right), \qquad \nu \neq 0,$$

$$c_0 = \frac{\gamma - 1}{\ln 2} - 1/2.$$

The theorem now follows, with $\phi(x) = -(\psi(2^x) - c_0)$, from (6), (16), (17) and (18).

The numerical bounds for $k \leq 4$ are obtained from (3); for small k, $\sum_1^\infty |c_\nu|$ is dominated by $|c_1|$ which is very small.

Acknowledgements

This work was supported in part by NSF Grants CCR-0208709, DBI-0321756 and NIH grant R01 GM068959-01.

References

1. Apostolico, A. The myriad virtues of suffix trees. In *Combinatorial Algorithms on Words*, A. Apostolico and Z. Galil, Eds., vol. 12 of *NATO Advanced Science Institutes, Series F*. Springer-Verlag, Berlin, 1985, pp. 85–96.
2. Apostolico, A., Bock, M. E., and Lonardi, S. Monotony of surprise and large-scale quest for unusual words. *J. Comput. Bio. 10*, 3-4 (2003), pp. 283–311.
3. Apostolico, A., Bock, M. E., Lonardi, S., and Xu, X. Efficient detection of unusual words. *J. Comput. Bio. 7*, 1/2 (2000), pp. 71–94.
4. Apostolico, A., and Szpankowski, W. Self-alignment in words and their applications. *J. Algorithms 13*, 3 (1992), pp. 446–467.
5. Barbour, A., Holst, K., Janson, S. *Poisson Approximation*. Oxford Press, Oxford, 1992.
6. Bell, T. C., Cleary, J. G., and Witten, I. H. *Text Compression*. Prentice Hall, 1990.
7. Burge, C., Campbell, A., and Karlin, S. Over- and under-representation of short oligonucleotides in DNA sequences. *Proc. Natl. Acad. Sci. U.S.A. 89* (1992), pp. 1358–1362.
8. Chang, W. I., and Lawler, E. L. Sublinear approximate string matching and biological applications. *Algorithmica 12*, 4/5 (1994), pp. 327–344.
9. Chen, M. T., and Seiferas, J. Efficient and elegant subword tree construction. In *Combinatorial Algorithms on Words*, A. Apostolico and Z. Galil, Eds., vol. 12 of *NATO Advanced Science Institutes, Series F*. Springer-Verlag, Berlin, 1985, pp. 97–107.
10. de Luca, A. On the combinatorics of finite words. *Theor. Comput. Sci. 218*, 1 (1999), pp. 13–39.
11. Farach, M. Optimal suffix tree construction with large alphabets. In *Proc. 38th Annual Symposium on Foundations of Computer Science* (1997), pp. 137–143.
12. Fickett, J. W., Torney, D. C., and Wolf, D. R. Base compositional structure of genomes. *Genomics 13* (1992), pp. 1056–1064.
13. Flaxman, A., Harrow, A. W., and Sorkin, G. B. Strings with maximally many distinct subsequences and substrings. *Electronic Journal of Combinatorics 11*, 1 (2004).
14. Ilie, L., Yu, S., and Zhang, K. Repetition Complexity of Words In *Proc. COCOON* (2002), pp. 320–329.
15. Karlin, S., Burge, C., and Campbell, A. M. Statistical analyses of counts and distributions of restriction sites in DNA sequences. *Nucleic Acids Res. 20* (1992), pp. 1363–1370.

16. Karlin, S., Mrazek, J., and Campbell, A. M. Frequent oligonucleotides and peptides of the *Haemophilus influenzae* genome. *Nucleic Acids Res. 24*, 21 (1996), pp. 4273–4281.

17. Kása, Z. On the *d*-complexity of strings. *Pure Math. Appl. 9*, 1-2 (1998), pp. 119–128.

18. Leve, F., and Seebold, P. Proof of a conjecture on word complexity. *Bulletin of the Belgian Mathematical Society 8*, 2 (2001).

19. Li, M., and Vitanyi, P. *Introduction to Kolmogorov Complexity and its Applications*. Springer-Verlag, Berlin, Aug. 1993.

20. Lothaire, M., Ed. *Algebraic Combinatorics on Words*. Cambridge University Press, 2002.

21. McCreight, E. M. A space-economical suffix tree construction algorithm. *J. Assoc. Comput. Mach. 23*, 2 (Apr. 1976), pp. 262–272.

22. Niederreiter, H. Some computable complexity measures for binary sequences, In *Sequences and Their Applications*, Eds. C. Ding, T. Hellseth and H. Niederreiter Springer Verlag, pp. 67-78, 1999.

23. Pevzner, P. A., Borodovsky, M. Y., and Mironov, A. A. Linguistics of nucleotides sequences II: Stationary words in genetic texts and the zonal structure of DNA. *J. Biomol. Struct. Dynamics 6* (1989), pp. 1027–1038.

24. Rahmann, S., and Rivals, E. On the distribution of the number of missing words in random texts. *Combinatorics, Probability and Computing* (2003), 12:73-87.

25. Reinert, G., Schbath, S., and Waterman, M. S. Probabilistic and statistical properties of words: An overview. *J. Comput. Bio. 7* (2000), pp. 1–46.

26. Shallit, J. On the maximum number of distinct factors in a binary string. *Graphs and Combinatorics 9* (1993), pp. 197–200.

27. Szpankowski, W. A generalized suffix tree and its (un)expected asymptotic behaviors, *SIAM J. Computing*, 22, (1993), pp. 1176–1198.

28. Szpankowski, W. *Average Case Analysis of Algorithms on Sequences*. John Wiley & Sons, New York, 2001.

29. Tomescu, I. On the number of occurrences of all short factors in almost all words. *Theoretical Computer Science* 290:3 (2003), pp. 2031-2035.

30. Tomescu, I. On Words Containing all Short Subwords. *Theoretical Computer Science* 197:1-2 (1998), pp. 235-240.

31. Troyanskaya, O. G., Arbell, O., Koren, Y., Landau, G. M., and Bolshoy, A. Sequence complexity profiles of prokaryotic genomic sequences: A fast algorithm for calculating linguistic complexity. *Bioinformatics 18*, 5 (2002), pp. 679–688.

32. Ukkonen, E. On-line construction of suffix trees. *Algorithmica 14*, 3 (1995), pp. 249–260.

33. Wang, M.-W. and Shallit, J. On minimal words with given subword complexity. *Electronic Journal of Combinatorics*, 5(1) (1998), #R35.

34. Weiner, P. Linear pattern matching algorithm. In *Proceedings of the 14th Annual IEEE Symposium on Switching and Automata Theory* (Washington, DC, 1973), pp. 1–11.

Approximate Point Set Pattern Matching on Sequences and Planes

Tomoaki Suga[1] and Shinichi Shimozono[2]

[1] Graduate School of Computer Science and Systems Engineering
[2] Department of Artificial Intelligence
Kyushu Institute of Technology
Kawazu 680–4, Iizuka, 820-8502 Japan
suga@daisy.ai.kyutech.ac.jp, sin@ai.kyutech.ac.jp

Abstract. The point set pattern matching problem is, given two sets "pattern" and "text" of points in Euclidean space, to find a linear transformation that maps the pattern to a subset of the text. We introduce an approximate point set pattern matching for axis-sorted point sequences that allows a translation, space insertions and deletions between points. We present an approximate pattern matching algorithm that runs with pattern size n and text size m in $O(nm^2)$ time in general, and in $O(nm)$ time if distances between two contiguous points in texts and patterns are finite. A variant of the four-Russian technique achieving $O(nm/\log n + n \log n)$ time is also provided. Furthermore, as a natural extension we present an approximate point set pattern matching on the plane, and give a polynomial-time algorithm that solves this problem.

Keywords: point set pattern matching, edit distance, approximate matching, musical sequence search

1 Introduction

The point set pattern matching problem by linear transformations asks, given a pair of a "pattern" set and a "text" set of points in Euclidean d-dimensional space, to find a linear transformation that exactly maps the pattern to a subset of the text. A rich class of problems, such as a motion detection of rigid bodies from images, relate to this fundamental problem. In [14], Rezende and Lee presented an exact point set pattern matching algorithm for the 1-dimensional case, and extended it to d-dimensional one with a circular scan technique that runs in $O(nm^d)$ time with the pattern size n and the text size m.

The idea of approximate matching is also a powerful tool for the point set pattern matching in practical applications. For example, in a two-dimensional gel electrophoresis analysis, we would like to determine whether a set of specific proteins, which is observed as a set of spots on a gel media plane, occurs among spots obtained from a sample material. Distances between spots depend on experimental conditions, which drift during an experiment. Hoffmann *et al.* [4], and Akutsu *et al.* [1] coped with this problem by approximate point set pattern matching schemes and heuristic algorithms. Unfortunately, their edit

S.C. Sahinalp et al. (Eds.): CPM 2004, LNCS 3109, pp. 89–101, 2004.
© Springer-Verlag Berlin Heidelberg 2004

distances by non-uniform transformations lead the matching problems to be NP-complete [1, 5] even for the 1-dimensional case.

Mäkinen and Ukkonen have discussed in [8, 11] a 1-dimensional approximate point pattern matching, which is an extension of the approximate string pattern matching, and applied it in a heuristic for the gel electrophoresis problem. Their algorithm runs in $O(nm)$ time with space insertions and deletions between points. Additionally substitutions for sets of points are handled in $O(nm^3 + n^2m^2 + n^3m)$ time. Also in [8] they have introduced an approximate 2-dimensional point matching problem, and have been proved that the problem is NP-hard.

The origin of our motivation is a melody search problem for digital musical scores such as internet contents, ringer melodies of mobile phones, streaming data for on–demand "Karaoke" systems, etc. So far, pattern matching and similarity measurement for musical score sequences have often been done by extending text search techniques (e.g. [6, 13]). However, string representations are not suitable in musical pattern search. For example, the same melodies in different keys are described as completely different strings. Thus in [13] the keys of scores are assumed to be normalized in advance. Kadota *et al.* [6] have prevented this by generating all possible transposes of the target passage. In recent works [7, 10], this difficulty is treated by string pattern matching algorithms that allow symbols in patterns to transpose, i.e. relative shifts for all symbols, by regarding symbols as numbers.

A digital musical score such as a MIDI sequence can be considered as a series of notes, where each note roughly represents a note-on time, duration and a note-number describing a pitch in half-steps. Then a score simply describes a set of points in a plane, which is like a cylinder of musical boxes or a piano-roll of a classic player piano [12]. In this representation, key transposes are naturally treated as exact translations in the pitch axis. Differences in tempo would be treated by a scaling factor along the time axis. Since a pattern provided by human may lack some notes, such as notes forming chords, we would like to find a pattern as a subset of a score. To attain these advantages, we formulate as in [15] a musical pattern matching as a point "set" pattern matching problem.

In this paper, we define an approximate point set pattern matching for axis-sorted point sequences that allows space insertions and deletions between points, and show a DP algorithm that runs in $O(nm^2)$ time and $O(nm)$ space with pattern size n and text size m. Then, we show an improvement of the DP algorithm, and prove that it achieves $O(nm)$ running time if point sequences have "finite resolution," i.e., the maximum distance between two contiguous points is fixed. It enables us to apply the algorithm for a huge size of text sequences. Also, this assumption allows us to apply ideas in approximate string matching: If point sequences have finite resolution, a tabular computation can be performed in $O(nm/\log n + n\log n)$ time by a speed-up technique analogous to the four-Russian technique [2, 3].

Furthermore, we introduce an edit distance and an approximate point set pattern matching for point sets in the plane, as a natural extension of our approximate point set pattern matching in sequences. Then a DP-like algorithm that runs in $O(n^2m^4)$ time and $O(n^2m^2)$ space is devised. To our knowledge, this is the first non-trivial 2-dimensional approximate point set matching providing a polynomial-time algorithm.

2 Point Sequences and Approximate Matching

As in [8], the 1-dimensional case is instructive and interesting itself. Furthermore, there are considerable amount of practical applications where points belong to two or higher dimensional space but are ordered by one dominating axis. For example, in musical sequence search, notes of scores must be treated as time-series sequences by "note-on time." In 3-D matching of protein molecule structure, a molecule is often abstracted as a folded chain of amino acids, and thus the order of amino acid residues is fixed as a sequence.

Also we assume that in a 1-D point sequence there are no two or more points that have the same value. This does not spoil our discussion since in such a case points have additional dimensions and thus we can choose an appropriate point by other attributes. The notions required for our problem can be formalized as follows.

A *1-dimensional point sequence*, or *point sequence* in short, is a finite sequence of strictly-increasing positive integers. For a point sequence p, the notion $|p|$ refers to the number of points. A point sequence $p' = (p'_1, \ldots, p'_r)$ is a *subsequence* of $p = (p_1, \ldots, p_n)$ if there is a subsequence $1 \leq k_1 < \cdots < k_r \leq n$ of indices such that $p'_i = p_{k_i}$ for all $1 \leq i \leq r$.

Let $p = (p_1, \ldots, p_n)$ and $t = (t_1, \ldots, t_m)$ be point sequences with $n \leq m$. We say p *matches* t if there is a subsequence $t' = (t_{k_1}, \ldots, t_{k_r})$ of t such that, for all $1 < i \leq n$, $p_i - p_{i-1} = t_{k_i} - t_{k_{i-1}}$. In this case, p and t are said to be a *pattern* and a *text*, respectively. Also we say that p *occurs in* t *at* k_1, and the index k_1 is an *occurrence position of* p *in* t.

This definition of matching is due to the point set matching by [14]. We extend this to an approximate version. Firstly, we introduce an edit distance between sequences having the same number of points.

Definition 1. Edit Distance between Point Sequences
Let p and q be point sequences of the same size n. The edit distance $d(p, q)$ *between p and q is $d(p, q) = 0$ if $n \leq 1$, and*

$$d(p, q) = \sum_{i=2}^{n} |p_i - p_{i-1} - (q_i - q_{i-1})| .$$

For the brevity of discussion, we define $d(p, q') = \infty$ for $|p| \neq |q'|$.

This edit distance does not satisfy the "identity" condition with the straightforward equivalence. It is due to the lack of the distance $|q_1 - p_1|$ between the first points. Anyway, the following fact supports that our formulation is mathematically well-defined.

For any point sequences p, q and r of size n, the edit distance satisfies

(a) $d(p, q) \geq 0$,
(b) $d(p, q) = d(q, p)$, and
(c) $d(p, q) \leq d(p, r) + d(r, q)$ for any point sequence r.

Now the criteria to find a subsequence of a text sequence that approximately matches to a pattern sequence is formalized as follows.

Definition 2. Approximate Point Set Pattern Matching
Let $p = (p_1, \ldots, p_n)$ be a pattern point sequence, $t = (t_1, \ldots, t_m)$ a text point sequence, and $\varepsilon \geq 0$ a positive integer. We say p approximately matches t at an index k within error ε if there is a subsequence t' of t such that t' begins with t_k and $d(p, t')$ is no more than ε.

Unlike the matching defined in [8], we must find a best subset of points from the given text of points. Fortunately, we can use recurrence relations to find a subsequence, and it allows us to apply a version of the "traditional" tabular computation method. In the following, we show the recurrence that holds for the minimum edit distances between a pattern point sequence and a subsequence of a text point sequences.

Let p be a pattern point sequence of size n, and t a text point sequence of size m. For all $1 \leq i \leq n$ and $1 \leq j \leq m$, we define the notion $D(i, j)$ as the minimum value of the edit distances between $(p_1, \ldots, p_i)$ and any subsequence t' of t such that $|t'| = i$ and t' ends with the point t_j. Then clearly $D(1, j) = 0$ for all $1 \leq j \leq m$, and by the definition of the edit distance $D(i, j) = \infty$ for $i > j$. Furthermore, the following recurrence holds.

Lemma 1. For all $1 \leq i \leq n$ and $1 \leq j \leq m$,

$$D(i, j) = \min_{i-1 \leq k < j} \{D(i - 1, k) + |p_i - p_{i-1} - (t_j - t_k)|\} .$$

Proof. It is clear that the left hand side is no greater than the right hand side. Also by assuming a sequence $1 \leq h_1 < \cdots < h_{i-1} < j$, it is immediate that

$$D(i, j) - |p_i - p_{i-1} - (t_j - t_{h_{i-1}})| \geq D(i - 1, h_{i-1})$$

and $i - 1 \leq h_{i-1}$. $\square$

This gives a straightforward DP algorithm. Given a pattern $p = (p_1, \ldots, p_n)$ and a text $t = (t_1, \ldots, t_m)$, we construct an $n \times m$ integer table D. According to the definition, all the cells in the first row of D are initialized to 0, and all $D(i, j)$ with $i > j$ are filled with ∞. Then the remaining cells are filled as follows.

Algorithm 1.
1. **For** i **from** 2 **to** n **do** the following:
   (a) **For** j **from** i **to** m **do** the following:
       i. $D(i, j) = \min_{i-1 \leq j' < j} \{D(i - 1, j') + |p_i - p_{i-1} - (t_j - t_{j'})|\}$;

For each cell $D(i, j)$, the algorithm looks up $j - i + 1$ cells in the $(i-1)$th row with $O(m)$ time, and thus the total tabular computation takes $O(nm^2)$ time and $O(nm)$ space. To obtain an occurrence, or a subsequence of the text, we invoke the "trace back" procedure which is exactly the same with that of approximate string matching (e.g., [3].) Enumeration of occurrences takes $O(m)$ time.

Corollary 1. *1-dimensional approximate point set pattern matching can be solved in $O(nm^2)$ time and $O(nm)$ space with pattern size n and text size m.*

3   Speed–Up for Practical Instances

Running time of the DP-algorithm presented in the previous section is polynomial but is quadratic with the size of text. To cope with this point, we utilize the following observations for practical instances in a particular application:

The resolution of positions of points in sequences are limited by accuracy of measuring devices, data format, human sense, etc. For example, in a MIDI sequence, a time delay (delta time) of a note from the previous one is often specified by a resolution 480 per quarter note. In gel electrophoresis technique, the size of gel media planes are limited, and positions of spots cannot be accurately determined since they appear as stains.

We summarize these observations and formalize as follows.

Definition 3. *Let L be a positive integer, and let C be a (possibly infinite) set of point sequences. We say C has finite resolution L if there is no two contiguous points in any sequence in C such that the distance between them is greater than L.*

In the following, we present Algorithm_2, an improved version of Algorithm_1. The modification eliminates redundant computations at Step 1-(a)-i of Algorithm_1 which searches for the minimum value, though the worst case time-complexity order remains same for general instance. However, if text and pattern have finite resolution, it becomes $O(nm)$ time algorithm.

Algorithm_2.
1. **For** i **from** 2 **to** n **do** the following:
   (a) $k \leftarrow i - 1$; /* The "pivot" index of the ith row. */
   (b) $D_- \leftarrow \infty$; /* The minimum value in the range $j' \in [i - 1, k - 1]$. */
   (c) **For** j **from** i **to** m **do** the following:
      i. $D_- \leftarrow D_- + (t_j - t_{j-1})$ if $j \neq i$;
      ii. $D_+ \leftarrow \infty$;
      iii. **For** j' **from** k **to** j **do** the following:
         A. **If** $p_i - p_{i-1} < t_j - t_{j'}$ **then**
             $k \leftarrow j' + 1$; /* Advancing the pivot. */
             $D_- \leftarrow \min\{D_-, D(i - 1, j') + |p_i - p_{i-1} - (t_j - t_{j'})|\}$;
         B. **else**
             $D_+ \leftarrow \min\{D_+, D(i - 1, j') + |p_i - p_{i-1} - (t_j - t_{j'})|\}$;
      iv. $D(i, j) \leftarrow \min\{D_-, D_+\}$;

At first, we show this algorithm is equivalent to Algorithm_1.

Lemma 2. *Algorithm 2 completes the tabular computation.*

Proof. Recall the recurrence relation of cells discussed in the previous section. By the notion $\Delta(i, j, j') = p_i - p_{i-1} - (t_j - t_{j'})$, it can be rewritten as follows.

$$D(i, j) = \min \left\{ \begin{array}{l} \min\limits_{i-1 \leq j' < j-1} \{D(i-1, j') + |\Delta(i, j, j')|\}, \\ D(i-1, j-1) + |\Delta(i, j, j-1)| \end{array} \right\}.$$

The idea of the algorithm is that the values considered in the right-hand side of this equation (except for $j' = j - 1$) are obtained by adding or subtracting $t_j - t_{j-1}$ to the values calculated in the computation of $D(i, j - 1)$. Assume that indices i and j are fixed. Since point sequences are strictly increasing, $\Delta(i, j, j')$ monotonically increases with j'. Thus there is the "pivot" index k with $i - 1 \leq k < j$ such that $\Delta(i, j, j') < 0$ for $i - 1 \leq j' < k$ and $\Delta(i, j, j') \geq 0$ for $k \leq j' < j$.

Let us consider the difference between $D(i-1, j-1)$ and $D(i-1, j)$. Let $k_{(j-1)}$ and $k_{(j)}$ be the pivots for $\Delta(i-1, j-1, j')$ and $\Delta(i-1, j, j')$, respectively. Note that $k_{(j-1)} \leq k_{(j)}$ because i is fixed. Since $\Delta(i, j-1, j') < 0$ for $i-1 \leq j' < k_{(j-1)}$, we can rewrite

$$D(i-1, j') + |\Delta(i, j, j')| = D(i-1, j') + |\Delta(i, j-1, j')| + (t_j - t_{j-1})$$

for all $i - 1 \leq j' < k_{(j-1)}$. This indicates that if $D(i-1, j') + |\Delta(i, j-1, j')|$ is minimum for some $j' < k_{(j-1)}$ then $D(i-1, j') + |\Delta(i, j, j')|$ is also the minimum in the same range. This value is stored in D_-. For $k_{(j-1)} \leq j' < k_{(j)}$, the sign of $\Delta(i, j, j')$ changes from plus to minus, and the minimum value is compared with and stored in D_-. This is carried out in Step 1-(c)-iii-A, as well as updates of the value k towards $k_{(j)}$. For the remaining range $k_{(j)} \leq j' < j$, we enumerate values as in Algorithm 1. The result of Step 1-(c) is the minimum value among all the indices $i - 1 \leq j' < j$. Thus the algorithm computes the same recurrences.   □

Theorem 1. *Let t and p be a text and a pattern point sequences, and let r be the ratio of the maximum distance between two contiguous points in p to the minimum distance between two contiguous points in t. Then, Algorithm 2 runs in $O(nm \cdot r)$ time with $n = |t|$ and $m = |p|$.*

Proof. The steps from 1-(c)-i to 1-(c)-iv of the algorithm takes $O(j - k)$ time. Let L be the maximum distance between two contiguous points in p. For any interval $I \subseteq [t_1, t_n]$ of length $2L$, the number of points of t in I is at most $2r$. Then the number of points in $[t_k, t_j]$ for some j and pivot k, at the beginning of Step 1-(c)-iii, is $O(r)$. In another words, k at Step 1-(c)-iii advances with the control index j in the loop at Step 1-(c). □

Corollary 2. *Algorithm 2 solves 1-D approximate point set pattern matching in $O(nm^2)$ time in general, and solves in $O(nm)$ time if a text and a pattern are chosen from a set having finite resolution.*

3.1 Computational Results

We briefly show by computational experiments that Algorithm_2 achieves drastic improvement. Table 1 shows execution time required to finish tabular computations by **Algorithm_1** and **Algorithm_2** for sequences of note-on time extracted from Standard MIDI Format (SMF) files on the internet. The algorithms are implemented by GNU C++ with STL, and tested on AMD Duron 800 MHz ($\approx$ Intel Celeron 800 MHz) based machine. A text with 4370 points is formed from a concatenation of multiple tracks in one SMF file, and texts with $*$ symbol are synthesized by concatenating tracks from multiple files of tunes to obtain huge inputs. The maximum distance between two contiguous points in patterns A and B is 480.

Execution time by **Algorithm_2** seems to be proportional to text size. The maximum number of iterations at Step 1-(c)-iii in **Algorithm_2** was 57 for these sequences.

Table 1. Execution time for tabular computations with patterns formed from 10 and 20 points.

Text	Pattern A ($n = 10$)		Pattern B ($n = 20$)	
(points)	**Algorithm_1**	**Algorithm_2**	**Algorithm_1**	**Algorithm_2**
4370	7.505 (sec.)	0.019 (sec.)	22.81 (sec.)	0.039 (sec.)
20309*	226.8	0.048	573.0	0.094
51173*	1489	0.107	3692	0.220
104579*	–	0.201	–	0.411
1154073*	–	2.238	–	4.577

3.2 Four–Russian Speed Up

The four–Russian speed up is a well-known technique (e.g. [3]) to finish a $n \times m$ tabular computation in $O(nm/\log^2 n + n \log^2 n)$ time with unit-cost RAM model. The technique divides a whole table into boundary-shared squares of a fixed size, and only their boundaries are filled by looking up values from a pre-computed table. The key is that the value of a cell depends on only 3 cells and two character symbols, and thus if an error bound is limited then the number of possible combinations of symbols and values is constant.

In general, this does not hold with point sequences, and applying the four–Russian technique to our approximate point set pattern matching is impossible. However, if point sequences have finite resolution, then a variant of the four-Russian speed up for the tabular computation in Algorithm_2 is available.

We define rectangular blocks of cells for filling up the DP-table as follows.

Definition 4. *Let C be a set of point sequences having finite resolution L, and let $p = (p_1, \ldots, p_n)$ and $t = (t_1, \ldots, t_m)$ be a pattern and a text sequence of*

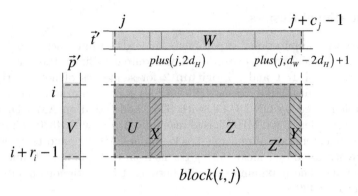

Fig. 1. Segments of a block on D and D_-, and the corresponding subsequences.

points in $\mathcal{C}$. We choose constants $d_H \geq L$ and $d_W \geq 4L$ appropriately. For any $1 \leq i \leq n$ and $1 \leq j \leq m$ satisfying $p_i + d_H \leq p_n$ and $t_j + d_W \leq t_m$, we define rectangle $block(i,j)$ of cells of D (and D_-) represented by the set of pairs of cell indices

$$block(i,j) = \{(i',j') \mid p_i \leq p_{i'} \leq p_i + d_H, t_j \leq t_{j'} \leq t_j + d_W\}.$$

The number of cells in a block depends on its position. We refer the number of rows and the number of columns of $block(i,j)$ to r_i and c_j, respectively. Also we define, for any $\delta \in \mathbb{Z}$, the notion $plus(i,\delta)$ of the text index such that $t_{plus(i,\delta)} \leq t_i + \delta$ and $plus(i,\delta)$ is maximum.

We place the first block, $block(i_1.j_1)$, at $i_1 = 1$ and $j_1 = n$. Then, $block(i_k, j_l)$, the i_kth from top and the j_lth from left, is placed at $i_k = i_{k-1} + r_{i_{k-1}} - 1$, $j_l = plus(j_{l-1}, d_W - 2d_H) + 1$. By this placement, the first row and the $(plus(j, 2d_H) - j + 1)$ left-aligned columns of $block(i,j)$, the area U (See Figure 1,) overlaps with the upper and the left neighbor blocks. Also, cells in the area Z' are shared with the right neighbor block and the lower neighbor block.

To determine values of cells in Z' and cells of D_- in Y, we only need values of cells in U, subsequences $(p_i, \ldots, p_{i+r_i-1})$ and $(t_j, \ldots, t_{j+c_j-1})$ of p and t, and the cells in D_- corresponding to the area X. We encode the subsequences of points by sequences of differences from the previous points, as $V = (0, p_{i+1} - p_i, \ldots, p_{i+r_i-1} - p_{i+r_i-2})$ and $W = (0, t_{j+1} - t_j, \ldots, t_{j+c_j-1} - t_{j+c_j-2})$.

Now we give our algorithm. Firstly, we directly fill the cells $(1,j)$ with $1 \leq j \leq m$ and (i,j) with $i > j$ of tables D and D_-, and compute the cells (i,j) with $i \leq j < n$ and with $n \leq j \leq plus(n, 2d_H)$. Secondly, we place blocks as presented above, as many as possible. At each placement, we fill the cells in Z' by looking up the pre-computed table that maps a combination of values of (U, V, W, X) to that of (Z', Y). The cell values larger than possible error ε are omitted as the positive infinity symbol ∞. Then, all the values in the last row of the DP-table D are determined.

The bottleneck of this algorithm is time complexity of preparing pre-computed table. For any block, the number of rows r_i is at most d_H, and the number of columns c_j is at most d_W. Thus U has at most $d_W + 2d_H(d_H - 1)$ cells, and X has at most d_H cells. A value of a cell is either $0, 1, \ldots, \varepsilon, \infty$, or the "no cell" symbol. Therefore, the number of possible combinations of values of U and X is $(\varepsilon + 3)^{d_W + 2d_H(d_H - 1)} \cdot (\varepsilon + 3)^{d_H - 1}$. Also, since sequences have finite resolution L, there are $L^{d_H} + L^{d_H - 1} + \cdots + L \leq (L + 1)^{d_H}$ possible combinations of values for V', and $(L + 1)^{d_W}$ possible combinations for W'. For each combination of quadruple (U, V', W', X), its result (Z, Y) can be computed within time $O(2d_H \cdot (d_W - 2d_H) \cdot (d_H - 1))$. By summarizing this, the total time complexity for preparing pre-computed table is $O\left(d_W \cdot d_H{}^2 \cdot (\varepsilon + 3)^{d_W + 2d_H{}^2} \cdot (L + 1)^{d_W + d_H}\right)$.

Therefore, the following holds.

Theorem 2. *If an error ε and $d_H \geq L$ can be regarded or chosen as constants, then for sufficiently large n and m, we can choose $d_W = \log_{(L+1) \cdot (\varepsilon + 3)} n \geq 4d_H$ and the pre-computation table can be prepared in $O(n \log n)$ time.*

Roughly, at most $\frac{Ln}{d_H} \cdot \frac{Lm}{d_W}$ blocks are placed in each table. By supposing a unit-cost RAM model that can look up the pre-computed table in $O(1)$ time, we obtain the following result.

Corollary 3. *If an error bound ε is regarded as a constant, and a pattern and a text are sufficiently large and drawn from a set C of point sequences having finite resolution, then a tabular computation of Algorithm 2 can be done in $O(\frac{nm}{\log n} + n \log n)$ time with unit-cost RAM model.*

4   Approximate Point Set Pattern Matching on the Plane

In this section, we introduce an edit distance and an approximate point set pattern matching on the plane, which is an extension from those of our approximate matching of point sequences. Then, we present a polynomial-time algorithm for the problem. To our knowledge, this is the first polynomial-time solvable approximate point set pattern matching on the plane.

A *point in the plane* is a pair $p = (p.x, p.y)$ of positive integers. We assume that, for any set $P \subseteq \mathbb{Z}^{+2}$ of points in the plane, any pair of points $p, p' \in P$ satisfies neither $p.x = p'.x$ nor $p.y = p'.y$. Also, we assume that a set P of points has an arbitrary but fixed order of points and can be regarded as a sequence $P = (p_1, \ldots, p_n)$ of points.

Let $P \subseteq \mathbb{Z}^{+2}$ be a set of points. We denote the indices of the right most point and the top point in P by $(P)_R$ and $(P)_T$. The *bounding rectangle of P* is the minimum axis-parallel rectangle that includes all the points in P. For any $1 \leq i, j \leq n$, we define the subset

$$P[i, j] = P[j, i] = \{p_k \in P \mid p_k.x \leq \max(p_i.x, p_j.x) \text{ and } p_k.y \leq \max(p_i.y, p_j.y)\}.$$

If the origin set of points is clear in a context, we identify $P[i, j]$ with its bounding rectangle and the pair $((P[i, j])_R, (P[i, j])_T)$ of indices, and simply denote it by $[i, j]$. The sets produced by removing the right-most point $p_{([i,j])_R}$ and the top point $p_{([i,j])_T}$ from $P[i, j]$ are denoted by $[i, j]_-$ and $[i, j]^-$, respectively.

Firstly, we define the edit distance between two sets of points.

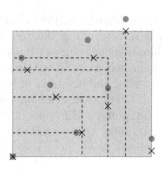 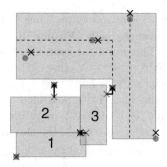

Fig. 2. The left-bottom aligned bounding rectangle of × points of P define axis-parallel space insertions and deletions (the first three ones, numbered in right) for matching to ∘ points of Q.

Definition 5. *Let P and Q be sets of n points in the plane. For any $1 \le i, j, k, l \le n$, the edit distance $d(i, j; k, l)$ between $P[i, j]$ and $Q[i, j]$ is*

1. *if $|P[i,j]| = |Q[k,l]| = 1$ then $d(i, j; k, l) = 0$,*
2. *if $|P[i,j]| \neq |Q[k,l]|$ then $d(i, j; k, l) = \infty$, and*
3. *otherwise, $d(i, j; k, l) =$*

$$\min \left\{ \begin{array}{l} d([i,j]_-; [k,l]_-) + |p_{([i,j])_R} - p_{([i,j]_-)_R} - (q_{([k,l])_R} - q_{([k,l]_-)_R})|, \\ d([i,j]^-; [k,l]^-) + |p_{([i,j])_T} - p_{([i,j]_-)_T} - (q_{([k,l])_T} - q_{([k,l]_-)_T})| \end{array} \right\},$$

where the distance $|\cdot|$ between points denotes L_1 norm on the plane.

This definition is naturally extending the edit distance of 1-dimensional point sequences to that of bi-axis sorted sequences. Figure 2 is exhibiting this idea. Firstly, the left-bottom points × of a set P and ∘ of a set Q are aligned, and are treated as bounding rectangles having only one point. This is the base case 1 of Definition 5. Then, each bounding rectangle is enhanced to have exactly one more point. It is implied by the case 2 of the definition. Among all possible pairs of such left-bottom aligned bounding rectangles, we choose a pair that minimizes the distance between the new added points. This defines the case 3 of the above definition. Note that the edit distance between P and Q of size n is always finite.

The matching is defined as follows.

Definition 6. *Let $P = \{p_1, \ldots, p_n\}$ and $T = \{t_1, \ldots, t_m\}$ be sets of points in the plane, and let $\varepsilon \ge 0$ be a positive integer. If there is a subset $T' \subseteq T$ such that the edit distance $d((P)_R, (P)_T; (T')_R, (T')_T)$ between P and T' is no more than ϵ, then we say that P approximately matches T at the position $(T')_R$.*

Let P be a set of n points. We denote by $[i, j]_{\le l}$ and $[i, j]^{\le l}$ with $1 \le l \le n$ the sets of the first l points of $P[i, j]$ in X-axis increasing order and the first l points in Y-axis increasing order, respectively.

Let P and T be sets of n and m points in the plane. For any $1 \le i, j \le n$ and any $1 \le k, l \le m$, we denote by $D(i, j; k, l)$ the minimum edit distance between

$P[i, j]$ and any subset $T' \subseteq T[k, l]$ that includes t_k and t_l, i.e. $T' \subseteq T$ such that $(T')_R = (T[k, l])_R$ and $(T')_T = (T[k, l])_T$. Then the base conditions are defined as follows.

1. If $|P[i, j]| = 1$ then $D(i, j; k, l) = 0$.
2. If $|T[k, l]| < |P[i, j]|$ then $D(i, j; k, l) = \infty$.
3. If $|P[i, j]| = 2$ and $([k, l])_R \neq ([k, l])_T$ then $D(i, j; k, l)$ is given by the edit distance between $P[i, j]$ and $\{t_{([k,l])_R}, t_{([k,l])_T}\}$, and thus $D(i, j; k, l) = |p_{([i,j])_R} - p_{([i,j]-)_R} - (t_{([k,l])_R} - t_{([k,l])_T})| = |p_{([i,j])_T} - p_{([i,j]-)_T} - (t_{([k,l])_T} - t_{([k,l])_R})|$.

In other cases, the following lemma holds.

Lemma 3. *For* $|P[i, j]| = n > 2$, *the following recurrences hold.*
(1) *If* $([k, l])_R = ([k, l])_T$, *then* $D(i, j; k, l) =$

$$
\min_{|P[i,j]|-1 \leq w \leq |T[k,l]|} \left\{
\begin{array}{l}
D([i, j]-; [k, l]_{\leq w}) \\
+|p_{([i,j])_R} - p_{([i,j]-)_R} - (t_{([k,l])_R} - t_{([k,l]_{\leq w})_R})|, \\
D([i, j]^-; [k, l]^{\leq w}) \\
+|p_{([i,j])_T} - p_{([i,j]-)_T} - (t_{([k,l])_T} - t_{([k,l]^{\leq w})_T})|
\end{array}
\right\}.
$$

(2) *If* $([k, l])_R \neq ([k, l])_T$, *then* $D(i, j; k, l) =$

$$
\min \left\{
\begin{array}{l}
\min_{|T[([k,l])_T,([k,l])_T]| \leq w < |T[k,l]|} \left\{
\begin{array}{l}
D([i, j]-; ([k, l]_{\leq w})_R, ([k, l])_T)+ \\
|p_{([i,j])_R} - p_{([i,j]-)_R} - \\
(t_{([k,l])_R} - t_{([k,l]_{\leq w})_R})|
\end{array}
\right\}, \\
\min_{|T[([k,l])_R,([k,l])_R]| \leq w < |T[k,l]|} \left\{
\begin{array}{l}
\{D([i, j]^-; ([k, l])_R, ([k, l]^{\leq w})_T)+ \\
|p_{([i,j])_T} - p_{([i,j]-)_R} - \\
(t_{([k,l])_T} - t_{([k,l]_{\leq w})_T})|
\end{array}
\right\}
\end{array}
\right\}.
$$

Proof. Similar to the proof of lemma 1.

Now we present an algorithm that finds occurrences at which P approximately matches T. Firstly, by enumerating pairs of indices $1 \leq i, j \leq n$, we prepare the list $List(P, r) = ((i_1, j_1), \ldots, (i_{n(P,r)}, j_{n(P,r)}))$ of all possible bounding rectangles of P with r points for $1 \leq r \leq n$, where $n(P, r)$ denotes the number of such rectangles in P. Also, we build $List(T, z)$ for T and $1 \leq z \leq m$, and denote the number of size z rectangles by $n(T, z)$. It takes $O(n^2 + m^2)$ time. Next, we construct a map from a quadruple (i, j, k, l) with $1 \leq i, j \leq n$ and $1 \leq k, l \leq m$ to $D(i, j; k, l)$, by the base conditions and the following tabular computation.

For r from 2 to n **do:**
    For z from r to m **do:**
        For each (i, j) in $List(P, r)$ **and for each** (k, l) in $List(T, z)$ **do:**
            Compute $D(i, j; k, l)$ by the recurrences.

The number of quadruples (i, j, k, l) is $O(n^2 m^2)$, and for each cell if we can look up the table D in constant time a computation for the recurrence requires $O(m^2)$ time. Finally, we find and construct all or one of occurrences by a back-tracking procedure.

Theorem 3. *The point set pattern matching on the plane with pattern size n and text size m is solvable in $O(n^2 m^4)$ time and $O(n^2 m^2)$ space.*

5   Concluding Remarks

In musical score sequence search, it is quite needed to regard differences of tempo between a pattern sequence and a text sequence. In exact point set pattern matching [14], global scaling is also considered in linear transformations. However, time insertions and deletions make impossible to applying the method same with [14]. An efficient algorithm that can deal with edit distance and global scaling is the primary open problem.

The technique utilized in Algorithm_2 that separates the difference value $\Delta(i, j, j')$ by its sign relates to "path–separability" in [9]. It is extensively used with data structures for minimum range queries on sequences to develop improved algorithms for edit distances with restrictions on gaps. Improving the running time of our algorithm by this scheme for instances without finite resolution, and its practical implementation should be our future work.

Acknowledgement

We would like to thank Ayumi Shinohara for suggesting the use of four-Russian technique. We are also grateful to the reviewers for their valuable suggestions and comments.

References

1. T. Akutsu, K. Kanaya, A. Ohyama and O. Fujiyama Matching of spots in 2D electrophoresis images. point matching under non-uniform distortions. In *Proc. 10th Annual Symposium on Combinatorial Pattern Matching, LNCS* 1645, 212–222, 1999.
2. V. L. Arlazarov, E. A. Dinic, M. A. Kronrod and I. A. Faradzev, On economic construction of the transitive closure of a directed graph, *Dokl. Acad. Nauk SSSR* **194**, 487–488, 1970.
3. D. Gusfield, Algorithms on strings, trees, and sequences computer science and computational biology, Cambridge University Press, 1997.
4. F. Hoffmann, K. Kriegel and C. Wenk, Matching 2D patterns of protein spots, In *Proc. 14th ACM Symposium on Computational Geometry*, 231–239, 1998.
5. S. Jokisch and H. Müller, Inter-point-distance-dependent approximate point set matching, Research Report No. 653, Dept. of Computer Science, University of Dortmund, 1997.
6. T. Kadota, M. Hirao, A. Ishino, M. Takeda, A. Shinohara and F. Matsuo, Musical sequence comparison for melodic and rhythmic similarities, In *Proc. 8th String Processing and Information Retrieval*, 111–122, 2001.
7. K. Lemström and V. Mäkinen, On minimizing pattern splitting in multi-track string matching, In *Proc. 14th Annual Symposium on Combinatorial Pattern Matching, LNCS* 2672, 237–253, 2003.

8. V. Mäkinen, Using edit distance in point-pattern matching, In *Proc. 8th String Processing and Information Retrieval*, 153–161, 2001.

9. V. Mäkinen, Parameterized approximate string matching and local-similarity-based point-pattern-matching, Ph.D. thesis, Report A–2003–6, Dept. of Computer Science, Univ. Helsinki, August 2003.

10. V. Mäkinen, G. Navarro and E. Ukkonen, Algorithms for transposition invariant string matching, In *Proc. 20th International Symposium on Theoretical Aspects of Computer Science, LNCS* 2607, 191–202, 2003.

11. V. Mäkinen and E. Ukkonen, Local similarity based point-pattern matching, In *Proc. 13th Annual Symposium on Combinatorial Pattern Matching, LNCS* 2373, 115–132, 2002.

12. D. Meredith, G. Wiggins and K. Lemström, Pattern induction and matching in polyphonic music and other multi-dimensional databasets, In *Proc. 5th World Multi-Conference on Systems, Cybernetics and Informatics*, Vol. X, 61 – 66, 2001.

13. M. Mongeau and D. Sankoff, Comparison of musical sequences, *Computers and the Humanities* **24** (3), 161–175, 1990.

14. P. J. de Rezende and D. T. Lee, Point set pattern matching in *d*-dimensions, *Algorithmica* **13** (4), 387–404, 1995.

15. E. Ukkonen, K. Lemström and V. Mäkinen, Geometric algorithms for transposition invariant content-based music retrieval, In *Proc. International Symposium on Music Information Retrieval*, 193–199, 2003.

Finding Biclusters by Random Projections

Stefano Lonardi[1], Wojciech Szpankowski[2], and Qiaofeng Yang[1]

[1] Dept. of Computer Science – University of California – Riverside, CA
[2] Dept. of Computer Sciences – Purdue University – West Lafayette, IN

Abstract. Given a matrix X composed of symbols, a bicluster is a submatrix of X obtained by removing some of the rows and some of the columns of X in such a way that each row of what is left reads the same string. In this paper, we are concerned with the problem of finding the bicluster with the largest area in a large matrix X. The problem is first proved to be **NP**-complete. We present a fast and efficient randomized algorithm that discovers the largest bicluster by random projections. A detailed probabilistic analysis of the algorithm and an asymptotic study of the statistical significance of the solutions are given. We report results of extensive simulations on synthetic data.

1 Introduction

Clustering refers to the problem of finding a partition of a set of input vectors, such that the vectors in each subset (cluster) are "close" to one another (according to some predefined distance). A common limitation to the large majority of clustering algorithms is their inability to perform on high dimensional spaces (see, e.g., [1, 2]).

Recent research has focused on the problem of finding hidden sub-structures in large matrices composed by thousands of high dimensional vectors (see, e.g., [3–10]). This problem is known as *biclustering*. In biclustering, one is interested in determining the similarity in a *subset* of the dimensions (subset that has to be determined as well). Although there exists several definitions of biclustering, it can be informally described as the problem of finding a partition of the vectors and a subset of the dimensions such that the projections along those directions of the vectors in each cluster are close to one another. The problem requires to cluster the vectors and the dimensions simultaneously, thus the name "biclustering".

Biclustering has important applications in several areas, such as data mining, machine learning, computational biology, and pattern recognition. Data arising from text analysis, market-basket data analysis, web logs, etc., is usually arranged in a contingency table or co-occurrence table, such as, a word-document table, a product-user table, a cpu-job table or a webpage-user table. Discovering a large bicluster in a product-user matrix indicates, for example, which users share the same preferences. Finding biclusters has therefore applications in recommender systems and collaborative filtering, identifying web communities, load balancing, discovering association rules, among others.

S.C. Sahinalp et al. (Eds.): CPM 2004, LNCS 3109, pp. 102–116, 2004.

In computational biology, this problem is associated with the analysis of gene expression data obtained from microarray experiments. Gene expression data is typically arranged in a table with rows corresponding to genes, and columns corresponding to patients, tissues, time points, etc. The classical approach to analyze microarray data is clustering. The process of clustering partitions genes into mutually exclusive clusters under the assumption that genes that are involved in the same genetic pathway behave similarly across all the testing conditions. The assumption might be true when the testing conditions are associated with time points. However, when the testing conditions are heterogeneous, such as patients or tissues, the previous assumption is not appropriate anymore. One would expect that a group of genes would exhibit similar expression patterns only in a subset of conditions, such as the subset of patients suffering from the same type of disease. Under this circumstance, biclustering becomes the alternative to the traditional clustering paradigm. The results of biclustering may enable one to discover hidden structures in gene expression data in which many genetic pathways might be embedded. It might also allow one to uncover unknown genetic pathways, or to assign functions to unknown genes in already known genetic pathways.

Biclustering is indeed, not a new problem. In fact, it is also known under several other names, namely "co-clustering", "two-way clustering" and "direct clustering". The problem was first introduced in the seventies in a paper by Hartigan [11]. Almost thirty years later, Cheng and Church [3] raised the interest on this problem for applications in gene expression data analysis.

Several other researchers studied the problem recently. Wang *et al.* propose the *pCluster* model that is capable of discovering shifting or scaling patterns from raw data sets [4]. Tanay *et al.* [5] combine a graph-theoretic approach with a statistical modeling of the data to discover biclusters in large gene expression datasets. Ben-Dor *et al.* [6] introduce a new notion of a bicluster called *order preserving submatrix*, which is a group of genes whose expression level induces a linear ordering across a subset of the conditions. Murali and Kasif [12] (see also [10]) propose the concept of *xmotif*, which is defined as a subset of genes whose expression is simultaneously conserved for a subset of samples.

As we were writing this document, we became aware of two other contributions to the subject, by Sheng *et al.* [8], and Mishra *et al.* [9], that use a randomized approach similar with the work described here. Sheng *et al.* [8] propose a randomized algorithm based on Gibbs sampling to discover large biclusters in gene expression data. Their model of a bicluster is probabilistic, that is, each entry of the matrix is associated with a probability. Mishra *et al.* [9] are concerned with the problem of finding ϵ-bicliques which maximizes the number of edges[1]. Given a bipartite graph (U, V, E), a subgraph (U', V') is ϵ-biclique if each vertex in U' is a neighbor of at least $(1 - \epsilon)$ fraction of vertices in V'. The authors give an efficient randomized algorithm that finds the largest ϵ-biclique, but no experimental results are reported.

---

[1] the connection between bicliques and bicluster will be explained in detail in Section 2

As shown in papers [12] and [8], the problem of biclustering gene expression data can be formulated on a discrete domain, by first discretizing the gene expression matrix into a matrix over a finite alphabet. The simplifying assumption is that the set of states in which each gene operates is finite, such as up-regulated, down-regulated or unchanged. Once the data is discretized into strings where each symbol corresponds to a state, the biclustering problem reduces to the problem of finding a subset of the rows and a subset of the columns such that the submatrix induced has the property that each row reads the same string. Such a submatrix would therefore correspond to a group of genes that exhibit a coherent pattern of states over a subset of conditions. This is indeed the formulation of the problem that we define in Section 2, which is first proved to be **NP**-complete. In Section 3 we present a randomized algorithm which is efficient and easy to understand and implement. Section 4 presents an asymptotic analysis that allows one to determine the statistical significance of the solution. Finally, in Section 5 we report simulation results on synthetic data.

2   Notations and Problem Definition

We use standard concepts and notation about strings. The set Σ denotes a nonempty *alphabet* of *symbols* and a *string* over Σ is an ordered sequence of symbols from the alphabet. We use the variable a as a shorthand for the cardinality of the set Σ, that is, $a = |\Sigma|$. Given a string x, the number of symbols in x defines the *length* $|x|$ of x.

Similarly, we can define a two-dimensional $n \times m$ string (or matrix) $X \in \Sigma^{n \times m}$ over the alphabet Σ. The element (i, j) of X is denoted by $X_{[i,j]}$. A *row selection* of size k of X is defined as the subset of the rows $R = \{i_1, i_2, \ldots, i_k\}$, where $1 \leq i_s \leq n$ for all $1 \leq s \leq k$. Similarly, a *column selection* of size l of X is defined as a subset of the columns $C = \{j_1, j_2, \ldots, j_l\}$, where $1 \leq j_t \leq m$ for all $1 \leq t \leq l$.

The submatrix $X_{(R,C)}$ *induced* by the pair (R, C) is defined as the matrix

$$
X_{(R,C)} = \begin{vmatrix} X_{[i_1,j_1]} & X_{[i_1,j_2]} & \cdots & X_{[i_1,j_l]} \\ X_{[i_2,j_1]} & X_{[i_2,j_2]} & \cdots & X_{[i_2,j_l]} \\ \cdots & \cdots & \cdots & \cdots \\ X_{[i_k,j_1]} & X_{[i_k,j_2]} & \cdots & X_{[i_k,j_l]} \end{vmatrix}
$$

Given a selection of rows R, we say that a column j, $1 \leq j \leq m$, is *clean* with respect to R if the symbols in the j-th column of X restricted to the rows R, are identical.

The problem addressed in this paper is defined as follows.

LARGEST BICLUSTER(f) problem
Instance: A matrix $X \in \Sigma^{n \times m}$ over the alphabet Σ.
Question: Find a row selection R and a column selection C such that the rows of $X_{(R,C)}$ are identical strings and the objective function $f(X_{(R,C)})$ is maximized.

Some examples of objective functions are the following.

- $f_1\left(X_{(R,C)}\right) = |R| + |C|$
- $f_2\left(X_{(R,C)}\right) = |R|$ provided that $|C| = |R|$
- $f_3\left(X_{(R,C)}\right) = |R|\,|C|$

The problem in general may have multiple solutions which optimize the objective function. The solutions may also "overlap", that is, they may share some elements of the original matrix.

The computational complexity of this family of problems depends on the objective function f. In the literature, the problem has been studied mostly from a graph-theoretical viewpoint which corresponds to the special case $\Sigma = \{0,1\}$. In fact, observe that a matrix $X \in \{0,1\}^{n \times m}$ is the adjacency matrix of a bipartite graph $G = (V_1, V_2, E)$ with $|V_1| = n$ and $|V_2| = m$. An edge $(i,j) \in E$ connects node $i \in V_1$ to node $j \in V_2$ if $X_{i,j} = 1$. Thus, a submatrix of 1's in X corresponds to a subgraph of G which is completely connected. Such a subgraph is called a *biclique*. Because of this relation, we use the terms "submatrix", "biclique", and "bicluster" interchangeably.

When the alphabet is binary and we are looking for the largest submatrix composed only by 1's [2], the LARGEST BICLUSTER reduces to well-known problems on bipartite graphs. More specifically, the LARGEST BICLUSTER problem associated with objective function f_1 is known as the MAXIMUM VERTEX BICLIQUE problem, and it can be solved in polynomial time because it is equivalent to the maximum independent set in bipartite graphs which, in turn, can be solved by a minimum cut algorithm (see, e.g., [13]). The same problem with objective function f_2 over a binary alphabet is called BALANCED COMPLETE BIPARTITE SUBGRAPH problem or BALANCED BICLIQUE problem and it is listed as GT24 among the **NP**-complete problems in Garey & Johnson's book [14] (see also [15]).

The LARGEST BICLUSTER problem with objective function f_3 and $\Sigma = \{0,1\}$ is called MAXIMUM EDGE BICLIQUE problem. The problem requires to find the biclique which has the maximum number of edges. The problem is proved to be **NP**-complete in [16] by reduction from 3SAT. The weighted version of this problem is shown **NP**-complete by Dawande *et al.* [17].

In [13] Hochbaum studies a problem related to MAXIMUM EDGE BICLIQUE, which is the problem of finding the number of edges that need to be deleted so that the resulting graph is a biclique. Hochbaum describes a 2-approximation algorithm based on LP-relaxation. According to Pasechnik [18] this approximation ratio does not hold for the original MAXIMUM EDGE BICLIQUE problem. Pasechnik shows a semidefinite relaxation, and claims that his relaxation is in general better than [13].

The following theorem establishes the hardness of the problem of finding the largest area bicluster over a general alphabet. For lack of space the proof is omitted.

---

[2] In general, a solution of the largest bicluster can contain a column of zeros, as long as they appear in all rows of the submatrix

Theorem 1. *The decision problem associated with* LARGEST BICLUSTER(f_3) *is* **NP**-*complete.*

By the same approach, LARGEST BICLUSTER(f_2) can also be proved to be **NP**-complete. In the rest of this paper we will concentrate our attention on the problem of finding the largest-area bicluster. For practical reasons that will become apparent in Section 3, the objective function that we are maximizing is

$$\tilde{f}_3\left(X_{(R,C)}, \hat{r}, \hat{c}\right) = |R|\,|C| \text{ provided that } |R| \geq \hat{r} \text{ and } |C| \geq \hat{c}$$

where $\hat{r}$ and $\hat{c}$ are two input parameters.

3   Randomized Search

Given that LARGEST BICLUSTER(f_3) problem is **NP**-complete, it is unlikely that a polynomial time algorithm could be found. In this paper, we present a randomized algorithm which finds a maximal solution with probability $1 - \epsilon$, where $0 < \epsilon < 1$.

Assume that we are given a large matrix $X \in \Sigma^{n \times m}$ in which a submatrix $X_{(R^*,C^*)}$ is implanted. Assume also that the submatrix $X_{(R^*,C^*)}$ is maximal. To simplify the notation, let $r^* \equiv |R^*|$ and $c^* \equiv |C^*|$.

The idea behind the algorithm comes from the following simple observation. Observe that if we knew R^*, then C^* could be determined by selecting the clean columns with respect to R^*. If instead we knew C^*, then R^* could be obtained by taking the maximal set of rows which read the same string. Unfortunately, neither R^* nor C^* is known. Our approach is to "sample" the matrix by random projections, with the expectation that at least some of the projections will overlap with the solution (R^*, C^*). Clearly, one can project either rows or columns. In what follows we describe how to retrieve the solution by sampling columns.

The algorithm works as follows. Select a random subset S of size k uniformly from the set of columns $\{1, 2, \ldots, m\}$. Assume for the time being that $S \cap C^* \neq \emptyset$. If we knew $S \cap C^*$, then (R^*, C^*) could be determined by the following three steps (1) select the string(s) w that appear exactly r^* times in the rows of $X_{[1:n, S \cap C^*]}$, (2) set R^* to be the set of rows in which w appears and (3) set C^* to be the set of clean columns corresponding to R^*.

The algorithm would work, but there are a few problems that are still unresolved. First, the set $S \cap C^*$ could be empty. The solution is to try different random projections S, relying on the argument that the probability that $S \cap C^* \neq \emptyset$ *at least once* will approach one with more and more projections. The second problem is that we do not really know $S \cap C^*$. But, certainly $S \cap C^* \subseteq S$, so our approach is to check all possible subsets $U \subseteq S$ such that $|U| \geq k_{\min}$, where $1 \leq k_{\min} \leq k$ is a user-defined parameter. The final problem is that we assumed that we knew r^*, but we do not. The solution is to introduce a *row threshold* parameter, called $\hat{r}$, that replaces r^*.

As it turns out, we need another parameter to avoid producing solutions with too few columns. The *column threshold* $\hat{c}$ is used to discard submatrices whose

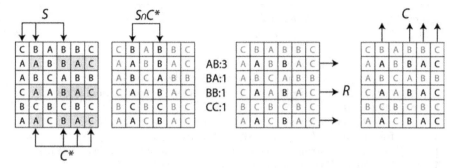

Fig. 1. An illustration of a recovery of the embedded matrix by random projections. C^* is the set of columns containing the embedded submatrix. S is a random selection of columns. By following the steps described in the text, the correct solution can be easily retrieved.

number of columns is smaller than $\hat{c}$. The algorithm considers all the submatrices which satisfy the user-defined row and column threshold as candidates. Among all candidate submatrices, only the ones that maximize the total area are kept. A sketch of the algorithm is shown in Figure 2. As noted in the introduction, a very similar strategy was developed independently and concurrently by Mishra et al. [9].

The algorithm depends on five key parameters, namely the projection size k, the minimum subset size k_{min}, the row threshold $\hat{r}$, the column threshold $\hat{c}$, and the number of iterations t. We discuss how to choose each of these in the rest of the section.

Parameter Selection. The projection size k is determined by a probabilistic argument. It is well-known that in a random string of size m over an alphabet of size a, the number of occurrences of substrings has two different probabilistic regimes (1) Gaussian distributed for strings shorter than $\log_a m$ and (2) Poisson distributed for strings longer than $\log_a m$ (see, e.g., [19]). Based on this observation, when $k_{min} = k$ we argue that $k = \log_a m$ is the optimal trade-off between generating too many trivial solutions (k too small) and potentially missing the solution (k too large). This value of k has been confirmed to be the optimal choice in our simulations. When $k_{min} = 1$, then k can be chosen significantly larger, but this will adversely affect the running time. An experimental comparison between $k_{min} = k$ (i.e., no subsets), and $k_{min} = 1$ (i.e., all subsets) is reported in Section 5.1.

The thresholds $\hat{r}$ and $\hat{c}$ are associated with the uncertainty on the size of the largest submatrix r^*, c^* for a particular input instance. There may be situations in which the user has already a good idea about r^*, c^*. If however r^* and c^* are completely unknown, then our target will be to find "statistically significant" biclusters. In Section 4 we will present a theorem (Thorem 2) which gives the expected number of columns of the largest submatrix in a random matrix, when the number of rows is fixed. Based on this, we propose the following trial-and-

---

LARGEST_BICLUSTER_C$(X, t, k, k_{\min}, \hat{r}, \hat{c})$
INPUT: X is a $n \times m$ matrix over Σ
    t is the number of iterations
    k is the projection size
    $k_{\min}$ is the size of the smallest subset of the projection
    $\hat{r}, \hat{c}$ are the "thresholds" on the number of rows and columns, resp.
1 **repeat** t **times**
2 **select** randomly a subset S of columns such that $|S| = k$
3 **for** all subsets $U \subseteq S$ such that $|U| \geq k_{\min}$ **do**
4   $D \leftarrow$ all strings induced by $X_{[1:n, U]}$ that appear at least $\hat{r}$ times
5   **for** each string w in D
6    $V \leftarrow$ rows corresponding to w
7    $Z \leftarrow$ all "clean" columns corresponding to V
8   **if** $|Z| \geq \hat{c}$ **then save** (V, Z)
9 **return** the (V, Z) that maximizes f

Fig. 2. A sketch of the algorithm that discovers large biclusters (sampling columns).

error strategy. Set $\hat{r}$ to some value between 1 and n, and use Theorem 2 to set the value $\hat{c}$. Run the algorithm. If the algorithm returns too many solutions, try to increase $\hat{r}$ and update $\hat{c}$ correspondingly. If there are no solutions, lower the value of $\hat{r}$ and repeat. Observe that the number of choices for $\hat{r}$ is finite since $\hat{r} \in [1, n]$. By using Theorem 2 to set the threshold $\hat{c}$, we are trying to filter out submatrices whose size is small enough that they could appear in totally random matrices.

Because of the randomized nature of the approach, there is no guarantee that the algorithm will find the solution after a given number of iterations. We therefore need to choose t so that the probability that the algorithm will recover the solution in at least one of the t trials is $1 - \epsilon$, where $0 < \epsilon < 1$ is a user-defined parameter.

Let $\alpha(n, m, k, r^*, c^*, a)$ be the probability of missing the solution in one of the trials assuming that r^* and c^* are known and that $k_{\min} = 1$. There are two disjoint cases in which the algorithm can miss (R^*, C^*). The first is when the random projection S misses completely C^*, i.e., $S \cap C^* = \emptyset$. The second is when $S \cap C^* = U \neq \emptyset$ but the string w chosen by the algorithm among the rows $X_{[1:n, U]}$ also appears in another row that does not belong to the set R^* of the real solution. In this case, the algorithm will select a set of rows larger than R^* and thus miss the solution. Hence, we have

$$\alpha(n, m, k, r^*, c^*, a) = \Pr\{S \cap C^* = \emptyset\} + \sum_{i=1}^{k} \Pr\{|S \cap C^*| = i \text{ and } |R| > r^*\}$$

$$= \Pr\{S \cap C^* = \emptyset\} + \sum_{i=1}^{k} \Pr\{|R| > r^* \text{ given } |S \cap C^*| = i\}\Pr\{|S \cap C^*| = i\}$$

Let Y be the random variable associated with the size of the set $S \cap C^*$, that is, $Y = |S \cap C^*|$. Since we are sampling without replacement, Y follows the hyper-geometric distribution.

$$\Pr\{Y = 0\} = \binom{m - c^*}{k} \Big/ \binom{m}{k} \quad \text{and} \quad \Pr\{Y = i\} = \binom{c^*}{i}\binom{m - c^*}{k - i} \Big/ \binom{m}{k}$$

In order to compute the probability of missing the solution given $|S \cap C^*| = i$, we have to estimate how likely a string w belonging to some of the rows of $X_{[1:n, U]}$ is more frequent than r^*. Assuming the symbols in the matrix X are generated by a symmetric Bernoulli i.i.d. model, the probability that w will never appear in the other $n - r^*$ rows is $\left(1 - \frac{1}{a^i}\right)^{n - r^*}$ and therefore

$$\Pr\{|R| > r^* \text{ given } |S \cap C^*| = i\} = 1 - \left(1 - \frac{1}{a^i}\right)^{n - r^*}$$

Combining all together, the probability of missing the solution in one iteration is given by

$$\alpha(n, m, k, r^*, c^*, a) = \frac{\binom{m - c^*}{k} + \sum_{i=1}^{k}\left(1 - \left(1 - \frac{1}{a^i}\right)^{n - r^*}\right)\binom{c^*}{i}\binom{m - c^*}{k - i}}{\binom{m}{k}}$$

Now suppose we want the probability of missing the solution to be smaller than a given ϵ, $0 < \epsilon < 1$. We can obtain the number of iterations t by solving the inequality $(\alpha(n, m, k, r^*, c^*, a))^t \leq \epsilon$, which gives

$$t \geq \frac{\log \epsilon}{\log \alpha(n, m, k, r^*, c^*, a)} \tag{1}$$

This bound on the number of iterations has been verified by our experimental results (compare Table 1 with our experimental results shown in Figure 3). For

Table 1. The estimated number of iterations for a matrix 256×256 with a submatrix 64×64, for different choices of ϵ, alphabet size a, and projection size k (sampling columns).

ϵ	$a = 2, k = 8$	$a = 4, k = 4$	$a = 8, k = 3$	$a = 16, k = 2$	$a = 32, k = 2$
0.005	18794	1342	306	179	99
0.05	10626	759	173	101	56
0.1	8168	583	133	78	43
0.2	5709	408	93	54	30
0.3	4271	305	70	41	23
0.4	3250	232	53	31	17
0.5	2459	176	40	23	13
0.6	1812	129	29	17	10
0.7	1265	90	21	12	7
0.8	792	57	13	8	4
0.9	374	27	6	4	2

example, by setting $a = 4$, $k = 4$, $\epsilon = 0.7$, equation (1) gives $t = 90$ iterations whereas the experimental results show that with 90 iterations we obtain a performance of $\epsilon = 0.689$.

The worst case time complexity of LARGEST_BICLUSTER_C is bounded by $O\left(t\sum_{j=k_{min}}^{k}\binom{k}{j}(kn+nm)\right)$. If $k_{min} = 1$, then the time complexity becomes $O\left(t2^k(kn+nm)\right)$. Although the complexity is exponential in k, choosing k to be $O(\log_a m)$ makes the algorithm run in $O\left(tm^{1/\log_2 a}(kn+nm)\right)$ time.

The probability of missing the solution changes significantly when we set $k_{min} = k$. In this case, we are not checking any of the subsets of S, but we simply rely on the fact that eventually one of the random projections S will end up completely contained in C^*, in which case we have a chance to find the solution.

Since we avoid checking the $O(2^k)$ subsets of S, the number of iterations t to achieve the same level of performance of the case $k_{min} = 1$ must be significantly larger. Indeed, by a similar argument as we did for $k_{min} = 1$, the probability of missing the solution when $k_{min} = k$ can be estimated by the following formula

$$\tilde{\alpha}(n,m,k,r^*,c^*,a) = \Pr\{|S \cap C^*| < k\} + \Pr\{|S \cap C^*| = k \text{ and } |R| > r^*\}$$
$$= 1 - \Pr\{|S \cap C^*| = k\} + \Pr\{|S \cap C^*| = k \text{ and } |R| > r^*\}$$
$$= 1 - \left(\binom{c^*}{k}/\binom{m}{k}\right) + \left(\left(1 - \left(1 - \frac{1}{a^k}\right)^{n-r^*}\right)\binom{c^*}{k}/\binom{m}{k}\right)$$
$$= 1 - \left(\left(1 - \frac{1}{a^k}\right)^{n-r^*}\binom{c^*}{k}/\binom{m}{k}\right)$$

As mentioned above, we also have the option to project the rows instead of the columns, which would result in a slightly different algorithm that we called LARGEST_BICLUSTER_R. The details and the analysis of this algorithm will be reported in the journal version of this manuscript.

Both strategies were implemented and tested extensively. Results are reported in Section 5.

4   Statistical Analysis

We now analyze the statistical significance of finding a large submatrix of size $r \times c$ hidden into a random $n \times m$ matrix over an alphabet of cardinality a. More specifically, we randomly generate a matrix $X \in \Sigma^{n \times m}$ using a memoryless source with parameters $\{p_1, \ldots, p_a\}$ where p_i is the probability of the i-th symbol in Σ. Given X, the goal is to characterize asymptotically the size of the largest submatrix in X.

For convenience of notation, let us call $P_r = p_1^r + p_2^r + \ldots + p_a^r$ the probability of observing a clean column over r rows, and let us define $H(x) = -x \ln x - (1 - x)\ln(1-x)$.

Table 2. The statistics of large submatrices in a random $\{0,1\}$-matrix of size 256×256. The second column reports the number of columns of the submatrices observed in a random matrix, whereas the third reports the prediction based on Theorem 2.

rows	columns observed	columns predicted
1	256	256
2	160	165.6771209
3	100	103.9626215
4	67	67.24371945
5	45	44.84053788
6	31	30.70906224
7	23	21.48364693
8	16	15.26873716

The first result characterizes the random variable associated with the number of columns of the largest bicluster, when we fix the number of rows. Both proofs are omitted due to lack of space.

Theorem 2. *Let $C_{n,m,r,a}$ be the random variable associated with the number of columns of the submatrix with the largest area in a matrix $X \in \Sigma^{n\times m}$ generated from a memoryless source, once the number of rows r is fixed. Then*

$$C_{n,m,r,a} \le mP_r + \sqrt{2P_r(1 - P_r)mF(n,r)} \equiv C_{\max}$$

with high probability and as $n \to \infty$, where

$$F(n,r) = \begin{cases} r\log n & \text{if } r = o(n) \\ nH(\alpha) & \text{if } r = \alpha n \text{ where } 0 < \alpha < 1 \end{cases}$$

When $r = o(n)$ the error term is $O(1/\log^d n)$ for some $d > 1$ that may depend on a, whereas the error becomes $O(1/\sqrt{n})$ when $r = \alpha n$. The prediction on random matrices is indeed quite accurate as reported in Table 2. We claim that the upper bound is actually an equality, that is, asymptotically and with high probability $C_{n,m,r,a} = C_{\max}$.

The practical implications of Theorem 2 are twofold. First, the expected number of columns can be used to set the column threshold parameter $\hat{c} \gg \max\{C_{\max}, 1\}$. That allows the algorithm to avoid considering statistically non-significant submatrices. Second, observe that when $\log n = o(m)$, then the dominant term of $C_{\max}$ is the average, say $\mathbf{E}[C]$, of the number of clean columns, that is, $\mathbf{E}[C] = mP_r$. This implies $C_{\max}/\mathbf{E}[C] \le 1 + o(1)$ for $\log n = o(m)$, and therefore with high probability any algorithm is asymptotically optimal. Clearly, this is not true for $r = \alpha n$. Finally, in passing we add that when we restrict the search to largest *squared* matrix (see objective function f_2 above), then its side is asymptotically equal to $2\log(n/(2\log n))/\log P_r^{-1}$.

The second result characterizes the random variable associated with the area of the solution. For convenience of notation, given a memoryless source with parameters $\{p_1, \ldots, p_a\}$ we define $p_{\max} = max_{1 \le i \le a} p_i$.

Theorem 3. *Let $A_{n,m,a}$ be the random variable associated with the area of the largest submatrix in a matrix $X \in \Sigma^{n \times m}$, $m \leq n$, generated from a memoryless source. Then, with high probability for any $\epsilon > 0$ and as $n \to \infty$*

$$A_{n,m,a} \leq (1 + \epsilon)rc$$

where $r = n/2$ and $c = 2\ln 2 / \ln p_{\max}^{-1}$.

The intuition behind Theorem 3 is that on random matrices one should expect the largest submatrix to be "skinny", that is, a few columns and lots of rows, or vice versa. For example, we expect the largest submatrix in a random $\{0, 1\}$-matrix of size 256×256 to be size 2×160 (see Table 2).

5   Implementation and Experiments

We implemented column- and row-sampling algorithms in C++ and tested the programs on a desktop PC with a 1.2GHz Athlon CPU and 1GB of RAM, under Linux. Although the algorithms do not require sophisticated data structures, in order to carry out step 4 in the algorithm of Figure 2, one needs a data structure to store the strings and their frequencies. Since k and a are usually not very large, our experience shows that a simple hash table (of size a^k) is a good choice. If a^k becomes too large, a trie would be a better data structure. If one uses the hash table, it is important to keep track of the non-zero entries in another balanced data structure. That would avoid the algorithm to spend $O(a^k)$ to search for the frequently occurring strings. Observe also that row-sampling algorithm does not require any hash table, or any other data structure. However, our experiments show that in order to get the same level of performance of the column sampling, the row sampling strategy needs a significantly larger projection k which adversely affects the running time.

Another issue is whether one should keep track of the projections generated so far to avoid generating duplicates. We studied this matter experimentally, and found that it is worthwhile to keep track of the projections in some balanced data structure only when k is small. If k is large, the overhead required to keep the data structure updated is much higher than the time wasted in processing the same projection multiple times.

5.1   Simulations

In order to evaluate the performance of the algorithms, we designed several simulation experiments. In these experiments we randomly generated one thousand 256×256 matrices of symbols drawn from an symmetric i.i.d. distribution over an alphabet of cardinality $a = 2, 4, 8, 16, 32$. Then, in each matrix we embedded a random 64×64 submatrix at random columns and random rows. We ran the algorithms for a few tens of iterations ($t = 5, \ldots, 100$), and for each choice of t we measured the number of successes out of the 1,000 distinct instances. Figure 3 summarizes the performance of LARGEST_BICLUSTER_C, for several choices of

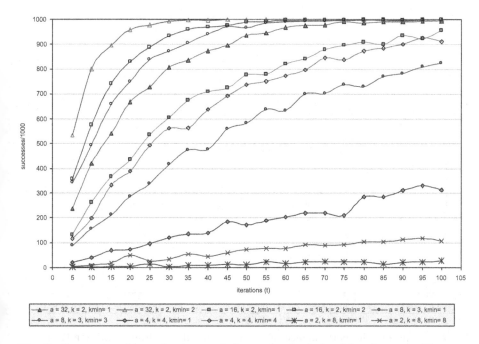

Fig. 3. Comparing the performance of the randomized algorithm LARGEST_BICLUS-TER_C when $k_{min} = k$ versus $k_{min} = 1$, for different choices of the alphabet size a. The projection size is $k = \log_a m$.

alphabet size a and projection size k, and minimum subset size k_{min}. Figure 4 summarizes the performance of LARGEST_BICLUSTER_R under the same conditions.

In order to make a fair comparison between $k_{min} = k$ and $k_{min} = 1$, the number of iterations for the case $k_{min} = k$ was multiplied by $2^k - 1$. Note that by doing so, we are assuming that one projection for $k_{min} = 1$ takes about the same time as one projection for $k_{min} = k$, which is not necessarily very accurate. Under this assumption, however, $k_{min} = k$ outperforms $k_{min} = 1$ (see Figure 3). This not necessarily true in the row sampling strategy (see Figure 4).

By comparing the performance of row sampling against column sampling, one can observe that if one uses the same set of parameters, column sampling always outperforms row sampling.

Unfortunately, we were unable to compare the performance of our randomized approach to other biclustering algorithms (e.g. [3, 4, 6, 5, 12, 8]), because their notion of bicluster is generally different from ours.

6   Conclusions

In this paper we have introduced the LARGEST BICLUSTER problem. This problem has a variety of applications ranging from computational biology to data

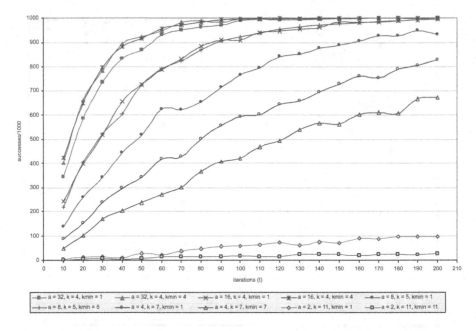

Fig. 4. Comparing the performance of the randomized algorithm LARGEST_BICLUS-TER_R for different choices of the alphabet size a and projection size k.

mining. As far as we know, the pattern matching community has not looked yet at this problem from a combinatorial perspective. Unfortunately, the problem is generally NP complete.

Here we presented a rather simple algorithm based on random projections. Its performance with respect to the number of projection was carefully analyzed. We have also presented a probabilistic analysis of the LARGEST BICLUSTER problem, which allows one to determine the statistical significance of a solution.

Our approach performs remarkably well on synthetic data. On large alphabets, thirty or so iterations are enough to give a performance close to 100%. With respect to other biclustering algorithms (see e.g., [3, 12, 8]), our algorithm simultaneously discovers multiple solutions which satisfy the user-defined parameters without masking or changing the original data. In addition to this, the algorithm will never report solutions which are completely contained in other solutions.

Acknowledgments

This research was supported in part by NSF Grants CCR-0208709, DBI-0321756 and the NIH Grant R01 GM068959-01.

References

1. Agrawal, R., Gehrke, J., Gunopulos, D., Raghavan, P.: Automatic subspace clustering of high dimensional data for data mining applications. In: Proceedings of the ACM SIGMOD International Conference on Management of Data (SIGMOD-98). Volume 27,2 of ACM SIGMOD Record., New York, ACM Press (1998) 94–105
2. Aggarwal, C.C., Procopiuc, C., Wolf, J.L., Yu, P.S., Park, J.S.: Fast algorithms for projected clustering. In: Proceedings of the ACM SIGMOD International Conference on Management of Data (SIGMOD-99). Volume 28,2 of SIGMOD Record., New York, ACM Press (1999) 61–72
3. Cheng, Y., Church, G.M.: Biclustering of expression data. In: Proceedings of the 8th International Conference on Intelligent Systems for Molecular (ISMB-00), Menlo Park, CA, AAAI Press (2000) 93–103
4. Wang, H., Wang, W., Yang, J., Yu, P.S.: Clustering by pattern similarity in large data sets. In: Proceedings of the 2002 ACM SIGMOD international conference on Management of data (SIGMOD-02), New York, ACM Press (2002) 394–405
5. Tanay, A., Sharan, R., Shamir, R.: Discovering statistically significant biclusters in gene expression data. In: Proceedings of the 10th International Conference on Intelligent Systems for Molecular Biology (ISMB'02), in Bioinformatics. Volume 18. (2002) S136–S144
6. Ben-Dor, A., Chor, B., Karp, R., Yakhini, Z.: Discovering local structure in gene expression data: The order-preserving submatrix problem. In: Proceedings of Sixth International Conference on Computational Molecular Biology (RECOMB 2002), ACM Press (2002) 45–55
7. Zhang, L., Zhu, S.: A new clustering method for microarray data analysis. In: Proceedings of the First IEEE Computer Society Bioinformatics Conference (CSB'02), IEEE Press (2002) 268– 275
8. Sheng, Q., Moreau, Y., Moor, B.D.: Biclustering microarray data by Gibbs sampling. In: Proceedings of European Conference on Computational Biology (ECCB'03). (2003) to appear
9. Mishra, N., Ron, D., Swaminathan, R.: On finding large conjunctive clusters. In: Proc. of the ACM Conference on Computational Learning Theory (COLT'03). (2003) to appear
10. Procopiuc, M., Jones, M., Agarwal, P., Murali, T.M.: A Monte-Carlo algorithm for fast projective clustering. In: Proceedings of the 2002 International Conference on Management of Data (SIGMOD'02). (2002) 418–427
11. Hartigan, J.A.: Direct clustering of a data matrix. Journal of the American Statistical Association **67** (1972) 123–129
12. Murali, T.M., Kasif, S.: Extracting conserved gene expression motifs from gene expression data. In: Proceedings of the Pacific Symposium on Biocomputing (PSB'03). (2003) 77–88
13. Hochbaum, D.S.: Approximating clique and biclique problems. Journal of Algorithms **29** (1998) 174–200
14. Garey, M.R., Johnson, D.S.: Computers and intractability: a guide to the theory of NP-completeness. Freeman, New York, NY (1979)
15. Grigni, M., Manne, F.: On the complexity of the generalized block distribution. In: In proceedings of Irregular'96, The third international workshop on parallel algorithms for irregularly structured problems. Volume 1117., Lecture Notes in Computer Science, Springer (1996) 319–326

16. Peeters, R.: The maximum-edge biclique problem is NP-complete. Technical Report 789, Tilberg University: Faculty of Economics and Business Adminstration (2000)

17. Dawande, M., Keskinocak, P., Swaminathan, J.M., Tayur, S.: On bipartite and multipartite clique problems. Journal of Algorithms **41** (2001) 388–403

18. Pasechnik, D.V.: Bipartite sandwiches. Technical report, available at `http://arXiv.org/abs/math.CO/9907109` (1999)

19. Reinert, G., Schbath, S., Waterman, M.S.: Probabilistic and statistical properties of words: An overview. J. Comput. Bio. **7** (2000) 1–46

20. Hastie, T., Tibshirani, R., Eisen, M., Alizadeh, A., Levy, R., Staudt, L., Chan, W., Botstein, D., Brown, P.: 'Gene shaving' as a method for identifying distinct sets of genes with similar expression patterns. Genome Biol. **1** (2000) 1–21

21. Lazzeroni, L., Owen, A.: Plaid models for gene expression data. Statistica Sinica **12** (2002) 61–86

22. Kluger, Y., Basri, R., Chang, J., Gerstein, M.: Spectral biclustering of microarray data: coclustering genes and conditions. Genome Res. **13** (2003) 703–16

23. Yang, J., Wang, H., Wang, W., Yu, P.S.: Enhanced biclustering on gene expression data. In: IEEE Symposium on Bioinformatics and Bioengineering (BIBE'03). (2003) to appear

24. Hanisch, D., Zien, A., Zimmer, R., Lengauer, T.: Co-clustering of biological networks and gene expression data. In: Proceedings of the 8th International Conference on Intelligent Systems for Molecular (ISMB-02), AAAI Press (2002) 145–154

25. Dhillon, I., Mallela, S., Modha, D.: Information-theoretic co-clustering. In: Proceedings of the Seventh ACM SIGKDD International Conference on Knowledge Discovery and Data Mining (KDD-03), ACM Press (2001) to appear

26. Gu, M., Zha, H., Ding, C., He, X., Simon, H.: Spectral relaxation models and structure analysis for k-way graph clustering and bi-clustering. Technical Report CSE-01-007, Department of Computer Science and Engineering, Pennsylvania State University (2001)

27. Dhillon, I.: Co-clustering documents and words using bipartite spectral graph parititioning. In: Proceedings of the Seventh ACM SIGKDD International Conference on Knowledge Discovery and Data Mining (KDD-01), New York, ACM Press (2001) 269–274

28. Szpankowski, W.: Average Case Analysis of Algorithms on Sequences. Wiley Interscience (2001)

Real-Time String Matching in Sublinear Space

Leszek Gąsieniec[1] and Roman Kolpakov[1]

Department of Computer Science, University of Liverpool, Liverpool L69 7ZF, UK
{leszek,roman}@csc.liv.ac.uk

Abstract. We study a problem of efficient utilisation of extra memory space in real-time string matching. We propose, for any constant $\varepsilon > 0$, a real-time string matching algorithm claiming $O(m^\varepsilon)$ extra space, where m is the size of a pattern. All previously known real-time string matching algorithms use $\Omega(m)$ extra space.

1 Introduction

A string matching problem is one of the most studied problems in algorithmic theory of computing. In standard string matching problem we are interested in finding all occurrences of one string P, called *a pattern*, in another (usually much longer) string T, called *a text*. Frequently, we denote the length of the pattern by $|P| = m$ and the length of the text by $|T| = n$. We say that the string matching algorithm is optimal if its total complexity is linear in n and m. A number of optimal linear time string matching algorithms has been proposed in the literature. Some of them perform well in the worst case, see, e.g., [12], and some of them work fast in the average case [1, 11]. In the context of some particular applications, we are interested in other aspects of the string matching problem. E.g., in design of optimal linear time algorithms requiring only constant extra space, see, e.g., [4, 7–9]. Note here that most of the string matching algorithms use $O(m)$ extra memory. In other applications we insist that the input text symbols are assessed consecutively, without possibility of backtracking, i.e., looking at symbols previously assessed. We call these *on-line* algorithms. A number of optimal (linear time) on-line algorithms have been proposed, in case where a linear extra space is available, see, e.g., [2, 3, 6]. Very recently, an optimal constant space on-line string matching algorithm was introduced in the context of compressed matching, see [10]. However, in some applications, e.g., when the text is a part of a large streamed data, we might be forced to assess input text symbols in real-time. I.e., the time difference between assessments of two consecutive text symbols must be bounded by a constant. This creates a need for real-time string matching algorithms. The first real-time string matching procedure was introduced in [5]. However, as most of the algorithms it uses and extra linear space. In this paper we discuss more efficient utilisation of memory in real-time string matching. We propose here a real-time algorithm that uses only $O(m^\varepsilon)$ extra space, for any constant $\varepsilon > 0$.

The paper is organised as follows. In section 2 we introduce a notion of a *partial next function* and we prove a number of its useful properties. In section 3 we present a crude version of our real-time string matching algorithm which is followed by a discussion on possible improvements in section 4.

S.C. Sahinalp et al. (Eds.): CPM 2004, LNCS 3109, pp. 117–129, 2004.

2   Partial Next Function

Let $W = w_1 \ldots w_n$ be a finite word. We will denote by $W[i..j]$ a subword $w_i \ldots w_j$ of W, and call the number i the *position* of this subword in W. For two subwords $W[i'..j']$, $W[i''..j'']$ of W, where $i' \leq i''$, the length $\max(0, \min(j', j'') - i'' + 1)$ of their intersection forms an *overlap* of the subwords. A period π of the word W is a natural number such that $w_i = w_{i+\pi}$, for any $i = 1, \ldots, n - \pi$. We denote by $\pi(W)$ the smallest period of W, and call the ratio $\frac{|W|}{\pi(W)}$ the *exponent* of W. A *repetition* in W is any subword of W with the exponent greater or equal to 2. A *cyclic root* of a repetition r is any subword in r of length $\pi(r)$. The following fact is well known.

Lemma 1. *Let* $r = W[h'..h'']$ *be a repetition in a word* W. *Two integers* $i, j \in \{h', \ldots, h'' - \pi(r) + 1\}$ *are positions of equal cyclic roots of* r *if and only if* $i - j \equiv 0 \pmod{\pi(r)}$.

A *maximal repetition* in W is a repetition $r = W[h'..h'']$, s.t.,

1. if $h' > 1$, then $w_{h'-1} \neq w_{h'-1+\pi(r)}$,
2. if $h'' < n$, then $w_{h''+1-\pi(r)} \neq w_{h''+1}$.

In other words, a repetition in W with the minimal period π is maximal if its one letter extension in W (to the left or to the right) results in a subword with the minimal period $> \pi$. We will also use the following known fact about maximal repetitions.

Proposition 1. *The overlap of two different maximal repetitions with the same minimal period* π *is smaller than* π

Let f be a integer function defined on a set $\{1, \ldots, m\}$, and $i_1 < i_2 < \ldots < i_q$ be all values, s.t., $f(i_j) \neq 0$, $j = 1, \ldots, q$. Then we can represent f as the list of pairs $(i_1, f(i_1)), (i_2, f(i_2)), \ldots, (i_q, f(i_q))$. We denote this list by $\text{list}[f]$. Moreover, we group consecutive pairs of $\text{list}[f]$ into sublists

$$\rho_1, \ldots, \rho_h \tag{1}$$

called *ranges*. The grouping process is done as follows. Initially the pair $(i_1, f(i_1))$ is placed in the first range ρ_1. Assume, that the pairs $(i_1, f(i_1)), \ldots, (i_j, f(i_j))$, for $1 \leq j < q$, are already placed appropriately in ranges $\rho_1, \ldots, \rho_s$. Now, the next available pair $(i_{j+1}, f(i_{j+1}))$ is settled in the range ρ_s if $f(i_j) = f(i_{j+1})$ and one of the following conditions is satisfied:

1. ρ_s contains only the pair $(i_j, f(i_j))$,
2. ρ_s contains also the pair $(i_{j-1}, f(i_{j-1}))$, and $i_{j+1} - i_j = i_j - i_{j-1}$.

Otherwise $(i_{j+1}, f(i_{j+1}))$ is settled in the next range ρ_{s+1}. Note that each range is composed of consecutive pairs $(i_{j'}, f(i_{j'})), (i_{j'+1}, f(i_{j'+1})), \ldots, (i_{j''}, f(i_{j''}))$, such that, $f(i_{j'}) = f(i_{j'+1}) = \ldots = f(i_{j''})$ and $i_{j''} - i_{j''-1} = \ldots = i_{j'+1} - i_{j'}$. Thus all these pairs can be represented with a help of four parameters: $f(i_{j'})$, $i_{j'}$, $i_{j''}$, and $i_{j'+1} - i_{j'}$. Note that the last parameter is not used if the range contains only one pair. For each range α, we will denote the four parameters by $\textbf{value}(\alpha)$, $\textbf{begin}(\alpha)$, $\textbf{end}(\alpha)$, and $\textbf{step}(\alpha)$, respectively. Thus the list $\text{list}[f]$ can be represented as the list (1) of ranges in $O(h)$ memory space. We will denote the double linked list (1) by $\text{list}^*[f]$.

Lemma 2. *Let g be an integer function defined on set $\{1, \ldots, m\}$, s.t., g has only one nonzero value. Then the number of ranges in $\text{list}^*[\max(f, g)]$ is larger by at most 2 comparing with the number of ranges in $\text{list}^*[f]$.*

Corollary 1. *Let g be an integer function defined on set $\{1, \ldots, m\}$, s.t., g has no more than k nonzero values. Then the number of ranges in $\text{list}^*[\max(f, g)]$ is larger by at most $2k$ comparing with the number of ranges in $\text{list}^*[f]$.*

Recall that in the string matching problem we are expected to find all occurrences of the pattern $P = p_1 \ldots p_m$ in the text $T = t_1 \ldots t_n$. Our small extra space real-time algorithm is based on the concept of the *next function* used in the Knuth-Morris-Pratt algorithm, see [12]. The next function, $\text{next} : \{1, \ldots, m\} \rightarrow \{0, 1, \ldots, m-1\}$, is computed during the preprocessing of the pattern P. Recall that $\text{next}(i)$ is defined as the largest positive integer k, s.t., $p_1 \ldots p_{k-1} = p_{i-k+1} \ldots p_{i-1}$ and $p_i \neq p_k$, if such k exists. Otherwise $\text{next}(i) = 0$. For a real $x > 1$ we define a function next_x as follows:

$$\text{next}_x(i) = \begin{cases} \text{next}(i) \text{ if } \text{next}(i) > x, \\ \qquad 0 \text{ otherwise.} \end{cases}$$

Lemma 3. *The list $\text{list}^*[\text{next}_x]$ contains $O(m/x)$ ranges.*

Proof: For $j = 2, \ldots, m$ denote by $\chi(j)$ the minimum number $i \geq j$, s.t., $p_i \neq p_{i-j+1}$, if such i exists. For $i = 2, \ldots, m$ let $U(i)$ be a set of all js, s.t., $\chi(j) = i$. Then

$$\text{next}(i) = \max(0, \quad \{i - j + 1 \mid j \in U(i)\}).$$

Therefore, if V_k is the set of all $j = 2, \ldots, m$, s.t., $\chi(j) - j \geq k$, we have

$$\text{next}_x(i) = \max(0, \quad \{i - j + 1 \mid j \in U(i) \cap V_{\lfloor x \rfloor}\}). \tag{2}$$

Denote the word $p_1 \ldots p_{\lfloor x \rfloor}$ by v and consider two possible cases:
CASE 1 Let $\pi(v) > |v'|/2$. Then the overlap of two occurrences of v in cannot be greater than $|v|/2$. So P contains no more than $2|P|/|v| = O(m/x)$ occurrences of v. On the other hand, note that $j \in V_{\lfloor x \rfloor}$ if and only if j is the position of an occurrence of v in P. Thus $V_{\lfloor x \rfloor}$ contains $O(m/x)$ elements. From formula 2 we can see that each nonzero value $\text{next}_x(i)$ is provided by a number j of $V_{\lfloor x \rfloor}$, s.t., $\chi(j) = i$. So the number of nonzero values of next_x is not greater that the number of elements in $V_{\lfloor x \rfloor}$. Thus next_x has $O(m/x)$ nonzero values. Hence $\text{list}^*[\text{next}_x]$ contains $O(m/x)$ ranges.
CASE 2 Let $\pi(v) \leq |v'|/2$. So v is a repetition (extended in P) preserving its minimal period to some maximal repetition $P[1..h]$. Consider any number $j \in V_{\lfloor x \rfloor}$. Assume that the occurrence of v at position j extends in P its minimal period to another maximal repetition $r = P[h'..h'']$. We call such a maximal repetition a *worthy* repetition. The position j is of a *first type* if $h + j \leq h''$. Otherwise, if we have $h'' < h + j \leq h'' + \pi(v) + 1$ when $h' \leq h'' - h + 1$ or $j < h' + \pi(v)$, then j is of a *second type*. We denote the subsets of positions of the first type and the second type in $V_{\lfloor x \rfloor}$ by $V'_{\lfloor x \rfloor}$ and $V''_{\lfloor x \rfloor}$ respectively. If j is neither of the first type nor the second type then j is a position of a *third type*. Let j be a position of the third type. Then $j \geq h' + \pi(v)$ and

$$h'' + 1 + \pi(v) - j < h. \tag{3}$$

Note that words $u' = P[\lfloor x \rfloor + 1..h]$ and $u'' = P[j + \lfloor x \rfloor..h'']$ are the maximal extensions of occurrences of v in P that preserve the period $\pi(v)$, and inequality (3). This implies

$$|u'| = h - \lfloor x \rfloor > h'' - (j + \lfloor x \rfloor) + 1 = |u''|.$$

Thus we have $u'' = P[\lfloor x \rfloor + 1..\lfloor x \rfloor + |u''|]$ and

$$p_{\lfloor x \rfloor + |u''| + 1} = p_{\lfloor x \rfloor + |u''| + 1 - \pi(v)} = p_{h'' + 1 - \pi(v)} \neq p_{h'' + 1}.$$

Thus $\chi(j) = h'' + 1$. It follows from $j \geq h' + \pi(v)$ that r contains another occurrence of v at position $j - \pi(v)$. So word $u''' = P[j + \lfloor x \rfloor - \pi(v)..h'']$ is the maximal extension of this occurrence of v that preserves the period $\pi(v)$. The inequality (3) implies also that

$$|u'| = h - \lfloor x \rfloor > h'' - (j + \lfloor x \rfloor - \pi(v)) + 1 = |u'''|.$$

Therefore, similarly as in case of the position j, the equality $\chi(j - \pi(v)) = h'' + 1$ holds for the position $j - \pi(v)$ too. Hence $\chi(j - \pi(v)) = \chi(j)$. Using equation (2), we obtain $\mathbf{next}(\chi(j)) \geq \chi(j) - (j - \pi(v)) + 1 > \chi(j) - j + 1$. Thus positions of the third type don't provide any values for $\mathbf{next}(i)$ in equation (2). And equation (2) can be rewritten in the form

$$\mathbf{next}_x(i) = \max(\mathbf{next}'_x(i), \mathbf{next}''_x(i))$$

where

$$\mathbf{next}'_x(i) = \max(0, \quad \{i - j + 1 \mid j \in U(i) \cap V'_{\lfloor x \rfloor}\}),$$
$$\mathbf{next}''_x(i) = \max(0, \quad \{i - j + 1 \mid j \in U(i) \cap V''_{\lfloor x \rfloor}\}).$$

Let now j be a position of the first type. In this case we have $h + j \leq h''$, and

$$|u'| = h - \lfloor x \rfloor < h'' - (j + \lfloor x \rfloor) + 1 = |u''|.$$

So we have $u' = P[j + \lfloor x \rfloor..j + \lfloor x \rfloor + |u'| - 1]$, and

$$p_{h+1} = p_{h+1-\pi(v)} = p_{j+\lfloor x \rfloor + |u'| - \pi(v)} \neq p_{j + \lfloor x \rfloor + |u'|}.$$

Thus, $\chi(j) = j + \lfloor x \rfloor + |u'| = j + h$, for any $j \in V'_{\lfloor x \rfloor}$. Therefore,

$$\mathbf{next}'_x(i) = \begin{cases} h+1 & \text{if } i - h \in V'_{\lfloor x \rfloor}; \\ 0 & otherwise. \end{cases}$$

Let $r_1 = P[h'_1..h''_1], \ldots, r_s = P[h'_s..h''_s]$ be all worthy repetitions in P, s.t., $h'_1 < h'_2 < \ldots < h'_s$, and for $k = 1, \ldots, s$ let J'_k be the set of all positions of the first type of occurrences of v contained in r_k. By Proposition 1, for any $k = 1, \ldots, s - 1$ we have $h''_k - \pi(v) < h'_{k+1}$. Hence any $j' \in J'_k$ and $j'' \in J'_{k+1}$ satisfy the inequalities $j' \leq h''_k - h < h'_{k+1} \leq j''$. Thus $\mathbf{list}[\mathbf{next}'_x]$ contains consecutive sublists $\mathcal{J}_1, \ldots, \mathcal{J}_s$ where $\mathcal{J}_k$ is the list of all pairs $(j + h, h + 1)$, s.t., $j \in J'_k$. Since all numbers of J'_k are positions of the cyclic root $p_1 \ldots p_{\pi(v)}$ in r_k, using Lemma 1, we conclude that these numbers form an arithmetic progression in the segment $[h'_k; h''_k - h]$ with the step $\pi(v)$. Therefore, for any consecutive pairs $(j' + h, h + 1), (j'' + h, h + 1)$ of J'_k we have $(j'' + h) - (j' + h) = j'' - j' = \pi(v)$. Thus we can see that the number of ranges representing the list consisting of sublists $\mathcal{J}_1, \ldots, \mathcal{J}_k$ is greater by at most 1 than the number of ranges representing the list consisting of sublists $\mathcal{J}_1, \ldots, \mathcal{J}_{k-1}$. Thus $\mathbf{list}^*[\mathbf{next}'_x]$ has no more than s ranges. Since $h'_{k+1} \geq h''_k + 1 - \pi(v) \geq h'_k + |v| - \pi(v) \geq h'_k + |v|/2$, we obtain that $s \leq \frac{2m}{|v|} + 1 = O(m/x)$. Thus $\mathbf{list}^*[\mathbf{next}'_x]$ contains $O(m/x)$ ranges. Consider now the set $V''_{\lfloor x \rfloor}$. Note that $V''_{\lfloor x \rfloor} = \bigcup_{k=1}^s J''_k$ where J''_k is the set of all positions of the second type of occurrences of v contained in r_k. Note that any $j \in J''_k$ satisfies the inequalities $\hat{h} \leq j\hat{h} + \pi(v)$, where $\hat{h} = \max(h'' - h + 1, h')$. On the other hand, by Lemma 1, for any two numbers $j', j'' \in J''_k$ which are different occurrences of the cyclic root $p_1 \ldots p_{\pi(v)}$ in r_k, we have $|j' - j''| \geq \pi(v)$. Thus J''_k contains no more than two different numbers. And $V''_{\lfloor x \rfloor}$ contains no more than $2s = O(m/x)$ different numbers. Hence $\mathbf{next}''_x$ has $O(m/x)$ nonzero values. Therefore, taking into account that $\mathbf{list}^*[\mathbf{next}'_x]$ has $O(m/x)$ ranges, by Corollary 1 we obtain that $\mathbf{list}^*[\mathbf{next}_x] = \mathbf{list}^*[\max(\mathbf{next}'_x, \mathbf{next}''_x)]$ has also $O(m/x)$ ranges.

3 Space Efficient Real-Time Text Search

Recall, that our main aim is to find all occurrences of the pattern $P = p_1 \ldots p_m$ in the text $T = t_1 \ldots t_n$ utilising the extra space as efficiently as possible. For this purpose we propose the following modification of the Knuth-Morris-Pratt algorithm.

Lemma 3 states that there exists a constant $C \geq 1$, s.t., for any pattern and $x > 1$, the list $\mathbf{list}^*[\mathbf{next}_x]$ has less than $C \ldots \frac{l}{x}$ ranges, where l is the length of the pattern. Consider any integer τ such that $2 \leq \tau \leq \log_{C+1} m$. For $i = 1, \ldots, \tau$ denote by m_i the number $\lfloor m^{i/\tau} \rfloor$, and by P_i the word $P[1..m_i]$. In our modification, instead of the next function for the pattern $P\$$ (where $\$$ is a symbol which does not occur in P) we use τ functions $h_0, h_1, \ldots, h_{\tau-1}$ where h_0 is the function $\mathbf{next}$ for the pattern $P_1\$$, and h_i is the function $\mathbf{next}_{m^{i/\tau}}$ for the pattern $P_{i+1}\$$, for $i = 1, \ldots, \tau-1$. Moreover, we assume that the functions $h_1, \ldots, h_{\tau-1}$ are represented as lists $\mathbf{list}^*[h_1], \ldots, \mathbf{list}^*[h_{\tau-1}]$ of ranges. For each list and range α we compute the pointer to the first range β in the list containing α, s.t., $\mathbf{end}(\beta) \geq \mathbf{value}(\alpha)$. We denote this pointer by $\mathbf{failrange}(\alpha)$. For $j = 1, \ldots, n$ we denote by $\mathbf{maxpref}_0(j)$ the largest $k \leq m^{1/\tau}$, s.t., $p_1 \ldots p_k = t_{j-k+1} \ldots t_j$; and if there is no such k, then $\mathbf{maxpref}_0(j) = 0$. Moreover for $i = 1, \ldots, \tau - 1$ we denote by $\mathbf{maxpref}_i(j)$ the largest k, s.t., $p_1 \ldots p_k = t_{j-k+1} \ldots t_j$ and $m^{i/\tau} < k \leq m^{(i+1)/\tau}$; and if is such k does not exist, then $\mathbf{maxpref}_i(j) = 0$.

The main idea of our algorithm is based on the following mechanism. During a test of the text symbol t_j, we compute simultaneously (with a possible short delay), for any $i = 0, 1, \ldots, \tau - 1$, all nonzero values $\mathbf{maxpref}_i(j)$, using functions COMP_MPREF$_i$. Since, for any $i = 1, \ldots, \tau-1$, a nonzero value $\mathbf{maxpref}_i(j)$ must be $> m^{i/\tau}$. And this happens only when there exists a close occurrence of P_i in T to the left of t_j. Thus the function COMP_MPREF$_i$ begins a computation of $\mathbf{maxpref}_i$ only after we have an occurrence of P_i in T provided by the function COMP_MPREF$_{i-1}$. Note that the delay of the output in the function COMP_MPREF$_{i-1}$ is not necessarily synchronized with the delay occurring in the function COMP_MPREF$_i$. Thus in order to communicate the positions of all occurrences of P_i for the function COMP_MPREF$_i$

we will use a queue (FIFO) $\mathbf{jqueue}_i$. In other words, the purpose of $\mathbf{jqueue}_i$ is to keep (temporarily) all positions j in T, s.t., $P_i = T[j - m_i..j - 1]$. The positions will be used in computation of $\mathbf{maxpref}_i$. The outline of our algorithm is shown below.

$\mathbf{jqueue}_1, \ldots, \mathbf{jqueue}_{\tau-1}$ are empty;
$j_0 := j_1 := \ldots j_{\tau-1} := 1$;
$\{j_i$ – current position for which the value $\mathbf{maxpref}_i$ is computed$\}$
$l_0 := 1$; $l_1, \ldots, l_{\tau-1}$ are undefined;
$\{l_i$ – position of pattern symbol compared with t_{j_i} in COMP_MPREF$_i\}$
for $j := 1$ **to** n
    begin read the next symbol t_j;
    for $i := \tau - 1$ **down to** 0
      COMP_MPREF$_i()$;
    if $j_i \leq j$ **then**
      COMP_MPREF$_i()$;
    end

Each time the function COMP_MPREF$_i$ is called, we start (or continue) to compute, with delay $j - j_i$, the value $\mathbf{maxpref}_i(j_i)$. The function COMP_MPREF$_0$ is essentially a single step of execution of the Knuth-Morris-Pratt algorithm for finding all occurrences of P_1 in T with only exception in case when $P_1 = T[j_0 - m_1..j_0 - 1]$. In this case we report the position j_0 as a potential value j_1 in $\mathbf{jqueue}_1$, if $j_0 > j_1$. Note that the inequality $j_0 \leq j_1$ indicates that the position j_0 is already processed by COMP_MPREF$_1$ during the execution of our algorithm.

function COMP_MPREF$_0()$
if $p_{l_0} = t_{j_0}$ **then begin**
    $j_0 := j_0 + 1$;
    if $l_0 = m_1$ **then begin**
      if $j_0 > j_1$ **then** put j_0 in $\mathbf{jqueue}_1$;
      $l_0 := h_0(m_1 + 1)$;
    end
    else $l_0 := l_0 + 1$;
    end
else $\{p_{l_0} \neq t_{j_0}\}$
    if $h_0(l_0) > 0$ **then** $l_0 := h_0(l_0)$;
    else $j_0 := j_0 + 1$; $l_0 := 1$;

The functions COMP_MPREF$_i$, for $i = 1, \ldots, \tau - 2$, differ from COMP_MPREF$_0$ in two principal elements. The first difference is in the fact that the function h_i is represented as a list of ranges. So, when we have a mismatch $p_{l_i} \neq t_{j_i}$ during the computation of $\mathbf{maxpref}_i(j_i)$, we cannot retrieve immediately the value $h_i(l_i)$, but we rather need to find in $\mathbf{list}^*[h_i]$ the range containing the pair $(l_i, h_i(l_i))$, if such a pair exists. For this purpose we maintain in COMP_MPREF$_i$ an additional variable $\mathbf{currange}_i$ pointing to the first range γ in $\mathbf{list}^*[h_i]$, s.t., $\mathbf{end}(\gamma) \geq l_i$. The second difference is that COMP_MPREF$_i$ computes $\mathbf{maxpref}_i(j)$ only for positions j, preceded by a match $p_1 \ldots p_k = t_{j-k} \ldots t_{j-1}$, s.t., $m_i \leq k < m_{i+1}$. So, if for some position j_i we have $\mathbf{maxpref}_i(j_i) = 0$, during the next call of COMP_MPREF$_i$ we don't continue to compute COMP_MPREF$_i$ for the next position $j_i + 1$, but we rather take as the next position to be processed by COMP_MPREF$_i$ the first to the right position j_i^{next}, s.t., $T[j_i^{\text{next}} - m_i .. j_i^{\text{next}} - 1] = P_i$. The position j_i^{next} is taken from $\mathbf{jqueue}_i$. If the queue $\mathbf{jqueue}_i$ is empty then COMP_MPREF$_i$ stops any computations till COMP_MPREF$_{i-1}$ delivers a new position to $\mathbf{jqueue}_i$. Note that for the correct functioning of our algorithm j_i^{next} have to be the first element in $\mathbf{jqueue}_i$. Thus we have to maintain $\mathbf{jqueue}_i$ properly. I.e., before the processing of each position j_i by COMP_MPREF$_i$ we have to remove j_i from $\mathbf{jqueue}_i$ if $j_i \in \mathbf{jqueue}_i$. The function COMP_MPREF$_{\tau-1}$ differs from other functions COMP_MPREF$_i$ for $i > 0$ only in case of detecting an occurrence of $P_\tau = P$ in T. In this case, instead of communicating the position of the occurrence to a queue, COMP_MPREF$_{\tau-1}$ reports the occurrence.

function COMP_MPREF$_i$() $\{i > 0\}$
if l_i is undefined **then**
    if jqueue$_i$ is not empty **then begin**
        $j_i :=$ the first element in $\mathbf{jqueue}_i$;
        remove the first element in $\mathbf{jqueue}_i$;
        $l_i := m_i + 1$; $\mathbf{currange}_i :=$ the first range of $\mathbf{list}^*[h_i]$;
    end
    else $\{$jqueue$_i$ is empty$\}$ **return;**
if $p_{l_i} = t_{j_i}$ **then begin**
    $j_i := j_i + 1$;
    if $j_i =$ the first element in $\mathbf{jqueue}_i$
        then remove the first element in $\mathbf{jqueue}_i$;
    if $l_i = m_{i+1}$ **then begin**
#if $i < \tau - 1$
        if $j_i > j_{i+1}$ **then** put j_i in $\mathbf{jqueue}_{i+1}$;
#else $\{i = \tau - 1\}$
        report the occurrence $T[j_i - m .. j_i - 1]$ of P;
#endif
        if the last range α of $\mathbf{list}^*[h_i]$ has the pair
        $(m_{i+1} + 1, h_i(m_{i+1} + 1))$ $\{h_i(m_{i+1} + 1) > m^{i/\tau}\}$
            then $l_i := h_i(m_{i+1} + 1)$; $\mathbf{currange}_i :=$ **failrange**(α);
        else undefine l_i;
    end
    else $\{l_i < m_{i+1}\}$ **begin**
        $l_i := l_i + 1$;
        if $l_i > \mathbf{end}(\mathbf{currange}_i)$
            then $\mathbf{currange}_i :=$ the next range in $\mathbf{list}^*[h_i]$;
    end
  end
else $\{p_{l_i} \neq t_{j_i}\}$
    if $\mathbf{currange}_i$ has the pair $(l_i, h_i(l_i))$ $\{h_i(l_i) > m^{i/\tau}\}$
        then $l_i := h_i(l_i)$; $\mathbf{currange}_i :=$ **failrange**$(\mathbf{currange}_i)$;
    else undefine l_i;

During the preprocessing stage, similarly as in the Knuth-Morris-Pratt algorithm, we apply our algorithm of computation of $\mathbf{maxpref}_i$, using the functions h_i, to the computation of the functions h_i themselves. During the computation of the lists $\mathbf{list}^*[h_i]$ we compute simultaneously for each range α the value $\mathbf{failrange}(\alpha)$. We present here, for the simplicity of presentation, a non-real-time version of the preprocessing which can be upgraded to the real-time by means of using queues similar to $\mathbf{jqueue}_i$. The outline of the preprocessing algorithm follows.

$h_0(1) = 0$; $l_0 := 1$; $l_1, \ldots, l_{\tau-1}$ are undefined;
$\mathbf{lastrange}_1, \ldots, \mathbf{lastrange}_{\tau-1}$ are undefined;
{$\mathbf{lastrange}_i$ points to the last created range in $\mathbf{list}^*[h_i]$}
for $j := 2$ **to** n
    for $i := \tau - 1$ **down to** 0
        if l_i is defined **then**
            if $j \leq m^{(i+1)/\tau}$ **then**
                COMP_NEXT$_i$();
            else {$j > m^{(i+1)/\tau}$}
                FIND_PATTERN$_i$();
        else continue the **for**-loop for i;

The function COMP_NEXT$_i$ computes the value $h_i(j)$ (and the value $h_i(m_{i+1} + 1)$ if $j = m_{i+1}$). The function FIND_PATTERN$_i$ searches for an occurrence of P_{i+1} ending at the position j in P, and, if this occurrence exists and the variable l_{i+1} is not defined, the execution of the functions COMP_NEXT$_{i+1}$ or FIND_PATTERN$_{i+1}$ is started from the next position $j + 1$. For $i > 0$ the functions COMP_NEXT$_i$ and FIND_PATTERN$_i$, as the functions COMP_MPREF$_i$, use the variable $\mathbf{currange}_i$ for fast retrieval of the value $h_i(l_i)$ from $\mathbf{list}^*[h_i]$.
The function COMP_NEXT$_0$ corresponds to a single execution of the Knuth-Morris-Pratt preprocessing for the pattern $P_1\$$.

function COMP_NEXT$_0$()
if $p_{l_0} = p_j$ **then** $h_0(j) := h_0(l_0)$;
else begin $h_0(j) := l_0$;
    repeat $l_0 := h_0(l_0)$;
    while $l_0 > 0$ **and** $p_{l_0} \neq p_j$;
    end
$l_0 := l_0 + 1$;
if $j = m_1$ **then** $h_0(j + 1) := l_0$;

The function FIND_PATTERN$_0$ is a slight modification of the Knuth-Morris-Pratt algorithm for finding all occurrences of P_1 in $P[2..n]$.

function FIND_PATTERN$_0$()
if $p_{l_0} = p_j$
    if $l_0 = m_1$
        then begin
        if l_1 is undefined
            then begin

$$l_1 = l_0 + 1;$$
if lastrange$_1$ is defined {list$^*[h_1]$ is not empty}
    then currange$_1$:= the first range of {list$^*[h_1]$;
else currange$_1$ is undefined;
end
$$l_0 := h_0(m_1 + 1);$$
end
else $l_0 := l_0 + 1;$
else $\{p_{l_0} \neq p_j\}$ **begin**
    repeat $l_0 := h_0(l_0);$
    while $l_0 > 0$ **and** $p_{l_0} \neq p_j;$
    $l_0 := l_0 + 1;$
end

The function COMP_NEXT$_i$, for $i > 0$ is presented below.

function COMP_NEXT$_i$() $\{i = 1, \ldots, \tau - 1\}$
if $p_{l_i} = p_j$ **then**
    if currange$_i$ has the pair $(l_i, h_i(l_i))$
        then put the pair $(j, h_i(l_i))$ in list$^*[h_i]$;
else begin
    place the pair (j, l_i) in list$^*[h_i]$;
    repeat
        if currange$_i$ has the pair $(l_i, h_i(l_i))$
            then $l_i := h_i(l_i);$ currange$_i$:= failrange(currange$_i$);
        else undefine l_i; **return**;
    while $l_i > 0$ **and** $p_{l_i} \neq p_j;$
    end
$l_i := l_i + 1;$
if currange$_i$ is defined **and** $l_i > $ end(currange$_i$)
    then currange$_i$:= the next range in list$^*[h_i]$;

The operation of settling the pair $(j, h_i(j))$ is implemented as follows. If **lastrange**$_i$ is not defined (**list**$^*[h_i]$ is empty), we create a new range α that contains this pair, and define **failrange**(α), **lastrange**$_i$, and **currange**$_i$ to be α. Let **lastrange**$_i$ be already defined. If $(j, h_i(j))$ belongs to **lastrange**$_i$ (i.e. **value**(**lastrange**$_i$)=$h_i(j)$, and either **lastrange**$_i$ contains only one pair or **step**(**lastrange**$_i$) = $j - ($end(**lastrange**$_i$)), we modify the parameters **step**(**lastrange**$_i$) and **end**(**lastrange**$_i$) appropriately. Otherwise we create a new range α for the pair $(j, h_i(j))$, and we define **failrange**(α) as **currange**$_i$, if $h_i(j) = l_i$, or **failrange**(**currange**$_i$), if $h_i(j) = h_i(l_i)$. We also set **lastrange**$_i$:= α. The functions FIND_PATTERN$_i$ make essentially the same computations as the functions COMP_NEXT$_i$.

function FIND_PATTERN$_i$() $\{i = 0, 1, \ldots, \tau - 2\}$
if $p_{l_i} = p_j$
    if $l_i = m_{i+1}$ **then begin**
        if l_{i+1} is undefined **then begin**

$l_{i+1} := l_i + 1;$
if lastrange$_{i+1}$ is defined {**list**$^*[h_{i+1}]$ is not empty}
    then currange$_{i+1} :=$ the first range of {**list**$^*[h_{i+1}]$;
else currange$_1$ is undefined;
end

#if $i > 0$

if the last range α of **list**$^*[h_i]$ has the pair
$(m_{i+1} + 1, h_i(m_{i+1} + 1))$
    then $l_i := h_i(m_{i+1} + 1);$ **currange**$_i :=$ **failrange**$(\alpha);$
else undefine $l_i;$

#else {$i = 0$}

$l_0 := h_0(m_1 + 1);$

#endif

    return;
else {$p_{l_i} \neq p_j$} **repeat**

#if $i > 0$

if currange$_i$ has the pair $(l_i, h_i(l_i))$
    then $l_i := h_i(l_i);$ **currange**$_i :=$ **failrange**(**currange**$_i);$
else undefine $l_i;$ **return**;

#else {$i = 0$}

$l_0 := h_0(l_0);$

#endif

    while $l_i > 0$ **and** $p_{l_i} \neq p_j;$
$l_i := l_i + 1;$

#if $i > 0$

if $l_i >$ **end**(**currange**$_i$)
    then currange$_i :=$ the next range in **list**$^*[h_i];$

#endif

In what follows we show that our algorithm works in real time and requires only $O(m^{1/\tau})$ additional space. First consider the function COMP_MPREF$_0$. Note that during the processing of the text symbol t_j the function is called for one or two times. In either case we consider the last call of COMP_MPREF$_0$. Let $l_0(j)$ and $j_0(j)$ be the values of l_0 and j_0 respectively before this call. It follows that:

Proposition 2. *For each* $j = 1, \ldots, n$ *we have* $m_1 - l_0(j) \geq 2(j - j_0(j)).$

Let $e_0(j)$ be the leftmost end position of an occurrence of P_i in T, s.t., $e_0(j) \geq j_0(j)$. Note that $e_0(j) \geq j_0(j) + (m_1 - l_0(j))$, so Proposition 2 implies $e_0(j) - j \geq j - j_0(j)$. From this observation we conclude that COMP_MPREF$_0$ finds each occurrence of P_i in T while our algorithm inspects the end symbol of this occurrence. Another conclusion from Proposition 2 is that the delay $j - j_0(j)$ is never larger than $m_1/2$ during the execution of our algorithm. Thus, for proper functioning of COMP_MPREF$_0$, we always have to save $O(m^{1/\tau})$ last input symbols of T.

Consider now the function COMP_MPREF$_i$ for $i > 0$. We say that a call of this function during the execution of our algorithm is *successful* if during this call the conditional operator "**if** $p_{l_i} = t_{j_i}$" is executed. Note that the call is successful if and only if after the execution of the previous operator the variable l_i is defined. Similarly as in the case of the function COMP_MPREF$_0$, we consider the last call of COMP_MPREF$_i$ when the text symbol t_j is inspected. If this call is successful, we denote by $l_i(j)$ and $j_i(j)$ the values of l_i and j_i respectively at the moment of execution of the conditional operator "**if** $p_{l_i} = t_{j_i}$" during this call. Let **rnumber**$_i(l_i)$ be the number of the last range α in **list**$^*[h_i]$, s.t., **begin**$(\alpha) \leq l_i$, if such a range exists; and **rnumber**$_i(l_i) = 0$ otherwise. We assume that **rnumber**$_i(m_{i+1}) > 0$; otherwise the analysis of COMP_MPREF$_i$ becomes trivial. Applying inductive reasoning on i, we prove the following statement.

Proposition 3. *If $l_i(j)$ is defined, then*

$$\mathbf{rnumber}_i(m_{i+1}) - \mathbf{rnumber}_i(l_i(j)) \geq 2(j - j_i(j)). \tag{4}$$

COMP_MPREF$_i$ finds each occurrence of P_{i+1} in T during the processing the end symbol of this occurrence.

Proof: Assume that this proposition is true for all values $\leq i - 1$. To prove it for i, we use induction on j. Note that the first j for which $l_i(j)$ is defined is preceded by an occurrence $T[j - m_i..j - 1]$ of P_i in T. This occurrence has to be found by COMP_MPREF$_{i-1}$ during successful inspection of the symbol t_{j-1}. So the position j has to be retrieved from **jqueue**$_i$ by COMP_MPREF$_i$ during the processing of the symbol t_j. Thus for such j we have $j_i(j) = j$, hence inequality (4) is trivially satisfied. There are three possible outcomes of the successful execution of COMP_MPREF$_i$.

1. Let $p_{l_i} = t_{j_i}$. In this case j_i is increased by one, and l_i is increased by at most one, so **rnumber**$_i(l_i)$ is also increased by at most one.
2. Let $p_{l_i} \neq t_{j_i}$ and $h_i(l_i) > 0$. Then **rnumber**$_i(l_i)$ is decreased by at least one, because in this case $h_i(l_i) = \mathbf{value}(\alpha) < \mathbf{begin}(\alpha)$, where α is the range with the number **rnumber**$_i(l_i)$ in **list**$^*[h_i]$.
3. Let $p_{l_i} \neq t_{j_i}$ and $h_i(l_i) = 0$. Then l_i becomes undefined. So at the next successful call of COMP_MPREF$_i$ we define $l_i = m_i + 1$ and take the value j_i from **jqueue**$_i$. Thus in this case **rnumber**$_i(l_i)$ becomes equal to 0, and j_i is increased by at least one.

Considering all possible outcomes of execution of two consecutive calls of the function COMP_MPREF$_i$, we can prove that the validity of inequality (4), for $j - 1$, implies its validity for j. The second statement of the proposition follows from inequality (4).

Now applying Proposition 3 to COMP_MPREF$_{\tau-1}$, we conclude that our algorithm finds all occurrences of $P_\tau = P$ in T. Moreover, from inequality (4) follows that during successfully completed calls of COMP_MPREF$_i$ the delay $j - j_i(j)$ is always no more than $(\mathbf{rnumber}_i(m_{i+1}) + 1)/2$. By Lemma 3 we have that **list**$^*[h_i]$ contains $O(m^{1/\tau})$ ranges. So **rnumber**$_i(m_{i+1}) = O(m^{1/\tau})$. Hence, for proper functioning of COMP_MPREF$_i$, we also have to save only $O(m^{1/\tau})$ last input symbols of T. Moreover note that **jqueue**$_i$ always contains positions between j_i and j. So **jqueue**$_i$ contains $O(m^{1/\tau})$ positions. We conclude that our string matching uses $O(m^{1/\tau} \cdot \tau)$ additional space. Finally, recall that the test of each new text character requires executions of COMP_MPREF$_i$, for every $i = 0, 1, \ldots, \tau - 1$. Each execution of these functions is performed in constant time, thus the total time required for the test of each new text symbol is bounded by $O(\tau)$. In particular, if τ is a constant integer, and $\varepsilon = 1/\tau$, the following theorem holds.

Theorem 1. *For any constant $\varepsilon > 0$ all occurrences of a pattern of length m can be found in any text in real time and $O(m^\varepsilon)$ additional space.*

4 Further Discussion

In previous section we show that during the execution of our algorithm for any i the delay $j - j_i$ is less than $Km^{1/\tau}$, where K is some constant not greater than C. Denote by m'_i, for $i = 1, \ldots, \tau - 1$ the number $\lceil m^{i/\tau} + Km^{1/\tau} \rceil$, and by P'_i the word $P[1..m'_i]$. Then we note that there exists an alternative algorithm in which for $i = 0, 1, \ldots, \tau - 2$ instead of $\mathbf{maxpref}_i(j)$ COMP_MPREF$_i$ computes $\mathbf{maxpref}'_i(j)$ which is the largest k, s.t., $p_1 \ldots p_k = t_{j-k+1} \ldots t_j$ and $m^{i/\tau} < k \leq m'_{i+1}$ ($k \leq m'_1$ for $i = 0$), if such k exists, and 0 otherwise. For this purpose we use a function h'_i which is the function $\mathbf{next}_{m^{i/\tau}}$ for the pattern $P'_{i+1}\$$. As in the crude version of our algorithm, for $i > 0$ the functions h'_i are represented as lists of ranges. Using Lemma 3, we can prove that all the lists require $O(m^{1/\tau})$ additional space. For any $i > 0$, the function COMP_MPREF$_i$ starts the computation of $\mathbf{maxpref}'_i$, only after COMP_MPREF$_{i-1}$ finds an occurrence of P'_i in T during the computation of $\mathbf{maxpref}'_{i-1}$. Assume we have an occurrence of P'_i in T which is found by COMP_MPREF$_{i-1}$ at the time when COMP_MPREF$_i$ already completed the computation of $\mathbf{maxpref}'_i$ (i. e., when l_i is defined). Since $j - j_i < Km^{1/\tau}$, the distance between the end position of this occurrence and the current position j_i is smaller than $Km^{1/\tau}$. So the prefix P_i of this occurrence is to the left of j_i. Therefore, COMP_MPREF$_i$ will find this occurrence with a help of the function h'_i. And we don't need to provide this occurrence to COMP_MPREF$_i$. And since COMP_MPREF$_{i-1}$ is obliged to provide an occurrence of P'_i in T to COMP_MPREF$_i$ only if l_i is undefined, we can avoid the use of the system of queues $\mathbf{jqueue}_i$.

References

1. R.S. Boyer, J.S. More, A Fast String Searching Algorithm, *Communication of the ACM*, 20(10), 1977, pp 762–772.
2. L. Colussi. Correctness and efficiency of string matching algorithms. *Information and Control*, 95, 1991, pp 225–251.
3. L. Colussi, Z. Galil, and R. Giancarlo. On the exact complexity of string matching. *Proc. 31st IEEE Symposium on Foundations of Computer Science*, (FOCS'90). pp. 135–143, 1990.
4. M. Crochemore and D. Perrin, Two-way string-matching. *J. Assoc. Comput. Mach.*, 38(3), p. 651–675, 1991.
5. Z. Galil, String Matching in Real Time, *Journal of the ACM*, 28(1), 1981, pp 134–149.
6. Z. Galil and R. Giancarlo. The exact complexity of string matching: upper bounds. *SIAM J. on Computing*, 21(3):407-437, 1992.
7. Z. Galil and J. Seiferas, Time-space-optimal string matching. *J. Comput. System Sci.*, 26, p. 280–294, 1983.
8. L. Gąsieniec, W. Plandowski and W. Rytter, The zooming method: a recursive approach to time-space efficient string-matching, *Theoret. Comput. Sci.*, 147, 1995, pp 19–30.

9. L. Gąsieniec, W. Plandowski and W. Rytter, Constant-space string matching with smaller number of comparisons: sequential sampling, Proc. *6th Annual Symposium on Combinatorial Pattern Matching*, (CPM'95), pp 78-89.
10. L. Gąsieniec and I. Potapov, Time/Space Efficient Compressed Pattern Matching, *Fundamenta Informaticae*, 56(1-2), 2003, pp. 137-154.
11. R.N. Horspool, Practical Fast Searching in Strings, *Software, Practice and Experience*, 10(6), 1980, pp 501–506.
12. D.E. Knuth, J.H. Morris and V.R. Pratt, Fast pattern matching in strings. *SIAM J. Comput.*, 6, p. 322–350, 1977.

On the k-Closest Substring
and k-Consensus Pattern Problems

Yishan Jiao[1], Jingyi Xu[1], and Ming Li[2]

[1] Bioinformatics lab, Institute of Computing Technology,
Chinese Academy of Sciences, 6#, South Road, Kexueyuan, Zhongguancun, Beijing,
P.R.China,
{jys,xjy}@ict.ac.cn
http://www.bioinfo.org.cn/Englishourlab.htm
[2] University of Waterloo
mli@uwaterloo.ca

Abstract. Given a set $S = \{s_1, s_2, \ldots, s_n\}$ of strings each of length m, and an integer L, we study the following two problems.

k-CLOSEST SUBSTRING problem: find k center strings $c_1, c_2, \ldots, c_k$ of length L minimizing d such that for each $s_j \in S$, there is a length-L substring t_j (closest substring) of s_j with $\min_{1 \leq i \leq k} d(c_i, t_j) \leq d$. We give a PTAS for this problem, for $k = O(1)$.

k-CONSENSUS PATTERN problem: find k median strings $c_1, c_2, \ldots, c_k$ of length L and a substring t_j (consensus pattern) of length L from each s_j minimizing the total cost $w = \sum_{j=1}^{n} \min_{1 \leq i \leq k} d(c_i, t_j)$. We give a PTAS for this problem, for $k = O(1)$.

Our results improve recent results of [10] and [16] both of which depended on the random linear transformation technique in [16]. As for general k case, we give an alternative and direct proof of the NP-hardness of $(2-\epsilon)$-approximation of the HAMMING RADIUS k-CLUSTERING problem, a special case of the k-CLOSEST SUBSTRING problem restricted to $L = m$.

Keywords: k-center problems, closest string and substrings, consensus pattern, polynomial time approximation scheme.

1 Introduction

While the original departure point of this study has been separating repeats in our DNA sequence assembly project, we have quickly realized that the problems we have abstracted relate to many widely studied in different areas from geometric clustering [4], [8], [3], [16] to DNA multiple motif finding [12], [7], [10].

In sequence assembly, the greatest challenge is to deal with repeats, when the shortest common superstring GREEDY algorithm [1] is applied. Given a collection of approximate repeats, if it is possible to separate them into original groups, the quality of sequence assembly algorithm will be improved at least for a class of repeats that are sufficiently different.

S.C. Sahinalp et al. (Eds.): CPM 2004, LNCS 3109, pp. 130–144, 2004.

Many classic computational problems such as clustering and common string find applications in a great variety of contexts related to molecular biology: finding conserved regions in unaligned sequences, genetic drug target identification, and classfying protein sequences. See [11], [12], [7] for comprehensive overviews of such applications.

Throughout the article, we use a fixed finite alphabet Σ. Let s and t be finite strings over Σ. Let $d(s,t)$ denote the Hamming distance between s and t, that is, the number of positions where s and t differ. $|s|$ is the length of s. $s[i]$ is the i-th character of s. Thus, $s = s[1]s[2]\ldots s[|s|]$.

In this article, we consider the following two problems:

k-CLOSEST SUBSTRING problem: Given a set $S = \{s_1, s_2, \ldots, s_n\}$ of strings each of length m, and an integer L, find k center strings $c_1, c_2, \ldots, c_k$ of length L minimizing d such that for every string $s_j \in S$, there is a length-L substring t_j (closest substring) of s_j with $\min_{1 \leq i \leq k} d(c_i, t_j) \leq d$. We call the solution $(\{c_1, c_2, \ldots, c_k\}, d)$ a k-clustering of S and call the number d the maximum cluster radius of the k-clustering.

k-CONSENSUS PATTERN problem: Given a set $S = \{s_1, s_2, \ldots, s_n\}$ of strings each of length m, and an integer L, find k median strings $c_1, c_2, \ldots, c_k$ of length L and a substring t_j (consensus pattern) of length L from each s_j minimizing the total cost $w = \sum_{1 \leq j \leq n} \min_{1 \leq i \leq k} d(c_i, t_j)$.

Some special-case versions of the above two problems have been studied by many authors including [6], [7], [10], [11], [12], [14], [13], [16]. Most related work focused on either the $k = 1$ case or the $L = m$ case. However, since protein biological functions are generally more related to local regions of protein sequence than to the whole sequence, both the k-CLOSEST SUBSTRING problem and the k-CONSENSUS PATTERN problem may find more applications in computational biology than their special-case versions.

In this article, we extend the random sampling strategy in [14] and [13] to give a deterministic PTAS for the $O(1)$-CLOSEST SUBSTRING problem. We also give a deterministic PTAS for the $O(1)$-CONSENSUS PATTERN problem. Using a novel construction, we give a direct and neater proof of the NP-hardness of $(2-\epsilon)$-approximation of the HAMMING RADIUS k-CLUSTERING problem other than the one in [7] which relied on embedding an improved construction in [3] into the Hamming metric .

2   Related Work

2.1   The Closest Substring Problem

The following problem was studied in [11], [6], [12], [13]:

CLOSEST STRING: Given a set $S = \{s_1, s_2, \ldots, s_n\}$ of strings each of length m, find a center string s of length m minimizing d such that for every string $s_i \in S$, $d(s, s_i) \leq d$.

Two groups of authors, [11] and [6], studied CLOSEST STRING problem and gave a ratio $4/3$ approximation algorithm independently. Finally, Li, Ma and Wang [12] gave a PTAS for the CLOSEST STRING problem.

Moreover, the authors in [14] and [13] generalized the above result and gave a PTAS for the following CLOSEST SUBSTRING problem:

CLOSEST SUBSTRING: Given a set $S = \{s_1, s_2, \ldots, s_n\}$ of strings each of length m, and an integer L, find a center string s of length L minimizing d such that for each $s_i \in S$ there is a length-L substring t_i (closest substrings) of s_i with $d(s, t_i) \leq d$.

The k-CLOSEST SUBSTRING problem degenerates into the CLOSEST SUBSTRING problem when $k = 1$.

2.2   The Hamming Radius k-Clustering Problem

The following is the problem studied by L. Gasieniec *et al* [7], [10]:

HAMMING RADIUS k-CLUSTERING problem: Given a set $S = \{s_1, s_2, \ldots, s_n\}$ of strings each of length m, find k center strings $c_1, c_2, \ldots, c_k$ of length m minimizing d such that for every string $s_j \in S$, $\min_{1 \leq i \leq k} d(c_i, s_j) \leq d$.

The HAMMING RADIUS k-CLUSTERING problem is abbreviated to HRC. The k-CLOSEST SUBSTRING problem degenerates into the HAMMING RADIUS k-CLUSTERING problem when $L = m$. The authors in [7] gave a PTAS for the $k = O(1)$ case when the maximum cluster radius d is small ($d = O(\log(n+m))$). Recently, J. Jansson [10] proposed a RPTAS for the $k = O(1)$ case. According to Jansson [10]: "We combine the randomized PTAS of Ostrovsky and Rabani [16] for the HAMMING p-MEDIAN CLUSTERING problem with the PTAS of Li, Ma, and Wang [13] for HRC restricted to $p = 1$ to obtain a randomized PTAS for HRC restricted to $p = O(1)$ that has a high success probability." In addition, Jansson wondered whether a deterministic PTAS can be constructed for the HAMMING RADIUS $O(1)$-CLUSTERING problem and whether a PTAS can be constructed for the $O(1)$-CLOSEST SUBSTRING problem.

It seems that the random linear transformation technique in [16] cannot be easily adapted to solve the $O(1)$-CLOSEST SUBSTRING problem. The Claim 5 in [16], which plays a key role, requires that different solutions of the $O(1)$-CLOSEST SUBSTRING problem should have the same n closest substrings (those which is nearest among all length-L substrings of some string to the center string which it is assigned to). It is clear that for the $O(1)$-CLOSEST SUBSTRING problem such a requirement is not guaranteed.

However, Lemma 1 in [16] is not strong enough to solve all possible cases. Therefore, since the requirement mentioned above is not guaranteed, some possible bad cases can not be excluded by using the triangle inequality as in [16].

Contrary to their method, we adopt the random sampling strategy in [14] and [13]. The key idea is to design a quasi-distance measure h such that for any cluster center c in the optimal solution and any substring t from some string in S, h approximates the Hamming distance between t and c very well. Using this measure h, we give a PTAS for the $O(1)$-CLOSEST SUBSTRING problem.

The random sampling strategy has found many successful applications in various contexts such as nearest neighbor search [9] and finding local similarities between DNA sequences [2].

As for the general k case, L. Gasieniec *et al* [7] showed that it is impossible to approximate HRC within any constant factor less than 2 unless P=NP. They improved a construction in [3] for the planar counterpart of HRC under the L_1 metric and then embedded the construction into the Hamming metric to prove the inapproximability result for HRC.

The geometric counterpart of HRC has a relatively longer research history than HRC itself.

In the geometric counterpart of HRC, the set of strings is replaced with the set of points in m-dimensional space. We call the problem *Geometric k-center problem*. When m=1, the problem is trival. For $m \geq 2$ case, it was shown to be NP-complete [4]. In 1988, the $(2-\epsilon)$-inapproximable result for the planar version of the *Geometric k-center problem* under the L_1 metric was shown [3]. Since a ratio 2 approximation algorithm in [8] assumed nothing beyond the triangle inequality, these bounds are tight. [3] adopted a key idea in [5] of embedding an instance of the vertex cover problem for planar graphs of degree at most 3 in the plane so that each edge e becomes a path p_e with some odd number of edges, at least 3. The midpoints of these edges then form an instance of the plannar k-center problem.

We adopt the similar idea to the one in [5] and give a novel construction purely in Hamming metric, which is used to prove the inapproximability result for HAMMING RADIUS k-CLUSTERING problem.

2.3   The Hamming p-Median Clustering Problem

The following problem was studied by Ostrovsky and Rabani [16]:

HAMMING p-MEDIAN CLUSTERING problem: Given a set $S = \{s_1, s_2, \ldots, s_n\}$ of strings each of length m, find k median strings $c_1, c_2, \ldots, c_k$ of length m minimizing the total cost $w = \sum_{j=1}^{n} \min_{1 \leq i \leq k} d(c_i, s_j)$.

The k-CONSENSUS PATTERN PROBLEM degenerates into the HAMMING p-MEDIAN CLUSTERING problem when $L = m$. Ostrovsky and Rabani [16] gave a RPTAS for the problem.

2.4   The Consensus Pattern Problem

The following is the problem studied in [12]:

CONSENSUS PATTERN: Given a set $S = \{s_1, s_2, \ldots, s_n\}$ of strings each of length m, and an integer L, find a median string c of length L and a substring t_j (consensus pattern) of length L from each s_j minimizing the total cost $w = \sum_{1 \leq j \leq n} d(c, t_j)$.

The k-CONSENSUS PATTERN problem degenerates into the CONSENSUS PATTERN problem when $k = 1$. [12] gave a PTAS for the CONSENSUS PATTERN problem. We extend it to give a PTAS for the $O(1)$-CONSENSUS PATTERN problem.

3   NP-Hardness of Approximating Hamming Radius k-Clustering

Firstly let us outline the underlying ideas of our proof of Theorem 1. Given any instance G of the vertex cover problem, with m' edges, we can construct an instance S of the HAMMING RADIUS k-CLUSTERING problem which has a k-clustering with the maximum cluster radius not exceeding 2 if and only if G has a vertex cover with $k - m'$ vertices. Such a construction is the key to our proof. Due to our construction, any two strings in S are at distance either at least 8 or at most 4 from each other. Thus finding an approximate solution within an approximation factor less than 2 is no easier than finding an exact solution. So if there is a polynomial algorithm for the HAMMING RADIUS k-CLUSTERING problem within an approximation factor less than 2, we can utilize it to solve the exact vertex cover number of any instance G. Thus a contradiction can be deduced.

Theorem 1. *If k is not fixed, the* HAMMING RADIUS k-CLUSTERING *problem can not be approximated within constant factor less than two unless P=NP.*

Proof. We reduce the vertex cover problem to the problem of approximating the HAMMING RADIUS k-CLUSTERING problem within any constant factor less than two.

Given any instance $G = (V, E)$ of the vertex cover problem with V containing n vertices $v_1, v_2, \ldots, v_n$ and E containing m' edges $e_1, e_2, \ldots, e_{m'}$. The reduction constructs an instance S of the HAMMING RADIUS k-CLUSTERING problem such that $k - m'$ vertices in V can cover E if and only if there is a k-clustering of S with the maximum cluster radius not exceeding 2. For each $e_i = v_{i_1} v_{i_2} \in E$, we denote $l(i) = \min(i_1, i_2)$ and $r(i) = \max(i_1, i_2)$. We denote a set $V' = \{u_{i_j} \mid 1 \le i \le m' \wedge 1 \le j \le 5\}$ of $5m'$ vertices.

We encode each $v_i \in V$ by a length-$6n$ binary string s_i of the form $0^{6(i-1)}1^6 0^{6(n-i)}$ and encode each $u_{i_j} \in V'$ by a length-$6n$ binary string t_{i_j} of the form $0^{6(l(i)-1)+j-1}1^{7-j}0^{6(r(i)-l(i)-1)}1^{j+1}0^{6(n-r(i))+5-j}$. See Fig. 1 for an illustration of such an encoding schema.

$$
\begin{array}{ll}
s_i : & \ldots 111111 \ldots 000000 \ldots \\
 & \text{(a)} \\
t_{i_1} : & \ldots 111111 \ldots 110000 \ldots \\
t_{i_2} : & \ldots 011111 \ldots 111000 \ldots \\
t_{i_3} : & \ldots 001111 \ldots 111100 \ldots \\
t_{i_4} : & \ldots 000111 \ldots 111110 \ldots \\
t_{i_5} : & \ldots 000011 \ldots 111111 \ldots \\
 & \text{(b)}
\end{array}
$$

Fig. 1. Illustration of the encoding schema (a) s_i ($1 \le i \le n$) (b) t_{i_j} ($1 \le i \le m', 1 \le j \le 5$)

So we have an instance S of the HAMMING RADIUS k-CLUSTERING problem with $S = \cup_{i=1}^{m'}\{t_{i_1}, t_{i_3}, t_{i_5}\}$.

We denote $c(v_i) = s_i$ for each $v_i \in V$ and $c(u_{i_j}) = t_{i_j}$ for each $u_{i_j} \in V'$. We denote $V_S = \cup_{i=1}^{m'}\{u_{i_1}, u_{i_3}, u_{i_5}\}$.

We define a graph $G' = (V \cup V', E')$ with $E' = \cup_{i=1}^{m'}\{v_{l(i)}u_{i_1}, u_{i_1}u_{i_2}, u_{i_2}u_{i_3}, u_{i_3}u_{i_4}, u_{i_4}u_{i_5}, u_{i_5}v_{r(i)}\}$.

Lemma 1. *Given any two adjacent vertices x, y in G', we have that $d(c(x), c(y)) = 2$.*

Proof. It is easy to check that there are only three possible cases and for each case the above conclusion holds.

  Case 1. There is some $1 \le i \le m'$ such that $x = v_{l(i)}$ and $y = u_{i_1}$.

  Case 2. There is some $1 \le i \le m'$ such that $x = u_{i_5}$ and $y = v_{r(i)}$.

  Case 3. There are some $1 \le i \le m'$ and $1 \le j \le 4$ such that $x = u_{i_j}$ and $y = u_{i_{j+1}}$. □

Lemma 2. *Given any two different vertices x, y in V_S satisfying that there is some $z \in V \cup V'$ which is adjacent to both x and y in G', we have that $d(c(x), c(y)) = 4$.*

Proof. It is easy to check that there are only five possible cases and for each case the above conclusion holds.

  Case 1. There are some $1 \le i, j \le m'$ and $l(i) = l(j)$ such that $x = u_{i_1}$ and $y = u_{j_1}$.

  Case 2. There are some $1 \le i, j \le m'$ and $l(i) = r(j)$ such that $x = u_{i_1}$ and $y = u_{j_5}$.

  Case 3. There are some $1 \le i, j \le m'$ and $r(i) = l(j)$ such that $x = u_{i_5}$ and $y = u_{j_1}$.

  Case 4. There are some $1 \le i, j \le m'$ and $r(i) = r(j)$ such that $x = u_{i_5}$ and $y = u_{j_5}$.

  Case 5. There are some $1 \le i \le m'$ and $j \in \{1, 3\}$ such that $x = u_{i_j}$ and $y = u_{i_{j+2}}$. □

Lemma 3. *Given any two vertices x, y in V_S, if no vertex in $V \cup V'$ is adjacent to both x and y in G', we have that $d(c(x), c(y)) \ge 8$.*

Proof. Let $x = u_{i_j}$, $y = u_{i'_{j'}}$, $1 \le i, i' \le m'$ and $j, j' \in \{1, 3, 5\}$. We consider three cases:

  Case 1. $i = i'$. Clearly, in this case, we have that $d(c(x), c(y)) = 8$.

  Case 2. $l(i), r(i), l(i')$ and $r(i')$ are all different from each other. Clearly, in this case, we have that $d(c(x), c(y)) = 16$.

  Case 3. Otherwise, exactly one of $l(i) = l(i')$, $r(i) = r(i')$, $l(i) = r(i')$ and $r(i) = l(i')$ holds. Without loss of generality, we assume that $l(i) = l(i')$. We denote $a = l(i) = l(i')$, $b = r(i)$, $c = r(i')$, $O = \{6a - 6 < l \le 6a \mid x(l) = 1\}$, $P = \{6a - 6 < l \le 6a \mid y(l) = 1\}$, $Q = \{6b - 6 < l \le 6b \mid x(l) = 1\}$ and $R = \{6c - 6 < l \le 6c \mid y(l) = 1\}$. Thus, we have the following inequality:

$$d(c(x), c(y)) \geq |Q| + |R| + ||O| - |P||$$
$$= |Q| + |R| + ||Q| - |R||$$
$$= 2\max(|Q|, |R|)$$
$$= 2\max(j, j') + 2 . \tag{1}$$

Clearly, in this case, $\max(j, j') \geq 3$. Therefore, by Formula (1), we have that $d(c(x), c(y)) \geq 8$. □

Lemma 4. *Given $k \leq 2m'$, $k - m'$ vertices in V can cover E if and only if there is a k-clustering of S with the maximum cluster radius equal to 2.*

Proof. If part: Considering the set $T = \{t_{i_3} \mid 1 \leq i \leq m'\}$ of strings. Any cluster can not contain two or more strings in T since otherwise the radius of this cluster would be at least 4. Therefore, there must be exactly $k - m'$ clusters, each of which doesn't contain any string in T. For each edge $e_i \in E$, either t_{i_1} or t_{i_5} would be in someone among those $k - m'$ clusters. On the other hand, for anyone (say, the l-th, $1 \leq l \leq k - m'$) among those $k - m'$ clusters, there must be some index $1 \leq j \leq n$ such that for each t_{i_1} (t_{i_5}) in the cluster, $l(i) = j$ ($r(i) = j$). We denote $f(l) = j$. Clearly, the $k - m'$ vertices $v_{f(1)}, v_{f(2)}, \ldots, v_{f(k-m')}$ can cover E.

Only if part: Since $k \leq 2m' < 3m' = |S|$ and for any two different strings $s, t \in S$ we have $d(s, t) \geq 4$, hence for any k-clustering of S, the maximum cluster radius is at least 2. Assuming that $k - m'$ vertices $v_{i_1}, v_{i_2}, \ldots, v_{i_{k-m'}}$ can cover E, we can introduce $k - m'$ strings $s_{i_1}, s_{i_2}, \ldots, s_{i_{k-m'}}$ as the first $k - m'$ center strings. The last m' center strings can be introduced as follows: since for each edge $e_i \in E$, either $s_{l(i)}$ or $s_{r(i)}$ is among the first $k - m'$ center strings, we introduce t_{i_4} as the $(k - m' + i)$-th center string if $s_{l(i)}$ is among the first $k - m'$ center strings and t_{i_2} otherwise. Clearly, this give a k-clustering of S with the maximum cluster radius equal to 2. □

Now we can conclude our proof.

First note that, by Lemmas 1, 2, 3, when $k \leq 2m'$ and there is a k-clustering of S with the maximum cluster radius equal to 2, finding an approximate solution within an approximation factor less than 2 is no easier than finding an exact solution.

Clearly, there is a critical $1 \leq k_c \leq 2m'$ satisfying that there is a k_c-clustering of S with the maximum cluster radius equal to 2 and for any $(k_c - 1)$-clustering of S, the maximum cluster radius is at least 4.

If there is a polynomial algorithm for the HAMMING RADIUS k-CLUSTERING problem within an approximation factor less than 2, we can apply it on each instance S in the decreasing order of k from $2m'$ downto 1. So once we find a k such that the maximum cluster radius of the instance S obtained by the algorithm is at least 4, we are sure that $k_c = k + 1$. Since when $k \geq k_c$, the algorithm can of course find a k-clustering of S with the maximum cluster radius equal to 2, and when $k = k_c - 1$, for any k-clustering of S, the maximum cluster radius is at least 4.

Therefore, from Lemma 4, the vertex cover number of the graph G is $k_c - m'$. That is, we can get the exact vertex cover number of any instance G of the vertex cover problem in polynomial time. However, the vertex cover problem is NP-hard. Contradiction, this completes our reduction.   □

4   A Deterministic PTAS
for the $O(1)$-Closest Substring Problem

In this section, we study the k-CLOSEST SUBSTRING problem when $k = O(1)$. Both the algorithm and the proof are based on the $k = 2$ case. We call it the 2-CLOSEST SUBSTRING problem. We make use of some results for the CLOSEST STRING problem and extend a random sampling strategy in [14] and [13] to give a deterministic PTAS for the 2-CLOSEST SUBSTRING problem. Finally, we give an informal statement explaining how and why it can be easily extended to the general $k = O(1)$ case. Since the HAMMING RADIUS $O(1)$-CLUSTERING problem is just a special-case version of the $O(1)$-CLOSEST SUBSTRING problem, the same algorithm also gives a deterministic PTAS which improves the RPTAS as in [10], for the HAMMING RADIUS $O(1)$-CLUSTERING problem.

4.1   Some Definitions

Let s, t be strings of length m. A multiset $P = \{j_1, j_2, \ldots, j_k\}$ such that $1 \leq j_1 \leq j_2 \leq \cdots \leq j_k \leq m$ is called a *position set*. By $s|_P$ we denote the string $s[j_1]s[j_2] \ldots s[j_k]$. We also write $d^P(s, t)$ to mean $d(s|_P, t|_P)$.

Let c, o be strings of length L, $Q \subseteq \{1, 2, \ldots, L\}$, $P = \{1, 2, \ldots, L\} \setminus Q$, $R \subseteq P$, we denote

- $h(o, c, Q, R) = d^Q(o, c) + \frac{|P|}{|R|} d^R(o, c)$,
- $f(s, c, Q, R) = \min_{\{\text{any length-}L \text{ substring } t \text{ of } s\}} h(t, c, Q, R)$,
- $g(s, c) = \min_{\{\text{any length-}L \text{ substring } t \text{ of } s\}} d(t, c)$.

The above function h is just the quasi-distance measure adopted by us. Also, the above function f and g are two new distance measure introduced by us. We refer to them as distance f or distance g from now on.

Let $S = \{s_1, s_2, \ldots, s_n\}$ be an instance of the 2-CLOSEST SUBSTRING problem, where each s_i is of length m ($m \geq L$). Let (c_A, c_B, A, B) be the solution so that the set S has been partitioned into two clusters A, B and c_A, c_B are the center strings of cluster A, B respectively.

We denote $d(c_A, c_B, A, B)$ the minimal d satisfying that $\forall_{s \in A} g(s, c_A) \leq d \wedge \forall_{s \in B} g(s, c_B) \leq d$.

Let $s_{i_1}, s_{i_2}, \ldots, s_{i_r}$ be r strings (allowing repeats) in S. Let $Q_{i_1, i_2, \ldots, i_r}$ be the set of positions where $s_{i_1}, s_{i_2}, \ldots, s_{i_r}$ agree, $P_{i_1, i_2, \ldots, i_r} = \{1, 2, \ldots, m\} \setminus Q_{i_1, i_2, \ldots, i_r}$.

4.2   Several Useful Lemmas

Lemma 5. *(Chernoff Bound) Let $X_1, X_2, \ldots, X_n$ be n independent random 0-1 variables, where X_i takes 1 with probability p_i, $0 < p_i < 1$. Let $X = \sum_{i=1}^{n} X_i$, and $\mu = E[X]$. Then for any $0 < \epsilon \leq 1$,*

(1) $\mathbf{Pr}(X > \mu + \epsilon n) < \exp\left(-\frac{1}{3}n\epsilon^2\right)$,
(2) $\mathbf{Pr}(X < \mu - \epsilon n) \leq \exp\left(-\frac{1}{2}n\epsilon^2\right)$.

Let $S = \{s_1, s_2, \ldots, s_n\}$ be an instance of the CLOSEST STRING problem, where each s_i is of length m. Let s, d_{opt} are the center string and the radius in the optimal solution. Let $\rho_0 = \max_{1 \leq i,j \leq n} \frac{d(s_i, s_j)}{d_{opt}}$.

Lemma 6. *For any constant r, $2 \leq r < n$, if $\rho_0 > 1 + \frac{1}{2r-1}$, then there are indices $1 \leq i_1, i_2, \ldots, i_r \leq n$ such that for any $1 \leq l \leq n$,*
$$d(s_l|_{Q_{i_1,i_2,\ldots,i_r}}, s_{i_1}|_{Q_{i_1,i_2,\ldots,i_r}}) - d(s_l|_{Q_{i_1,i_2,\ldots,i_r}}, s|_{Q_{i_1,i_2,\ldots,i_r}}) \leq \frac{1}{2r-1}d_{opt}.$$

Lemma 7. $|P_{i_1,i_2,\ldots,i_r}| \leq rd_{opt}$.

Lemma 5 is Lemma 1.2 in [13]. Lemma 6 is Lemma 2.1 in [13]. Lemma 7 is Claim 2.4 in [13]. The following is a lemma which plays a key role in the proof of our main result.

Lemma 8. *For any constant r, $2 \leq r < n$, there are r strings $s_{i_1}, s_{i_2}, \ldots, s_{i_r}$ (allowing repeats) in S and a center string c' such that for any $1 \leq l \leq n$, $d(c', s_l) \leq (1 + \frac{1}{2r-1})d_{opt}$ where $c'|_{Q_{i_1,i_2,\ldots,i_r}} = s_{i_1}|_{Q_{i_1,i_2,\ldots,i_r}}$, $c'|_{P_{i_1,i_2,\ldots,i_r}} = s|_{P_{i_1,i_2,\ldots,i_r}}$.*

Proof. If $\rho_0 \leq 1 + \frac{1}{2r-1}$, we can choose s_1 r times as the r strings $s_{i_1}, s_{i_2}, \ldots, s_{i_r}$. Otherwise, by Lemma 6, such r strings $s_{i_1}, s_{i_2}, \ldots, s_{i_r}$ do exist. ☐

4.3   A PTAS for the 2-Closest Substring Problem

How to choose n closest substrings and partition n strings into two sets accordingly are the only two obstacles on the way to the solution. We use the random sampling strategy in [14] and [13] to handle both the two obstacles. This is a further application of the random sampling strategy in [14] and [13].

Now let us outline the underlying ideas. Let $S = \{s_1, s_2, \ldots, s_n\}$ be an instance of the 2-CLOSEST SUBSTRING problem, where each s_i is of length m. Suppose that in the optimal solution, $t_1, t_2, \ldots, t_n$ are the n closest substrings (those which is nearest among all length-L substrings of some string to the center string which it is assigned to) which form two instances of the CLOSEST STRING problem based on the optimal partition. Thus, if we can obtain the same partition together with choice as one in the optimal solution, we can solve the CLOSEST STRING problem on such two instances respectively. Unfortunately, we don't know the exact partition together with choice in the optimal solution. However, by virtue of the quasi-distance measure h which approximates the Hamming distance d very well, we can get a "not too bad" partition together with choice as compared with the one in the optimal solution. That is, even if some string is partitioned into "wrong" cluster, its distance g from "wrong" center will not exceed $(1 + \epsilon)$ times its distance g from the "right" center in the optimal solution with ϵ as small as we desire. The detailed algorithm (Algorithm 2-Closest Substring) is given in Fig. 2. We prove Theorem 2 in the rest of the section.

Algorithm 2-Closest Substring

Input n strings $s_1, s_2, \ldots, s_n \in \Sigma^m$, integer L.
Output   two center strings c_1'', c_2''.

1. **for** each r length-L substrings $t_{i_1}, t_{i_2}, \ldots, t_{i_r}$ (allowing repeats, but if t_{i_p} and t_{i_q} are both chosen from the same s_i then $t_{i_p} = t_{i_q}$) of the n input strings **do**
   (1) $Q_1 = \{1 \leq k \leq L \mid t_{i_1}[k] = t_{i_2}[k] = \ldots = t_{i_r}[k]\}$, $P_1 = \{1, 2, \ldots, L\} \setminus Q_1$.
   (2) Let R_1 be a multiset containing $\lceil 4\frac{1}{\epsilon^2} \log(mn) \rceil$ uniformly random positions from P_1.
   (3) **for** each r length-L substrings $t_{j_1}, t_{j_2}, \ldots, t_{j_r}$ (allowing repeats, but if t_{j_p} and t_{j_q} are both chosen from the same s_j then $t_{j_p} = t_{j_q}$) of the n input strings **do**
      (a) $Q_2 = \{1 \leq k \leq L \mid t_{j_1}[k] = t_{j_2}[k] = \ldots = t_{j_r}[k]\}$, $P_2 = \{1, 2, \ldots, L\} \setminus Q_2$.
      (b) Let R_2 be a multiset containing $\lceil 4\frac{1}{\epsilon^2} \log(mn) \rceil$ uniformly random positions from P_2.
      (c) **for** each string y_1 of length $|R_1|$ and each string y_2 of length $|R_2|$ **do**
         (i) Let c_1' be the string such that $c_1'|_{R_1} = y_1$ and $c_1'|_{Q_1} = t_{i_1}|_{Q_1}$, and c_2' be the string such that $c_2'|_{R_2} = y_2$ and $c_2'|_{Q_2} = t_{j_1}|_{Q_2}$.
         (ii) **for** l from 1 to n **do**
            Assign s_l into set C_1' if $f(s_l, c_1', Q_1, R_1) \leq f(s_l, c_2', Q_2, R_2)$ and set C_2' otherwise. We denote $c'(l) = c_1'$, $Q'(l) = Q_1$, $R'(l) = R_1$ if $s_l \in C_1'$ and $c'(l) = c_2'$, $Q'(l) = Q_2$, $R'(l) = R_2$ otherwise. let t_l' be the length-L substring (if several such substrings exist, we choose one of them arbitrarily) of s_l such that $h(t_l', c'(l), Q'(l), R'(l)) = f(s_l, c'(l), Q'(l), R'(l))$.
         (iii) Using the method in [13], solve the optimization problem defined by Formula (9) approximately to get a solution (x_1, x_2) within error $\max(\epsilon|P_1|, \epsilon|P_2|)$.
         (iv) Let c_1'' be the string such that $c_1''|_{P_1} = x_1$ and $c_1''|_{Q_1} = t_{i_1}|_{Q_1}$, and c_2'' be the string such that $c_2''|_{P_2} = x_2$ and $c_2''|_{Q_2} = t_{j_1}|_{Q_2}$.
         (v) Let $d = \max_{1 \leq l \leq n} \min(g(s_l, c_1''), g(s_l, c_2''))$.
2. Output the (c_1'', c_2'') with minimum d in step 1(3)(c).

Fig. 2. The PTAS for the 2-CLOSEST SUBSTRING problem

Theorem 2. *Algorithm 2-Closest Substring is a PTAS for the* 2-CLOSEST SUB-STRING *problem.*

Proof. Let ϵ be any small positive number and $r \geq 2$ be any fixed integer. Let $S = \{s_1, s_2, \ldots, s_n\}$ be an instance of the 2-CLOSEST SUBSTRING problem, where each s_i is of length m. Let $T = \{u_j \mid \exists_{s \in S} u_j$ is a length-L substring of $s\}$.

Suppose that in the optimal solution, $t_1, t_2, \ldots, t_n$ are the n closest substrings, C_1, C_2 are the two clusters which S is partitioned into and c_1, c_2 are the length-L center strings of the two clusters. We denote $T_1 = \{t_j \mid s_j \in C_1\}, T_2 = \{t_j \mid s_j \in C_2\}$ and $d_{opt} = \max_{1 \le j \le n} \min(d(t_j, c_1), d(t_j, c_2))$. Both T_1 and T_2 form instances of the CLOSEST STRING problem, so by trying all possibilities, we can assume that $t_{i_1}, t_{i_2}, \ldots, t_{i_r}$ and $t_{j_1}, t_{j_2}, \ldots, t_{j_r}$ are the r strings that satisfy Lemma 8 by considering T_1 and T_2 respectively. Let Q_1 (Q_2) be the set of positions where $t_{i_1}, t_{i_2}, \ldots, t_{i_r}$ $(t_{j_1}, t_{j_2}, \ldots, t_{j_r})$ agree and $P_1 = \{1, 2, \ldots, L\} \setminus Q_1$ $(P_2 = \{1, 2, \ldots, L\} \setminus Q_2)$. Let c'_1 be the string such that $c'_1|_{P_1} = c_1|_{P_1}$ and $c'_1|_{Q_1} = t_{i_1}|_{Q_1}$, and c'_2 be the string such that $c'_2|_{P_2} = c_2|_{P_2}$ and $c'_2|_{Q_2} = t_{j_1}|_{Q_2}$. By Lemma 8, the solution (c'_1, c'_2, T_1, T_2) is a good approximation of the optimal solution (c_1, c_2, T_1, T_2), that is,

$$d(c'_1, c'_2, T_1, T_2) \le (1 + \frac{1}{2r - 1}) d_{opt} . \tag{2}$$

As for P_1 (P_2) where we know nothing about c'_1 (c'_2), we randomly pick $\lceil 4\frac{1}{\epsilon^2} \log(mn) \rceil$ positions from P_1 (P_2). Suppose that the multiset of these random positions is R_1 (R_2). By trying all possibilities, we can assume that we can get $c'_1|_{Q_1 \cup R_1}$ and $c'_2|_{Q_2 \cup R_2}$ at some point. We then partition each $s_j \in S$ into set C'_1 if $f(s_j, c'_1, Q_1, R_1) \le f(s_j, c'_2, Q_2, R_2)$ and set C'_2 otherwise.

For each $1 \le j \le n$, we denote $c'(j) = c'_1, Q'(j) = Q_1, R'(j) = R_1, P'(j) = P_1$ if $s_j \in C'_1$ and $c'(j) = c'_2, Q'(j) = Q_2, R'(j) = R_2, P'(j) = P_2$ otherwise.

For each $1 \le j \le n$, let t'_j be the length-L substring (if several such substrings exist, we choose one of them arbitrarily) of s_j such that $h(t'_j, c'(j), Q'(j), R'(j)) = f(s_j, c'(j), Q'(j), R'(j))$. We denote $T'_1 = \{t'_j \mid s_j \in C'_1\}$ and $T'_2 = \{t'_j \mid s_j \in C'_2\}$.

We can prove the following Lemma 9.

Lemma 9. *With high probability, for each $u_j \in T$,*

$$|d(u_j, c'_1) - h(u_j, c'_1, Q_1, R_1)| \le \epsilon |P_1| \text{ and } |d(u_j, c'_2) - h(u_j, c'_2, Q_2, R_2)| \le \epsilon |P_2| . \tag{3}$$

Proof. Let $\lambda = \frac{|P_1|}{|R_1|}$. It is easy to see that $d^{R_1}(u_j, c'_1)$ is the sum of $|R_1|$ independent random 0-1 variables $\sum_{i=1}^{|R_1|} X_i$, where $X_i = 1$ indicates a mismatch between c'_1 and u_j at the i-th position in R_1. Let $\mu = E[d^{R_1}(u_j, c'_1)]$. Obviously, $\mu = d^{P_1}(u_j, c'_1)/\lambda$. Therefore, by Lemma 5 (2), we have the following inequality:

$$\begin{aligned}
&Pr(d(u_j, c'_1) - h(u_j, c'_1, Q_1, R_1) \ge \epsilon |P_1|) \\
&= Pr(d^{R_1}(u_j, c'_1) \le (d(u_j, c'_1) - d^{Q_1}(u_j, c'_1))/\lambda - \epsilon |R_1|) \\
&= Pr(d^{R_1}(u_j, c'_1) \le d^{P_1}(u_j, c'_1)/\lambda - \epsilon |R_1|) \\
&= Pr(d^{R_1}(u_j, c'_1) \le \mu - \epsilon |R_1|) \\
&\le \exp(-\frac{1}{2} \epsilon^2 |R_1|) \\
&\le (mn)^{-2} , \tag{4}
\end{aligned}$$

where the last inequality is due to the setting $|R_1| = \lceil 4\frac{1}{\epsilon^2}\log(mn)\rceil$ in step 1(2) of the algorithm. Similarly, using Lemma 5 (1), we have

$$Pr(d(u_j, c_1') - h(u_j, c_1', Q_1, R_1) \leq -\epsilon|P_1|) \leq (mn)^{-\frac{4}{3}} . \tag{5}$$

Combining Formula (4) with (5), we have that for any $u_j \in T$,

$$Pr(|d(u_j, c_1') - h(u_j, c_1', Q_1, R_1)| \geq \epsilon|P_1|) \leq 2(mn)^{-\frac{4}{3}} . \tag{6}$$

Similarly, we have that for any $u_j \in T$,

$$Pr(|d(u_j, c_2') - h(u_j, c_2', Q_2, R_2)| \geq \epsilon|P_2|) \leq 2(mn)^{-\frac{4}{3}} . \tag{7}$$

Summing up over all $u_j \in T$, we have that with probability at least $1 - 4(mn)^{-\frac{1}{3}}$, Formula (3) holds. $\qquad\square$

Now we can conclude our proof.

For each $1 \leq j \leq n$, we denote $c(j) = c_1'$, $Q(j) = Q_1$, $R(j) = R_1$, $P(j) = P_1$ if $t_j \in T_1$ and $c(j) = c_2'$, $Q(j) = Q_2$, $R(j) = R_2$, $P(j) = P_2$ otherwise. Therefore, we have that Formula (8) holds with probability at least $1 - 4(mn)^{-\frac{1}{3}}$.

$$
\begin{aligned}
d(c'(j), t_j') &\leq h(t_j', c'(j), Q'(j), R'(j)) + \epsilon|P'(j)| \\
&\leq h(t_j, c(j), Q(j), R(j)) + \epsilon|P'(j)| \\
&\leq d(c(j), t_j) + \epsilon|P(j)| + \epsilon|P'(j)| \\
&\leq d(c_1', c_2', T_1, T_2) + \epsilon|P(j)| + \epsilon|P'(j)| \\
&\leq (1 + \frac{1}{2r - 1})d_{opt} + \epsilon|P(j)| + \epsilon|P'(j)| \\
&\leq (1 + \frac{1}{2r - 1} + 2\epsilon r)d_{opt}, \quad \forall 1 \leq j \leq n ,
\end{aligned}
\tag{8}
$$

where the first and the third inequality is by Formula (3), the second inequality is due to the definition of the partition, the fifth inequality is by Formula (2) and the last inequality is due to Lemma 7.

By the definition of c_1' and c_2', the following optimization problem has a solution $(x_1, x_2) = (c_1|_{P_1}, c_2|_{P_2})$ such that $d \leq (1 + \frac{1}{2r-1} + 2\epsilon r)d_{opt}$.

$$
\begin{cases}
\min \ d ; \\
d(x_1, t_j'|_{P_1}) \leq d - d^{Q_1}(t_{i_1}, t_j') , \quad \forall t_j' \in T_1' ; |x_1| = |P_1| . \\
d(x_2, t_j'|_{P_2}) \leq d - d^{Q_2}(t_{j_1}, t_j') , \quad \forall t_j' \in T_2' ; |x_2| = |P_2| .
\end{cases}
\tag{9}
$$

We can solve the optimization problem within error $\epsilon r d_{opt}$ by applying the method for the CLOSEST STRING problem [13] to set T_1' and T_2' respectively. Let (x_1, x_2) be the solution of the optimization problem. Then by Formula (9) we have

$$d(x_1, t_j'|_{P_1}) \leq (1 + \frac{1}{2r - 1} + 2\epsilon r)d_{opt} - d^{Q_1}(t_{i_1}, t_j') + \epsilon|P_1| , \quad \forall t_j' \in T_1' .$$

$$d(x_2, t_j'|_{P_2}) \leq (1 + \frac{1}{2r - 1} + 2\epsilon r)d_{opt} - d^{Q_2}(t_{j_1}, t_j') + \epsilon|P_2| , \quad \forall t_j' \in T_2' . \tag{10}$$

Let (c_1'', c_2'') be defined as in step $1(3)(c)(iv)$, then by Formula (10), we have

$$d(c_1'', t_j') = d(x_1, t_j'|_{P_1}) + d^{Q_1}(t_{i_1}, t_j')$$
$$\leq (1 + \frac{1}{2r-1} + 2\epsilon r)d_{opt} + \epsilon|P_1|$$
$$\leq (1 + \frac{1}{2r-1} + 3\epsilon r)d_{opt}, \quad \forall t_j' \in T_1',$$
$$d(c_2'', t_j') = d(x_2, t_j'|_{P_2}) + d^{Q_2}(t_{j_1}, t_j')$$
$$\leq (1 + \frac{1}{2r-1} + 2\epsilon r)d_{opt} + \epsilon|P_2|$$
$$\leq (1 + \frac{1}{2r-1} + 3\epsilon r)d_{opt}, \quad \forall t_j' \in T_2'. \tag{11}$$

we can get a good approximate solution $(c_1'', c_2'', T_1', T_2')$ with $d(c_1'', c_2'', T_1', T_2') \leq (1 + \frac{1}{2r-1} + 3\epsilon r)d_{opt}$. It is easy to see that the algorithm runs in polynomial time for any fixed positive r and ϵ. For any $\delta > 0$, by properly setting r and ϵ such that $\frac{1}{2r-1} + 3\epsilon r \leq \delta$, the algorithm outputs in polynomial time a solution (c_1'', c_2'') with high probability such that $\min(g(s_j, c_1''), g(s_j, c_2'')) \leq (1 + \delta)d_{opt}$ for each $1 \leq j \leq n$. The algorithm can be derandomized by standard methods [15]. □

4.4   The General $k = O(1)$ Case

Theorem 2 can be trivially extended to the $k = O(1)$ case.

Theorem 3. *There is a PTAS for the* $O(1)$-CLOSEST SUBSTRING *problem.*

We only give an informal explanation of how such an extension can be done. Unlike Jansson's RPTAS in [10] for the HAMMING RADIUS $O(1)$-CLUSTERING problem, which made use of the apex of tournament to get a good assignment of strings to clusters, our extension is trivial. (Due to space limitation, the explanation is omitted and given in supplementary material.)

5   The $O(1)$-Consensus Pattern Problem

It is relatively straightforward to extend the techniques in [12] for the CONSENSUS PATTERN problem to give a PTAS for the $O(1)$-CONSENSUS PATTERN problem. Since the algorithm for the $k = O(1)$ case is tedious, we only give an algorithm for the $k = 2$ case. The detailed algorithm (Algorithm 2-Consensus Pattern) is described in Fig. 3.

Theorem 4. *Algorithm 2-Consensus Pattern is a PTAS for the* 2-CONSENSUS PATTERN *problem.*

Again, since the proof is tedious and does not involve new ideas, we only give an informal explanation about how such an extension can be done. (Due to space limitation, the explanation is omitted and given in supplementary material.)

Algorithm 2-Consensus Pattern

Input n strings $s_1, s_2, \ldots, s_n \in \Sigma^m$, integer L.

Output   two median strings c_1'', c_2''.

1. **for** each r length-L substrings $t_{i_1}, t_{i_2}, \ldots, t_{i_r}$ (allowing repeats, but if t_{i_p} and t_{i_q} are both chosen from the same s_i then $t_{i_p} = t_{i_q}$) of the n input strings **do**

    (1) Let c_1' be the column-wise majority string of $t_{i_1}, t_{i_2}, \ldots, t_{i_r}$.

    (2) **for** each r length-L substrings $t_{j_1}, t_{j_2}, \ldots, t_{j_r}$ (allowing repeats, but if t_{j_p} and t_{j_q} are both chosen from the same s_j then $t_{j_p} = t_{j_q}$) of the n input strings **do**

        (a) Let c_2' be the column-wise majority string of $t_{j_1}, t_{j_2}, \ldots, t_{j_r}$.

        (b) **for** j from 1 to n **do**

             Assign s_j into set C_1' if $g(s_j, c_1') \leq g(s_j, c_2')$ and set C_2' otherwise. We denote $c'(j) = c_1'$ if $s_j \in C_1'$ and $c'(j) = c_2'$ otherwise. let t_j' be the length-L substring (if several such substrings exist, we choose one of them arbitrarily) of s_j such that $d(t_j', c'(j)) = g(s_j, c'(j))$.

        (c) We denote $T_1' = \{t_j' \,|\, s_j \in C_1'\}$, $T_2' = \{t_j' \,|\, s_j \in C_2'\}$. Let c_1'' be the column-wise majority string of all strings in T_1' and c_2'' be the column-wise majority string of all strings in T_2'.

        (d) Let $w = \sum_{1 \leq j \leq n} \min(g(s_j, c_1''), g(s_j, c_2''))$.

2. Output the (c_1'', c_2'') with minimum w in step 1(2).

Fig. 3. The PTAS for the 2-CONSENSUS PATTERN problem

We have presented a PTAS for the $O(1)$-CONSENSUS PATTERN problem. Since the MIN-SUM HAMMING MEDIAN CLUSTERING problem (i.e., HAMMING p-MEDIAN CLUSTERING problem) discussed in [16] is just a special-case version of the $O(1)$-CONSENSUS PATTERN problem, we also present a deterministic PTAS for that problem.

Acknowledgement

We would like to thank Dongbo Bu, Hao Lin, Bin Ma, Jingfen Zhang and Zefeng Zhang for various discussions. This work is partially supported by the NSFC major research program "Basic Theory and Core Techniques of Non Canonical Knowledge".

References

1. A. Blum, T. Jiang, M. Li, J. Tromp, and M. Yannakakis: Linear Approximation of Shortest Superstrings. Journal of the ACM. **41(4)** (1994) 630–647
2. J. Buhler: Efficient large-scale sequence comparison by locality-sensitive hashing. Bioinformatics **17** (2001) 419–428
3. T. Feder and D. H. Greene: Optimal algorithms for approximate clustering. In Proceedings of the 20th Annual ACM Symposium on Theory of Computing (1988) 434–444
4. R.J. Fowler, M.S. Paterson and S.L. Tanimoto: Optimal packing and covering in the plane are NP-complete. Inf. Proc. Letters **12(3)** (1981) 133–137

5. M.R. Garey and D. S. Johnson: The rectilinear Steiner tree problem is NP-complete. SIAM J.Appl.Math. **32** (1977) 826–834

6. L. Gasieniec, J. Jansson, and A. Lingas: Efficient approximation algorithms for the Hamming center problem. Proc. 10th ACM-SIAM Symp. on Discrete Algorithms (1999) S905–S906

7. L. Gasieniec, J. Jansson, and A. Lingas: Approximation Algorithms for Hamming Clustering Problems. Proc. 11th Symp. CPM. (2000) 108–118

8. T. F. Gonzalez: Clustering to minimize the maximum intercluster distance. Theoretical Computer Science **38** (1985) 293–306

9. P. Indyk, R. Motwani: Approximate Nearest Neighbors: Towards Removing the Curse of Dimensionality. Proc. 30th Annual ACM Symp. Theory Comput. (1998) 604–613

10. J. Jansson: Consensus Algorithms for Trees and Strings. Doctoral dissertation (2003)

11. K. Lanctot, M. Li, B. Ma, L. Wang and L. Zhang: Distinguishing string selection problems. Proc. 10th ACM-SIAM Symp. Discrete Algorithms (1999) 633–642

12. M. Li, B. Ma, and L. Wang: Finding similar regions in many strings. In Proceedings of the 31st Annual ACM Symposium on Theory of Computing (1999) 473–482

13. M. Li, B. Ma, and L. Wang: On the Closest String and Substring Problems. Journal of ACM. **49(2)** (2002) 157–171

14. B. Ma: A polynomial time approximation scheme for the Closest Substring problem. In Proc. 11th Annual Symposium on Combinatorial Pattern Matching (2000) 99–107

15. R. Motwani and P.Raghavan: Randomized Algorithms. Cambridge Univ. Press (1995)

16. R. Ostrovsky and Y. Rabani: Polynomial-Time Approximation Schemes for Geometric Min-Sum Median Clustering. Journal of ACM. **49(2)** (2002) 139–156

A Trie-Based Approach
for Compacting Automata

Maxime Crochemore[1], Chiara Epifanio[2],
Roberto Grossi[3], and Filippo Mignosi[2]

[1] Institut Gaspard-Monge, Université de Marne-la-Vallée,
France and King's College (London), Great Britain
mac@univ-mlv.fr
[2] Dipartimento di Matematica e Applicazioni, Università di Palermo, Italy
{epifanio,mignosi}@math.unipa.it
[3] Dipartimento di Informatica, Università di Pisa, Italy
grossi@di.unipi.it

Abstract. We describe a new technique for reducing the number of
nodes and symbols in automata based on tries. The technique stems
from some results on anti-dictionaries for data compression and does not
need to retain the input string, differently from other methods based on
compact automata. The net effect is that of obtaining a lighter automaton than the directed acyclic word graph (DAWG) of Blumer et al., as
it uses less nodes, still with arcs labeled by single characters.

Keywords: Automata and formal languages, suffix tree, factor and suffix
automata, index, text compression.

1 Introduction

One of the seminal results in pattern matching is that the size of the minimal
automaton accepting the suffixes of a word (DAWG) is linear [4]. This result
is surprising as the maximal number of subwords that may occur in a word is
quadratic according to the length of the word. Suffix trees are linear too, but
they represent strings by pointers to the text, while DAWGs work without the
need of accessing it.

DAWGs can be built in linear time. This result has stimulated further work.
For example, [8] gives a compact version of the DAWG and a direct algorithm
to construct it. In [13] and [14] it is given an algorithm for online construction
of DAWGs. In [11] and [12] space-efficient implementations of compact DAWGs
are designed. For comparisons and results on this subject, see also [5].

In this paper we present a new compaction technique for shrinking automata
based on antifactorial tries of words. In particular, we show how to apply our
technique to factor automata and DAWGs by compacting their spanning tree
obtained by a breadth-first search. The average number of nodes of the structure
thus obtained can be sublinear in the number of symbols of the text, for highly
compressible sources. This property seems new to us and it is reinforced by

S.C. Sahinalp et al. (Eds.): CPM 2004, LNCS 3109, pp. 145–158, 2004.

the fact the number of nodes for our automata is always smaller than that for DAWGs.

We build up our finding on "self compressing" tries of antifactorial binary sets of words. They were introduced in [7] for compressing binary strings with antidictionaries, with the aim of representing in a compact way antidictionaries to be sent to the decoder of a static compression scheme. We present an improvement scheme for this algorithm that extends its functionalities to any chosen alphabet for the antifactorial sets of words M. We employ it to represent compactly the automaton (or better the *trim*) $\mathcal{A}(M)$ defined in [6] for recognizing the language of all the words avoiding elements of M (we recall that a word w avoids $x \in M$ if x does not appear in w as a factor).

Our scheme is general enough for being applied to any index structure having a failure function. One such example is that of (generalized) suffix tries, which are the uncompacted version of well-known suffix trees. Unfortunately, their number of nodes is $O(n^2)$ and this is why researchers prefer to use the $O(n)$-node suffix tree. We obtain compact suffix tries with our scheme that have a linear number of nodes but are different from suffix trees. Although a compact suffix trie has a bit more nodes than the corresponding suffix tree, all of its arcs are labeled by single symbols rather than factors (substrings). Because of this we can completely drop the text, as searching does not need to access the text contrarily to what is required for the suffix tree. We exploit suffix links for this kind of searching. As a result, we obtain a family of automata that can be seen as an alternative to suffix trees and DAWGs.

This paper is organized as follows. Section 2 contains our generalization of some of the algorithms in [7] so as to make them work with any alphabet. Section 3 presents our data structure, the compact suffix trie and its connection to automata. Section 4 contains our new searching algorithms for detecting a pattern in the compact tries and related automata. Finally, we present some open problems and further work on this subject in Section 5.

2   Compressing with Antidictionaries and Compact Tries

In this section we describe a non-trivial generalization of some of the algorithms in [7] to any alphabet A, in particular with ENCODER and DECODER algorithms described next. We recall that if w is a word over a finite alphabet A, the set of its factors is called $F(w)$. For instance, if $w = $ aeddebc, then $F(w) = \{\varepsilon, \mathtt{a}, \mathtt{b}, \ldots, \mathtt{aeddebc}\}$.

Let us take some words in the complement of $F(w)$, *i.e.*, let us take some words that are not factors of w, call these *forbidden*. This set of such words AD is called an *antidictionary* for the language $F(w)$. Antidictionaries can be finite as well as infinite. For instance, if $w = $ aeddebc the words aa, ddd, and ded are forbidden and the set $\{\mathtt{aa}, \mathtt{ddd}, \mathtt{ded}\}$ is an antidictionary for $F(w)$. If $w_1 = 001001001001$, the infinite set of all words that have two 1's in the i-th and $i + 2$-th positions, for some integer i, is an antidictionary for w_1.

We want to stress that an antidictionary can be any subset of the complement of $F(w)$. Therefore an antidictionary can be defined by any property concerning words.

The compression algorithm in [7] treats the input word in an on-line manner. Let us suppose to have just read the word v, proper prefix of w. If there exists any word $u = u'a$, where $a \in \{0, 1\}$, in the antidictionary AD such that u' is a suffix of v, then surely the letter following v cannot be a, i.e., the next letter is b, with $b \neq a$. In other words, we know in advance the next letter b that turns out to be "redundant" or predictable. As remarked in [7], this argument works only in the case of binary alphabets.

We show how to generalize the above argument to any alphabet A, i.e., any cardinality of A. The main idea is that of eliminating redundant letters with the compression algorithm ENCODER. In what follows the word to be compressed is noted $w = a_1 \cdots a_n$ and its compressed version is denoted by $\gamma(w)$.

ENCODER (antidictionary AD, word $w \in A^*$)

1. $v \leftarrow \varepsilon; \gamma \leftarrow \varepsilon$;
2.   **for** $a \leftarrow$ first to last letter of w
3.   **if** there exists a letter $b \in A, b \neq a$ such that
          for every suffix u' of v, $u'b \notin AD$ **then**
4. $\gamma \leftarrow \gamma.a$;
5. $v \leftarrow v.a$;
6.   **return** $(|v|, \gamma)$;

As an example, let us run this algorithm on the string $w =$ aeddebc, with $AD = \{$aa, ab, ac, ad, aeb, ba, bb, bd, be, da, db, dc, ddd, ea, ec, ede, ee$\}$.

The steps of the execution are described in the next array by the current values of the prefix $v_i = a_1 \cdots a_i$ of w that has been just considered and of the output $\gamma(v_i)$. In the case of a positive answer to the query to the antidictionary AD, the array indicates the value of the corresponding forbidden word u, too. The number of times the answer is positive in a run corresponds to the number of bits erased.

ε	$\gamma(\varepsilon) = \varepsilon$	
$v_1 =$ a	$\gamma(v_1) =$ a	
$v_2 =$ ae	$\gamma(v_2) =$ a	aa, ab, ac, ad $\in AD$
$v_3 =$ aed	$\gamma(v_3) =$ a	ea, ec, ee, aeb $\in AD$
$v_4 =$ aedd	$\gamma(v_4) =$ a	da, db, dc, ede $\in AD$
$v_5 =$ aedde	$\gamma(v_5) =$ a	da, db, dc, ddd $\in AD$
$v_6 =$ aeddeb	$\gamma(v_6) =$ ab	
$v_7 =$ aeddebc	$\gamma(v_7) =$ ab	ba, bb, bd, be $\in AD$

Remark that γ is not injective. For instance, $\gamma($aed$) = \gamma($ae$) =$ a.

In order to have an injective mapping we consider the function $\gamma'(w) = (|w|, \gamma(w))$. In this case we can reconstruct the original word w from both $\gamma'(w)$ and the antidictionary.

Remark 1. Instead of adding the length $|w|$ of the whole word w other choices are possible, such as to add the length $|w'|$ of the last encoded fragment w' of w. In the special case in which the last letter in w is not erased, we have that $|w'| = 0$ and it is not necessary to code this length. We will examine this case while examining the algorithm DECOMPACT.

The decoding algorithm works as follows. The compressed word is $\gamma(w) = b_1 \cdots b_h$ and the length of w is n. The algorithm recovers the word w by predicting the letter following the current prefix v of w already decompressed. If there exists a unique letter a in the alphabet A such that for any suffix u' of v, the concatenation $u'a$ does not belong to the antidictionary, then the output letter is a. Otherwise we have to read the next letter from the input γ.

DECODER (antidictionary AD, word $\gamma \in A^*$, integer n)
1. $v \leftarrow \varepsilon$;
2.   **while** $|v| < n$
3.   **if** there exists a unique letter $a \in A$ such that for any u' suffix of v
         $u'a$ does not belong to AD **then**
4. $v \leftarrow v \cdot a$;
5.   **else**
6. $b \leftarrow$ next letter of γ;
7. $v \leftarrow v \cdot b$;
8.   **return** (v);

The antidictionary AD must be structured in order to answer, for a given word v, whether there exist $|A| - 1$ words $u = u'b$ in AD, with $b \in A$ and $b \neq a$, such that u' is a suffix of v. In case of a positive answer the output should also include the letter a.

Languages avoiding finite sets of words are called *local* and automata recognizing them are ubiquitously present in Computer Science (cf [2]).

Given an antidictionary AD, the algorithm in [6], called L-AUTOMATON, takes as input the trie $\mathcal{T}$ that represents AD, and gives as output an automaton recognizing the language $L(AD)$ of all words avoiding the antidictionary. This automaton has the same states as those in trie $\mathcal{T}$ and the set of labeled edges of this automaton includes properly the one of the trie. The transition function of automaton $\mathcal{A}(AD)$ is called δ. This automaton is complete, i.e., for any letter a and for any state v, the value of $\delta(v, a)$ is defined.

If AD is the set of the minimal forbidden words of a text t, then it is proved in [6] that the trimmed version of automaton $\mathcal{A}(AD)$ is the *factor automaton* of t. If the last letter in t is a letter \$ that does not appear elsewhere in the text, the factor automaton coincides with the DAWG, apart from the set of final states. In fact, while in the factor automaton every state is final, in the DAWG the only final state is the last one in every topological order. Therefore, if we have a technique for shrinking automata of the form $\mathcal{A}(AD)$, for some antidictionary (AD), this technique will automatically hold for DAWG, by appending at the end of the text a symbol \$ that does not appear elsewhere. Actually the trie that we compact is the spanning tree obtained by a breadth-first search of the DAWG, as this spanning tree is the trie obtained by pruning all leaves from the trie of all minimal forbidden words of text t (cf. [6]). The appendix presents an example of this procedure.

Self-compressing tries. Let us analyze now a technique to achieve a better compression ratio than the one obtained with the simple application of algorithm ENCODER.

If AD is an antifactorial antidictionary for a text t then for any word $v \in AD$ the set $AD \setminus \{v\}$ is an antidictionary for v. This last property lets us to compress v using $AD \setminus \{v\}$ or a subset of it. Hence the new technique, that is analogous to the one in [7], will consist in *self-compressing* the antidictionary of the text t that is not necessarily binary.

Let us go in details and consider a generic word v in AD. Words u in $AD \setminus \{v\}$ of length smaller than or equal to v's one will be used to compress v and to obtain its compressed version $\gamma_1(v)$. If a word u in AD, such that $|u| = |v|$, is used for compressing v, then u and v have the same prefix x of length $|u| - 1$ of v. Further $\gamma_1(v) = \gamma_1(u) = \gamma_1(x)$.

In this case we loose information on last letters of u and v. We have two choices. The first one is to compress a word v in AD by using as an antidictionary the subset of AD including all words having length strictly smaller than $|v|$. The second choice consists in adopting some strategy to let only one of the two words being compressed using the other one. We will proceed along the first choice, i.e. while compressing a given word in the antidictionary we would not use words of its same length.

Definition 1. *A r-uple of words $(v_1, \ldots, v_r)$, being r the cardinality of the alphabet A, is called stopping r-uple if $v_1 = u_1 a_1, \ldots v_r = u_r a_r$, where a_i are distinct symbols in A and u_i is a suffix of u_{i+1}, for any i such that $1 \leq i \leq r - 1$.*

Proposition 1. *Let t be a text and AD be an antifactorial antidictionary of t. If there exists a stopping r-uple $(v_1, \ldots, v_r)$, with $v_i = u_i a_i, a_i \in A, 1 \leq i \leq r$, then u_r is a suffix of t and does not appear elsewhere in t. Moreover there exists at most one such r-uple of words.*

Proposition 2. *Let $(v_1, \ldots, v_r)$ be the only possible stopping r-uple in the antidictionary AD of a text t. If the prefix u_{r-1} of length $|v_{r-1}| - 1$ of the penultimate word v_{r-1} in the lexicographic word is a proper suffix of the prefix u_r of length $|v_r| - 1$ of the last word v_r, then last letter in v_r is never erased.*

Let us now examine another algorithm, similar to that presented in [7], apart from some details, that is very important in our case. The idea of this algorithm is to "erase" states with only one outgoing edge whose label is predictable by the previous compression algorithm that uses as antidictionary the leaves of the trie having strictly smaller depth. In other words, consider a state p that has a unique outgoing edge labeled by a such that, for any other symbol $b \neq a$, the longest suffix of pb stored in the trie (i.e., $\delta(p, b)$) is a leaf. Then we erase state p together with its outgoing edge and we replace p by its unique child as described in SELF-COMPRESS. This algorithm works in the case when in the antidictionary there are no stopping r-uples.

```
SELF-COMPRESS (trie T, function δ)
  1.   i ← root of T;
  2.   create root i';
  3.   add (i, i') to empty queue Q;
  4.   while Q ≠ ∅
  5.        extract (p, p') from Q;
  6.        if p has more than one child then
  7.             for each child qⱼ of p
  8.                  create q'ⱼ as child of p';
  9.                  add (qⱼ, q'ⱼ) to Q;
 10.        else if q is a unique child of p and
                 q = δ(p, a), a ∈ A then
 11.             if δ(p, b) is a leaf ∀b ∈ A, b ≠ a then
 12.                  add + sign to q';
 13.                  add (q, p') to Q;
 14.             else create q' as a-child of p';
 15.                  add (q, q') to Q;
 16.   return trie having root i';
```

In this case we do not delete the child, but the parent of a compressed edge, and we add a + sign to the remaining node. Note that with this new algorithm there is no guarantee that function δ is always defined on the nodes belonging to the compacted trie (this property was guaranteed with previous SELF-COMPRESS algorithm). To avoid this drawback, δ can be recovered by the relation $s(v)$ (usually called suffix link), the longest suffix of v that is a node in the trie, where the left-hand δ is the new version defined in terms of the old version, the right-hand δ,

$$\delta(p,a) = \begin{cases} \delta(p,a) & \text{if it is defined} \\ \delta(s^n(p),a) & \text{if not, where } n = \min\{m \mid \delta(s^m(p),a) \text{ is defined}\} \end{cases}$$

Hence, the output of this algorithm represents a compact version of the input trie and, if we include the new function δ, the compact version of the automaton $\mathcal{A}(AD)$.

To deal with the case when the antidictionary contains a stopping r-uple, we have to add some more information to the output of the algorithm. This added information can be to keep a pointer to the leaf corresponding to the longest element of the stopping r-uple together with the length of the compressed ingoing edge to this leaf. Details are left to the reader.

Compact tries present some very useful and important properties. In particular we are able to check whether a pattern p is recognized by the automaton $\mathcal{A}(AD)$, by only "decompacting" a small part of the compact trie, output of previous algorithm, and not by "decompacting" the whole trie described in [7].

The algorithm SEARCH that realizes this task is described in Section 4. Before that, we introduce compact suffix tries in Section 3, because we make use of their properties in the next section.

Recall that if the text ends with a symbol $ the compacted version of automa-
ton $\mathcal{A}(AD)$ can be seen as a compacted version of the DAWG, that is different
from the classical one, called CDAWG.

Proposition 3. *Every node v that is in our compacted DAWG and is not in
the CDAWG is such that it has only one outgoing edge and its suffix link points
to a node of the CDAWG.*

3   Compact Suffix Tries

We want to extend our technique to suffix tries that do not represent antifactorial
sets of words.

We adopt the notation defined in [15], and assume that the reader is familiar
with suffix trees. Given a word $w = a_1a_2 \ldots a_n$ of n symbols drawn from an
alphabet A, called letters, we denote by $w[j \ldots j + k - 1]$ the factor of w of
length k that appears in w at position j. The length of w is denoted by $|w| = n$.
The empty word ϵ is a factor of any word. A factor of the form $w[j, |w|]$ (resp.
$w[1, j]$) is called a suffix (resp. prefix) of w.

The suffix trie $ST(w)$ of a word w is a trie where the set of leaves is the set
of suffixes of w that do not appear previously as factors in w.

We can identify a node of the trie with the label of the unique path from the
root of the trie to the given node. Sometimes we can skip this identification and,
given a node v, we call σ_v the label word that represents the unique path from
the root to v.

In this section we define our compacted version $CST(w)$ of the suffix trie.
Basically we use the same approach of the suffix tree, but we compact a bit less,
i.e., we keep all nodes of the suffix tree $S(w)$ and we keep some more nodes of
the trie. In this way we can avoid to keep the text aside.

In order to describe this new data structure, we have to define first of all its
nodes, namely,

1. all the nodes of suffix tree $S(w)$, and
2. all the nodes v of trie $ST(w)$ such that $s(v)$ is a node of suffix tree $S(w)$.

Recall that in the suffix tree, for any suffix $w[j, |w|]$ of w there is a node v such
that $\sigma_v = w[j, |w|]$, even if $w[j, |w|]$ has appeared as factor in another position.

At this point one may wonder why are we keeping the nodes of type 2. What
is the relation between these nodes and minimal forbidden words? The answer to
the question is given by next proposition, that can be seen as a kind of converse
of Proposition 3.

Proposition 4. *For any node v described in point 2, there exists a letter a such
that $\sigma_v a$ is a minimal forbidden word of w.*

A suffix trie is not a trie of minimal forbidden words, but if we add to the
trie the minimal forbidden words, we can describe a compacting algorithm that
is analogous to that presented in Section 2. The nodes that are neither of type 1,

nor of type 2, and that are not leaves can be erased. More precisely, we do not need to introduce explicitly the minimal forbidden words in the trie.

Let us now define the arcs of $CST(w)$. Since the nodes of $CST(w)$ are all nodes of the trie $ST(w)$, for any node v we assign the same number of outgoing arcs, each of them labeled by a different letter. Hence, for any arc of the form (v, v') with label $l((v, v')) = a$ in $ST(w)$, we have an arc (v, x) with label $l((v, x)) = a$ in $CST(w)$, where the node x is

(i) v' itself, in the case it is still present in $CST(w)$;
(ii) the first node in $CST(w)$ that is a descendant of v' in $ST(w)$, when v' is not a node of $CST(w)$.

In case (ii), we consider as this arc (v, x) represents the whole path from v to x in the trie $Tr(w)$, and we say that this arc has length greater than one. We further add a $+$ sign to node x, in order to record this information.

To complete the definition of $CST(w)$ we keep the suffix link function over these nodes. Notice that, by definition, for any node v of $CST(w)$, the node pointed by a suffix link, $s(v)$, is always a node of the suffix tree $S(w)$ and hence it also belongs to $CST(w)$.

We now show that the number of nodes of $CST(w)$ is still linear in the length $|w|$ of w, independently from the alphabet size. Indeed, since the number of nodes of the suffix tree $S(w)$ is linear, we only have to prove that the number of nodes that we add to it to obtain the nodes of $CST(w)$ is linear in the length $|w|$ of w. This is what we prove in the lemma.

Lemma 1. *The number of nodes v of the trie $ST(w)$ such that v does not belong to the tree $S(w)$ and $s(v)$ is a node of the tree $S(w)$ is smaller than the text size.*

Notice that Lemma 1 states a result that is "alphabet independent", i.e., the result does not depend on the alphabet size. By definition, since the number of nodes in the tree is smaller than $2|w|$, it is straightforward to prove that the number of added nodes is smaller than this quantity multiplied by the cardinality of the alphabet.

The power of our new data structure is that it still has some nice features when compared to the suffix tree. In fact, we possibly reduce the amount of memory required by arc labels. While suffix trees arcs are labeled by substrings, compact suffix tries ones are labeled by single symbols. So they can be accessed "locally" without jumping to positions in the text, which is no more needed. Also, this property is "alphabet independent", as stated in next lemma.

Lemma 2. *Let A be any alphabet and w be a word on it. The number of nodes of the compact suffix trie $CST(w)$ is linear in the length n of w, whichever cardinality A has.*

We remark here that, if we add terminal states to the DAWG of a word w, the structure thus obtained is the minimal automaton accepting the suffixes of w. This implies that if we reduce the size of this automaton, we consequently reduce that of the DAWG.

4   Searching in Compact Tries and Related Automata

In this section we show how to check whether a pattern p occurs as a factor in a text w, by just "decompacting" a small part of the compact suffix trie $CST(w)$ associated with w (not the whole trie). Note that the arcs in $CST(w)$ are labeled with single letters instead of substrings, so we must recover them somehow without accessing the text.

Generally speaking, with the same technique we can check whether a pattern p is recognized by an automaton $\mathcal{A}(AD)$. In this section we center our description around the algorithms for compact suffix tries, but they work on automata with minor modifications.

Let us begin by examining a function that returns the string associated with the path in the trie between two nodes u and v.

```
DECOMPACT (Compacted trie CST, function δ, function s, arc (u, v))
 1.  w ← ε;
 2.  a ← label of (u, v);
 3.  if v has not a + then
 4.      w ← a;
 5.  else
 6.      z ← s(u);
 7.      while z ≠ s(v)
 8.          w ← w concat DECOMPACT(CST, δ, s,(z, δ(z, a)));
 9.          z ← δ(z, a);
10.         if z has only one child x in CST then
11.             a ← label of (z, x);
12.  return w
```

Proposition 5. *Let u be a node in the compacted trie. If v is a child of u, then the path from $s(u)$ to $s(v)$ in the compacted trie has no forks.*

An easy variation of this algorithm can be used to check whether a pattern occurs in a text. More precisely, algorithm SEARCH takes as input a node u and a pattern p. If there is no outgoing path in the decompacted trie from u labeled p, it returns "null." If such a path exists and v is its incoming node, it returns v. The pattern p is represented either as a static variable or as a pointer to pointers. We denote by $succ(p)$ the operation that, received as input a pointer to a letter, points to next letter and $p[0]$ points to the first letter of p. The proof of the correctness of this algorithm follows by induction and by Proposition 5.

```
SEARCH (Compacted trie CST, function δ, function s, node u, pattern p)
 1.  if p[0] = end of string then
 2.      return(u)
 3.  else
 4.      if ∄v such that p[0] is label of (u, v) then
 5.          return (null)
 6.      else
 7.          let v such that (u, v) has label p[0];
 8.          if v has not a + then
 9.              SEARCH(CST, δ, s, v, succ(p))
10.         else
11.             z ← s(u);
12.             while z ≠ s(v)
13.                 z' ← δ(z, p[0]);
14.                 SEARCH(CST, δ, s,z,p)
15.                 z ← z'
```

Notice that, since SEARCH returns a pointer, we can use this information to detect all the occurrences in an index. The drawback of DECOMPACT and SEARCH is that they may require more then $O(m)$ time to decompact m letters and this does not guarantee a searching cost that is linear in the pattern length. To circumvent this drawback, we introduce a new definition, that we use for a linear-time version of DECOMPACT (hence, SEARCH).

Definition 2. *Let s be the suffix link function of a given trie. We define* **super-suffix link** *for a given node v in the trie, the function*

$$s^* : v \to s^k(u),$$

where u is a node in the trie and k is the greatest integer such that $s^k(u)$ and $s^k(v)$ are connected by a single edge (i.e., of path length 1), while the path from $s^{k+1}(u)$ to $s^{k+1}(v)$ has a length strictly greater than 1.

In order to appreciate how this new definition is useful for our goal, consider the task of decompacting the edge (u, v) in CST to reconstruct the substring originally associated with the uncompacted tries. We do this by considering repeatedly $(s(u), s(v))$, $(s^2(u), s^2(v))$, and so on, producing symbols only when the path from the two nodes at hand has at least one more edge. Unfortunately, we cannot guarantee that $(s(u), s(v))$ satisfies this property. With super-suffix links, we know that this surely happens with $(s(s^*(u)), s^*(s(v)))$ by Definition 2. Now, DECOMPACT works in linear time with the aid of super-suffix links.

```
SUPERDECOMPACT (Compacted trie CST, function δ, function s, arc (u,v))
  1.  w ← ε;
  2.  a ← label of (u,v);
  3.  if v has not a + then
  4.       w ← a;
  5.  else
  6.       z ← s(s*(v));
  7.       while z ≠ s(δ(s*(v), a))
  8.            w ← w concat SUPERDECOMPACT(CST, δ, s,(z, δ(z, a)));
  9.            z ← δ(z, a);
 10.            if z has only one child x in CST then
 11.                 a ← label of (z, x);
 12.  return w
```

As a result of the linearity of SUPERDECOMPACT, we also have:

Lemma 3. *Algorithm SEARCH for a pattern p takes $O(|p|)$ time by using super-suffix links instead of suffix links.*

Full-Text Indexing. A full-text index over a fixed text t is an abstract data type based on the set of all factors of t, denoted by $Fact(t)$. Such data type is equipped with some operations that allow it to answer the following query: given $x \in A^*$, find the list of all occurrences of x in t. If the list is empty, then x is not a factor of t.

Suffix trees and compacted DAWGs (CDAWGs) can answer to previous query in time proportional to the length of x plus the size of the list of the occurrences. By superposing the structure of CDAWGs to our compacted DAWGs we can obtain the same performance. More precisely, for any edge q of our compacted DAWG, let us define recursively $final(q)$

$$final(q) = \begin{cases} q & \text{if } q \text{ has more than two children or } q \text{ is the} \\ & \text{only final state} \\ \\ final(\delta(q,a)) & \text{if not, and } (\delta(q,a)) \text{ is the only outgoing edge} \\ & \text{from } q. \end{cases}$$

For any edge (p, q) we add another edge $p, final(q)$ labelled by the length of the path from p to $final(q)$ in the DAWG. These new edges allow to simulate CDAWGs.

We have examples of infinite sequences of words where our compacted DAWGs have size exponentially smaller than the text size. This is the first time to our knowledge that full-text indexes show this property.

5 Conclusions and Further Work

In this paper we have presented a new technique for compacting tries and their corresponding automata. They have arcs labeled with single characters; they do not need to retain the input string; also, they have less nodes than similar automata. We are currently performing some experiments for testing the effective gain in terms of final size of the automata. For highly compressible sources, our automata seem to be sublinear in space. We do not known if algorithm SEARCH strictly needs super-suffix links to work in linear time on the average (in the worst case, they are needed). Finally, we are investigating methods for a direct construction of our automata.

References

1. A. V. Aho and M. J. Corasick. Efficient string matching: an aid to bibliographic search, *Comm. ACM* **18:6** (1975) 333–340.
2. M.-P. Béal. Codage Symbolique, Masson, 1993.
3. M.-P. Béal, F. Mignosi, and A. Restivo. Minimal Forbidden Words and Symbolic Dynamics, in (*STACS'96*, C. Puech and R. Reischuk, eds., LNCS **1046**, Springer, 1996) 555–566.
4. A. Blumer, J. Blumer, D. Haussler, A. Ehrenfeucht, M. T. Chen, and J. Seiferas. The Smallest Automaton Recognizing the Subwords of a Text, Theoretical Computer Science, **40**, 1, 1985, 31–55.
5. M. Crochemore, Reducing space for index implementation, in Theoretical Computer Science, **292**, 1, 2003, 185–197.
6. M. Crochemore, F. Mignosi, A. Restivo, Automata and forbidden words, in Information Processing Letters, **67**, 3, 1998, 111–117.

7. M. Crochemore, F. Mignosi, A. Restivo, S. Salemi, Data compression using anti-dictonaries, in Special issue *Lossless data compression*, J. Storer ed., Proceedings of the IEEE, **88**, 11, 2000), 1756–1768.

8. M. Crochemore, R. Vérin, Direct Construction of Compact Directed Acyclic Word Graphs, in CPM97, A. Apostolico and J. Hein, eds., LNCS **1264**, Springer-Verlag, 1997, 116–129.

9. V. Diekert, Y. Kobayashi. *Some identities related to automata, determinants, and Möbius functions,* Report 1997/05, Fakultät Informatik, Universität Stuttgart, 1997, in (*STACS'96*, C. Puech and R. Reischuk, eds., LNCS **1046**, Springer-Verlag, 1996) 555–566.

10. B. K. Durgan. *Compact searchable static binary trees,* in Information Processing Letters, **89**, 2004, 49–52.

11. J. Holub. *Personal Communication,* 1999.

12. J. Holub, M. Crochemore. *On the Implementation of Compact DAWG's,* in Proceedings of the *7th Conference on Implementation and Application of Automata,* University of Tours, Tours, France, July 2002, LNCS **2608**, Springer-Verlag, 2003, 289–294.

13. S. Inenaga, H. Hoshino, A. Shinohara, M. Takeda, S. Arikawa, G. Mauri, G. Pavesi. *On-Line Construction of Compact Directed Acyclic Word Graphs,* To appear in Discrete Applied Mathematics (special issue for CPM'01).

14. S. Inenaga, H. Hoshino, A. Shinohara, M. Takeda, S. Arikawa, G. Mauri, G. Pavesi. *On-Line Construction of Compact Directed Acyclic Word Graphs,* Proceedings of *CPM 2001*, LNCS **2089**, Springer-Verlag, 2001, 169–180.

15. M. Lothaire. *Algebraic Combinatorics on Words.* Encyclopedia of Mathematics and its Applications, **90**, Cambridge University Press (2002).

Appendix

Let us examine, as an example, the text $t = abaababaaba\$$. Minimal words of text t are: aaa, $aabaa$, $aa\$$, $babaabab$, $babab$, $baba\$$, bb, $b\$$, $\$a$, $\$b$, $\$\$$.

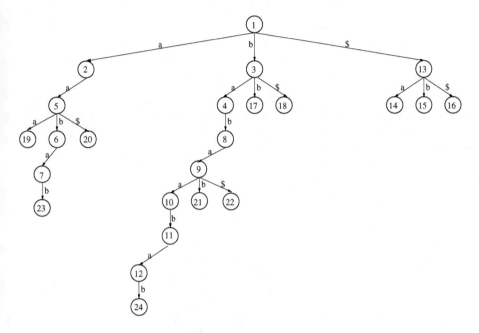

Fig. 1. Trie of the minimal forbidden factors.

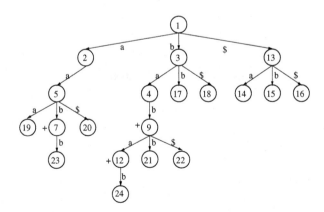

Fig. 2. Self-Compressed trie of the minimal forbidden factors.

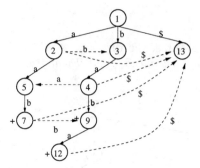

Fig. 3. Compacted Self-Compressed DAWG of the text t. It has 9 states. It has been obtained by adding the function δ whenever undefined (dotted arcs) to the self-compressed trie and by trimming it, i.e. by pruning the leaves and the corresponding ingoing and outgoing arcs. Suffix links and super-suffix links are not drawn.

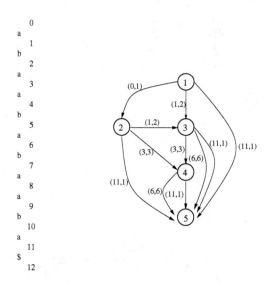

Fig. 4. CDAWG of the text t.

A Simple Optimal Representation
for Balanced Parentheses

Richard F. Geary[1], Naila Rahman[1], Rajeev Raman[1], and Venkatesh Raman[2]

[1] Department of Computer Science, University of Leicester, Leicester LE1 7RH, UK
{r.geary,naila,r.raman}@mcs.le.ac.uk
[2] Institute of Mathematical Sciences, Chennai, India 600 113
vraman@imsc.res.in

Abstract. We consider *succinct*, or highly space-efficient, representations of a (static) string consisting of n pairs of balanced parentheses, that support natural operations such as finding the matching parenthesis for a given parenthesis, or finding the pair of parentheses that most tightly enclose a given pair. This problem was considered by Jacobson, [*Proc. 30th FOCS*, 549–554, 1989] and Munro and Raman, [*SIAM J. Comput.* **31** (2001), 762–776], who gave $O(n)$-bit and $2n + o(n)$-bit representations, respectively, that supported the above operations in $O(1)$ time on the RAM model of computation. This data structure is a fundamental tool in succinct representations, and has applications in representing suffix trees, ordinal trees, planar graphs and permutations.

We consider the practical performance of parenthesis representations. First, we give a new $2n + o(n)$-bit representation that supports all the above operations in $O(1)$ time. This representation is conceptually simpler, its space bound has a smaller $o(n)$ term and it also has a simple and uniform $o(n)$ time and space construction algorithm.

We implement our data structure and Jacobson's, and evaluate their practical performance (speed and memory usage), when used in a succinct representation of large static XML documents. As a baseline, we compare our representations against a widely-used implementation of the standard DOM (Document Object Model) representation of XML documents. Both succinct representations use orders of magnitude less space than DOM, but are slightly slower and provide less functionality.

1 Introduction

Given a static balanced string of $2n$ parentheses, we want to represent it *succinctly* or space-effficiently, so that the following operations are supported in $O(1)$ time on the RAM model:

- FINDOPEN(x), FINDCLOSE(x): To find the index of the opening (closing) parenthesis that matches a given closing (opening) parenthesis x.
- ENCLOSE(x): To find the opening parenthesis of the pair that most tightly encloses x.

S.C. Sahinalp et al. (Eds.): CPM 2004, LNCS 3109, pp. 159–172, 2004.
© Springer-Verlag Berlin Heidelberg 2004

By counting the number of balanced parenthesis strings, one can see that the string requires $2n - O(\lg n)$ bits in the worst case, so a naive representation of the string is very close to optimal in terms of space usage. However, the above operations would essentially take linear time to support. One way to support $O(1)$-time operations is to note that the string is static and precompute and store answers for all possible arguments, but this uses $O(n \lg n)$ bits, $\Theta(\lg n)$ times more space than necessary. Jacobson [14] and Munro and Raman [16] gave $O(n)$-bit and $2n + o(n)$-bit representations, respectively, that supported the above operations in $O(1)$ time on the RAM model of computation[1]. Parenthesis representations are fundamental to succinct data structures, and have applications to suffix trees [20, 18], *ordinal* trees [2, 1, 16, 11], k-page graphs [14, 16] and stack-sortable permutations [17]. A topical motivation, and the starting point of our work, is the use of this data structure in the representation of (large, static) XML documents. The correspondence between XML documents and ordinal trees is well-known (see e.g. Fig. 1). In this paper we consider simplified

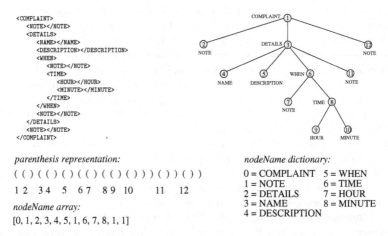

Fig. 1. Top left: Small XML fragment (only tags shown). Top right: Corresponding tree representation. Bottom: Succinct representation of document.

XML documents, where we ignore a number of secondary features[2], and also assume that the document consists purely of markup (i.e. there is no free text).

The XML Document Object Model (DOM) [8] is a standard interface through which applications can access XML documents. DOM implementations store an entire XML document in memory, with its tree structure preserved. At the heart of DOM is the Node interface, which represents a single node in the tree. The node interface contains attributes such as nodeName, nodeValue and nodeType to store information about the node, as well as parentNode, firstChild, lastChild,

---

[1] Jacobson's result was stated for the bit-probe model, but it can be modified to run in $O(1)$ time on the RAM model [15].

[2] Such as: attributes and their values, namespace nodes, comments, etc.

`previousSibling` and `nextSibling`, which act as a means to access other related nodes. The usual, but naive, way of implementing the DOM is to store with each node a pointer to the parent, the first/last child, and the previous/next sibling. Unfortunately, this can take up many times more memory than the raw XML file. This 'XML bloat' significantly impedes the scalability and performance of current XML query processors [23], especially if the DOM representation does not fit in main memory (which can happen for fairly modest-sized documents).

To represent XML documents succinctly, while providing the essential features of the `Node` interface, we store the tree as a sequence of parentheses, identifying nodes with the position of their open parentheses. This is augmented with a standard data structure that, for any position i in the sequence, gives the number of open parentheses in positions $1, \ldots, i$, and occupies $o(n)$ bits [15], this gives the pre-order number of the node numbered i. We also store a sequence of values $\sigma_1, \ldots, \sigma_n$, where σ_i is the tag of the i-th node in pre-order (see Fig. 1).

The viability of such a representation depends crucially on the speed and space-efficiency of the parenthesis data structure. A good implementation must find the right trade-off between storing pre-computed data – the "insignificant" $o(n)$ terms can easily dominate space usage – and computation time. There has been work on implementations of space-efficient trees, including k-ary trees, where each edge from a node to its children is labelled with a distinct letter from an alphabet [7] and Patricia trees [6] among others. Chupa, in unpublished work, described an implementation of a restricted static binary tree representation [5]. Compressed self-indexing dictionaries have been implemented in [9, 12]. We are not aware of any implementations of a parenthesis data structure.

We begin by giving a new, conceptually simple, $2n + o(n)$-bit parenthesis data structure. Our new data structure uses no complex subroutines (e.g. [16] use perfect hash tables) and it has a lower order term in the space usage of $O(n \lg \lg n / \lg n)$ bits versus $\Theta(n \lg \lg \lg n / \lg \lg n)$ bits in [16]. It also has a simple and uniform $o(n)$-time and space construction algorithm, which is not known of the data structure of [16]. Indeed, to achieve $O(n)$ construction time, [16] need to use either randomisation, or a recent complex algorithm [13, Theorem 1.1] for constructing perfect hash tables.

We implement a version of Jacobson's data structure as well as the new one, evaluating their space usage and speed. As a baseline, we also compare with CenterPoint XML [3] which is an open-source C++ XML DOM library. The standard test we perform with an XML document is to perform a DFS traversal of the tree, both in a standard DOM implementation and in our representation, counting the number of nodes of a given type (this is a fairly canonical operation in manipulating XML documents). As expected, both succinct schemes use orders of magnitude less space than Centerpoint XML – it is surprising how modest the computational overhead of the succinct schemes is.

2   A Simple Parenthesis Representation

Both Jacobson's and Munro and Raman's representations divide the given string of parentheses into equal-sized blocks of B parentheses, and identify a set of

$O(n/B)$ parentheses as *pioneers*. They explicitly keep the position of the match-ing parenthesis of the pioneer parentheses. They also store enough other infor-mation with blocks and/or with individual parentheses to detect pioneer paren-theses, as well as to find the match of any parenthesis, from the position of the match of its closest pioneer parenthesis. They also store a small number of tables, typically, to find answers within a block.

Jacobson takes $B = \Theta(\lg n)$ and so the number of pioneer parentheses is $O(n/\lg n)$. He stores (essentially) the location of the matching parenthesis for each pioneer explicitly. He uses a bit vector (along with $O(n)$ bits of auxiliary storage) to detect pioneer parentheses, and keeps the *excess* – the number of open minus the number of closing parenthesis – at each block boundary. Each of the above takes $\Theta(n)$ bits, and so the overall space bound is also $\Theta(n)$ bits. In order to reduce the space bound to $2n + o(n)$ bits, Munro and Raman employ a three level blocking scheme (big, small and tiny), using blocks of $\Theta(\lg^2 n)$, $\Theta((\lg \lg n)^2)$ and $\Theta(\lg \lg n)$ respectively, storing auxiliary data at each level. In particular, they store the positions of $\Theta(n/(\lg n)^2)$ pioneer parentheses (with respect to big blocks) in a perfect hash table. Constructing this perfect hash table takes $O(n)$ expected time and space [10] or $O(n)$ time using the rather complex algorithm of [13][3]. The need to store (slightly different) auxiliary information at different block sizes contributes both to the implementation complexity and the lower-order term in the space bound (the latter is important in determining space usage in practice).

Our representation also divides the given parenthesis string into blocks of size $\Theta(\lg n)$. We modify the definition of a pioneer so that the sequence of pi-oneer parenthesis is itself a balanced string of $O(n/\lg n)$ parentheses. Our rep-resentation is based on three important observations. First, the positions of the sequence of pioneer parentheses can be stored using $o(n)$ bits using a *fully index-able dictionary* (FID) [19] (see below). Second, representing the string of pioneer parenthesis recursively gives enough information to support the basic operations in constant time. (Recursing at most twice, we have a set of $O(n/\lg^2 n)$ pioneer parentheses, which is small enough that it can be stored using the trivial rep-resentation.) Third, looking closely at the requirements of the FID, we are able to replace the FID of [19] by a very simple data structure. We now discuss the new parenthesis structure, following the above outline.

Fully Indexable Dictionaries. Given a bit-vector of length M which has 1s at a set of positions $S \subseteq [M]$ [4], $|S| = N$, and zeros elsewhere, we define the operations:

RANK(x, S): Given $x \in [M]$, return $|\{y \in S | y \leq x\}|$.
SELECT(i, S): Given $i \in [N]$, return the $i + 1$-st smallest element in S.

---

[3] One needs to use the result of [13, Section 4] rather than the main result (Theorem 1.1), in order to get a uniform algorithm.
[4] For nonnegative integers M, $[M] = \{0, \dots, M - 1\}$.

We call a representation of S that supports the above two operations in $O(1)$ time a *nearest neighbour dictionary (NND)*, as the operations below are also supported in $O(1)$ time:

PRED(x, S): Given $x \in [M]$, return x if $x \in S$ and $\max\{y \in S | y < x\}$ otherwise;
SUCC(x, S): Given $x \in [M]$, return x if $x \in S$ and $\min\{y \in S | y > x\}$ otherwise;

An NND that supports RANK and SELECT, on S and $\bar{S}$ simultaneously, where $\bar{S}$ is the complement of S, in $O(1)$ time has been called a *fully indexable dictionary* (FID) [19]. The following is known about FID (and hence about NND) representations:

Theorem 1 ([19, Lemma 4.1]). *There is an FID for a set $S \subseteq [M]$ of size N using at most $\left\lceil \lg \binom{M}{N} \right\rceil + O(M \lg \lg M / \lg M)$ bits.*

In particular, we have, from Theorem 1:

Corollary 1. *There is a NND for a set $S \subseteq [M]$ of size $N = O(M/\lg M)$ that uses $O(M \lg \lg M / \lg M) = o(M)$ bits.*

2.1   The New Representation

We now assume that we are given a balanced string of $2n$ parentheses and our goal is to support FINDOPEN, FINDCLOSE and ENCLOSE operations in constant time. We now describe the new data structure to store any balanced string of parenthesis of length $2N \leq 2n$.

If N is $O(n/\lg^2 n)$, then we represent the sequence using the trivial structure which stores the pre-computed answer for each of the operations above, for every parenthesis. This takes $O(N \lg N) = o(n)$ bits. Otherwise, we divide the parenthesis string into equal-sized *blocks* of size $B = \lceil (\lg N)/2 \rceil$. We number these blocks $1, \ldots, \beta \leq 4N/\lg N$, and by $b(p)$ we denote the block in which the parenthesis p lies. The matching parenthesis of p is denoted by $\mu(p)$. We call a parenthesis p *far* if $\mu(p)$ is not in the same block as p (and note that $\mu(p)$ is itself a far parenthesis). At any position i, we call the number of open parenthesis minus the number of closing parenthesis in positions $1, \ldots, i$ as the *left excess* at i. Similarly, we call the number of closing parenthesis minus the number of open parenthesis in positions $i, \ldots, 2N$ as the *right excess* at i.

We now define a pioneer parenthesis, in the spirit of Jacobson. Consider an opening far parenthesis p, and let q be the far opening parenthesis that most closely precedes p in the string. We say that p is an *opening pioneer* if $b(\mu(p)) \neq b(\mu(q))$. The definition of a closing pioneer p is as above, except that q would be the far parenthesis immediately *after* p. A *pioneer* is either an opening or closing pioneer. Note that the match of a pioneer may not be a pioneer itself.

Lemma 1 ([14, Theorem 1]). *The number of opening pioneers in a balanced string divided into β blocks is at most $2\beta - 3$. The same holds for the number of closing pioneers.*

Proof. The *pioneer graph* which has nodes $1, \ldots, \beta$ and edges $(b(p), b(\mu(p)))$, for all opening pioneers p, is outerplanar and has no parallel edges. Therefore, it has at most $2\beta - 3$ edges.

For a given block size B, we define the *pioneer family* as the set of all pioneers, together with all their matching parentheses (recall that if p is a pioneer, $\mu(p)$ need not be one). Clearly, the substring comprising only the parentheses in the pioneer family is balanced. We now bound the size of the pioneer family.

Proposition 1. *The size of the pioneer family is at most $4\beta - 6$.*

Proof. The pioneer family graph, defined analogously to the pioneer graph, is itself outerplanar, allowing us to conclude that the pioneer family is of size at most $2 \cdot (2\beta - 3)$ or $4\beta - 6$.

Remark 1. The upper bound on the number of pioneers in Jacobson's original definition and in the pioneer family is the same. We note that the pioneer family is in fact the set of all pioneers, if we modify the definition of a pioneer parenthesis slightly to include the leftmost (rightmost) far opening (closing) parenthesis from every block in addition to the original pioneers.

Our structure has the following four parts.

1. The original parenthesis string π of length $2N$,
2. A NND (Corollary 1) that stores the set $P \subseteq [2N]$ of the positions in π that belong to the pioneer family,
3. a recursive parenthesis data structure for the pioneer family, and
4. A constant number of tables that allow us to operate on blocks in $O(1)$ time. For example, a table that stores for for every block b, and for every $i = 1, \ldots, B$, the position of the matching parenthesis of the parenthesis at position i, if the match is inside the block (the table stores 0 if the match is not inside the block). Such tables take at most $O(\sqrt{N}(\lg N)^2) = o(N)$ bits.

We now calculate the space usage. The tables take $O(\sqrt{N} \lg N \lg \lg N)$ bits. Since $|P| \le 16(N/\lg N)$, the NND for pioneers take $O(N \lg \lg N / \lg N)$ bits by Corollary 1. Thus, if $S(N)$ is the space used by the structure, then $S(N)$ satisfies:

$$S(N) = O(N \lg N) \text{ if } N \text{ is } O(n/\lg^2 n) \text{ and}$$
$$S(N) = 2N + S(8N/\lg N) + O(N \lg \lg N / \lg N) \text{ otherwise}$$

It is easy to see that $S(n) = 2n + O(n \lg \lg n / \lg n) = 2n + o(n)$ bits.

2.2   Operations

Now we describe how the operations are implemented.

FINDCLOSE(p): Let p be the position of an open parenthesis. First determine by a table lookup whether it is far. If not, the table gives the answer. If it is, use PRED(p, P) to find the previous pioneer p^*. We can show that this will be an open parenthesis. Find its position in the pioneer family using RANK(p^*, P) and find its match in the pioneer family using the recursive structure for P; assume that this match is the j-th parenthesis in the pioneer family. We then use SELECT(j, P) to find the position of $\mu(p^*)$ in π. Now observe that since the first far parenthesis in each block is a pioneer, p^* and p are in the same block. Compute i, the change in left excess between p and p^*, using a table lookup. Noting that $\mu(p)$ is the leftmost closing parenthesis in $b(\mu(p^*))$ starting from $\mu(p^*)$, with right excess i relative to $\mu(p^*)$, we locate $\mu(p)$ using a table. FINDOPEN is similar.

ENCLOSE(c): Let $p =$ ENCLOSE(c) such that p and c are both open parentheses. From one (or two) table lookup(s) determine whether either of $\mu(p)$ or p is in the same block as c. If so, we can return p using, if necessary, one call to FINDOPEN. If not, we proceed as follows. Let $c' =$ SUCC(c, P). If c' is a closing parenthesis then let $p' =$ FINDOPEN(c'). Otherwise find the position of c' in the pioneer family using RANK, find the parentheses enclosing c' in the pioneer family and using SELECT translate the result into a parenthesis p' in π. We claim that in both cases $(p', \mu(p'))$ is the pair of pioneer family parentheses that most tightly encloses c. Let $q =$ SUCC(p', P). If q is in the same block as p' then p is the first far parenthesis to the left of q. Otherwise, p is the rightmost far parenthesis in the block containing p'. In either case, the answer is obtained from a table.

To prove the correctness of the algorithm, we observe that if p or $\mu(p)$ are in the same block as c, then we can find p using table lookup (and possibly FINDOPEN($\mu(p)$)). Otherwise since both p and $\mu(p)$ are in different blocks to c, $b(p) < b(c) < b(\mu(p))$ and hence both p and $\mu(p)$ must be far parentheses.

From the definition of a pioneer, there must exist exactly one pair of pioneers $(p', \mu(p'))$ such that $b(p') = b(p)$ and $b(\mu(p')) = b(\mu(p))$; and the pair $(p', \mu(p'))$ is the tightest enclosing pioneer pair of c. If there were a tighter enclosing pioneer pair, this pair would be enclosed by p and hence p would not be the tightest enclosing parenthesis. That the algorithm correctly computes p' is seen from the following:

1. if c' is a closing parenthesis, then it must enclose c. It must be the tightest enclosing pioneer because it is the first pioneer to the right of c. Therefore $p' =$ FINDOPEN(c').
2. if c' is an opening parenthesis, then c and c' must share the same tightest enclosing pioneer parenthesis. Hence $p' =$ ENCLOSE(c').

Now, note that there are a number of (1 or more) far parentheses in $b(p)$ that have their matching parentheses in $b(\mu(p))$; the left-most of these far parentheses is p' and the rightmost is p. As has been observed before, there is only 1 pioneer in $b(p)$ that points to $b(\mu(p))$, and from the definition of a pioneer this means that there is no pioneer that occurs between p' and p.

Therefore, if q is the next pioneer in $b(p)$ to the right of p', then p must be the last far parenthesis in $b(p)$ before q, and if there are no pioneers to the right of p' in $b(p)$ then p must be the rightmost far parenthesis in the block. This is indeed what the above algorithm computes. We thus have:

Theorem 2. *A balanced string of $2n$ parenthesis can be represented using $2n + O(n \lg \lg n / \lg n)$ bits so that the operations* FINDOPEN, FINDCLOSE *and* ENCLOSE *can be supported in $O(1)$ time.*

2.3   Simplifying the NND

Our structure, although conceptually simple, uses the the (fairly complex) data structure of Theorem 1 as a subroutine. We now greatly simplify this subroutine as well, by modifying the definition of the pioneer family. Call a block *near* if it has no pioneer (and hence no far) parenthesis. We add to the pioneer family (as defined above) *pseudo*-pioneers consisting of the first and the last parenthesis of every near block (it is is easy to see that the string corresponding to the modified pioneer family is balanced and has size $O(N/\lg N)$).

We now argue that pseudo-pioneers do not affect the operations FINDOPEN, FINDCLOSE and ENCLOSE. For FINDOPEN(x) (FINDCLOSE(x)), where x is a near parenthesis, the answer will be obtained by a table lookup. If x is a far parenthesis, the answer is obtained by first searching for the previous (next) pioneer p. Since p will always be in $b(x)$, and $b(x)$ is not a near block, p cannot be a pseudo-pioneer and the earlier procedures go through.

When we perform ENCLOSE(c) on an open parenthesis, where c is in a block that does not contain pseudo-pioneers, we first check to see if either the opening or the closing enclosing parenthesis is in the block using table lookup; if it is then we have computed ENCLOSE(c) correctly (with possibly one call to FINDOPEN). Otherwise, we locate the next pioneer c' after c and check to see if c' is an opening or closing parenthesis. It is possible that c' is a pseudo-pioneer that is an opening parenthesis, but if this is the case, the closest enclosing pioneer parenthesis pair of c is the same as that of c', and hence we get a valid result by performing ENCLOSE(p) on the pioneer bitvector. If we were to perform ENCLOSE(c) where c is in a near block we will always be able to compute ENCLOSE(c) using table lookup, unless c is the first position in the near block. In the latter case, we can still compute ENCLOSE(c) correctly using this method because the closest enclosing pioneer pair of c is the same, even with the pseudo-pioneers.

Since every block has at least a (pseudo-)pioneer, the gap between the positions of two successive pioneers in the modified pioneer family is at most $2B = O(\lg N)$. This allows us to simplify the NND(s) in item (2) as follows.

A Simple NND for Uniformly Sparse Sets. We now consider the problem of creating an NND for a bit-vector of length M with 1s in a uniformly sparse set $S \subseteq [M]$ of positions. Specifically, we assume that $N = |S| = O(M/\lg M)$ and further that if $S = \{x_1, \ldots, x_N\}$ and $x_1 < \ldots < x_N$, then for $i = 1, \ldots, N$, $x_i - x_{i-1} \le (\lg M)^c$ for some constant c (take $x_0 = 0$). Our scheme uses four arrays of $O(M \lg \lg M / \lg M)$ bits each and three tables of $O(M^{2/3})$ bits each.

Let $t = \lg M/(2c \lg \lg M)$ and $S_t = \{x_t, x_{2t}, \ldots\}$. In the array A_1, we list the elements of S_t explicitly: A_1 thus takes $N \lceil \lg M \rceil / t = O(M \lg \lg M / \lg M)$ bits. In array A_2, we store the differences between consecutive elements of S, i.e., we let $A_2[i] = x_i - x_{i-1}$ for $i \geq 1$ (take $x_0 = 0$). Since all values in A_2 are $O((\lg M)^c)$ by assumption, each entry can be stored using $c \lg \lg M$ bits, and array A_2 takes $O(N \lg \lg M)$ or $O(M \lg \lg M / \lg M)$ bits in all. A table T_1 contains, for every bit string of length $ct \lg \lg M = (\lg M)/2$, the sum of the t values obtained by treating each group of consecutive $c \lg \lg M$ bits as the binary encoding of an integer. The table takes $O(\sqrt{M} \lg \lg M)$ bits.

Now $\text{SELECT}(i, S)$ can be obtained in $O(1)$ time as follows: let $i' = \lfloor (i+1)/t \rfloor$ and $i'' = (i+1) \bmod t$. Let $x = A_1[i']$. Obtain y as the concatenation of the values in $A_2[i'+1], A_2[i'+2], \ldots, A_2[i'+i'']$; these $ci'' \lg \lg M < ct \lg \lg M$ bits are padded with trailing zeroes to make y be $(\lg M)/2$ bits long (this is done in $O(1)$ time by reading at most two $\lg M$-bit words from A_2 followed by masks and shifts). Return $\text{SELECT}(i, S) = x + T_1[y] - 1$ (recall that $[M]$ starts with 0).

To support the RANK operation, we store two more arrays. We (conceptually) divide $[M]$ into blocks of t consecutive values, where the i-th block b_i is $\{(i-1)t, \ldots, it-1\}$, and let $A_3[i] = |b_i \cap S_t|$, for $i = 1, \ldots, M/t$. Noting that $A_3[i] \in \{0, 1\}$, we conclude that A_3 takes $O(M/t) = O(M(\lg \lg M)/\lg M)$ bits. Viewing A_3 as the bitvector of a set, the following standard auxiliary information permits $O(1)$-time rank queries (details omitted) on A_3: an array A_4 containing, for $i = 1, \ldots, 2M/\lg M$, $A_3[1] + \ldots + A_3[i \cdot (\lg M)/2]$, and a table T_2 containing, for every bit-string of $(\lg M)/2$ bits, the number of 1s in that bit-string. Clearly A_4 occupies $O(M/t)$ bits and T_2 takes $O(\sqrt{M} \lg \lg M)$ bits.

Finally, we have another table T_3, which contains, for every bit string of length $ct \lg \lg M = (\lg M)/2$, interpreted as a sequence of t non-negative integers of $c \lg \lg M$ bits each, and a value $i \leq t(\lg M)^c$, the largest $l \leq t$ such that the sum of the first l integers is less than i. T_3 takes $O(\sqrt{M}(\lg M)^{c+2})$ bits.

Now $\text{RANK}(i, S)$ is implemented as follows. Let $i' = \lfloor i/t \rfloor$ and let $r = \text{RANK}(i', A_3)$. Observe that $A_2[r] = x_{rt}$ is the largest element in S_t that is $\leq i$. Let y be the concatenation of the values in $A_2[rt+1], A_2[rt+2], \ldots, A2[(r+1)t]$, and return $\text{RANK}(i, S) = rt + T_3[y, i - x_{rt}]$. Thus we have:

Theorem 3. *Let $S \subseteq M$ be a subset of size $O(M/\lg M)$ and let the difference between two consecutive values of S be $O((\lg M)^c)$ for some constant c. Then there is a simple representation for S (using four arrays and three tables) taking $O(M \lg \lg M / \lg M) = o(M)$ bits in which the operations $\text{RANK}(x, S)$ and $\text{SELECT}(i, S)$ can be supported in constant time.*

Remark 2. Using Theorem 3 in place of Corollary 1, we get a parenthesis data structure that uses $2n + O(n \lg \lg n / \lg n) = 2n + o(n)$ bits, and is manifestly simple. Note that most applications involving succinct data structures (including the parenthesis one) would anyway require the table T_2.

Remark 3. The construction of this data structure is both simple and efficient: given a parenthesis string π of length $2n$, all auxiliary data structures can be constructed in $O(n/\lg n)$ time using additional $O(n \lg \lg n / \lg n)$ bits of workspace, as follows. We first determine the pioneer family of π. This is done in two passes over π, to determine the lists of closing and opening pioneers, respectively. By merging the two lists we produce the array A_2 of the NND.

We determine the closing pioneers by processing each block in turn. We assume that when processing the i-th block, there is a temporary stack that contains, for every block among the blocks $1, \ldots, i - 1$ that has one or more currently unmatched (far) open parenthesis, the number of such parentheses. Clearly the space used by the stack is $O(n \lg \lg n / \lg n)$ bits. We use table lookup on the i-th block to determine the number j of far closing parentheses in this block. If $j > 0$ then, using the stack and table lookup, it is easy to determine which of the far parentheses are pioneers. If there are no far parenthesis at all, we designate the last parenthesis as a (pseudo)-pioneer. Using table lookup, we determine the number of far open parentheses and push this on to the stack. If any closing pioneers are found we write their positions down in a temporary array A_c, again storing differences in positions of successive closing pioneers, rather than the positions themselves (since closing pioneers may be more than poly-log positions apart, A_c needs to be represented with a little care to fit in $O(n \lg \lg n / \lg n)$ bits). We calculate A_o, the positions of open pioneer parentheses, similarly, and merge A_o with A_c to give A_2. It is fairly easy to see that the entire process takes $O(n / \lg n)$ time.

3   Implementation and Experiments

An important difference between the implementation and theoretical description is that we do not constrain the size of a block to be $\lceil (\lg n)/2 \rceil$. Indeed, larger block sizes are both possible and necessary, and the size of the tables is determined by the need for them to fit into the cache of a computer. For this reason, table lookups are done using 8-bit or 16-bit "chunks" and blocks contain more than one chunk (e.g. we consider $B = 32, 64, \ldots, 256$). For simplicity (and to make it easier to share tables) we use the same block and chunk sizes for the recursive data structures. The two data structures we implemented are the new $2n + o(n)$ recursive data structure and a minor variant of Jacobson's data structure[5].

We implemented the data structures in C++, and ran some tests on a Pentium 4 machine and a Sun UltraSparc-III machine. The Pentium 4 has a 512MB RAM, a 2.4GHz CPU and a 512KB cache, running Debian Linux. The compiler was g++ 2.95 with optimisation level 2. The Sun UltraSparc-III has an 8GB RAM, a 1.2GHz CPU and a 8MB cache, running SunOS 5.8. The compiler was g++ 3.3 with optimisation level 2.

In order for us to be able to understand the space and time requirements of our representations in real-life situations, we gathered statistics from some real-world XML files. For this task we compiled a list of 12 XML files: `treebank.xml`, `sprot.xml`, `elts.xml`, `w3c1.xml`, `pcc2.xml`, `tpc.xml`, `tal1.xml`, `stats1.xml` [4] `mondial-3.0.xml`, `orders.xml`, `nasa.xml` [21] and `votable.xml` [22]. The files come from different applications and have different characteristics. Fig. 2 shows the sizes of the files and number of nodes.

Running times were evaluated against a baseline of CenterPoint XML's DOM implementation, release 2.1.7. The test in each case was to traverse the tree in

---

[5] Whereas Jacobson's data structure could never go below $4n$ bits, the new one can.

treebank	sprot	elts	w3cl	pcc2	tpc	stats1	tal1	mondial	orders	votable	nasa
7	11	114	216	247	294	657	718	1,743	5,253	15,536	24,463
.8	.8	.6	10.4	17.8	35.2	56.4	49.9	57.4	300	1,991	1,425

Fig. 2. Test file names, file sizes in KB, XML nodes in thousands.

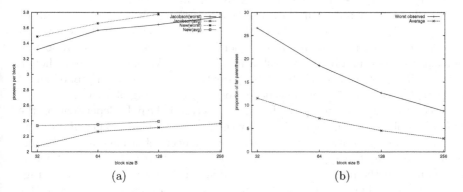

(a)   (b)

Fig. 3. Statistics on XML files. The x-axis has the block size B and the y-axis has: (a) Number of pioneers per block and (b) The proportion of parentheses that are far.

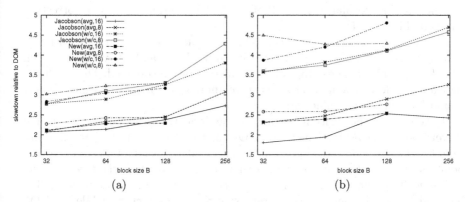

(a)   (b)

Fig. 4. Average slowdown relative to DOM overall files and worst instance over all files. The x-axis has the block size B and the y-axis has slowdown relative to DOM: (a) on Sun UltraSparc-III, (b) on Pentium 4. Plot marks in (b) are as in (a).

Table 1. Space used by Jacobson implementation and new DS, excluding tables, assuming a pioneer density in the original bitstring of 2.4 per block and 4 per block.

	$PD = 4$		$PD = 2.4$	
	Jacob	New	Jacob	New
32	16.00	9.31	12.80	6.09
64	9.00	4.89	7.40	3.81
128	5.50	3.30	4.70	2.89
256	3.75		3.35	

DFS order and count the number of nodes of a particular type. We performed multiple tree walks in order to get more stable running times; repetitions varied from 50 (for the large files) to 100000 (for the smaller files). All files fit in memory, even using DOM, to keep the test fair.

We observed that there were several parameters that influenced performance, some that were data-dependent and others that were program parameters.

Two key parameters of the data are the proportion of far parentheses, and the number of pioneers, for a given block size. The former is important, as the computation time for many operations is higher for a far parenthesis. The proportion of far parentheses in a bitvector is shown in Fig. 3(b). We note that even though it is possible to have essentially $2n$ far parentheses, the observed proportion of far parentheses is very small, and decreases with increasing blocksize.

Secondly, the space usage of the data structure is heavily dependent on the number of pioneers (Fig. 3(a)). We show the number of pioneers per block using both Jacobson's definition, as well as using our pioneer definition plus the dummy pioneers required to support the simple NND. Note that the theoretical worst-case bound for the number of pioneers per block is just under 4, but on average there were less than 60% of the theoretical maximum number of pioneers; this held regardless of which definition was used. In both graphs, the worst-case values displayed are the values associated with pcc2.xml which almost always has the worst performance of our 12 files; this file contains a proof of correctness and is highly nested.

We calculate the space bounds (excluding tables) used by our data structure, taking both a worst-case pioneer density of 4 per block and an average-case density of 2.4 per block. The details of these calculations are omitted. The results are shown in Table 1. The new data structure approaches the information-theoretic bound for quite small values of B.

In general on the Pentium 4 machine the data structures were 1.7 to 4 times slower than the DOM implementation. On the UltraSparc-III the data structures were 1 to 2.5 times slower. This is very good considering that DOM simply follows pointers. The key benefit however, is that the space to store the tree structure is drastically reduced. For example, with $B = 128$ we can choose between $4.7n$ bits and $2.89n$ bits, on average, to represent the tree structure, while the DOM implementation takes 96 bits (verifed by looking at source code), and suffer an average slowdown of 2.5 or 2.7 on a Pentium 4. Of course, we do not expect a $30 - 35\times$ reduction in the overall size of an in-memory XML representation but pointers would typically be half the cost of the node. The DOM implementation we consider stores pointers to parent, first child and next sibling, presumably to reduce space usage. Operations such as getLastChild or getPreviousSibling require traversal over all child nodes or many siblings. It would be relatively easy to modify the tree walk experiment to visit child nodes in reverse order and this would have a significant impact on the performance of the DOM implementation.

Fig. 5 shows for our new data structure slowdown versus percentage of far parentheses on the Pentium 4 using a blocksize of 64 bits. The clear trend is that slowdown increases with the proportion of far parentheses.

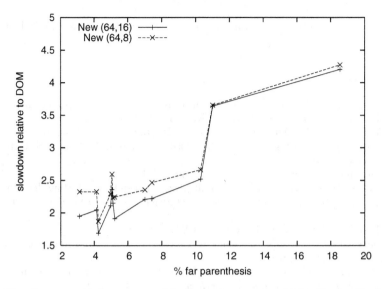

Fig. 5. Slowdown versus percentage of far parentheses using 64-bit blocks. Each point represents data for one of our sample XML files.

It is interesting to note that at least on these data sets the new DS is no better than Jacobson. For example, roughly similar space and time performance is obtained by Jacobson with $B = 256$ and the new DS with $B = 64$. The new DS can be expected to be slower for a given B because it has a more complex algorithm for dealing with matches of pioneers, and also because of the NND in the middle. However, as B increases, the proportion of far parentheses (at least on these real-world example) decreases. Indeed, we have observed that the percentage increase in running times is similar to the proportion of far parentheses.

4 Conclusions and Further Work

We have given a conceptually simple succinct representation for balanced parenthesis that supports natural parenthesis operations in constant time. This immediately gives a simpler optimal representation for all applications of these data structures. The new representation has theoretical advantages as well, such as a simple sublinear-time and space construction algorithm, and an improved lower-order term in the space bound. We believe the data structure can be further simplified.

We note that on the Pentium 4, which has a relatively small cache, chunk size 8 and 16 perform slightly differently on "average" and "difficult" data sets. This may be due to the fact that for "average" data most of the running time is dominated by the total number of operations. For "difficult" data we make accesses to a larger number of tables, when these tables no longer fit in cache the running time begins to be effected by memory access costs. In general, optimising the cache performance of table lookup is an interesting problem.

We would like to thank R Ramani for coding versions of the rank/select data structures.

References

1. D. A. Benoit, E. D. Demaine, J. I. Munro and V. Raman. Representing trees of higher degree. In *Proc. 6th WADS*, LNCS 1663, 169-180, 1999.
2. D. A. Benoit, E. D. Demaine, J. I. Munro, R. Raman, V. Raman and S. S. Rao. Representing trees of higher degree. TR 2001/46, Dept. of Maths & CS, University of Leicester, 2001.
3. Centerpoint XML, http://www.cpointc.com/XML.
4. J. Cheney. Compressing XML with multiplexed hierarchical PPM models. In *Proc. Data Compression Conference (DCC 2001)*, IEEE Computer Society, pp. 163–172, 2001.
5. K. Chupa. MMath Thesis, University of Waterloo (1997).
6. D. Clark and J. I. Munro. Efficient suffix trees on secondary storage. In *Proc. 7th ACM-SIAM SODA*, pp. 383–391, 1996.
7. J. J. Darragh, J. G. Cleary and I. H. Witten. Bonsai: a compact representation of trees. In *Software-Practice and Experience*, **23** (1993), pp. 277–291.
8. A. Le Hors, P. Le Hégaret, L. Wood, G. Nicol, J. Robie, M. Champion, S. Byrne. Document Object Model (DOM) Level 2 Core Specification Version 1.0. W3C Recommendation 13 November, 2000. `http://www.w3.org/TR/DOM-Level-2-Core`. W3C Consortium, 2000.
9. P. Ferragina, G. Manzini. An experimental study of a compressed index. *Information Sciences*, **135** (2001) pp. 13–28.
10. M. L. Fredman, J. Komlós, and E. Szemerédi. Storing a sparse table with $O(1)$ worst case access time. *J. ACM*, **31** (1984), pp. 538–544.
11. R. F. Geary, R. Raman and V. Raman. Succinct ordinal trees with level-ancestor queries. In *Proc. 15th ACM-SIAM SODA*, pp. 1–10, 2004.
12. Roberto Grossi, Ankur Gupta and Jeffrey Scott Vitter. When indexing equals compression: Experiments on suffix arrays and trees. In *Proc. 15th ACM-SIAM SODA*, pp. 629–638, 2004.
13. T. Hagerup, P. B. Miltersen and R. Pagh. deterministic dictionaries. *Journal of Algorithms* **41** (1): 69-85 (2001).
14. G. Jacobson. Space-efficient static trees and graphs. In *Proc. 30th FOCS*, 549-554, 1989.
15. J. I. Munro. Tables. In *Proc. 16th FST&TCS conference*, LNCS 1180, 37-42, 1996.
16. J. I. Munro and V. Raman. Succinct representation of balanced parentheses and static trees. *SIAM J. Computing*, **31** (2001), pp. 762–776.
17. I.Munro, R. Raman, V. Raman and S. S. Rao. Succinct representation of Permutations. In *Proc. 30th ICALP*, LNCS 2719, 345-356, 2003.
18. I.Munro, V. Raman and S. S. Rao. Space Efficient Suffix Trees In *J. of Algorithms*, **39** (2001), pp. 205–222.
19. R. Raman, V. Raman and S. S. Rao. Succinct indexable dictionaries with applications to encoding k-ary trees and multisets. In *Proc. 13th ACM-SIAM SODA*, pp. 233–242, 2002.
20. K. Sadakane. Succinct representations of lcp information and improvements in the compressed suffix arrays. In *Proc. 13th ACM-SIAM SODA*, pp. 225–232, 2002.
21. UW XML Repository. `http://www.cs.washington.edu/research/xmldatasets/`.
22. `http://cdsweb.u-strasbg.fr/doc/VOTable/`.
23. http://xml.apache.org/xindice/FAQ.

Two Algorithms
for LCS Consecutive Suffix Alignment

Gad M. Landau[1,2,*], Eugene Myers[3], and Michal Ziv-Ukelson[4,**]

[1] Dept. of Computer Science, Haifa University, Haifa 31905, Israel
landau@cs.haifa.ac.il
[2] Department of Computer and Information Science, Polytechnic University, Six
MetroTech Center, Brooklyn, NY 11201-3840, USA
landau@poly.edu
[3] Div. of Computer Science, UC Berkeley, Berkeley, CA 94720-1776, USA
gene@eecs.berkeley.edu
[4] Dept. of Computer Science, Technion – Israel Institute of Technology,
Haifa 32000, Israel
michalz@cs.technion.ac.il

Abstract. The problem of aligning two sequences A and B to determine
their similarity is one of the fundamental problems in pattern matching.
A challenging, basic variation of the sequence similarity problem is the
incremental string comparison problem, denoted **Consecutive Suffix
Alignment**, which is, given two strings A and B, to compute the alignment solution of each suffix of A versus B.

Here, we present two solutions to the Consecutive Suffix Alignment Problem under the LCS metric. The first solution is an $O(nL)$ time and space
algorithm for constant alphabets, where n is the size of the compared
strings and $L \leq n$ denotes the size of the LCS of A and B.

The second solution is an $O(nL + n \log |\Sigma|)$ time and $O(L)$ space algo-
rithm for general alphabets, where Σ denotes the alphabet of the com-
pared strings. (Note that $|\Sigma| \leq n$.)

1 Introduction

The problem of comparing two sequences A of size n and B of size m to determine their similarity is one of the fundamental problems in pattern matching.
Standard dynamic programming sequence comparison algorithms compute an
$(m + 1) \times (n + 1)$ matrix DP, where entry $DP[i, j]$ is set to the best score for
the problem of comparing A^i with B^j, and A^i is the prefix, $a_1, a_2, \ldots, a_i$ of A.
However, there are various applications, such as Cyclic String Comparison [8,

* Research supported in part by NSF grant CCR-0104307, by NATO Science Pro-
gramme grant PST.CLG.977017, by the Israel Science Foundation grant 282/01, by
the FIRST Foundation of the Israel Academy of Science and Humanities, and by
IBM Faculty Partnership Award.
** Research supported in part by the Aly Kaufman Post Doctoral Fellowship and by
the Bar-Nir Bergreen Software Technology Center of Excellence.

S.C. Sahinalp et al. (Eds.): CPM 2004, LNCS 3109, pp. 173–193, 2004.
© Springer-Verlag Berlin Heidelberg 2004

14], Common Substring Alignment Encoding [9–11], Approximate Overlap for DNA Sequencing [8] and more, which require the computation of the solution for the comparison of B with progressively longer suffixes of A, as defined below.

Definition 1. *The* **Consecutive Suffix Alignment Problem** *is, given two strings A and B, to compute the alignment solution of each suffix of A versus B.*

By *solution* we mean some encoding of a relevant portion of the DP matrix computed in comparing A and B. As will be seen in detail later, the data-dependencies of the fundamental recurrence, used to compute an entry $DP[i, j]$, is such that it is easy to extend DP to a matrix DP' for B versus Aa by computing an additional column. However, efficiently computing a solution for B versus aA given DP is much more difficult, in essence requiring one to work against the "grain" of these data-dependencies. The further observation that the matrix for B versus A, and the matrix for B versus aA can differ in $O(n^2)$ entries suggests that the relationship between such adjacent problems is non-trivial. One might immediately suggest that by comparing the reverse of A and B, prepending symbols becomes equivalent to appending symbols, and so the problem, as stated, is trivial. But in this case, we would ask for the delivery of a solution for B versus Aa. To simplify matters, we will focus on the core problem of computing a solution for B versus aA, given a "forward" solution for B versus A. A "forward" solution of the problem contains an encoding of the comparison of all (relevant) prefixes of B with all (relevant) prefixes of A. It turns out that the ability to efficiently prepend a symbol to A when given all the information contained in a "forward" solution allows one to solve the applications mentioned above with greater asymptotic efficiency then heretofore possible.

There are known solutions to the Consecutive Suffix Alignment problem for various string comparison metrics. For the LCS and Levenshtein distance metrics, the best previously published algorithm [8] for incremental string comparison computes all suffix comparisons in $O(nk)$ time, provided the number of differences in the alignment is bounded by parameter k. When the number of differences in the best alignment is not bounded, one could use the $O(n(n+m))$ results for incremental Levenshtein distance computation described in [8, 7]. Schmidt [14] describes an $O(nm)$ incremental comparison algorithm for metrics whose scoring table values are restricted to the interval $[-S, M]$. Here, we will focus on incremental alignment algorithms for the LCS metric.

The simplest form of sequence alignment is the problem of computing the *Longest Common Subsequence* (LCS) between strings A and B [1]. A *subsequence* of a string is any string obtained by deleting zero or more symbols from the given string. A *Common Subsequence* of A and B is a subsequence of both, and an LCS is one of greatest length. Longest Common Subsequences have many applications, including sequence comparison in molecular biology as well as the widely used *diff* file comparison program. The LCS problem can be solved in $O(mn)$ time, where m and n are the lengths of strings A and B, using dynamic programming [5]. More efficient LCS algorithms, which are based on the observation that the LCS solution space is highly redundant, try to limit the computation only to those entries of the DP table which convey essential information, and exploit in various ways the *sparsity* inherent to the LCS problem. Sparsity allows us to relate algorithmic performances to parameters other than the lengths of the input strings. Most LCS algorithms that exploit sparsity have their natural predecessors in either Hirshberg [5] or Hunt-Szymanski [6]. All Sparse LCS algorithms are preceded by an $O(n \log |\Sigma|)$ preprocessing [1]. The Hirshberg algorithm uses $L = |LCS[A, B]|$ as a parameter, and achieves an $O(nL)$ complexity. The Hunt-Szymanski algorithm utilizes as parameter the number of matches between A and B, denoted r, and achieves an $O(r \log L)$ complexity. Apostolico and Guerra [2] achieve an $O(L \cdot m \cdot \min(\log |\Sigma|, \log m, \log(2n/m))$ algorithm, where $m \leq n$ denotes the size of the shortest string among A and B, and another $O(m \log n + d \log(nm/d))$ algorithm, where $d \leq r$ is the number of dominant matches (as defined by Hirschberg [5]). This algorithm can also be implemented in time $O(d \log \log \min(d, nm/d))$ [4]. Note that in the worst case both d and r are $\Omega(n^2)$, while L is always bounded by n.

Note that the algorithms mentioned in the above paragraph compute the LCS between two strings A and B, however the objective of this paper is to compute all LCS solutions for each of the n suffixes of A versus B, according to Definition 1.

1.1   Results

In this paper we present two solutions to the Consecutive Suffix Alignment Problem under the LCS metric. The first solution (Section 3) is an $O(nL)$ time and space algorithm for constant alphabets, where n is the size of A, m is the size of B and $L \leq n$ denotes the size of the LCS of A and B. This algorithm computes a representation of the Dynamic Programming matrix for the alignment of each suffix of A with B.

The second solution (Section 4) is an $O(nL + n \log |\Sigma|)$ time, $O(L)$ space incremental algorithm for general alphabets, that computes the comparison solutions to $O(n)$ "consecutive" problems in the same asymptotic time as its standard counterpart [5] solves a single problem. This algorithm computes a representation of the last row of each of the Dynamic Programming matrices that are computed during the alignment of each suffix of A with B.

Both algorithms are extremely simple and practical, and use the most naive data structures.

Note that, due to lack of space, all proofs are omitted. A full version of the paper, including proofs to all lemmas, can be found in:

$$http : //www.cs.technion.ac.il/ \sim michalz/lcscsa.pdf$$

2   Preliminaries

An *LCS graph* [14] for A and B is a directed, acyclic, weighted graph containing $(|A| + 1)(|B| + 1)$ nodes, each labeled with a distinct pair $(x, y)(0 \leq x \leq |A|, 0 \leq y \leq |B|)$. The nodes are organized in a matrix of $(|A| + 1)$ rows and $(|B| + 1)$

columns. An index pair (x, y) in the graph where $A[x] = B[y]$ is called a *match*. The LCS graph contains a directed edge with a weight of zero from each node (x, y) to each of the nodes $(x, y+1)$, $(x+1, y)$. Node (x, y) will contain a diagonal edge with a weight of one to node $(x+1, y+1)$, if $(x+1, y+1)$ is a match.

Maximal-score paths in the LCS graph represent optimal alignments of A and B, and can be computed in $O(n^2)$ time and space complexity using dynamic programming. Alternatively, the LCS graph of A versus B can be viewed as a sparse graph of matches, and the alignment problem as that of finding highest scoring paths in a sparse graph of matches. Therefore, paths in the LCS Graph can be viewed as chains of matches.

Definition 2. *A k-sized chain is a path of score k in the LCS graph, going through a sequence of k matches $(x_1, y_1)(x_2, y_2) \ldots (x_k, y_k)$, such that $x_j < x_{j+1}$ and $y_j < y_{j+1}$ for successive matches (x_j, y_j) and (x_{j+1}, y_{j+1}).*

Both algorithms suggested in this paper will execute a series of n iterations numbered from n down to 1. At each iteration, an increased sized suffix of string A will be compared with the full string B. Increasing the suffix of A by one character corresponds to the extension of the LCS graph to the left by adding one column. Therefore, we define the growing LCS graph in terms of generations, as follows (see Figure 2).

Definition 3. Generation k *(G_k for short) denotes the LCS graph for comparing B with A_k^n. Correspondingly, L_k denotes $LCS[B, A_k^n]$, and reflects the size of the longest chain in G_k.*

We define two data structures, to be constructed during a preprocessing stage, that will be used by the consecutive suffix alignment algorithms for the incremental construction and navigation of the representation of the LCS graph for each generation (see Figure 1).

Definition 4. *MatchList(j) stores the list of indices of match points in column j of DP, sorted in increasing row index order.*

MatchLists can be computed in $O(n \log |\Sigma|)$ preprocessing time.

Definition 5. *NextMatch(i, A[j]) denotes a function which returns the index of the next match point in column j of DP with row index greater than i, if such a match point exists. If no such match point exists, the function returns $NULL$.*

A $NextMatch[i, \alpha]$ table, for all $\alpha \in \Sigma$, can be constructed in $O(n|\Sigma|)$ time and space. When the alphabet is constant, a $NextMatch$ table can be constructed in $O(n)$ time and space.

3   The First Algorithm

The first algorithm consists of a preprocessing stage that is followed by a main stage. During the preprocessing stage, the $NextMatch$ table for strings A and B is constructed.

MatchList:
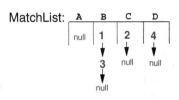

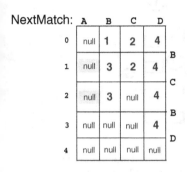

Fig. 1. The MatchList and NextMatch data structures.

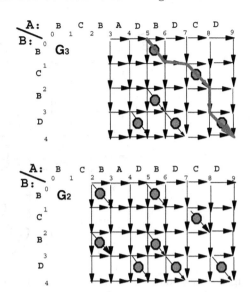

Fig. 2. The LCS Graphs G_3 and G_2 for the comparison of strings $A =$ "$BCBADBCDC$" versus $B =$ "$BCBD$". Grey octagons denote match points. A chain of size 3 is demonstrated in G_3, corresponding to the common subsequence "BCD".

During the main stage, the first algorithm will interpret the LCS graph for each generation as a dynamic programming graph, where node $[i, j]$ in G_k stores the value of the longest chain from the upper, leftmost corner of the graph up to node $[i, j]$. Therefore, we will formally define the graph which corresponds to each generation, as follows (see Figure 3).

Definition 6. DP^k *denotes the dynamic programming table for comparing string B with string A_k^n, such that $DP^k[i, j]$, for $i = 1 \ldots m$, $j = k \ldots m$, stores $LCS[B_1^i, A_k^j]$. DP^k corresponds to G^k as follows. $DP^k[i, j] = v$ if v is the size of the longest chain that starts in some match in the upper, left corner of G^k and ends in some match with row index $\leq i$ and column index $\leq j$.*

Using Definition 5, the objective of the first Consecutive Alignments algorithm could be formally defined as follows: **compute DP^k for each $k \in [1, n]$.**

Applying the dynamic programming algorithm to each of the n problems gives an $O(n^3)$ algorithm. It would be better if one could improve efficiency by incrementally computing DP^k from either DP^{k-1} or DP^{k+1}. At first glance this appears impossible, since computing DP^k from DP^{k+1} may require recomputing every value. Thus, attempting to compute the DP table for each problem incrementally appears doomed because the *absolute* value of as many as $O(n^2)$ elements can change between successive problem instances. However, based on

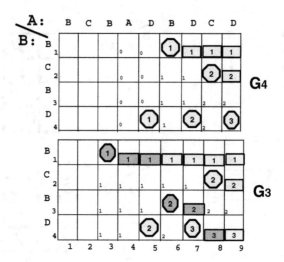

Fig. 3. The update operations applied by the first Consecutive Alignments algorithm, during the computation of the partition points of generation G_3 from the partition points of generation G_4. Partition points are indicated as rectangles and hexagons, and the numbers inside stand for their value. The hexagons represent partition points that are both partition points and match points. The grey rectangles and hexagons represent the new partition points in generation G_3.

the observation that each column of the LCS DP is a monotone staircase with unit-steps, we will apply *partition encoding* [5] to the DP table, and represent each column of DP^k by its $O(L)$ partition points (steps), defined as follows (see Figure 4).

Definition 7. P^k *denotes the set of partition points of* DP^k, *where partition point* $P^k[j, v]$, *for* $k = 1 \ldots n, j = k \ldots n, v = 0 \ldots L_k$, *denotes the first entry in column* j *of* DP^k *which bears the value of* v.

In terms of chains in the LCS graph, $P^k[j, v] = i$ if i is the lowest row index to end a chain that is contained in the first j columns of G_k. It is now clear that instead of computing DP^k it suffices to compute P^k for $k = n \ldots 1$.

3.1   Computing P^k from P^{k+1}

The consecutive alignments algorithm consists of n stages. The LCS graph for comparing strings B and A is grown from right to left in n stages. At stage k, column k is appended to the considered LCS graph. Correspondingly, P^k is obtained by inheriting the partition points of P^{k+1} and updating them as follows. The first, newly appended column of P^k has only one partition point - which is the first match point $[i, k]$ in column k (see column 3 of G_3 in Figure 3). This match point corresponds to a chain of size 1, and indicates the index i such that all entries in column k of DP^k of row index smaller than i are zero, and all

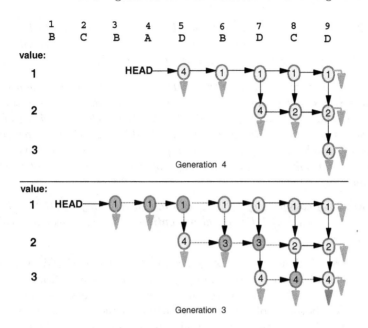

Fig. 4. The implementation of the partition-point data structure as a doubly-linked list. The grey circles represent the new partition points in generation G_3.

entries from index i and up are one. Therefore, stage k of the algorithm starts by computing, creating and appending the one partition point, which corresponds to the newly appended column k, to the partition points of P^k.

Then, the columns inherited from P^{k+1} are traversed in a left-to-right order, and updated with new partition points.

We will use two important observations in simplifying the update process. First, in each traversed column of P^k, at most one additional partition point is inserted, as will be shown in Lemma 2. We will show how to efficiently compute this new partition point. The second observation, which will be asserted in Conclusion 1, is that once the leftmost column j is encountered, such that no new partition point is inserted to column j of P^k, the update work for stage k of the algorithm is complete. Therefore, the algorithm will quit the column traversal and exit stage k when it hits the first, leftmost column j in P^k that is identical to column j of P^{k+1}.

The incremental approach applied in the first algorithm is based in the following lemma, which analyzes the differences in a given column from one generation of DP to the next.

Lemma 1. *Column j of DP^k is column j of DP^{k+1} except that all elements that start in some row I_j are greater by one. Formally, for column j of DP^k there is an index I_j such that $DP^k[i,j] = DP^{k+1}[i,j]$ for $i < I_j$ and $DP^k[i,j] = DP^{k+1}[i,j] + 1$ for $i \geq I_j$.*

The next Lemma immediately follows.

Lemma 2. *Column j in P^k consists of all the partition points which appear in column j of P^{k+1}, plus at most one new partition point. The new partition point is the smallest row index I_j, such that $delta[I_j] = DP^k[I_j, j] - DP^{k+1}[I_j, j] = 1$.*

Claim 3. *For any two rectangles in a DP table, given that the values of the entries in vertices in the upper and left border of the rectangles are the same and that the underlying LCS subgraphs for the rectangles are identical - the internal values of entries in the rectangles will be the same. Furthermore, adding a constant c to each entry of the left and top borders of a given rectangle in the DP table would result in an increment by c of the value of each entry internal to the rectangle.*

Conclusion 1: *If column j of DP^k is identical to column j of DP^{k+1}, then all columns greater than j of DP^k are also identical to the corresponding columns of DP^{k+1}.*

The correctness of Conclusion 1 is immediate from Claim 3. Given that the structure of the LCS graph in column $j + 1$ does not change from DP^{k+1} to DP^k, that the value of the first entry in the column remains zero, and that all values in its left border (column j of DP^k) remain the same as in DP^{k+1}, it is clear that the dynamic programming algorithm will compute the exact same values for column $j+1$ of DP^{k+1} and for column $j+1$ of DP^k. The same claim follows inductively when computing the values of column $j+2$ of DP^k from the values of column $j + 1$, and so on.

The suggested algorithm will traverse the columns of P^k from left to right. In each of the traversed columns it will either insert a new partition point or halt according to Conclusion 1.

3.2   Computing the New Partition Points of Column j of P^k

In this section we will show how to compute the new partition points of any column $j > k$ of P^k, using the partition points of column j of P^{k+1}, the partition points of column $j - 1$ of P^k, and the match points of column j of P^k. We start by constraining the range of row indices of column j in which the new partition point will be searched.

Lemma 3. *Let $I_{j-1} = P^k_{j-1}[v]$ denote the new partition point in column $j - 1$ of P^k, and let I_j denote the index of the new partition point in column j of P^k. $P^k_{j-1}[v] \leq I_j \leq P^k_{j-1}[v + 1]$.*

We will next show that there are two possible cases to consider when computing the new partition point of column j, as specified in the lemma below (see Figure 5).

Lemma 4. *Let $I_{j-1} = P^k_{j-1}[v]$ denote the row index of the new partition point in column $j - 1$ of P^k. Let I_j denote the row index of the new partition point in column j of P^k. I_j can assume one of two values, according to the following two cases.*

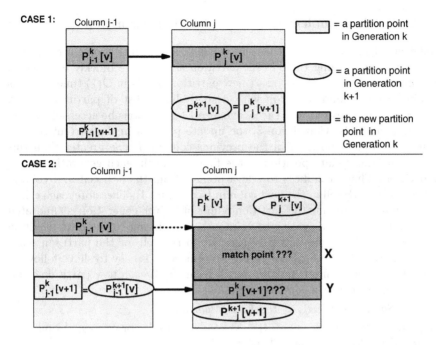

Fig. 5. The three possible scenarios to be considered when computing the new partition point of column j in generation G_k.

case 1. $I_{j-1} = P^k_{j-1}[v] \leq P^{k+1}_j[v]$, *in which case* $I_j = P^k_j[v] = I_{j-1}$.

case 2. $I_{j-1} = P^k_{j-1}[v] > P^{k+1}_j[v]$, *in which case*

$$I_j = P^k_j[v+1] = min\{P^k_{j-1}[v+1], NextMatch(I_{j-1} = P^k_{j-1}[v], j)\}.$$

Conclusion 2: *At each of the columns traversed by the algorithm, during the computation of P^k from the partition points of P^{k+1}, except for the last column that is considered for update, a single partition point is inserted. As for the last column considered in generation G_k, the algorithm quits the update of P^k, following Conclusion 1, upon realizing that there is no partition point to insert to this column, and it is therefore similar to the previous column.*

Conclusion 3: *The new partition point in column j of P^k, if such exists, is one of four options:*

1. *The new partition point of column $j - 1$.*
2. *The partition point that immediately follows the new partition point of column $j - 1$.*
3. *Some match point at an index that falls between the new partition point of column $j - 1$ and the match point that immediately follows in column j.*
4. *Some match point at an index that falls between the last partition point of column $j - 1$ and index $m + 1$.*

3.3   An Efficient Implementation of the First Algorithm

An efficient algorithm for the consecutive suffix alignments problem requires a data structure modelling the current partition that can be quickly updated in accordance with Lemma 4. To insert new partition points in $O(1)$ time suggests modelling each column partition with a singly-linked list of partition points. However, it is also desired that successive insertion locations be accessed in $O(1)$ time. Fortunately, by Conclusion 3, the update position in the current column is either the update position in the previous column or one greater than this position, and the update position in the first column in each generation is the first position. Thus, it suffices to add a pointer from the i-th cell in a column partition to the i-th cell in the next column (see Figure 4). Therefore, each cell in the mesh which represents the partition points of a given generation is annotated with its index, as well as with two pointers, one pointing to the next partition point in the same column and the other set to the cell for the partition point of the same value in the next column. Furthermore, it is easy to show, following Lemma 4, that the pointer updates which result from each new partition-point insertion can be correctly completed in $O(1)$ time.

Time and Space Complexity of the First Algorithm.

During the preprocessing stage, the $NextMatch$ table for strings A and B is constructed in $O(n|\Sigma|)$ time and space.

By conclusion 2, the number of times the algorithm needs to compute and insert a new partition point is linear with the final number of partition points in P^1. Given the $NextMatch$ table which was prepared in the preprocessing stage, the computation of the next partition point, according to Lemma 4, can be executed in constant time. Navigation and insertion of a new partition point can also be done in constant time according to Conclusion 3 (see Figure 4).

This yields an $O(nL)$ time and space complexity algorithm for constant alphabets.

4   The Second Algorithm

The second algorithm takes advantage of the fact that many of the Consecutive Suffix Alignment applications we have in mind, such as Cyclic String Comparison [8, 14], Common Substring Alignment Encoding [9–11], Approximate Overlap for DNA Sequencing [8] and more, actually require the computation of the last row of the LCS graph for the comparison of each suffix of A with B. Therefore, the objective of the second algorithm is to compute the partition encoding of the last row of the LCS graph for each generation. This allows to compress the space requirement to $O(L)$. Similarly to the first algorithm, the second algorithm also consists of a preprocessing stage and a main stage. This second algorithm performs better than the first algorithm when the alphabet size is not constant. This advantage is achieved by a main stage that allows the replacement of the $NextMatch$ table with a $MatchList$ data structure (see Figure 1). The $MatchList$ for strings A and B is constructed during the preprocessing stage.

4.1   An $O(L_k)$ Size $TAILS$ Encoding of the Solution for G_k

In this section we will examine the solution that is constructed from all the partition-point encodings of the last rows of DP^k, for $k = n \ldots 1$. We will apply some definitions and point out some observations which lead to the conclusion that the changes in the encoded solution, from one generation to the next, are constant. The main output of the second algorithm will be a table, denoted $TAILS$, that is defined as follows.

Definition 8. $TAILS[k, j]$ *is the column index of the j-th partition point in the last row of G_k. In other words, $TAILS[k, j] = t$ if t is the smallest column index such that $LCS[B, A_k^t] = j$.*

Correspondingly, the term *tail* is defined as follows.

Definition 9. *Let t denote the value at entry j of row k of $TAILS$.*

1. *t is considered a **tail** in generation G_k (see Figures 6, 7).*
2. *The **value** of tail t in generation G_k, denoted val_t, is j. That is, $LCS[A_k^t, B] = j$.*

It is easy to see that, in a given generation, tails are ordered in left to right *column* order and increasing *size*.

   In the next lemma we analyze the changes in the set of values from row $k+1$ to row k of $TAILS$, and show that this change is $O(1)$.

Lemma 5. *If column k of the LCS graph contains at least one match, then the following changes are observed when comparing row $k + 1$ of $TAILS$ to row k of $TAILS$:*

1. *$TAILS[k, 1] = k$.*
2. *All other entries from row $k + 1$ are inherited by row k, except for at most one entry which could be lost:*
   Case 1. *All entries are passed from row $k + 1$ to row k of tails and shifted by one index to the right. In this case $LCS[B, A_k^n] = LCS[B, A_{k+1}^n] + 1$.*
   Case 2. *One entry value, which appeared in row $k + 1$ disappears in row k. In this case $LCS[B, A_k^n] = LCS[B, A_{k+1}^n]$.*
      – *All values from row $k + 1$ of $TAILS$ up to the disappearing entry are shifted by one index to the right in row k of $TAILS$.*
      – *All values from row $k + 1$ of $TAILS$ which are greater than the disappearing entry remain intact in row k of $TAILS$.*

From the above lemma we conclude that, in order to compute row k of $TAILS$, it is sufficient to find out whether or not column k of G contains at least one match point, and if so to compute the entry which disappeared from row $k + 1$ of $TAILS$. Hence, from now on the algorithm will focus only on columns where there is at least one match point and on discovering, for these columns, which entry (if at all) disappears in the corresponding row of $TAILS$.

   From now on we will focus on the work necessary for updating the set of L_k values from row $k + 1$ of $TAILS$ to row k of $TAILS$. Therefore, we simplify the

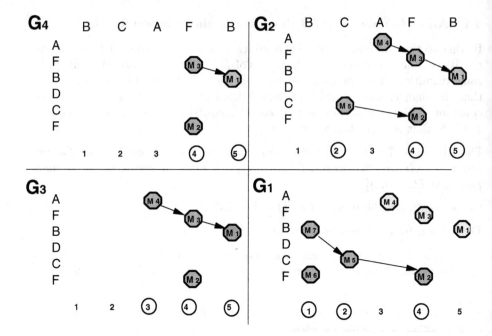

Fig. 6. The evolution of leftmost chains from chains that are not necessarily leftmost. For each generation, the dark circles around column indices directly below the bottom row of the graph mark active tails.

notation to focus on the L_k values in row k of $TAILS$. We note that these L_k values denote column indices of leftmost-ending chains of sizes $1 \ldots L_k$ in G_k. We will refer to these values from now on as the set of *tails* of generation G_k.

4.2 The $O(L^2)$ Active Chains in a Given Generation

In this section we will describe the new data structure which is the core of our algorithm. Note that $TAILS[k, j] = t$ if t is the index of the smallest column index to end a j-sized chain in G_k. So, in essence, in iteration k of the algorithm we seek all leftmost-ending chains of sizes $1 \ldots L_k$ in the LCS graph G_k.

Recall that, in addition to the output computation for G_k, we have to prepare the relevant information for the output computation in future generations. Therefore, in addition to the $O(L_k)$ leftmost ending chains we also wish to keep track of chains which have the potential to become leftmost chains in some future generation. Note that a leftmost chain of size j in a given generation does not necessarily evolve from a leftmost chain of size $j-1$ in some previous generation (see Figure 6). This fact brings up the need to carefully define the minimal set of chains which need to be maintained as candidates to become leftmost chains in some future generation.

By definition, each chain starts in a match (the match of smallest row index and leftmost column index in the chain) and ends in a match. At this stage it is already clear that an earlier (left) last-match is an advantage in a chain, according to the tail definition. It is quite intuitive that a lower first-match is an advantage as well, since it will be easier to extend it by matches in future columns. Hence, a chain of size k is redundant if there exists another chain of size k that starts lower and ends to its left. Therefore, we will maintain as candidates for expansion only the non-redundant chains, defined as follows.

Definition 10. *A chain c_1 of size j is an* **active chain** *in generation G_k, if there does not exist another chain c_2 of size j in G_k, such that both conditions below hold:*

1. *c_2 starts lower than c_1.*
2. *c_2 ends earlier than c_1.*

For the purpose of tail computation, it is sufficient to maintain the row index of the first match in each chain and the column index of the last match in each chain.

- The row number of the first match in an active chain is denoted a *head*.
- The column index of a last match in an active chain is denoted an *end-point*.

Note that two or more different chains could share the same head in a given generation. For example, match m_7, corresponding to a head of row index 3, is the origin of active chains of sizes $2-3$ in generation G_1 of Figure 6. Based on this observation, we decided to count the number of different matches which serve as heads and end-points in a given generation. To our surprise, we discovered that in a given generation G_k, the number of distinct heads is only L_k (see Conclusion 4), and the number of distinct end-points in G_k is only L_k (see Lemma 6 which comes up next). This observation is the key to the efficient state encoding in our algorithm.

Lemma 6. *Each active chain ends in a tail.*

We have shown in Lemma 6 that each end-point is a tail. Therefore, from now on we will use the term *tail* when referring to end-points of active chains. We consider two active chains of identical sizes which have the same head and the same tail as one.

4.3   An $O(L_k)$ *HEADS* Representation
of the State Information for G_k

In this section we will show that the number of distinct heads in generation G_k is exactly L_k. In order to count the distinct heads, we associate with each tail a set of relevant heads, as follows.

Definition 11. *H_t denotes the set of heads of active chains that end in tail t.*

The active heads in G_k are counted as follows. The tails are traversed left to right, in increasing size and index, and the new heads contributed by each tail

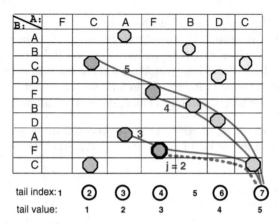

Fig. 7. The set of chains H_7 for the tail with value 5 and index 7 in generation G_2 of the consecutive suffix alignment of strings $A =$ "$BCBADBCDC$" versus $B =$ "$BCBD$". The head which is new_7 is highlighted with a thicker border, and the corresponding shortest chain of size 2 is dotted. The dark circles around column indices directly below the bottom row of the graph mark the active tails in G_2.

are noted (a head h is contributed by tail t if $h \in H_t$ and $h \notin H_{t_1}$ for any $t_1 < t$). The counting of the active heads which are contributed by each tail t will be based on the following two observed properties of H_t. These properties, given below, will be proven in the rest of this section.

Property 1 of H_t. Let j denote the size of the smallest chain in H_t. The chains headed by H_t form a consecutive span, ordered by increasing head height and increasing chain size, starting with the lowest head which is the origin of the j-chain of H_t, and ending with the highest head which is the origin of the val_t-chain which ends in t (see Figure 7).

Property 2 of H_t. The head which is the origin of the smallest chain (size j) of H_t is the one and only new head in H_t. All other heads are included in H_{t_1} for some $t_1 < t$.

The following Lemmas 7 to 9 formally assert the two observed properties of H_t.

Lemma 7. *The heads of H_t are ordered in increasing height and increasing chain size.*

Lemma 8. *For any tail t, the sizes of active chains which correspond to the heads in H_t form a consecutive span.*

Lemma 9. *The head h_1 of the smallest chain in H_t is new. That is, there is no active chain that originates in h_1 and ends in some tail to the left of t.*

Conclusion 4: *From Lemmas 8 and 9 we conclude that as we scan the tails for generation G_k from left to right, each tail contributes exactly one new head to the expanding list of active heads. Therefore, there are exactly L_k different row*

*indices which serve as active heads (to one or more active chains) in genera-
tion G_k.*

The new head that is contributed by each tail t is a key value which will
represent the full H_t set for tail t in our algorithm, and therefore it is formally
defined and named below.

Definition 12. *new_t is the head of the smallest chain in H_t.*

We have found one more property of H_t which will be relevant to our algorithm,
as proven in the next lemma.

Lemma 10. *H_t includes all heads that are higher than new_t and start at least
one active chain which ends in some tail $t_3 < t$.*

Up till now we have analyzed the set H_t of heads that start all active chains
which end in tail t. Next, we will symmetrically analyze the set of tails that end
chains which originate in a given active head h.

We associate with each head a set of relevant tails, as follows.

Definition 13. *T_h denotes the set of tails of active chains that start in head h.*

Lemma 11. *For any head h, the sizes of active chains which correspond to the
tails in T_h form a consecutive span.*

4.4 Changes in *HEADS* and *TAILS*
from One Generation to the Next

In this section we discuss the changes in the sets of active heads and tails as the
poset of matches for generation G_{k+1} is extended with the matches of column k.
Following the update, some changes are observed in the set of active heads, in
the set of active tails, and in the head-to-tail correspondence which was analyzed
in the previous section. (When we say head-to-tail correspondence, we mean the
pairing of head h_i with a tail t_i such that $h_i = new_{t_i}$.)

Throughout the algorithm, the relevant state information will be represented
by a dynamic list *HEADS* of active heads, which is modified in each generation
G_k, based on the match points in column k of *DP*.

Definition 14. *$HEADS_k$ denotes the set of active heads in generation G_k,
maintained as a list which is sorted in increasing height (decreasing row index).
Each head $h_{first} \in HEADS_k$ is annotated with two values. One is its height,
and the second is the tail t such that $h_{first} = new_t$.*

In iteration k of the algorithm, two objectives will be addressed.

1. The first and main objective is to compute the tail that dies in G_k. In Lemma
   12 we will show that, for any tail t that was active in G_{k+1}, the size of H_t
   can only decrease by one in G_k. Therefore, the tail to disappear in G_k is the
   tail t such that the size of H_t decreases from one to zero in G_k.
2. The second objective is to update the state information ($HEADS_{k+1}$ list)
   so it is ready for the upcoming computations of G_{k-1} ($HEADS_k$ list).

In this section we will show that both of the objectives above can be achieved by first merging $HEADS_{k+1}$ with the heights of matches in column k, and then traversing the list of heads once, in a bottom-up order, and modifying, in constant time, the head-to-tail association between active tails and their *new* head representative, if such an association indeed changes in G_k. (That is, if a given head h was new_t for some tail t in G_{k+1} and h is no longer new_t in G_k).

Lemma 12. *From one generation to the next, the number of active heads in H_t can only decrease by one. Furthermore, of all the chains that start in some head in H_t and end in t, only the shortest chain, the one headed by new_t in G_{k+1}, could be de-activated in G_k without being replaced by a lower head of a similar-sized active chain to t.*

From this we conclude that the tail to disappear from row k of $TAILS$ is the tail t such that the number of heads in H_t went down from one to zero in generation G_k. It remains to show how this dying tail t can be identified during the single, bottom up traversal of the list $HEADS_{k+1}$, following the merge with the matches of column k.

We are now ready to address the merging of the match points from column k with $HEADS_{k+1}$. The discussion of how the matches that are merged with $HEADS_{k+1}$ affect its transformation into $HEADS_k$ will be partitioned into four cases. First we discuss the first (lowest) match and the heads which fall below it. We then explain what happens to two consecutive matches with no head in between. The third case deals with the matches above the highest head in $HEADS_{k+1}$, if such exist. The fourth and main part of our discussion, deals with the changes to the "slice" of heads in $HEADS_{k+1}$ which fall either between two consecutive new matches in column k, or above the highest match in column k.

Case 1: The lowest match in column k. The first match in column k is a new head. It is the first chain, of size 1, of the tail k, and therefore is new_k. All heads below this match are unaffected, since no new chain that starts lower than these heads could have possibly been created in G_k.

Case 2: Two consecutive matches with no heads in between. For any sequence of consecutive matches in column k with no head in between, all match points, except for the lowest match point in the sequence, are redundant.

Case 3. Matches above the highest head in $HEADS_{k+1}$. The lowest match in column k which is above the highest active head in $HEADS_{k+1}$, if such a match exists, becomes a new head. Consider the longest chain in G_{k+1}, of size L_{k+1}, that ends in tail L_{k+1}. Clearly, this chain's head is the highest head in the list. This chain will be extended by the new match to a lowest, leftmost $L_k = L_{k+1} + 1$ chain, and therefore this match is a new head.

Case 4. The series of heads that fall between two consecutive matches. This case, which includes the series of remaining heads above the highest match in column k, is the most complex case and covers the identification of the disappearing tail. It will therefore be discussed in detail in the next subsection.

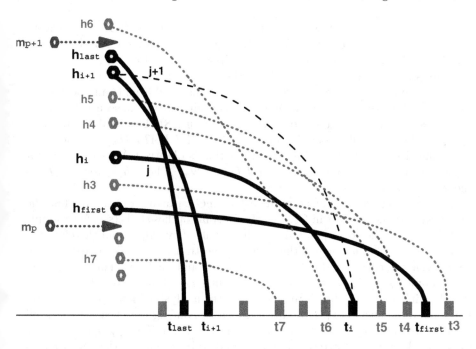

Fig. 8. The $HEADS$ list traversal in increasing height during its update with the match points of column k in generation G_k. The modified heads, as well as their corresponding tails and *new* chains are highlighted in black.

4.5   Heads That Fall between Two Matchpoints in Column k

Throughout this subsection we will use the following notation, as demonstrated in Figure 8.

- Let m_p, m_{p+1} denote two consecutive matches in column k, such that m_{p+1} is higher than m_p.
- Let h_{first} denote the first head in $HEADS_{k+1}$ which is higher than or equal to m_p and lower than m_{p+1}.
- Let $UPDATED_{m_p,k}$ denote the series of heads $h_i = h_{first} \ldots h_{last}$ which fall between m_p and m_{p+1} and whose head-to-tail association changed in generation k, ordered in increasing height. Let $t_i = t_{first} \ldots t_{last}$ denote the series of corresponding tails.
- Let h_{i+1} and h_i denote any pair of consecutive heads in $UPDATED_{m_p,k}$.

Consider the set of heads from $HEADS_{k+1}$ that fall between m_p and m_{p+1}. In this subsection we will show that some of the heads in this set remain unmodified in $HEADS_{k+1}$ (see, for example, heads h_3, h_4 and h_5 in Figure 8) while others change (see, heads h_{first}, h_i, h_{i+1} and h_{last} in Figure 8).

Lemma 14 shows that all heads in $HEADS_{k+1}$ that fall between m_p and m_{p+1} and are not in $UPDATED_{m_p,k}$ remain unchanged in G_k. Lemma 13 claims that $h_{first} \in UPDATED_{m_p,k}$, since it dies in generation G_k and is replaced with m_p. In Conclusion 8 we assert that all heads in $UPDATED_{m_{p,k}}$, except for h_{first}, remain active in G_k and survive the transformation of $HEADS_{k+1}$ to $HEADS_k$. Additional characteristics of the heads in the set $UPDATED_{m_{p,k}}$ are then investigated. In Lemma 14 we learn that the heads in $UPDATED_{m_{p,k}}$ and their corresponding tails form a series of increasing head heights and decreasing tail column indices, such that the *new* chains from any two consecutive heads in the series must cross. (See the dark chains in Figure 8). The challenge of modifying the head-to-tail association of the heads in $UPDATED_{m_{p,k}}$ is addressed in Lemma 16, and an interesting chain-reaction is observed. We show that, for any two consecutive heads $h_i, h_{i+1} \in UPDATED_{m_{p,k}}$, head h_{i+1} replaces h_i as the new new_{t_i} in G_k.

Next, we consider the tails that are active in G_{k+1} in order to try and find the tail which becomes extinct in G_k. Clearly, for some of these tails (see, for example t_6 in Figure 8), the corresponding *new* head falls above m_{p+1} and therefore the decision as to whether or not they survive the transition to G_k is delayed till later when the corresponding span is traversed and analyzed. For others (see, for example t_7 in Figure 8), the corresponding *new* head falls below m_p and therefore it has already been treated during the analysis of some previous span. For some tails (such as t_3, t_4 and t_5), the corresponding *new* heads indeed fall between m_p and m_{p+1} but are not included in $UPDATED_{m_{p,k}}$, and therefore these tails keep their shortest chain as is, and will also survive the transition to G_k. In Conclusion 8 we assert that the tails which correspond to all heads in $UPDATED_{m_{p,k}}$, except for t_{last}, are kept alive in G_k. As for t_{last}, this is the only candidate for extinction in G_k, and in Lemma 17 we show that in the last span of traversed heads, if the highest match in column k falls below the highest head in $HEADS_{k+1}$, then t_{last} of this last span will finally be identified as the dying tail in G_k.

Lemma 13. *h_{first} is no longer an active head in G_k. Instead, the height of match m_p replaces h_{first} in $HEADS_k$.*

We have shown that h_{first} dies in generation G_k and is replaced with the height of m_p in $HEADS_k$. From this we conclude that h_{first} is the first head in $UPDATED_{m_{p,k}}$. The next lemma will help further separate the heads which participate in $UPDATED_{m_{p,k}}$ from the heads that remain unmodified from $HEADS_{k+1}$ to $HEADS_k$.

Lemma 14. *Consider two heads $h_1, h_2 \in HEADS_{k+1}$, such that h_2 is higher than h_1, and given that there is no match point in column k which falls between h_1 and h_2. Let $h_1 = new_{t_1}$ and $h_2 = new_{t_2}$. If $t_1 < t_2$, then the chain from h_2 to t_2 remains active in G_k.*

The above Lemma immediately leads to the following conclusion.

Conclusion 5 *The heads in $UPDATED_{m_{p,k}}$ and their corresponding tails form a series of increasing head heights and decreasing tail column indices.*

We have shown that all heads which are not in $UPDATED_{m_{p,k}}$ remain unmodified. We will next show that all heads in $UPDATED_{m_{p,k}}$, even though modified, survive the transformation from $HEADS_{k+1}$ to $HEADS_k$. In order to do so we first prove the following two lemmas, which lead to the conclusion that the modifications in the head-to-tail associations of heads in $UPDATED_{m_p,k}$ consist of a chain reaction in which each head is re-labelled with the tail of the head below.

Lemma 15. *Let h_i and h_{i+1} denote two heads in $HEADS_{k+1}$, such that $h_i = new_{t_i}$, $h_{i+1} = new_{t_{i+1}}$, and h_{i+1} is the first head above h_i such that its corresponding new tail t_{i+1} falls to the left of t_i. Let j denote the size of the chain from h_i to t_i. If the chain from h_i to t_i becomes de-activated in G_k, then all chains that originate in h_{i+1} and are of sizes smaller than or equal to j will be de-activated in G_k.*

The following conclusion is immediate from the above Lemma 15 and the definition of $UPDATED_{m_p,k}$.

Conclusion 6. *Let h_i and h_{i+1} denote two heads in $HEADS_{k+1}$, such $h_i = new_{t_i}$, $h_{i+1} = new_{t_{i+1}}$, and h_{i+1} is the first head above h_i such that its corresponding new tail t_{i+1} falls to the left of t_i. If $h_i \in UPDATED_{m_p,k}$, then it follows that $h_{i+1} \in UPDATED_{m_p,k}$.*

Observation 1.
For any tail t_i, if the new chain from new_{t_i} to t_i becomes de-activated in G_k, and let j denote the size of the active chain from new_{t_i} to t_i in G_{k+1}. In generation G_k H_{t_i} will no longer include any head of a j-sized chain to t_i.

The above observation is correct since, by definition, an active chain to a given tail can only evolve in G_k by extending a chain shorter by one to the same tail, a chain that was active in G_{k+1}, with a new match from column k. However the fact that the j-sized chain to t_i was the *new* chain of H_{t_i} in G_{k+1} implies, by definition, that there was no shorter active chain to t_i in G_{k+1}. Therefore, H_{t_i} no longer includes any head of a j-sized chain to t_i in G_k.

Lemma 16. *[Chain Reaction.] For any two consecutive heads $h_i, h_{i+1} \in UPD\text{-}ATED_{m_p,k}$, such that $h_i = new_{t_i}$ in G_{k+1}. In generation G_k, h_{i+1} becomes new_{t_i}.*

The above lemma implies that, for any two consecutive heads $h_i, h_{i+1} \in UPD\text{-}ATED_{m_p,k}$, there is, in G_k, an active chain from t_i to h_{i+1}. Since all it takes is one active chain per generation to keep the head that starts this chain and the tail that ends this chain alive, this leads to the the following conclusion.

Conclusion 7.

- The heads $h_{first+1} \ldots h_{last} \in UPDATED_{m_p,k}$ remain active in G_k.
- The corresponding tails $t_{first} \ldots t_{last-1}$ remain active in G_k.

The only two critical indices in any $UPDATED$ series, which are not included in the above lists of heads and tails that remain active in G_k, are h_{first} and t_{last}. Since we already know, by Lemma 13, that any head that serves as h_{first} to some

$UPDATED$ series becomes extinct in G_k and is replaced with the height of the highest match point below, the only remaining issue to settle is what happens to tails that serve as t_{last} in G_k. This is done in the next Lemma, which claims that all tails that serve as t_{last} for some $UPDATED$ series between two match points remain active in G_k, and the tail that serves as t_{last} for the last span of heads in $HEADS_{k+1}$, if there is indeed no match point in column k above the highest head in this span, is the tail that becomes extinct in G_k.

Lemma 17. *The tail to disappear from row k of $TAILS$, i.e. the tail which becomes inactive in generation G_k, is the t_{last} of the last $UPDATED$ series of $HEADS_{k+1}$.*

The second algorithm computes the rows of $TAILS$ incrementally, in decreasing row order. Row k of $TAILS$ will be computed from row $k + 1$ of $TAILS$ by inserting the new tail k, (if such exists) and by removing the "disappearing" tail (if such exists). The algorithm maintains a dynamic list $HEADS$ of active heads. Each head is annotated with two fields: its height and a label associating it with one of the active tails t for which it is new_t. Upon the advancement of the computation from row $k + 1$ of the $TAILS$ table to row k, the poset of matches is extended by one column to the left to include the matches of column k of the LCS graph for A versus B. Given the list $HEADS_{k+1}$, sorted by increasing height, the algorithm computes the new list $HEADS_k$, obtained by merging and applying the matches of column k to $HEADS_{k+1}$, and the "disappearing entry" for row k of $TAILS$ is finally realized.

Lemma 18. $r \leq nL$

Time and Space Complexity of the Second Algorithm.

Since $r \leq nL$, the total cost of merging r matches with n lists of size L each is $O(nL)$. In iteration k, up to L_{k+1} new head height values may be updated, and up to one new head created. The linked list of L_{k+1} heads is then traversed once, and for each item on the list up to one, constant time, swap operation is executed. Therefore, the total work for n iterations is $O(nL)$. There is an additional $O(n \log |\Sigma|)$ preprocessing term for the construction of Match Lists. (Note that we only need to create match lists for characters appearing in B, and that $|\Sigma| \leq n$). Thus, the second algorithm runs in $O(nL + n \log |\Sigma|)$ time, and requires $O(L)$ space.

References

1. A. Apostolico, String editing and longest common subsequences. In G. Rozenberg and A. Salomaa, editors, *Handbook of Formal Languages*, Vol. 2, 361–398, Berlin, 1997. Springer Verlag.
2. Apostolico A., and C. Guerra, The longest common subsequence problem revisited. *Algorithmica*, **2**, 315–336 (1987).
3. Carmel, D.,N. Efraty , G.M. Landau, Y.S. Maarek and Y. Mass, An Extension of the Vector Space Model for Querying XML Documents via XML Fragments, *ACM SIGIR'2002 Workshop on XML and IR*, Tampere, Finland, Aug 2002.

4. Eppstein, D., Z. Galil, R. Giancarlo, and G.F. Italiano, Sparse Dynamic Programming I: Linear Cost Functions, *JACM*, **39**, 546–567 (1992).

5. Hirshberg, D.S., "Algorithms for the longest common subsequence problem", *JACM*, **24**(4), 664–675 (1977).

6. Hunt, J. W. and T. G. Szymanski. "A fast algorithm for computing longest common subsequences." *Communications of the ACM* , **20** 350–353 (1977).

7. Kim, S., and K. Park, "A Dynamic Edit Distance Table.", *Proc. 11th Annual Symposium On Combinatorial Pattern Matching*, 60–68 (2000).

8. Landau, G.M., E.W. Myers and J.P. Schmidt, Incremental string comparison, *SIAM J. Comput.*, **27**, 2, 557–582 (1998).

9. Landau, G.M. and M. Ziv-Ukelson, On the Shared Substring Alignment Problem, *Proc. Symposium On Discrete Algorithms*, 804–814 (2000).

10. Landau, G.M., and M. Ziv-Ukelson, On the Common Substring Alignment Problem, *Journal of Algorithms*, **41**(2), 338–359 (2001)

11. G. M. Landau, B. Schieber and M. Ziv-Ukelson, Sparse LCS Common Substring Alignment, *CPM 2003*, 225-236

12. Myers, E. W., "Incremental Alignment Algorithms and their Applications," *Tech. Rep. 86-22, Dept. of Computer Science, U. of Arizona.* (1986).

13. Myers, E. W., "An O(ND) Difference Algorithm and its Variants," *Algorithmica*, **1**(2): 251-266 (1986).

14. Schmidt, J.P., All Highest Scoring Paths In Weighted Grid Graphs and Their Application To Finding All Approximate Repeats In Strings, *SIAM J. Comput*, **27**(4), 972–992 (1998).

15. Sim, J.S., C.S. Iliopoulos and K. Park, "Approximate Periods of Strings." *Proc. 10th Annual Symposium On Combinatorial Pattern Matching*, 132-137 (1999).

Efficient Algorithms for Finding Submasses in Weighted Strings

Nikhil Bansal[1], Mark Cieliebak[2,3], and Zsuzsanna Lipták[4]

[1] IBM Research, T.J. Watson Research Center
P.O. Box 218, Yorktown Heights, NY 10598
nikhil@us.ibm.com
[2] ETH Zurich, Institute of Theoretical Computer Science
Clausiusstr. 49, CH-8092 Zurich
cieliebak@inf.ethz.ch
[3] Center for Web Research, Department of Computer Science, University of Chile
[4] Universität Bielefeld, AG Genominformatik, Technische Fakultät
Postfach 10 01 31, D-33592 Bielefeld
zsuzsa@cebitec.uni-bielefeld.de

Abstract. We study the Submass Finding Problem: Given a string s over a weighted alphabet, i.e., an alphabet Σ with a weight function $\mu : \Sigma \to \mathbb{N}$, decide for an input mass M whether s has a substring whose weights sum up to M. If M is indeed a submass, then we want to find one or all occurrences of such substrings. We present efficient algorithms for both the decision and the search problem. Furthermore, our approach allows us to compute efficiently the number of different submasses of s. The main idea of our algorithms is to define appropriate polynomials such that we can determine the solution for the Submass Finding Problem from the coefficients of the product of these polynomials. We obtain very efficient running times by using Fast Fourier Transform to compute this product. Our main algorithm for the decision problem runs in time $\mathcal{O}(\mu_s \log \mu_s)$, where μ_s is the total mass of string s. Employing standard methods for compressing sparse polynomials, this runtime can be viewed as $\mathcal{O}(\sigma(s) \log^2 \sigma(s))$, where $\sigma(s)$ denotes the number of different submasses of s. In this case, the runtime is independent of the size of the individual masses of characters.

1 Introduction

Over the past few years, interest in the area of weighted strings has received increasing attention. A *weighted string* is defined over an alphabet $\Sigma = \{a_1, \ldots, a_{|\Sigma|}\}$ with a weight function $\mu : \Sigma \to \mathbb{N}$, which assigns a specific weight (or mass) to each character of the alphabet. The weight of a string s is just the sum of the weights of all characters in s.

Several applications from bioinformatics can be formalized as problems on strings over a weighted alphabet; most notably, mass spectrometry experiments, which constitute an experimentally very efficient and therefore promising alternative method of protein identification and de-novo peptide sequencing, and are also increasingly being used for DNA.

S.C. Sahinalp et al. (Eds.): CPM 2004, LNCS 3109, pp. 194–204, 2004.

For our purposes, proteins are strings over the 20-letter amino acid alphabet. The molecular masses of the amino acids are known up to high precision. In order to enforce that the masses be positive integers, we assume that non-integer masses have been scaled. One of the main applications of protein mass spectrometry is database lookup. Here, a protein is cut into substrings (*digestion*), the molecular mass of the substrings is determined, and the list of masses is compared to a protein database [1–3].

In the following, we skip some of the proofs due to space limitations.

1.1 Definitions and Problem Statements

Before we define the problems we are going to solve in this paper, we first need to fix some notation for weighted strings. Given a finite alphabet Σ and a mass function $\mu : \Sigma \to \mathbb{N}$, where $\mathbb{N}$ denotes the set of positive integers (excluding 0). We denote by $\mu_{\max} = \max \mu(\Sigma)$, the largest mass of a single character. For a string $s = s_1 \ldots s_n$ over Σ, define $\mu(s) := \sum_{i=1}^{n} \mu(s_i)$. We denote the length n of s by $|s|$. We call $M > 0$ a *submass* of s if there exists a substring t of s with mass M, or, equivalently, if there is a pair of indices (i, j) such that $\mu(s_i \ldots s_j) = M$. We call such a pair (i, j) a *witness* of M in s, and we denote the number of witnesses of M in s by $\kappa(M) = \kappa(M, s)$. Note that $\kappa(M) \leq n$. We want to solve the following problems:

Submass Query Problem. Fix a string s over Σ. Let $|s| = n$.
INPUT: k masses $M_1, \ldots, M_k \in \mathbb{N}$.
OUTPUT: A subset $I \subseteq \{1, \ldots, k\}$ such that $i \in I$ if and only if M_i is a submass of s.

Submass Witness Problem. Fix a string s over Σ. Let $|s| = n$.
INPUT: k masses $M_1, \ldots, M_k \in \mathbb{N}$.
OUTPUT: A subset $I \subseteq \{1, \ldots, k\}$ such that $i \in I$ if and only if M_i is a submass of s, and a set $\{(b_i, e_i) : i \in I, (b_i, e_i) \text{ is witness of } M_i\}$.

Submass All Witnesses Problem. Fix a string s over Σ. Let $|s| = n$.
INPUT: k masses $M_1, \ldots, M_k \in \mathbb{N}$.
OUTPUT: A subset $I \subseteq \{1, \ldots, k\}$ such that $i \in I$ if and only if M_i is a submass of s, and for each $i \in I$, the set of all witnesses $W_i := \{(b, e) : (b, e) \text{ is witness of } M_i \text{ in } s\}$.

The three problems above can be solved by a simple algorithm, which we refer to as LINSEARCH. It moves two pointers along the string, one pointing to the potential beginning and the other to the potential end of a substring with mass M. The right pointer is moved if the mass of the current substring is smaller than M, the left pointer, if the current mass is larger than M. The algorithm solves each problem in $\Theta(kn)$ time and $\mathcal{O}(1)$ space in addition to the storage space required for the input string and the output.

Another algorithm, BINSEARCH, computes all submasses of s in a preprocessing step and stores the submasses in a sorted array, which can then be queried in time $\mathcal{O}(\log n)$ for an input mass M for the SUBMASS QUERY PROBLEM and the SUBMASS WITNESS PROBLEM. The storage space required is pro-

portional to $\sigma(s)$, the number of different submasses of string s. For the SUBMASS
ALL WITNESSES PROBLEM, we need to store in addition all witnesses, requir-
ing space $\Theta(n^2)$; in this case, the query time becomes $\mathcal{O}(k \log n + K)$, where
$K = \sum_{i=1}^{k} \kappa(M_i)$ is the number of witnesses for the query masses. Note that
any algorithm solving the SUBMASS ALL WITNESSES PROBLEM will have runtime $\Omega(K)$.

In this paper, we present a novel approach to the problems above which often
outperforms the naïve algorithms. The main idea is similar to using generating
functions for counting objects, which have been applied, for instance, in attacking
the Coin Change Problem [4]. We apply similar ideas using finite polynomials
rather than infinite ones as follows. We define appropriate polynomials such
that we can determine the solution for the three problems above from the coefficients of the product of these polynomials. We will obtain very efficient running
times by using Fast Fourier Transform to compute this product. More precisely,
ALGORITHM 1 solves the SUBMASS QUERY PROBLEM with preprocessing time
$\mathcal{O}(\mu_s \log \mu_s)$, query time $\mathcal{O}(k \log n)$ and storage space $\Theta(\sigma(s))$, where μ_s denotes
the total mass of the string s. For the SUBMASS WITNESS PROBLEM, we present
a Las Vegas algorithm, ALGORITHM 2, with preprocessing time $\mathcal{O}(\mu_s \log \mu_s)$, expected query time $\mathcal{O}(\mu_s \log^3 \mu_s + k \log n)$, and storage space $\Theta(\sigma(s))$. Finally, we
present ALGORITHM 3, a deterministic algorithm for the SUBMASS ALL WIT-
NESSES PROBLEM with preprocessing time $\mathcal{O}((K n \mu_s \log \mu_s)^{\frac{1}{2}})$ and running time
$\mathcal{O}((K n \mu_s \log \mu_s)^{\frac{1}{2}})$, where K is the output size, i.e., the total number of witnesses.

Many algorithms for weighted strings, such as BINSEARCH, have a space
complexity which is proportional to $\sigma(s)$, the number of submasses of s. For this
reason, we define the following problem:

Number of Submasses Problem. Given string s of length n, how many
different submasses does s have?

This problem is of interest because we can use $\sigma(s)$ to choose between algorithms whose complexity depends on this number. It is open how the number
of submasses of a given string can be computed efficiently. It can, of course, be
done in $\Theta(n^2 \log \sigma(s))$ time by computing the masses of all substrings $s_i \ldots s_j$,
for all pairs of indices $1 \le i \le j \le n$, and counting the number of different
masses. We show how ALGORITHM 1 can be adapted to solve the NUMBER OF
SUBMASSES PROBLEM in time $\mathcal{O}(\mu_s \log \mu_s)$, outperforming the naïve algorithm
for small values of μ_s.

Throughout the paper, we present our runtimes as a function of μ_s, the total
mass of the string s. However, we can use the technique of Cole and Hariharan [5]
in a straightforward way to give a Las Vegas algorithm with expected running
time in terms of the number of distinct submasses. This transformation loses a
factor of at most $O(\log n)$ in the running time.

1.2   Related Work

In [3], several algorithms for the SUBMASS QUERY PROBLEM were presented,
including LINSEARCH, BINSEARCH, and an algorithm with $\mathcal{O}(n)$ storage space

and query time $\mathcal{O}(\frac{kn}{\log n})$, using $\mathcal{O}(n)$ time and space for preprocessing. However, this is an asymptotic result only, since the constants are so large that for a 20-letter alphabet and realistic string sizes, the algorithm is not applicable. Another algorithm was presented in [3] which solves the SUBMASS QUERY PROBLEM for binary alphabets with query time $\mathcal{O}(\log n)$ and $\mathcal{O}(n)$ space but does not produce witnesses.

Edwards and Lippert [1] considered the SUBMASS ALL WITNESSES PROBLEM and presented an algorithm that preprocesses the database by compressing witnesses using suffix trees. However, they work under the assumption that the queries are limited in range.

The study of weighted strings and their submasses[1] has further applications in those problems on strings over an un-weighted alphabet where the focus of interest are not substrings, but rather equivalence classes of substrings defined by multiplicity of characters. One examines objects of the form $(n_1, \ldots, n_{|\Sigma|})$ which represent all strings $s_1 \ldots s_n$ such that the cardinality of character a_i in each string is exactly n_i, for all $1 \leq i \leq |\Sigma|$. These objects have been referred to in recent publications variously as *compositions* [6], *compomers* [7,8], *Parikh-vectors* [9], *multiplicity vectors* [3], and *π-patterns* [10]. A similar approach has been referred to as *Parikh-fingerprints* [11,12]. Here, Boolean vectors are considered of the form $(b_1, \ldots, b_{|\Sigma|})$, where $b_i = 1$ if and only if a_i occurs in the string. Applications range from identifying gene clusters [12] to pattern recognition [11], alignment [6] or SNP discovery [8].

2   Searching for Submasses Using Polynomials

In this section, we introduce the main idea of our algorithms, the encoding of submasses via polynomials. We first prove some crucial properties, and then discuss algorithmic questions.

Let $s = s_1 \ldots s_n$. In the rest of the paper, we denote by μ_s the total mass of the string s, and the empty string by ε. Define, for $0 \leq i \leq n$, $p_i := \sum_{j=1}^{i} \mu(s_j) = \mu(s_1 \ldots s_i)$, the i'th prefix mass of s. In particular, $p_0 = \mu(\varepsilon) = 0$. We define two polynomials

$$P_s(x) := \sum_{i=1}^{n} x^{p_i} = x^{\mu(s_1)} + x^{\mu(s_1 s_2)} + \ldots + x^{\mu_s}, \tag{1}$$

$$Q_s(x) := \sum_{i=0}^{n-1} x^{\mu_s - p_i} = x^{\mu_s} + x^{\mu_s - \mu(s_1)} + \ldots + x^{\mu_s - \mu(s_1 \ldots s_{n-1})} \tag{2}$$

Now consider the product of $P_s(x)$ and $Q_s(x)$,

$$C_s(x) := P_s(x) \cdot Q_s(x) = \sum_{m=0}^{2\mu_s} c_m x^m. \tag{3}$$

Since any submass of s with witness (i, j) can be written as a difference of two prefix masses, namely as $p_j - p_{i-1}$, we obtain the following

---

[1] Note that we use the expressions "weight" and "mass" synomymously, hence "weighted string" but "submass."

Lemma 1. *Let $P_s(x), Q_s(x)$ and $C_s(x)$ from Equations (1) through (3). Then for any $m \le \mu_s$, $\kappa(m) = c_{m+\mu_s}$, i.e., the coefficient $c_{m+\mu_s}$ of $C_s(x)$ equals the number of witnesses of m in s.*

Lemma 1 immediately implies the following facts. For a proposition $\mathcal{P}$, we denote by $[\mathcal{P}]$ the Boolean function which equals 1 if $\mathcal{P}$ is true, and 0 otherwise. Then $\sum_{m=\mu_s+1}^{2\mu_s}[c_m \ne 0] = \sigma(s)$, the number of submasses of s. Furthermore, $\sum_{m=\mu_s+1}^{2\mu_s} c_m = \frac{n(n+1)}{2}$. Thus, polynomial C_s also allows us to compute the number of submasses of s.

2.1   Algorithm and Analysis

The algorithm simply consists of computing $C_s(x)$.

ALGORITHM 1

1. Preprocessing step:
   Compute μ_s, compute $C_s(x)$, and store in a sorted array all numbers $m - \mu_s$ for exponents $m > \mu_s$ where $c_m \ne 0$.
2. Query step:
   (a) For the SUBMASS QUERY PROBLEM: Search for each query mass M_i for $1 \le i \le k$, and return **yes** if found, **no** otherwise.
   (b) For the NUMBER OF SUBMASSES PROBLEM: Return size of array.

The polynomial $C_s(x)$ can be computed with Fast Fourier Transform (FFT) [13], which runs in time $\mathcal{O}(\mu_s \log \mu_s)$, since $\deg C_s = 2\mu_s$. As mentioned in the Introduction, we can employ methods from [5] for sparse polynomials and reduce $\deg C_s$ to $\mathcal{O}(\sigma(s))$, the number of non-zero coefficients. However, for the rest of this paper, we will refer to the running time as proportional to $\mu_s \log \mu_s$.

Theorem 1. ALGORITHM 1 *solves the* SUBMASS QUERY PROBLEM *in time* $\mathcal{O}(\mu_s \log \mu_s + k \log n)$, *or in time* $\mathcal{O}(\mu_s \log \mu_s + k)$, *depending on the storage method.* ALGORITHM 1 *solves the* NUMBER OF SUBMASSES PROBLEM *in time* $\mathcal{O}(\mu_s \log \mu_s)$.

Proof. The preprocessing step takes time $\mathcal{O}(\mu_s \log \mu_s)$. The query time for the SUBMASS QUERY PROBLEM is $\mathcal{O}(k \cdot \log \sigma(s)) = \mathcal{O}(k \log n)$. Instead of using a sorted array, we can instead store the submasses in an array of size μ_s (which can be hashed to $\mathcal{O}(\sigma(s))$ size) and allow for direct access in constant time, thus reducing query time to $\mathcal{O}(k)$.   □

Along the same lines, for the NUMBER OF SUBMASSES PROBLEM, our algorithm allows computation of $\sigma(s)$ in $\mathcal{O}(\mu_s \log \mu_s) = \mathcal{O}(n \cdot \mu_{\max} \log(n \cdot \mu_{\max}))$ time. The naïve solution of generating all submasses requires $\Theta(n^2 \log n)$ time and $\Theta(\sigma(s))$ space (with sorting), or $\Theta(n^2)$ time and $\Theta(\mu_s)$ space (with an array of size μ_s). Our algorithm thus outperforms this naïve approach as long as $\mu_{\max} = o(\frac{n}{\log n})$.

3   A Las Vegas Algorithm for Finding Witnesses

We now describe how to find a witness for each submass of the string s in time $\mathcal{O}(\mu_s \, \text{polylog}(\mu_s))$.

Our high level idea is the following: We first note that given a mass M, if we know the ending position j of a witness of M, then, using the prefix masses $p_1, \ldots, p_n$, we can easily find the beginning position of this witness. To do this, we simply do a binary search amongst the prefix masses $p_1, \ldots, p_{j-1}$ for $p_j - M$. Below we will define two suitable polynomials of degree at most μ_s such that the coefficient of $x^{M+\mu_s}$ in their product equals the sum of the ending positions of substrings that have mass M.

Now, if we knew that there was a unique witness of mass M, then the coefficient would equal the ending position of this witness. However, this need not always be the case. In particular, if there are many witnesses with mass M, then we would need to check all partitions of the coefficient of $x^{M+\mu_s}$, which is computationally far too costly. To get around this problem, we look for the witnesses of M in the string s, where we do not consider all pairs of positions but instead random subsets of these.

By using the definition of $Q(x)$ from (2), set

$$R_s(x) := \sum_{i=1}^{n} i \cdot x^{p_i} \quad \text{and} \tag{4}$$

$$F_s(x) := R_s(x) \cdot Q_s(x) = \sum_{m=0}^{2\mu_s} f_m x^m. \tag{5}$$

In the following lemma, we use the definition of c_m from (3).

Lemma 2. Let $m > \mu_s$. If $c_m = 1$, then f_m equals the ending position of the (sole) witness of $m - \mu_s$.

Proof. By definition, $f_m = \sum_{(i,j) \text{ witness of } m} j$ for any $m > \mu_s$. If $c_m = 1$, by Lemma 1, $m - \mu_s$ has exactly one witness (i_0, j_0). Thus, $f_m = j_0$. □

3.1   The Algorithm

We first run a procedure which uses random subsets to try and find witnesses for the query masses. It outputs a set of pairs (m, j_m), where m is a submass of s, and j_m is the ending position of one witness of m. For all query masses which are in this set, we find the beginning positions with binary search within the prefix masses, as described above, to find the witness in time $\mathcal{O}(\log n)$. For any remaining query masses, we run LINSEARCH. In the following, let $[x^i]A(x)$ denote[2] the coefficient a_i of x^i of the polynomial $A(x) = \sum_j a_j x^j$.

---

[2] Incidentally, our two different uses of "[ ]" are both standard, for generating functions and logical expressions, respectively. Since there is no danger of confusion, we have chosen to use both rather than introduce new ones.

ALGORITHM 2

1. Compute $C_s(x)$ from Equation (3), and check which of the queries are sub-masses of s.
2. Procedure TRY-FOR-WITNESS
    (i) For a from 1 to $2 \log n$, do:
        (ii) Let $b = 2^{-a/2}$. Repeat $24 \log n$ times:
            (iii) – Generate a random subset I_1 of $\{1, 2, \ldots, n\}$, and a random subset I_2 of $\{0, 1, 2, \ldots, n-1\}$, where each element is chosen independently with probability b.
            – Compute $P_{I_1}(x) = \sum_{i \in I_1} x^{p_i}$, $Q_{I_2}(x) = \sum_{i \in I_2} x^{\mu_s - p_i}$ and $R_{I_1}(x) = \sum_{i \in I_1} i \cdot x^{p_i}$.
            – Compute $C_{I_1, I_2}(x) = P_{I_1}(x) \cdot Q_{I_2}(x)$ and $F_{I_1, I_2}(x) = R_{I_1}(x) \cdot Q_{I_2}(x)$.
            – Let $c_i = [x^i] C_{I_1, I_2}(x)$ and $f_i = [x^i] R_{I_1, I_2}(x)$.
            – For $i > \mu_s$, if $c_i = 1$ and if i has not yet been successful, then store the pair $(i - \mu_s, f_i)$. Mark i as successful.
3. For all submasses amongst the queries M_ℓ, $1 \le \ell \le k$, if an ending position was found in this way, find the beginning position with binary search amongst the prefix masses.
4. If there is a submass M_ℓ for which no witness was found, find one using LINSEARCH.

3.2   Analysis

We first give an upper bound on the failure probability of procedure TRY-FOR-WITNESS for a particular query mass M.

Lemma 3. *(1) For a query mass M with $\kappa(M) = \kappa$, and $a = \lfloor \log_2 \kappa \rfloor$. Consider the step 2.iii of ALGORITHM 2. The probability that the coefficient $c_{M+\mu_s}$ of $C_{I_1, I_2}(x)$ for a (as defined above) is not 1 is at most $\frac{7}{8}$.*
(2) Procedure TRY-FOR-WITNESS does not find a witness for a given submass M with probability at most $1/n^3$. Moreover, the probability that the procedure fails for some submass is at most $1/n$.

Theorem 2. ALGORITHM 2 *solves the* SUBMASS WITNESS PROBLEM *in expected time* $\mathcal{O}(\mu_s \log^3 \mu_s + k \log n)$.

Proof. Denote the number of distinct submasses amongst the query masses by k'. By Lemma 3, the probability that the algorithm finds a witness for each of the $k' = O(n^2)$ submasses is at least $1 - 1/n$. In this case, the expected running time is the time for running the procedure for finding witness ending positions, plus the time for finding the k' witnesses:

$$\mathcal{O}(\underbrace{\mu_s \log \mu_s}_{\text{Step 1.}} + \underbrace{2 \log n}_{\text{Step 2.i}} \cdot \underbrace{24 \log n}_{\text{Step 2.ii}} \cdot \underbrace{\mu_s \log \mu_s}_{\text{Steps 2.iii}}) + k \cdot \mathcal{O}(\log n)$$

In the case when the algorithm does not output all witnesses, we simply run LINSEARCH search for all the submasses in time $O(kn)$. However, since the probability of this event is at most $1/n$, the excepted time in this case is at most $O(k)$. This implies the required bound on the running time.   □

4   A Deterministic Algorithm for Finding All Witnesses

Recall that, given the string s of length n and k query masses $M_1, \ldots, M_k$, we are able to solve the SUBMASS ALL WITNESSES PROBLEM in $\Theta(k \cdot n)$ time and $\mathcal{O}(1)$ space with LINSEARCH, or in $\Theta(n^2 \log n + k \log n)$ time and $\Theta(n^2)$ space with BINSEARCH. Thus, the two naïve algorithms yield a runtime of $\Theta(\min(kn, (n^2 + k) \log n))$.

Our goal here is to give an algorithm which outperforms the bound above, provided certain conditions hold. Clearly, in general it is impossible to beat the bound $\min(kn, n^2)$ because that might be the size of the output, K, the total number of witnesses to be returned. Our goal will be to produce something good if $K << kn$.

Now consider two strings s and t. We are interested in submasses of $s \cdot t$ with a witness which spans or touches the border between s and t. More precisely, we refer to a witness (i, j) of m as a *border-spanning* witness if and only if $i \leq |s| \leq j$. We can encode such witnesses again in a polynomial, using the definition of $P(x)$ from (1). The idea is that the mass of a border-spanning witness can be written as the sum of a prefix mass of s^r, the reverse string of s, and a prefix mass of t. Note that here, we also allow 0 as a submass.

Lemma 4. *For two strings s, t, and the polynomial*

$$D_{s,t}(x) := (x^0 + P_{s^r}(x)) \cdot (x^0 + P_t(x)) = \sum_{m=0}^{\mu(s)+\mu(t)} d_m x^m, \qquad (6)$$

the coefficient d_m equals the number of border-spanning witnesses of m in $s \cdot t$.

4.1   The Algorithm

The algorithm combines the polynomial method with LINSEARCH in the following way: We divide the string s into g substrings of approximately equal length. We then use polynomials to identify, for each query mass M and each witness (b, e) of M, which substrings the beginning and end index lie in. Then we use LINSEARCH on these substings to actually find the witnesses. The crucial observation is given in Lemma 5. We now describe the details.

We divide the string s into g substrings of approximately equal length: $s = t_1 \cdot t_2 \cdots t_g$ (where we will choose g later), and denote by $M_{i,j} = \sum_{m=i+1}^{j-1} \mu(t_m)$. In particular, if $j \leq i + 1$, then $M_{i,j} = 0$.

In order to have a good choice for g, we need to know the total size of the output, $K = \sum_{\ell=1}^{k} \kappa(M_\ell)$. This we can do by computing $C_s(x)$ and then adding up the coefficients c_{M_ℓ} for $1 \leq \ell \leq k$. We now set $g = \lceil (\frac{Kn}{\mu_s \log \mu_s})^{\frac{1}{2}} \rceil$. Observe that if $Kn \leq \mu_s \log \mu_s$, then $g = 1$, in which case we are better off running LINSEARCH. So let $Kn > \mu_s \log \mu_s$.

In step 2.(b) of the following algorithm, we modify LINSEARCH to only return border-spanning submasses. This can be easily done by setting the second pointer at the start of the algorithm to the last position of the first string, and by breaking when the first pointer moves past the first position of the second string.

ALGORITHM 3

1. Preprocesssing step:
   (a) Compute μ_s and $C_s(x)$ as defined in (3), and compute $K = \sum_{\ell=1}^{k} c_{M_\ell}$.
       Set $g = \lceil (\frac{Kn}{\mu_s \log \mu_s})^{\frac{1}{2}} \rceil$.
   (b) For each $1 \leq i \leq g$, compute $C_{t_i}(x)$.
   (c) For each $1 \leq i < j \leq g$, compute $D_{t_i, t_j}(x)$ as defined in (6).
2. Query step: For each $1 \leq \ell \leq k$,
   (a) For each i such that $[x^{M_\ell + \mu(t_i)}]C_{t_i}(x) \neq 0$, run LINSEARCH on t_i for M_ℓ and return all witnesses.
   (b) For each pair (i, j) such that $[x^{M_\ell - M_{i,j}}]D_{t_i, t_j}(x) \neq 0$, run LINSEARCH on $t_i \cdot t_j$ for submass $M_\ell - M_{i,j}$ and return all border-spanning witnesses.
   (c) If M_ℓ was not a submass of any of the above strings, return no.

4.2   Analysis

The following lemma shows the correctness of ALGORITHM 3.

Lemma 5. *For $1 \leq M \leq \mu_s$,*

$$\kappa(M) = \sum_{i=1}^{g}[x^M + \mu(t_i)]C_{t_i} + \sum_{1 \leq i < j \leq g}[x^{M - M_{i,j}}]D_{t_i, t_j}(x).$$

Proof. Observe that for any witness (b, e) of M, there is exactly one pair (i, j) such that b lies in string t_i and e in t_j. If $i = j$, then M is a submass of t_i and by Lemma 1 contributes exactly one to the coefficient $[x^{M + \mu(t_i)}]C_{t_i}(x)$. Otherwise, $i < j$, and $M - M_{i,j}$ is a submass of the concatenated string $t_i \cdot t_j$ with the witness (b', e'), where (b', e') is shifted appropriately (i.e., $b' = b - \sum_{i' < i}|t_{i'}|$ and $e' = |t_i| + \sum_{i' < j}|t_{i'}|$). Moreover, (b', e') is a border-spanning submass of $t_i \cdot t_j$. Thus, by Lemma 4, (b', e') contributes exactly one to $[x^{M - M_{i,j}}]D_{t_i, t_j}(x)$.   □

Using FFT for computing the polynomials, the preprocessing step of ALGORITHM 3 has runtime $\mathcal{O}(g\mu_s \log \mu_s)$. The query time is $\mathcal{O}(g\mu_s \log \mu_s + K\frac{n}{g})$. With $g = \lceil (\frac{Kn}{\mu_s \log \mu_s})^{\frac{1}{2}} \rceil$, we obtain the following

Theorem 3. ALGORITHM 3 *solves the* SUBMASS ALL WITNESSES PROBLEM *in time* $\mathcal{O}((Kn\mu_s \log \mu_s)^{\frac{1}{2}})$, *where μ_s is the mass of the string, and K is the total number of witnesses, i.e., the output size.*

To better understand this result, let $\bar{\kappa}$ denote the average size of the output, so $\bar{\kappa} = K/k$. Then the runtime is $(k\bar{\kappa}n\mu_s \log \mu_s)^{1/2}$. Note that the running time of the combination of the naïve algorithms for the submass all witnesses problem is $O(\min(kn, n^2 \log n))$. Thus, our algorithm beats the running time of the naïve algorithms above if $\bar{\kappa}\mu_s \log \mu_s = o(kn)$ and $(\bar{\kappa}k\mu_s \log \mu_s) = o(n^3 \log^2 n)$.

5   Discussion

In this paper we gave algorithms for several variants of finding substrings with particular submasses in a given weighted string. Our algorithms are most interesting when the masses of the individual characters are small compared to the length of the string, or more generally, when the number of different possible submasses is small compared to n^2.

Most of our algorithms have running time complexity dependent (up to polylog factors) on the number of different submasses in the given weighted string. While this may not be the best possible running time, it seems that improving this significantly will be hard. For example, consider the problem of finding the number of different submasses $\sigma(s)$. Our algorithm for this problem has runtime $\mathcal{O}(\sigma(s) \log \sigma(s))$. It is not hard to see that the easier problem of deciding whether $\sigma(s)$ is exactly equal to $n(n+1)/2$ or not is at least as hard as the 4-Sum problem. The 4-Sum problem is conjectured to have a run time complexity of $\Omega(n^2)$ [14, 15] and is one of the major problems in computational geometry. So, it is unlikely that even the number of different submasses can be determined in time $o(n^2)$ in the general case.

References

1. Edwards, N., Lippert, R.: Generating peptide candidates from amino-acid sequence databases for protein identification via mass spectrometry. In: Proc. of 2nd WABI. LNCS (2002) 68–81
2. Lu, B., Chen, T.: A suffix tree approach to the interpretation of tandem mass spectra: Applications to peptides of non-specific digestion and post-translational modifications. Bioinformatics Suppl. 2 (ECCB) (2003) II113–II121
3. Cieliebak, M., Erlebach, T., Lipták, Z., Stoye, J., Welzl, E.: Algorithmic complexity of protein identification: Combinatorics of weighted strings. DAM (2004) 27–46
4. Wilf, H.: generatingfunctionology. Academic Press (1990)
5. Cole, R., Hariharan, R.: Verifying candidate matches in sparse and wildcard matching. In: Proc. of 34th STOC. (2002)
6. Benson, G.: Composition alignment. In: Proc. of 3rd WABI. LNCS (2003) 447–461
7. Böcker, S.: Sequencing from compomers: Using mass spectrometry for DNA de-novo sequencing of 200+ nt. In: Proc. of 3rd WABI. LNCS (2003) 476–497
8. Böcker, S.: SNP and mutation discovery using base-specific cleavage and MALDI-TOF mass spectrometry. Bioinformatics, Suppl. 1 (ISMB) (2003) i44–i53
9. Salomaa, A.: Counting (scattered) subwords. EATCS **81** (2003) 165–179
10. Eres, R., Landau, G.M., Parida, L.: A combinatorial approach to automatic discovery of cluster-patterns. In: Proc. of 3rd WABI. LNCS (2003) 139–150

11. Apostolico, A., Landau, G., Satta, G.: Efficient text fingerprinting via Parikh mapping. J. of Discrete Algorithms (to appear)
12. Didier, G.: Common intervals of two sequences. In: Proc. of 3rd WABI. LNCS (2003) 17–24
13. Cooley, J.W., Tukey, J.W.: An algorithm for the machine calculation of complex Fourier series. Mathematics of Computation **19(90)** (1965) 297–301
14. Demaine, E.D., Mitchell, J.S.B., O'Rourke, J.: The open problems project. http://cs.smith.edu/ orourke/TOPP/ (2004)
15. Erickson, J.: Lower bounds for linear satisfiability problems. In: Proc. of 6th SODA. (1995) 388–395

Maximum Agreement
and Compatible Supertrees*
(Extended Abstract)

Vincent Berry and François Nicolas

Équipe *Méthodes et Algorithmes pour la Bioinformatique - L.I.R.M.M.*
{vberry,nicolas}@lirmm.fr

Abstract. Given a collection of trees on n leaves with identical leaf set, the MAST, resp. MCT, problem consists in finding a largest subset of leaves such that all input trees restricted to these leaves are isomorphic, resp. have a common refinement. For MAST, resp. MCT, on k rooted trees, we give an $O(min\{3^p kn, 2.27^p + kn^3\})$ exact algorithm, where p is the smallest number of leaves to remove from input trees in order for these trees to be isomorphic, resp. to admit a common refinement. This improves on [14] for MAST and proves fixed-parameter tractability for MCT. We also give an approximation algorithm for (the complement of) MAST similar to the one in [2], but with a better ratio and running time, and extend it to MCT.

We generalize MAST and MCT to the case of supertrees where input trees can have non-identical leaf sets. For the resulting problems, SMAST and SMCT, we give an $O(N + n)$ time algorithm for the special case of two input trees (N is the time bound for solving MAST, resp. MCT, on two $O(n)$-leaf trees). Last, we show that SMAST and SMCT parameterized in p are W[2]-hard and cannot be approximated in polynomial time within a constant factor unless P = NP, even when the input trees are rooted triples.

We also extend the above results to the case of unrooted input trees.

1 Introduction

Given a set of leaf-labelled trees with identical leaf sets, the *maximum agreement subtree* problem (MAST) consists in finding a subtree homeomorphically included in all input trees and with the largest number of leaves [33, 15, 2, 19, 26, 12]. This pattern matching problem on trees arises in various areas, among which the reconstruction of *evolutionary trees* (or *phylogenies*) whose leaves represent living species and internal nodes represent ancestral species. The shape of the tree describes the speciation pattern from which current species originated.

---

* Supported by the *Action Incitative Informatique-Mathématique-Physique en Biologie Moléculaire* [ACI IMP-Bio].

S.C. Sahinalp et al. (Eds.): CPM 2004, LNCS 3109, pp. 205–219, 2004.

Motivation for extending the MAST and MCT problems to supertrees.
A recent problem in phylogenetics is to infer trees from a collection of input trees
on overlapping, but different, sets of leaves. Each input tree is built from a separate data set which does not include all studied species for various reasons. These
trees are then given as input to a method that proposes a tree, called a *supertree*,
i.e., a tree including all (or most) species according to their relative positions in
the input trees. For various reasons, the input trees can disagree on the position
of several species. Depending on the way they handle such conflicts, supertree
methods can be divided into two main categories [34]: *(i)* optimization methods
which tend to resolve conflicts according to a specified optimization criterion [3,
30, 32, 28, 11] ; *(ii)* consensus methods which output a supertree displaying only
parts of the species' history on which the input trees agree. The drawback of approach *(i)* is that output supertrees sometimes contain undesirable or unjustified
resolutions of conflicts [29, 31, 6, 36, 28].Approach *(ii)* has been less investigated
in the context of supertrees, the two proposed methods being the strict consensus [18], sometimes criticized because of the poor amount of information of
the produced supertree [36, 7], and the reduced consensus [34], which usually
proposes a set of complementary supertrees as output (see [34, Sect. 4]). Both
these methods focus on the *clusters* (sets of leaves under internal nodes). Several
authors remarked that an alternative would be to focus on *leaves* themselves,
because in many cases removing a few species on the position of which the input
trees disagree, could enable to produce a single informative supertree [18, 36, 7].
Here we follow this proposition by extending the MAST problem to the case of
input trees on overlapping sets of leaves. We call SMAST the resulting problem
concerned with the inference of a supertree. We also extend a variant of MAST
called *maximum compatible tree* (MCT) which is of interest when input trees
are non-binary [21, 23, 16], to obtain the SMCT problem. MCT and SMCT allow
multifurcating nodes (high degree nodes) of input trees to be resolved (split into
several nodes) according to other input trees.

Apart from inferring an estimate of the species' history, SMAST and SMCT
can play the same role as MAST in the context of supertrees, i.e., measuring the
similarity of input trees or identifying species that could be implied in horizontal
transfers of genes. Moreover, the supertree they produce most likely contains
leaves from most input trees (see Lem. 5) and, by definition, agrees topologically
with all of them. It is thus a good candidate to strengthen the results of the
popular *matrix representation with parsimony* (MRP) method [3, 30]. E.g., [8]
explicitly recommended to use such a *seed* tree (i.e., a tree with leaves spanning
most input trees) to improve the relatively low accuracy observed for MRP when
the input trees overlap moderately.

Note that [19] designed a variant of MAST that builds a tree they call a
"supertree", but with a different meaning from that considered in phylogenetics
and here.

Previous work on MAST and MCT. MAST is NP-hard on only three rooted
trees of unbounded degree [2], and MCT on two rooted trees of unbounded degree
[23]. Efficient polynomial time algorithms have been recently proposed for MAST

on two rooted n-leaf trees: $O(n \log n)$ for binary trees [12], and $O(\sqrt{d}n \log \frac{2n}{d})$ for trees of degree bounded by d [26]. When the two input trees are unrooted and of unbounded degree, one can use the $O(n^{1.5})$ algorithm of [25]. When k rooted trees are given as input and have maximum degree d, MAST can be solved in $O(n^d + kn^3)$[15, 2, 9] and MCT in $O(2^{2kd}n^k)$ [16]. MAST is known to be *fixed-parameter tractable* (*FPT*) in p, the smallest number of leaves to remove from the input set of leaves such that the input trees agree: [14] describe an algorithm in $O(3^p kn \log n)$ and [1] give an algorithm in $O(2.27^p + kn^3)$. The MAST problem (maximizing the number of leaves in an agreement subtree) is likely to be hard to approximate [23, 17], whereas the *complement* of MAST (i.e., minimizing the number of leaves to remove to obtain an agreement between the input trees) can be 4-approximated in polynomial time [23].

Results. Following the work of [14], we obtain an $O(min\{3^p kn, 2.27^p + kn^3\})$ algorithm for both MCT and MAST. This improves on the bound for the MAST problem w.r.t. [14] and is the first result of fixed-parameter tractability for MCT. Moreover, from this standpoint, MCT has the same complexity as MAST which was not expected (on two general trees, the former is NP-hard while efficient algorithms exist for the second, and on bounded degree trees, less efficient algorithms exist for MCT). We also give an approximation algorithm for the complement of MAST similar to the one in [2], but with a better ratio and running time, and extend it to MCT.

Then, we show how to extend MAST and MCT in a natural way to obtain the problems SMAST and SMCT on supertrees. SMAST and SMCT are NP-hard in general, as they are equivalent to MAST, resp. MCT, in the case of input trees with identical leaf sets. For both SMAST and SMCT, we give an $O(N + n)$ algorithm for the case of two input trees, where N is the time bound for solving MAST, resp. MCT, on two $O(n)$-leaf trees.

Finally, by reduction from the HITTING SET problem, we show that SMAST and SMCT are more difficult than MAST and MCT, as they are W[2]-hard for p. Thus, there is little hope to obtain efficient exact algorithms for these problems on more than two trees, suggesting to resort on heuristic algorithms to solve the problem. However, it will be difficult to prove tight approximation results for such heuristics, as we also show that no polynomial time algorithm can approximate SMAST and SMCT within a constant factor, unless P = NP, even when the input trees are rooted triples (binary trees on only three leaves).

All above results can be extended to the case of unrooted trees (with an extra n factor in the case of the FPT and approximation algorithms). Sect. 2 below presents definitions, then Sect. 3 and 4 give results for MAST and MCT, then Sect. 5 those for SMAST and SMCT.

2   Definitions and Preliminaries

Trees we consider are *evolutionary trees* (also called *phylogenies*). Such a tree T has its leaf set $L(T)$ in bijection with a label set and is either *rooted* (at a node denoted $r(T)$), in which case all internal nodes have at least two children, or *unrooted*, in which case internal nodes have degree at least three. When there is no ambiguity, we identify leaves with their labels. The size $\#T$ of a tree T is the number of its leaves. Let u be a node in a rooted tree, we denote by $S(u)$ the subtree rooted at u (i.e., u and its descendants) and by $L(u)$ the leaves of this subtree. If C is a set of nodes in a tree, then define $L(C) := \bigcup_{u \in C} L(u)$. Given a rooted tree T and a set of leaves $L \subseteq L(T)$, we denote $lca_T(L)$ the lowest common ancestor of leaves L in T.

Definition 1 *Given a set L of labels and a tree T, the* restriction *of T to L, denoted $T|L$, is obtained by taking the smallest subtree of T which connects leaves with label in L and making it homeomorphically irreducible (by suppressing non-root nodes of degree two and identifying the two edges to which they were connected).*

Definition 2 *Given two trees T, T', we write $T \sqsubseteq T'$ iff $T = T'|L(T)$. Given a collection $\mathcal{T} = \{T_1, \ldots, T_k\}$ of trees with a common leaf set L, an* agreement subtree *of $\mathcal{T}$ is any tree T with leaves in L s.t. $\forall T_i \in \mathcal{T}, T \sqsubseteq T_i$. The* Maximum Agreement Subtree *problem (*MAST*) consists in finding an agreement subtree of $\mathcal{T}$ with the largest number of leaves. We denote $MAST(\mathcal{T})$ such a tree.*

In phylogenetic analysis, this definition is sometimes considered too stringent when input trees can have *multifurcations* (nodes with more than 2 children). Indeed, such a node can either represent a multi-speciation event, or an uncertainty (irresolution) concerning the relative branching of the child subtrees of the node. The MAST problem is best suited for the first interpretation, as the presence of a multifurcation in a input tree, will impede to resolve the node according to other input trees. For the second interpretation, [21] introduced the MCT problem, a variant of MAST that allows multifurcations to be resolved in the output tree:

Definition 3 *A tree T* refines *a tree T', and we write $T \trianglerighteq T'$, if by contracting some internal edges of T one can obtain T' (contracting an edge means merging its extremities). More generally, a tree T refines a collection $\mathcal{T} = \{T_1, \ldots, T_k\}$, denoted $T \trianglerighteq \mathcal{T}$, whenever T refines all T_i's in $\mathcal{T}$. Given a collection $\mathcal{T} = \{T_1, \ldots, T_k\}$ of input trees with identical leaf set L, a tree T with leaves in L is said to be* compatible *with $\mathcal{T}$ iff $\forall T_i \in \mathcal{T}, T \trianglerighteq T_i|L(T)$. The* MCT *problem consists in finding a tree compatible with $\mathcal{T}$ having the largest number of leaves. We denote $MCT(\mathcal{T})$ such a tree.*

Note that $\#MCT(\mathcal{T}) \geq \#MAST(\mathcal{T})$ and that MCT is equivalent to MAST when input trees are binary.

Definition 4 *Let T be a rooted tree. For on any three leaves $a, b, c \in L(T)$, there are only three possible binary shapes for $T|\{a, b, c\}$, denoted $a|bc$, resp. $b|ac$, resp. $c|ab$, depending on their innermost grouping of two leaves (bc, resp. ac, resp ab). These trees are called* rooted triples *(or resolved triples). Alternatively $T|\{a, b, c\}$ can be a* fan *(also called* unresolved triple*), connecting the three leaves* to a same internal node, which is denoted (a, b, c). We define $tr(T)$, resp. $f(T)$, to be the set of rooted triples, resp. fans, induced by the leaves of a tree T. We extend these definitions to define rooted triples and fans of a collection of trees $\mathcal{T} = \{T_1, \ldots, T_k\}$: $tr(\mathcal{T}) := \bigcap_{T_i \in \mathcal{T}} tr(T_i)$ and $f(\mathcal{T}) := \bigcap_{T_i \in \mathcal{T}} f(T_i)$. Let T, T' be two rooted trees, a set of three leaves $\{a, b, c\} \subseteq L(T) \cap L(T')$ is a hard conflict, resp. a soft conflict, between T, T' whenever $a|bc \in tr(T)$ and $b|ac \in tr(T')$, resp. whenever $a|bc \in tr(T)$ and $(a, b, c) \in f(T')$.

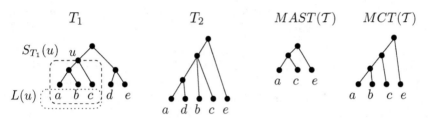

Fig. 1. A collection $\mathcal{T} = \{T_1, T_2\}$, one of the $MAST(\mathcal{T})$ trees, and the $MCT(\mathcal{T})$. T_1, T_2 have a hard conflict on $\{a, c, d\}$ (since $d|ac \in tr(T_1)$ while $c|ad \in tr(T_2)$) and in soft conflict on a, b, c (since $c|ab \in tr(T_1)$ while $(a, b, c) \in f(T_2)$).

Lemma 1 *Let $\mathcal{T}$ be a collection of trees with identical leaf set L and T, T' two rooted trees.*

(i) T is an agreement subtree of $\mathcal{T}$ iff $tr(T) \subseteq tr(\mathcal{T})$ and $f(T) \subseteq f(\mathcal{T})$.
(ii) T is compatible with $\mathcal{T}$ iff $\forall T_i \in \mathcal{T}, tr(T_i) \subseteq tr(T)$ and $L(T) \subseteq L$.
(iii) T is isomorphic to T' iff $tr(T) = tr(T')$ and $f(T) = f(T')$.
(iv) T refines T' iff $tr(T') \subseteq tr(T)$ and $L(T) = L(T')$.

Proof. (i) is [9, Lem. 6.6], (ii) is [10, Thm. 1], (iii) derives from (i), and (iv) results from (ii). □

3   An FPT Algorithm for MAST and MCT

The FPT algorithm described in [14] solves the MAST problem by resorting extensively to an $O(n \log n)$ subroutine that checks whether two input trees are isomorphic or alternatively that identifies a (soft or hard) conflict. We show below an equivalent subroutine but with $O(n)$ running-time, which then leads to improve the complexity of the FPT algorithm.

Sketch of the algorithm **Check-Isomorphism-or-Find-Conflict.** Our subroutine is based on the linear-time algorithm of [20, 35] to which we add a simple postprocessing in the case where non-isomorphism is detected. The algorithm of [20, 35] examines recursively and simultaneously the two input trees T_1, T_2, using the following property on *cherries* (i.e., nodes having only leaves as children):

Lemma 2 ([20]) *Let T_1, T_2 be two isomorphic trees and v_1 a cherry in T_1. Then, there exists a cherry $v_2 \in T_2$ s.t. $L(v_1) = L(v_2)$.*

The two input trees are processed bottom-up, repeatedly replacing the cherry of each tree ($v_1 \in T_1$ and $v_2 \in T_2$) by a new leaf with a same new label. Non-isomorphism is detected whenever the algorithm finds a cherry $v_1 \in T_1$ and a leaf $l \in L(v_1)$ such that the parent node v_2 of l in T_2 is not a cherry or has leaf set $L(v_2) \neq L(v_1)$. In this case, a simple linear-time search of $S(v_2)$ and $L(v_1)$ suffices to find leaves $l', l'' \in L(v_1) \cup L(v_2)$ such that $\{l, l', l''\}$ is a conflict between the two input trees (see [5] for full details).

Theorem 1 *Let T_1, T_2 be two rooted trees with identical leaf sets, in time $O(n)$ algorithm* CHECK-ISOMORPHISM-OR-FIND-CONFLICT *either concludes that the trees are isomorphic or otherwise identifies a hard or soft conflict between T_1, T_2.*

Concerning the MCT problem, Lem. 1 (ii) implies that any tree compatible with $\mathcal{T}$ will have to eliminate at least one leaf of every *hard* conflict $\{a, b, c\}$ between the input trees (soft conflicts do not preclude the existence of a tree compatible with $\mathcal{T}$). If trees in $\mathcal{T}$ have no hard conflict, then there is a tree T refining $\mathcal{T}$ and thus $T := MCT(\mathcal{T})$ (because T includes all leaves of $\mathcal{T}$). We are then interested in an subroutine that either identifies a hard conflict or returns a refinement of $\mathcal{T}$. As above, this subroutine is designed for two input trees $\mathcal{T} = \{T_1, T_2\}$. For solving MCT on $k > 2$ input trees, we repeatedly apply the subroutine on the output refinement and on an input tree not yet examined (and so on, until the k input trees have been processed or a conflict is found). For this reason we need the subroutine to output a *minimum* refinement T of $\{T_1, T_2\}$, i.e. such that $\forall T' \trianglerighteq \{T_1, T_2\}, T' \trianglerighteq T$. Otherwise, a too refined tree T, could induce artificial hard conflict with input trees examined afterwards.

Sketch of algorithm Find-Refinement-or-Conflict. As for MAST, given two input trees T_1, T_2 with same leaf set L, the algorithm gradually prunes parts of T_1 and T_2, repeatedly eating cherries in T_1 and corresponding parts in T_2 until the trees are reduced to a single leaf or a (hard) conflict is found. However, for solving MCT, we need a different result than the one of Lem. 2:

Lemma 3 *Let T_1, T_2 be two trees admitting a common refinement, v_1 a cherry in T_1 and $v_2 := lca_{T_2}(L(v_1))$. Then there exists a subset C of (at least two) children of v_2 such that $L(v_1) = L(C)$.*

The search of T_1 and T_2 is guided by cherries $v_1 \in T_1$. Given v_1, the algorithm identifies $v_2 := lca_{T_2}(L(v_1))$ and the set C of v_2's children such that $L(v_1) \subseteq L(C)$. Then either:

(i) $L(v_1) = L(C) = L(v_2)$ (i.e., C is the set of all children of v_2), then subtrees $S(v_1)$ and $S(v_2)$ are pruned;

(ii) $L(v_1) = L(C) \subset L(v_2)$, then subtree $S(v_1)$ and the set S of subtrees $S(u)$, for $u \in C$, are pruned;

(iii) $L(v_1) \neq L(C)$, then a conflict involving two leaves $l, l' \in L(v_1)$ and a leaf $l'' \in L(C) - L(v_1)$ is identified.

Pruning a subtree $S(v_1), S(v_2)$ or a set S of subtrees means deleting all its nodes and replacing it (the subtree or the whole set S under v_2) by a single leaf. Both in T_1 and T_2 this new leaf is given a new label, l^*.

To build the refinement of T_1 and T_2, a forest of growing trees is maintained (initially containing a leaf-tree for each leaf in L). These trees are gradually assembled until one tree remains (i.e., if T_1, T_2 have no conflict), that is the output refinement of T_1, T_2. In cases (i) and (ii), some trees of the forest are assembled to reproduce the shape of subtrees pruned from T_1, T_2 [1]. This guarantees the minimality of the computed refinement. Correctness of this refinement follows from the fact that pruned subtrees of T_1, T_2 have no conflict. The pruning of subtrees guarantees that the leaf sets of T_1, T_2 are always identical to one another through the whole process: at each step, the same set of leaves are deleted, and a same new leaf l^* is added. If the repeated processing of T_1, T_2 does not meet any conflict, it stops when both trees have been reduced to a single leaf. See [5] for the pseudo-code of the subroutine.

Theorem 2 *Let T_1, T_2 be two rooted trees with identical leaf sets, in time $O(n)$ algorithm* FIND-REFINEMENT-OR-CONFLICT(T_1, T_2) *either returns a tree T minimally refining T_1 and T_2, or otherwise identifies a hard conflict between T_1, T_2.*

Proof. The algorithm is traversing T_1, T_2 a constant number of times, spending a constant time at each of the $O(n)$ nodes and edges. Nodes v_2 are identified in $O(n)$ globally (i.e., other the whole process) by exploring each time a different subtree of T_2 (or using dynamic data structures proposed by [13]). The list of cherries in T_1 is maintained in $O(n)$ globally, sets of subtrees $S(v_l)$ corresponding to processed cherries of T_1 are identified and removed in $O(n)$ globally.   □

Thus, in the case of both MAST and MCT we have a $O(n)$ time subroutine that can be used over and over in the FPT algorithm described in [14] in order to solve MAST and MCT time $O(3^p kn)$.

An alternative approach to solve MAST and MCT (as remarked by [14, 1] for the former problem) is to reduce these problems to the 3-HITTING-SET problem (3HS), which is well-known for admitting an efficient FPT algorithm [27] and a 3-approximation algorithm. Indeed, in time $O(kn^3)$ it is possible to know the triples and fans induced by all input trees, hence to know the set $\mathcal{C}$ of three-leaves sets on which the input trees conflict. This set $\mathcal{C}$ can be seen as an instance of 3HS. Any hitting set of $\mathcal{C}$ is a set of leaves whose removal eliminates all conflicts in the input trees. This reasoning is used in [1] to obtain a $O(2.27^p + kn^3)$ algorithm for MAST, where $p = \#(L - L(T_M))$ for a maximum agreement subtree T_M.

Theorems 1 and 2 and the above reasoning imply:

Theorem 3 *Given a collection $\mathcal{T} = \{T_1, \ldots, T_k\}$ of rooted trees on an identical set L of n leaves, a tree $T_M := MAST(\mathcal{T})$ (resp. $T_M := MCT(\mathcal{T})$) can be found in time $O(\min\{3^p kn, 2.27^p + kn^3\})$, where $p = \#(L - L(T_M))$.*

This improves on the result of [14] for MAST by a $\log n$ factor. Concerning MCT, this is the first time that the problem is shown to be FPT. More precisely,

---

[1] In case (ii) though, only subtrees in S are exactly reproduced, then they are connected to a new root node, representing the cluster $L(v_1)$.

this means that the burden of the complexity can depend only on the level of disagreement between the input trees. When considering a collection of trees disagreeing on few species, we obtain an efficient algorithm, whatever the number and degree of the input trees.

When considering a collection $\mathcal{T}$ of unrooted trees, a repeated use of the above algorithms suffices to solve MAST, resp. MCT. Given $l \in L$, define $\mathcal{T}^l$ to be the collection of rooted trees on $L - \{l\}$ obtained by rooting all trees of $\mathcal{T}$ at l and then removing this leaf (and its incident edge). Then:

Lemma 4 *Let $\ell \in L$ s.t. #$MAST(\mathcal{T}^\ell)$, resp. #$MCT(\mathcal{T}^\ell)$, is maximum, and let $T_M := MAST(\mathcal{T}^\ell)$, resp. , $T_M := MCT(\mathcal{T}^\ell)$. Then unrooting T_M by grafting the leaf ℓ at its root gives an unrooted tree $MAST(\mathcal{T})$, resp. $MCT(\mathcal{T})$.*

Thus, given a collection of unrooted trees on n leaves, MAST, resp. MCT, can be solved in n runs of the above algorithms (one run for each rooted collection $\mathcal{T}^l$, $l \in L$). This only adds an extra n factor to the complexity bound.

4   Approximating MAST and MCT

The *complement* of MAST problem, resp. MCT problem, denoted CMAST, resp. CMCT, is minimizing the number of input leaves to remove from the trees such that they are isomorphic, resp. admit a common refinement. This alternative formulation of the problem is of interest only in the context of approximation algorithms, which may have different approximation ratio depending on which version of the problem is considered. To obtain a 4-approximation algorithm for solving CMAST on *unrooted* trees, [2] use implicitly a reduction to 4Hs, which is similar to the reduction to 3Hs described above. Using the latter reduction gives a 3-approximation algorithm for solving CMAST on *rooted* trees. The same ideas applies for the CMCT problem. Moreover, in the case of a collection $\mathcal{T}$ of *unrooted* trees, by trying successively all rooted collections $\mathcal{T}^l, l \in L$, one can improve the approximation ratio given in [2] (mainly using Lem. 4). Then:

Theorem 4 *CMAST and CMCT on rooted or unrooted collections of trees can be 3-approximated in polynomial time.*

The complexity of our algorithm is $O(kn^3)$ for rooted trees and $O(kn^4)$ for unrooted trees, which also improves on [2] (see full paper for details).

5   Extending Problems to the Supertree Context

We now consider the case of supertree inference, where input trees are allowed to have different (but overlapping) sets of leaves. We first show how to extend MAST and MCT to this context.

Definition 5 *Given a collection $\mathcal{T} = \{T_1, \ldots, T_k\}$ of input trees, we denote by $L(\mathcal{T}) := \bigcup_{T_i \in \mathcal{T}} L(T_i)$ the set of all input leaves. Leaves appearing in only one input tree are called* specific.

In the following we assume w.l.o.g. that any input tree shares at least two leaves with other input trees.

Definition 6 *Given a collection $T = \{T_1, \ldots, T_k\}$ of trees on overlapping sets of leaves, an* agreement supertree *of T is a tree T with $L(T) \subseteq L(T)$ and s.t. $\forall T_i \in T, T|L(T_i) = T_i|L(T)$. The* Maximum Agreement Supertree *problem (SMAST) consists in finding an agreement supertree of T with the largest number of leaves. Such a tree is denoted $SMAST(T)$. In a similar way, we define the* Maximum Compatible Supertree *problem (SMCT) as that of finding a largest tree T with $L(T) \subseteq L(T)$ s.t. $\forall T_i \in T, T|L(T_i) \trianglerighteq T_i|L(T)$. Such a tree is denoted $SMCT(T)$.*

The two problems stated above are natural extensions of the problems defined in Sect. 2: SMAST, resp. SMCT, is equivalent to MAST, resp. MCT, when all input trees have an identical leaf set.

The following observation states that the supertrees defined above are likely to contain a non-trivial number of leaves, and moreover, leaves from many, if not all, input trees. This suggests that these supertrees might be good *seed* trees for the MRP method [3, 30] or its MRF variant [11].

Lemma 5 *Let T be a collection of trees with overlapping sets of leaves, all specific leaves of T appear in any tree $SMAST(T)$ and in any tree $SMCT(T)$.*

5.1 Solving SMAST and SMCT on 2 Trees

Given two trees T_1, T_2 on overlapping sets of leaves, let $L_\cap := L(T_1) \cap L(T_2)$ be the set of leaves common to both trees.

Lemma 6

- *Any tree $MAST(T_1|L_\cap, T_2|L_\cap)$ (resp. $MCT(T_1|L_\cap, T_2|L_\cap)$) is the restriction to $L_\cap$ of some tree $SMAST(T_1, T_2)$ (resp. $SMCT(T_1, T_2)$);*
- *the restriction to $L_\cap$ of any tree $SMAST(T_1, T_2)$ (resp. $SMCT(T_1, T_2)$) is a tree $MAST(T_1|L_\cap, T_2|L_\cap)$ (resp. $MCT(T_1|L_\cap, T_2|L_\cap)$).*

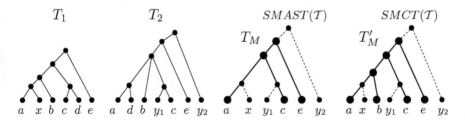

Fig. 2. A collection $T = \{T_1, T_2\}$ of input trees ($L_\cap = \{a, b, c, d, e\}$), the trees $T_M :=$ $MAST(T|L_\cap)$ and $T'_M := MCT(T|L_\cap)$ (in bold lines) and the supertrees $SMAST(T)$ and $SMCT(T)$.

Note that this result does not hold for three or more input trees. E.g., for $\mathcal{T} = \{(((((a, b), c), d), e), f), (((((d, e), f), b), c), a), ((a, c), b)\}$, no leaf common to all trees in $\mathcal{T}$ appears in a tree $SMAST(\mathcal{T})$ or $SMCT(\mathcal{T})$.

We now describe the algorithm to compute a tree $T := SMAST(\mathcal{T})$ for the pair of trees $\mathcal{T} = \{T_1, T_2\}$. Computing $SMCT(\mathcal{T})$ requires only small modifications. Define $L_S := L(\mathcal{T}) - L_\cap$ to be the set of specific leaves of $\mathcal{T}$ and a *specific subtree* to be a subtree $S(v)$ of T_1 (or $S(v)$ of T_2) such that $L(v) \subseteq L_S$.

Sketch of the algorithm. Let T_M be a precomputed $MAST(T_1|L_\cap, T_2|L_\cap)$. Lemmas 5 and 6 imply that there exists a tree $SMAST(\mathcal{T})$ on $L(T_M) \cup L_S$ admitting T_M as a restriction. Once T_M is computed, T_1 and T_2 are restricted to $L(T_m) \cup L_S$, then a tree $MAST(\mathcal{T})$ is built by grafting specific subtrees of T_1 and T_2 to a correct place in T_M. The approach is similar to the $O(n^3)$ algorithm of [18] for computing a strict consensus supertree, which attaches one by one specific leaves to a common *backbone* tree (sensu [18]). However here, the algorithm proceeds by successively attaching whole specific subtrees at a same time to T_M, leading to an $O(n)$ running time.

Before grafting specific subtrees in T_M, two data structures have to be precomputed. First, for each node $n_M \in T_M$, we need to store its *twin* node in T_1, defined as $\mathsf{Twin}_1(n_M) := lca_{T_1}(L(n_M))$, and its twin node in T_2, $\mathsf{Twin}_2(n_M) := lca_{T_2}(L(n_M))$. To initialize the Twin structures we need to preprocess the trees T_M, T_1, T_2 so that least common ancestor (lca) queries can be answered in $O(1)$ time (this preprocess can be done in $O(n)$ [22]). Then, by a simple traversal of T_M we can easily determine for each node $n_M \in T_M$ two leaves $l, l' \in L(n_M)$ s.t. $lca(l, l') = lca(L(n_M))$ (the equality holds jointly in trees T_M, T_1, T_2). Thus, each Twin_i structure ($i = 1, 2$) is initialized in $O(n)$ time, i.e., the cost of traversing T_M plus for each node n_M the cost of a lca queries in T_i. Apart from the Twin structures, the algorithm also needs to know for each node $n_i \in T_i$ ($i = 1, 2$), the list $\mathsf{SpecificChild}(n_i)$ of its child nodes v_i s.t. $S(v_i)$ is a specific subtree. These lists are precomputed in $O(n)$ time by traversing once T_1 and T_2.

Using these data structures, the pseudo-code $\textsc{GraftSpecific}(T_1, T_2)$ describes precisely how a tree $SMAST(\mathcal{T})$ is obtained by grafting specific subtrees of T_1 and T_2 into T_M. In this pseudo-code, an *initial* node or edge means one that is present in T_M when it is first computed (line 1).

Processing once each node and edge of T_M, as well as making one copy of each specific subtree of T_1 and T_2 clearly costs $O(n)$. Hence we have:

Theorem 5 *Let $\mathcal{T} = \{T_1, T_2\}$ be a pair of rooted trees with overlapping leaf sets, drawn from a set of n labels. The algorithm $\textsc{GraftSpecific}(T_1, T_2)$ returns a tree $SMAST(\mathcal{T})$ in time $O(N + n)$, where N is the time needed to compute the maximum agreement subtree of two $O(n)$-leaf rooted trees.*

Currently, $N = O(\sqrt{d}n \log \frac{2n}{d})$ where d is the maximum degree of the input trees [26].

Minor Modifications for Solving Related Problems. The case of unrooted input trees is handled in the same way, as Lemmas 5 and 6 are also valid in that

---

Algorithm 1: GRAFTSPECIFIC(T_1, T_2)

Input: Two rooted trees T_1, T_2 on overlapping sets of leaves.

Result : A tree $SMAST(T_1, T_2)$.

1 $T_M := MAST(T_1|L_\cap, T_2|L_\cap)$; $T_1 \leftarrow T_1|(L(T_M) \cup L_S)$; $T_2 \leftarrow T_2|(L(T_M) \cup L_S)$
   Compute the Twin and SpecificChild data structures
   for *each initial node* $u \in T_M$ **do**
   $\quad$ **for** *each* $s_i \in$ SpecificChild(Twin$_1(u)$) $\cup$ SpecificChild(Twin$_2(u)$) **do**
   $\quad\quad$ Add a copy of $S(s_i)$ as a new child of u

2 **for** *each initial edge* $(u, v) \in T_M$ *(with* $u =$parent(v)*)* **do**
   $\quad$ **for** $i \leftarrow 1$ *to* 2 **do**
   $\quad\quad$ $n_i \leftarrow$parent(Twin$_i(v)$) ; $v_i \leftarrow v$
   $\quad\quad$ **while** $n_i \neq$Twin$_i(u)$ **do**
   $\quad\quad\quad$ Insert a node m_i between v_i and its parent node in T_M
   $\quad\quad\quad$ **for** *each* $s_i \in$ SpecificChild(Twin$_i(n_i)$) **do**
   $\quad\quad\quad\quad$ Add a copy of $S(s_i)$ as a new child subtree of v_i
   $\quad\quad\quad$ $v_i \leftarrow m_i$; $n_i \leftarrow$ parent(n_i)

   /* Grafting specific subtrees lying above $r(T_M)$ */
3 **for** $i \leftarrow 1$ *to* 2 **do**
   $\quad$ $n_i \leftarrow$ Twin$_i(r(T_M))$
   $\quad$ **while** parent$(n_i) \neq \emptyset$ **do**
   $\quad\quad$ $n_i \leftarrow$parent(n_i)
   $\quad\quad$ Create a node m_i having T_M as child subtree
   $\quad\quad$ **for** *each* $s_i \in$ SpecificChild(Twin$_i(n_i)$) **do**
   $\quad\quad\quad$ Add a copy of $S(s_i)$ as a child subtree of m_i
   $\quad\quad$ Redefine T_M to be the tree rooted at m_i

---

setting. Changes in the pseudo-code consist in computing an *unrooted* tree T_M (line 1), then, before line 2, choosing a common leaf l and rooting T_1, T_2, T_M at the internal node to which l is connected. This gives an $O(N' + n)$ algorithm for solving SMAST on two unrooted trees, where N' is the time required to compute an unrooted maximum agreement subtree of two $O(n)$ leaf-trees. Currently, $N' = O(n^{1.5})$ [25].

In a similar way, to solve the SMCT we obtain T_M (line 1) by resolving MCT instead of MAST. Then, we restrict input trees to leaves of $L(T_M)$ and L_S. However, before starting to graft specific subtrees, we need to refine the input trees to reproduce the structure of T_M (this is done in $O(n)$ time). Thus, MCT is solved in $O(N'' + n)$ time, where N'' is the time required to solve MCT on two $O(n)$-leaf trees. Currently, N'' is $O(min\{2^{4d}n^2, 3^p n, 2.27^p + n^3\})$ ([16] and Thm. 3) for rooted trees. Considering unrooted trees adds an extra n factor.

When aiming at producing an estimate of a phylogeny underlying the input trees, the fact that the supertree may contain some arbitrary edges is not satisfactory. This happens whenever a same edge in T_M corresponds to paths in T_1 and T_2 from which specific subtrees are hanging in both trees. Instead of

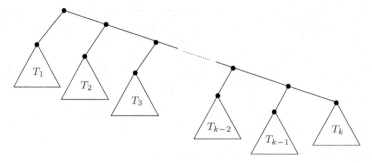

Fig. 3. The tree $rake(T_1, T_2, \ldots, T_k)$ for a sequence $T_1, T_2, \ldots, T_k$ of trees.

grafting subtrees of T_2 above those of T_1 (loops 2 and 3), we can transpose the lack of information of the input collection concerning the relative order of these subtrees by connecting them (or their leaves[2]) to a single multifurcating node m_i in T. This implies a minor modification which does not change the running time of the algorithm. Note that the produced tree might not be a maximum agreement supertree anymore, as some internal edges of the input trees can be collapsed to m in T.

5.2 Intractability of SMAST and SMCT

As MAST is the special case of SMAST where input trees have identical sets of leaves, the NP-hardness result for only three input trees obtained by [2] for MAST also holds for SMAST, as the inapproximability results of [23, 17]. Similarly, SMCT in NP-hard for 2 trees from the result of [23].

We now show that there is an important gap in complexity between MAST, resp. MCT, and SMAST, resp. SMCT, on more than two trees, even of bounded size. More precisely, we prove, by reduction from the general HITTING SET problem (HS), that SMAST is NP-complete and W[2]-hard for the parameter p, defined as $\#L(T) - \#SMAST(T)$. This holds even for instances consisting only of rooted triples and for instances consisting of unrooted 4-leaf trees (*quartets*). In comparison, MAST on k input trees of bounded size can be solved in time $O(k)$ and MAST is FPT for parameter p in the general case.

If not specified, trees in this section are to be considered rooted.

Definition 7 *We recursively define the function* rake *associating a tree to a non-empty ordered sequence of trees with non-intersecting leaf sets:*

- $rake(T) = T$ *for any tree* T *and*
- $rake(T_1, T_2, \ldots, T_k) = (T_1, rake(T_2, \ldots, T_k))$ *for any sequence of trees* T_1, $T_2, \ldots, T_k$ *of length* $k \geq 2$ *s.t.* $L(T_i) \cap L(T_j) = \emptyset, \forall i, j \in [1, k], i \neq j$.

Figure 3 illustrates the previous definition. The second item above uses the parenthetical notation for trees. We now describe the gadget that is used to reduce HS to SMAST:

---

[2] Though, when aiming at a *seed* tree for the *MRP* method, we guess it might be better to conserve the topology of the specific subtrees.

Gadget: *Let $m \geq 1$ and a set of distinct labels x^1, x^2, ..., x^m, y^1, y^2, ..., y^m, we define $\mathcal{G}$ to be the following collection of rooted triples:*

$$\left\{ y^h | y^{h+1} x^{h+1}, \, y^h | x^{h+1} x^{h+2} \right\}_{h \in [1,m]}$$

where we set $x^{m+1} := x^1$, $x^{m+2} := x^2$ et $y^{m+1} := y^1$.

$\mathcal{G}$ has the following properties:

1. *there is no agreement supertree of $\mathcal{G}$ on $L(\mathcal{G})$*
2. *the following tree is an agreement supertree of $\mathcal{G}$:*

$$\text{rake}(y^j, y^{j+1}, \ldots, y^m, y^1, y^2, \ldots, y^{j-1}, T^*)$$

where $j \in [1,m]$ and T^* is any tree with leaves $\left\{ x^1, x^2, \ldots, x^m \right\} - \left\{ x^j \right\}$.

Theorem 6 *The* SMAST *problem is* NP*-complete and* W[2]*-hard for the parameter p, even for instances were trees in $\mathcal{T}$ are rooted triples, resp quartets.*

The proof relies on a reduction from the general HITTING SET problem (HS) [5].

Let CSMAST, resp. CSMCT, be the *complement* problem of SMAST, resp. SMCT, defined as minimizing the number of input leaves to remove to obtain an agreement supertree, resp. a compatible supertree. The reduction from HS used in Thm. 6 is an L-reduction, hence using the result of [4] for HS we obtain:

Theorem 7 CSMAST *is not approximable within a constant factor unless* P = NP.

The two above results of intractability and inapproximability also hold for SMCT and CSMCT, as the reduction uses binary trees, case in which they are equivalent to SMAST, resp. CSMAST. Note that these results hold as well both for the rooted and the unrooted versions of the problems.

Final note: after acceptation of this paper, we learnt that a paper with a work similar to material in Sect. 5.1 appears independently at the same time [24].

References

1. J. Alber, Gramm J., and R. Niedermeier. Faster exact algorithms for hard problems: a parameterized point of view. *Disc. Math.*, 229:3–27, 2001.
2. A. Amir and D. Keselman. Maximum agreement subtree in a set of evolutionary trees: metrics and efficient algorithm. *SIAM J. on Comp.*, 26(3):1656–1669, 1997.
3. B.R. Baum. Combining trees as a way of combining data sets for phylogenetic inference, and the desirability of combining gene trees. *Taxon*, 41:3–10, 1992.
4. M. Bellare, S. Goldwasser, C. Lund, and A. Russeli. Efficient probabilistically checkable proofs and applications to approximations. In *Proceedings of the Twenty-Fifth Annual A.C.M. Symposium on Theory of Computing*, pages 294–304, 1993.
5. V. Berry and F. Nicolas. Maximum agreement and compatible supertrees. (available from http://www.lirmm.fr/~vberry) 04045, LIRMM, 2004.

6. O.R.P. Bininda-Edmonds and H.N. Bryant. Properties of matrix representation with parsimony analyses. *Syst. Biol.*, 47:497–508, 1998.
7. O.R.P. Bininda-Edmonds, J.L. Gittleman, and M.A. Steel. The (super)tree of life: procedures, problems, and prospects. *Ann. Rev. Ecol. Syst.*, 2002.
8. O.R.P. Bininda-Edmonds and M.J. Sanderson. Assessment of the accuracy of matrix representation with parsimony analysis supertree construction. *Syst. Biol.*, 50(4):565–579, 2001.
9. D. Bryant. *Building trees, hunting for trees and comparing trees*. PhD thesis, University of Canterbury, Department of Math., 1997.
10. D. Bryant and M.A. Steel. Extension operations on sets of leaf-labelled trees. *Adv. Appl. Math.*, 16:425–453, 1995.
11. D. Chen, L. Diao, O. Eulenstein, and D. Fernandez-Baca. Flipping: a supertree construction method. *DIMACS Series in Disc. Math. and Theor. Comp. Sci.*, 61:135–160, 2003.
12. R. Cole, M. Farach, R. Hartigan, Przytycka T., and M. Thorup. An $O(n \log n)$ algorithm for the maximum agreement subtree problem for binary trees. *SIAM J. on Computing*, 30(5):1385–1404, 2001.
13. R. Cole and R. Hariharan. Dynamic lca queries on trees. In *Proc. of the 10th ann. ACM-SIAM symp. on Disc. alg. (SODA'99)*, pages 235 – 244, 1999.
14. R.G. Downey, M.R. Fellows, and U. Stege. Computational tractability: The view from mars. *Bull. of the Europ. Assoc. for Theoret. Comp. Sci.*, 69:73–97, 1999.
15. M. Farach, T. Przytycka, and M. Thorup. Agreement of many bounded degree evolutionary trees. *Inf. Proc. Letters*, 55(6):297–301, 1995.
16. G. Ganapathysaravanabavan and T. Warnow. Finding a maximum compatible tree for a bounded number of trees with bounded degree is solvable in polynomial time. In O. Gascuel and B.M.E. Moret, editors, *Proc. of the Workshop on Algorithms for Bioinformatics (WABI'01)*, volume 2149 of *LNCS*, pages 156–163, 2001.
17. L. Gasieniec, J. Jansson, A. Lingas, and A. Ostlin. On the complexity of constructing evolutionary trees. *J. of Combin. Optim.*, 3:183–197, 1999.
18. A.G. Gordon. Consensus supertrees: the synthesis of rooted trees containing overlapping sets of labelled leaves. *J. of Classif.*, 3:335–346, 1986.
19. A. Gupta and N. Nishimura. Finding largest subtrees and smallest supertrees. *Algorithmica*, 21(2):183–210, 1998.
20. D. Gusfield. Efficient algorithms for inferring evolutionary trees. *Networks*, 21:19–28, 1991.
21. A.M. Hamel and M.A. Steel. Finding a maximum compatible tree is NP-hard for sequences and trees. *Appl. Math. Lett.*, 9(2):55–59, 1996.
22. D. Harel and R.E. Tarjan. Fast algorithms for finding nearest common ancestor. *Computer and System Science*, 13:338–355, 1984.
23. J. Hein, T. Jiang, L. Wang, and Zhang K. On the complexity of comparing evolutionary trees. *Disc. Appl. Math.*, 71:153–169, 1996.
24. J. Jansson, J. H.-K. Ng, K. Sadakane, and W.-K. Sung. Rooted maximum agreement supertrees. In *Proceedings of the Sixth Latin American Symposium on Theoretical Informatics (LATIN)*, 2004. (in press).
25. M.Y. Kao, T.W. Lam, W.K. Sung, and H.F. Ting. A decomposition theorem for maximum weight bipartite matchings with applications to evolutionary trees. In *Proc. of the 8th Ann. Europ. Symp. Alg. (ESA)*, pages 438–449. Springer-Verlag, New York, NY, 1999.
26. M.Y. Kao, T.W. Lam, W.K. Sung, and H.F. Ting. An even faster and more unifying algorithm for comparing trees via unbalanced bipartite matchings. *J. of Algo.*, 40:212–233, 2001.

27. R. Niedermeier and P. Rossmanith. An efficient fixed parameter algorithm for 3-Hitting Set. *Journal of Discrete Algorithms*, 1:89–102, 2003.
28. R. Page. Modified mincut supertrees. In O. Gascuel and M.-F. Sagot, editors, *Proc. of the Workshop on Algorithms for Bioinformatics (WABI'02)*, LNCS, pages 538–551. Springer-Verlag, 2002.
29. A. Purvis. A modification to Baum and Ragan's method for combining phylogenetic trees. *Syst. Biol.*, 44:251–255, 1995.
30. M.A. Ragan. Matrix representation in reconstructing phylogenetic relationships among the eukaryots. *Biosystems*, 28:47–55, 1992.
31. F. Ronquist. Matrix representation of trees, redundancy, and weighting. *Syst. Biol.*, 45:247–253, 1996.
32. C. Semple and M.A. Steel. A supertree method for rooted trees. *Disc. Appl. Math.*, 105:147–158, 2000.
33. M.A. Steel and T. Warnow. Kaikoura tree theorems: Computing the maximum agreement subtree. *Information Processing Letters*, 48:77–82, 1993.
34. J.L. Thorley and M. Wilkinson. A view of supertrees methods. In *Bioconsensus, DIMACS*, volume 61, pages 185–194. Amer. Math. Soc. Pub., 2003.
35. T. J. Warnow. Tree compatibility and inferring evolutionary history. *Journal of Algorithms*, 16:388–407, 1994.
36. M. Wilkinson, J. Thorley, D.T.J. Littlewood, and R.A. Bray. *Interrelationships of the Platyhelminthes*, chapter 27, Towards a phylogenetic supertree of Platyhelminthes. Taylor and Francis, London, 2001.

Polynomial-Time Algorithms for the Ordered Maximum Agreement Subtree Problem

Anders Dessmark, Jesper Jansson, Andrzej Lingas, and Eva-Marta Lundell

Department of Computer Science, Lund University, Box 118, 221 00 Lund, Sweden
{andersd,jj,andrzej,emj}@cs.lth.se

Abstract. For a set of rooted, unordered, distinctly leaf-labeled trees, the NP-hard maximum agreement subtree problem (MAST) asks for a tree contained (up to isomorphism or homeomorphism) in all of the input trees with as many labeled leaves as possible. We study the ordered variants of MAST where the trees are uniformly or non-uniformly ordered. We provide the first known polynomial-time algorithms for the uniformly and non-uniformly ordered homeomorphic variants as well as the uniformly and non-uniformly ordered isomorphic variants of MAST. Our algorithms run in time $O(kn^3)$, $O(n^3 \min\{nk, \ n + \log^{k-1} n\})$, $O(kn^3)$, and $O((k+n)n^3)$, respectively, where n is the number of leaf labels and k is the number of input trees.

1 Introduction

The basic combinatorial problem of finding a *largest common subsequence* for a set of sequences (LCS) and the well known problem of finding a *maximum agreement subtree* for a set of trees with distinctly labeled leaves (MAST) fall in the general category of problems of finding the *largest common subobject* for an input set of combinatorial objects.

In [7], Fellows *et al.* in particular studied the largest common subobject problem constrained to the so called p-sequences, i.e., sequences where each element occurs at most once. A p-sequence can be seen as an ordered, distinctly leaf-labeled star tree. In this paper, we study a natural generalization of the largest common subobject problem for p-sequences in which the objects are allowed to be arbitrary rooted, ordered trees with distinctly labeled leaves. Since this problem can be also regarded as a restriction of the MAST problem where tree ordering is required, we term it the *ordered maximum agreement subtree problem*.

For an extensive literature and motivations for the LCS and MAST problems, the reader is referred to [3,13,17] and [2,5,6,11,15,16], respectively. The NP-hardness [2] and approximation NP-hardness [4,10,11] of the general MAST problem is one of the motivations for studying its ordered variants in this paper.

1.1 Variants of the Maximum Agreement Subtree Problem

A tree whose leaves are labeled by elements belonging to a finite set S so that no two leaves have the same label is said to be *distinctly leaf-labeled by S*. Throughout this paper, each leaf in such a tree is identified with its corresponding element

S.C. Sahinalp et al. (Eds.): CPM 2004, LNCS 3109, pp. 220–229, 2004.

in S. Let T be a rooted tree distinctly leaf-labeled by a given finite set S. For any subset S' of S, $T|S'$ denotes the tree obtained by first deleting from T all leaves which are not in S' and all internal nodes without any descendants in S' along with their incident edges, and then contracting every edge between a node having just one child and its child. Similarly, $T||S'$ denotes the tree obtained by deleting from T all leaves which are not in S' and all internal nodes without any descendants in S' along with their incident edges.

In *the maximum homeomorphic agreement subtree problem* (MHT), the input is a finite set S and a set $\mathcal{T} = \{T_1, ..., T_k\}$ of rooted, unordered trees, where each $T_i \in \mathcal{T}$ is distinctly leaf-labeled by S and no $T_i \in \mathcal{T}$ has a node of degree 1, and the goal is to find a subset S' of S of maximum cardinality such that $T_1|S' = ... = T_k|S'$. In *the maximum isomorphic agreement subtree problem* (MIT), the input is a finite set S and a set $\mathcal{T} = \{T_1, ..., T_k\}$ of rooted, unordered trees, where each $T_i \in \mathcal{T}$ is distinctly leaf-labeled by S, and the goal is to find a subset S' of S of maximum cardinality such that $T_1||S' = ... = T_k||S'$.

An *ordered* tree is a rooted tree in which the left-to-right order of the children of each node is significant. The *leaf ordering* of an ordered, leaf-labeled tree is the sequence of labels obtained by scanning its leaves from left to right. A set $\mathcal{T}$ of ordered trees distinctly leaf-labeled by S is said to be *uniformly ordered* if all trees in $\mathcal{T}$ have the same leaf ordering.

We study the following four ordered variants of MHT and MIT:

- The ordered maximum homeomorphic agreement subtree problem (OMHT)
- The ordered maximum isomorphic agreement subtree problem (OMIT)
- The uniformly ordered maximum homeomorphic agreement subtree problem (UOMHT)
- The uniformly ordered maximum isomorphic agreement subtree problem (UOMIT)

OMHT and OMIT are defined in the same way as MHT and MIT except that $\mathcal{T}$ is required to be a set of *ordered* trees. UOMHT and UOMIT are the special cases of OMHT and OMIT in which $\mathcal{T}$ is required to be uniformly ordered. Note that OMHT and OMIT are the natural generalization of the largest common subobject problem for p-sequences studied by Fellows *et al.* in [7].

From here on, n and k denote the cardinalities of S and $\mathcal{T}$, respectively.

1.2   Motivations

In certain evolutionary tree construction situations, one can determine or accurately estimate the leaf ordering of a planar embedding of the true tree by taking into account other kinds of data such as the geographical distributions of the species or data based on some measurable quantitative characteristics (average life span, size, *etc*). The ordered variants of MHT and MIT might also arise in graphical representation of evolutionary trees where additional restrictions are placed on the leaves (e.g., that they must be ordered alphabetically) for ease of presentation. In the context of the approximation NP-hardness of

MHT and MIT, their ordered restrictions are of theoretical interest in their own rights. Does the leaf ordering restriction make the problems computationally feasible? In [9], Gąsieniec et al. showed that an analogous ordering restriction on an NP-hard optimization problem occurring in the construction of evolutionary trees admits a polynomial-time algorithmic solution. Our results on the ordered variants of MHT and MIT will further confirm the power of ordering.

1.3   Related Results

Fellows, Hallet, and Stege studied the LCS problem for p-sequences among many other problems in their interesting article [7], and claimed that they could solve this problem for k p-sequences with n symbols in time $O(kn(k + \log n))$ [1].

Steel and Warnow [15] presented the first exact polynomial-time algorithms to solve MHT and its unrooted counterpart UMHT[2] when $k = 2$ and the degrees are unbounded. Since then, many improvements have been published, the fastest currently known being an algorithm for MHT with $k = 2$, invented by Kao, Lam, Sung, and Ting [12], that runs in $O(\sqrt{D} n \log(2n/D))$ time, where D is the maximum degree of the two input trees. Note that this is $O(n \log n)$ for trees with maximum degree bounded by a constant and $O(n^{1.5})$ for trees with unbounded degrees. Finally, for two rooted, *ordered* trees, a maximum agreement subtree can be computed in $O(n \log^2 n)$ time [16].

Amir and Keselman [2] considered the more general case $k \geq 2$. They proved that MHT is NP-hard already for three trees with unbounded degrees, but solvable in polynomial time for three or more trees if the degree of at least one of the trees is bounded by a constant. For the latter case, Farach, Przytycka, and Thorup [6] gave an algorithm with improved efficiency running in $O(kn^3 + n^d)$ time, where d is an upper bound on at least one of the input trees degrees; Bryant [5] proposed a conceptually different algorithm with the same running time.

Hein, Jiang, Wang, and Zhang [11] proved that MHT with three trees with unbounded degrees cannot be approximated within a factor of $2^{\log^\delta n}$ in polynomial time for any constant $\delta < 1$, unless NP $\subseteq$ DTIME$[2^{\mathrm{polylog}\, n}]$. This inapproximability result also holds for UMHT [11]. Bonizzoni, Della Vedova, and Mauri [4] showed that it can be carried over to MIT restricted to three trees with unbounded degrees as well, and that even stronger bounds can be proved for MIT in the general case. Gąsieniec, Jansson, Lingas, and Östlin [10] proved that MHT is hard to approximate in polynomial time even for instances containing only trees of height 2, and showed that if the number of trees is bounded by a constant and all of the input trees' heights are bounded by a constant, then MHT can be approximated within a constant factor in $O(n \log n)$ time.

---

[1] Their algorithm, allowing different interpretations, seems to fail already for $x_1 = 1234$ and $x_2 = 1342$, producing either 12 or 34 instead of 134. Therefore, we were forced to use a weaker upper time-bound for this problem, given in Lemma 5 in this paper, in order to derive one of our main results.

[2] UMHT is defined like MHT except that all trees are unrooted and $T|S'$ now denotes the tree obtained by first deleting from T all nodes (and their incident edges) not on any path between two leaves in S', and then contracting every node with degree 2.

1.4   Our Results and Organization of the Paper

We present the first known polynomial-time algorithms for the uniformly and non-uniformly ordered homeomorphic variants (UOMHT and OMHT) as well as the uniformly and non-uniformly ordered isomorphic variants (UOMIT and OMIT) of the maximum agreement subtree problem. They run in time $O(kn^3)$, $O(n^3 \min\{nk, n + \log^{k-1} n\})$, $O(kn^3)$, and $O((k+n)n^3)$, respectively. Our results have been obtained by utilization of deep structural properties of ordered agreement subtrees which allowed for significant pruning of the otherwise unfeasible number of combinations of subproblems needed for the exact solution.

In Section 2, we introduce some common notation for our algorithms. In Section 3, we present the algorithm for UOMHT. Section 4 is devoted to the algorithm for OMHT. Section 5 describes the algorithms for UOMIT and OMIT.

2   Notation

We use the following notation. For any $a, b \in S$, denote by $OMHT_{a,b}$ the problem OMHT under the additional constraint that for any valid solution S', the leaf ordering of $T_1|S'$ must begin with a and end with b. Define $OMIT_{a,b}$, $UOMHT_{a,b}$, and $UOMIT_{a,b}$ analogously. Furthermore, let $UOMHT_{(a),(b)}$ be UOMHT restricted to the subset of S consisting of all elements in the interval $[a, b]$ in the leaf ordering. Note that while a and b are required to belong to any solution to $UOMHT_{a,b}$, they are not necessarily included in a solution to $UOMHT_{(a),(b)}$.

We also find it convenient to write $S = \{a_1, ..., a_n\}$ and in the uniformly ordered case let a_l precede a_r in the leaf ordering if $l < r$.

3   A Polynomial-Time Algorithm for UOMHT

In this section, we present an algorithm called *All-Pairs* that solves the uniformly ordered maximum homeomorphic agreement subtree problem. Our algorithm employs dynamic programming to build a table of solutions for $UOMHT_{(a_l),(a_r)}$ for all pairs of leaves a_l and a_r with $l \leq r$, using a procedure called *One-Pair*. *All-Pairs* and *One-Pair* are listed in Fig. 1 and Fig. 2, respectively.

Given indices l and r, *One-Pair* first computes $UOMHT_{a_l, a_r}$ and then finds the solution to $UOMHT_{(a_l),(a_r)}$ by taking the largest of $UOMHT_{a_l, a_r}$ and the two previously computed optimal solutions to subproblems $UOMHT_{(a_l),(a_{r-1})}$ and $UOMHT_{(a_{l+1}),(a_r)}$, both stored in the dynamic programming table.

When computing $UOMHT_{a_l, a_r}$, the algorithm considers the path P between a_l and a_r and the positions along P where the intervening leaves have their *lowest ancestors*. By *the lowest ancestor of leaf a_j*, we mean the node which is the lowest common ancestor of a_j and either a_l or a_r, i.e., the position on P where the path from a_j to the root joins P. By *the lowest ancestor edge of leaf a_j*, we mean the first edge on the path leading from the lowest ancestor of a_j to a_j.

For each input tree T_i, let v_i be the lowest common ancestor of a_l and a_r. In Step 1, the intervening leaves are divided into three classes depending on whether

---

Algorithm   *All-Pairs*

Input: An instance of UOMHT.

Output: The subset of leaves in a maximum agreement subtree of $\mathcal{T}$.

    for *length* = 1 to n **do**
        for *left* = 1 to $n - length + 1$ **do**
            Compute $UOMHT_{(a_{left}),(a_{left+length-1})}$ by a subroutine call to *One-Pair(left, left + length − 1)* and enter the result in the table.
        endfor
    endfor
    return $UOMHT_{(a_1),(a_n)}$

End *All-Pairs*

Fig. 1. A dynamic programming algorithm for UOMHT.

their lowest ancestor is inside the path from v_i to a_l, equal to v_i itself, or inside the path from v_i to a_r. Any leaf that does not belong to the same class in all input trees can be discarded, since it can not exist in a solution together with a_l and a_r (Step 2). Next, in Step 3, the remaining intervening leaves are further processed and divided into sets according to their lowest ancestors so that two leaves having the same lowest ancestor belong to the same set. The divisions created by the different trees are merged into sets $S_1, ..., S_m$ so that two leaves belong to the same set S_p if and only if they have the same lowest ancestor in every tree in $\mathcal{T}$. The leaves in any such set S_p clearly form a consecutive sequence.

Each set S_p is further divided into subsets in Step 4. As in the previous division, every input tree is traversed to find the lowest ancestor edge for each leaf which connects it to the path between a_l and a_r. The resulting divisions are then merged to create the subsets $S_{p,1}, S_{p,2}, ...$ with the property that two leaves are in the same subset if and only if they have the same lowest ancestor edge in every tree in $\mathcal{T}$ (see Fig. 3). As above, the leaves in such a subset form a consecutive sequence. It also holds for any subset $A \subseteq S_{p,q}$ that if A induces a homeomorphic agreement subtree then the set $A \cup \{a_l, a_r\}$ also induces a homeomorphic agreement subtree. Such a maximum subset A can be looked up in the table (Step 5). However, two members of two different $S_{p,q}$ subsets may have the same lowest ancestor edge in some tree even though no such pair of leaves can belong to the same solution. Since members of a subset share a lowest ancestor edge in all trees, a conflict in one tree must hold for all pairs of leaves induced by the two subsets concerned. Thus, we can represent such a conflict as an edge in a special graph having the $S_{p,q}$ sets as vertices and the sizes of UOMHT solutions looked up in Step 5 as vertex weights (Step 6), and then compute a maximum weighted independent set in this graph (Step 7).

The maximum subset of leaves, connected to the same lowest ancestor in all trees, that can be included together with a_l and a_r in a homeomorphic agreement subtree are now known. Next, we choose the sets to be included in $UOMHT_{a_l, a_r}$. This is done in Steps 8 and 9 similarly as for the subsets of a single set. We build a

---

Algorithm *One-Pair*

Input: An instance of UOMHT and the index of two leaves a_l and a_r.

Output: The subset of leaves in a maximum agreement subtree of T restricted to the set of leaves in the range from a_l to a_r (i.e., $UOMHT_{(a_l),(a_r)}$).

1 For each $T_i \in T$, divide the leaves $\{a_{l+1}, ..., a_{r-1}\}$ into sets according to the location of their lowest ancestor on the path between a_l and a_r.

2 Remove every leaf a_j, such that all subtrees induced by $\{a_l, a_j, a_r\}$ are not homeomorphic, from further consideration.

3 Construct the sets $S_1, ..., S_m$ of leaves such that two leaves are in the same set if and only if they are in the same set constructed in Step 1 for all trees in T, and furthermore, for any two leaves $a_i \in S_p$ and $a_j \in S_q$, if $p < q$ then $i < j$.

   for $p = 1$ to m **do**

4   Construct the sets $S_{p,1}, ..., S_{p,m_p}$ of leaves such that two leaves are in the same set if and only if they are connected by the same edge to the path between a_l and a_r for all trees in T, and furthermore, for any two leaves $a_i \in S_{p,s}$ and $a_j \in S_{p,t}$, if $s < t$ then $i < j$.

5   **for** $q = 1$ to m_p **do**

           $S_{p,q} = UOMHT_{(a_i),(a_j)}$ where i and j are the smallest and largest index, respectively, such that $a_i \in S_{p,q}$ and $a_j \in S_{p,q}$.

      endfor

6   Construct a weighted graph $G_p = (V_p, E_p)$, where $V_p = \{S_{p,1}, ..., S_{p,m_p}\}$ with weights according to their sizes, and E_p contains every pair of subsets of leaves such that they share a lowest ancestor edge in some $T_i \in T$.

7   Compute the maximum weighted independent set of G_p and let S_p be the union of the sets in the solution.

   endfor

8 Construct a weighted graph $G = (V, E)$ where V is the set $\{S_1, ..., S_m\}$ with weights according to their sizes and E contains every pair of sets of leaves such that they are in the same set for some $T_i \in T$.

9 Compute the maximum weighted independent set of G and let W be the union of the sets in the solution.

10 return The largest of W, $UOMHT_{(a_l),(a_{r-1})}$ and $UOMHT_{(a_{l+1}),(a_r)}$.

End *One-Pair*

Fig. 2. The subroutine for computing one entry of the dynamic programming table.

graph with the S_p sets as the vertices, with weights according to the solutions to the maximum weighted independent set problems for the subsets of the sets. The conflict edges now represent two separate sets sometimes connected to the same lowest ancestor (regardless of lowest ancestor edge). The maximum weighted independent set for this graph provides us with the solution to $UOMHT_{a_l, a_r}$.

Lemma 1. *One-Pair solves* $UOMHT_{(a_l),(a_r)}$ *in* $O(km)$ *time, with* $m = r - l + 1$.

Proof. The correctness follows from the discussion above. As for the time complexity, Steps 1 to 3 can be done in one traversal of each tree in T (this takes $O(km)$ time). Whenever the ancestor changes, this is recorded in the present

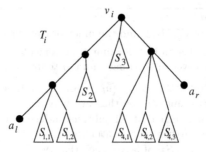

Fig. 3. An input tree T_i where the leaves $\{a_{l+1}, .., a_{r-1}\}$ are divided into sets according to lowest ancestor and lowest ancestor edge.

leaf. After all trees have been traversed, the union of these recorded changes will divide the leaves into the sought sequence of sets. Another traversal of the trees in $\mathcal{T}$ is sufficient to make the further division in Step 4.

Computing the weights for the subsets is done in $O(m)$ time, as there are $O(m)$ subsets. The edge sets can also be computed in the course of the traversal, by noting the lowest index of leaf that so far has shared a lowest ancestor edge with the current leaf. Since the leaves are uniformly ordered, a conflict with one leaf must also result in a conflict with all the intervening leaves. Solving the maximum weighted independent set problems in Step 7 for the constructed graphs can be done in linear time by the following dynamic programming procedure:

Process the vertices in the order of leaves contained. Keep a table of the maximum weight subset of vertices computed thus far. For every new entry (corresponding to the set $S_{p,q}$), compare the entry for set $S_{p,q-1}$ with the weight of the current vertex ($|S_{p,q}|$) added to the entry corresponding to $S_{p,s}$ where s is the largest index $s < q$ such that there is no conflict between $S_{p,s}$ and $S_{p,q}$, and choose the larger of the two values.

The edge set in Step 8 can be computed during the initial traversal. The final maximum weighted independent set problem in Step 9 can be solved in linear time with a dynamic programming procedure analogous to that of Step 7.   □

Theorem 1. *Algorithm All-Pairs solves UOMHT in $O(kn^3)$ time.*

Proof. There are $O(n^2)$ pairs; each pair requires $O(kn)$ time by Lemma 1.   □

4 An Algorithm for OMHT

In this section we present an algorithm for the ordered maximum homeomorphic agreement subtree problem running in time $O(n^3 \min\{nk, \ n + \log^{k-1} n\})$.

For each $a_l, a_r \in S$ where a_l precedes a_r in the leaf ordering of all trees in $\mathcal{T}$, or $a_l = a_r$, we consider the subproblem $OMHT_{a_l,a_r}$ (if $a_l = a_r$, the solution to $OMHT_{a_l,a_r}$ is trivially the singleton $\{a_l\}$). Our algorithm for $OMHT_{a_l,a_r}$ is similar to that for $UOMHT_{a_l,a_r}$, again focusing on the path between a_l and a_r. Let v_i be the lowest common ancestor of a_l and a_r in T_i. As before, the intervening leaves are divided into three classes depending on whether their lowest ancestor is inside the path from v_i to a_l, equal to v_i, or inside the path from v_i to a_r. Any leaf that does not belong to the same class in all trees is immediately discarded as it cannot exist in a solution together with a_l and a_r.

With each of the remaining intervening leaves b, we associate the point $p(b) = (p(b)_1, ..., p(b)_k)$ in $\mathbb{N}^k$, where for $i = 1, ..., k$, $p(b)_i$ is defined as the distance between a_l and the lowest ancestor of b on the path from a_l to a_r in T_i. With each of the points p, we associate the set S_p of remaining leaves b for which $p = p(b)$. In turn, with each leaf $b \in S_p$, we associate another point $q(b) = (q(b)_1, ..., q(b)_k)$ in $\mathbb{N}^k$, where for $i = 1, ..., k$, $q(b)_i$ is the rank of the lowest ancestor edge of b on the path from a_l to a_r in T_i. We say that a point x in $\mathbb{N}^k$ *strictly dominates* a point y in $\mathbb{N}^k$ if each coordinate of x is greater than the corresponding one of y.

Lemma 2. *A pair of remaining leaves b, d, where $p(b) \neq p(d)$, can jointly with a_l and a_r be a subset of a leaf set inducing a homeomorphic agreement subtree for $\mathcal{T}$ if and only if $p(b)$ strictly dominates $p(d)$ or vice versa.*

Lemma 3. *A pair of remaining leaves b, $d \in S_p$, where $q(b) \neq q(d)$, can jointly with a_l and a_r be a subset of a leaf set inducing a homeomorphic agreement subtree for $\mathcal{T}$ if and only if $q(b)$ strictly dominates $q(d)$ or vice versa.*

Following the idea of double set subdivision from the previous section, we associate with each of the points q a set $S_{p,q}$ of remaining leaves $b \in S_p$ for which $q = q(b)$. We let the weight of each such point q be the weight the maximum cardinality of a solution to $OMHT_{c,d}$, where c, $d \in S_{p,q}$ and c and d occur in the same order in the leaf ordering of all trees in $\mathcal{T}$.

The points q can be interpreted as degenerate k-trapezoids [8]. By Lemma 3, it suffices to find a maximum weighted independent set in the k-trapezoid graph induced by the points q, i.e., in the intersection graph of the k-trapezoids, in order to find a largest subset of S_p that can jointly with a_l and a_r be a subset of a leaf set inducing a homeomorphic agreement subtree for $\mathcal{T}$.

Lemma 4. *[8] A maximum weighted independent set in a k-trapezoid graph on n vertices given with its box representation can be found in $O(n \log^{k-1} n)$ time.*

Consequently, we assign as the weight of p the cardinality of such a largest subset. Analogously, the points p can be interpreted as degenerate k-trapezoids. By Lemma 2, it is sufficient to find a maximum weighted independent set in the k-trapezoid graph induced by the points p to solve $OMHT_{a_l,a_r}$.

When k is large, the above method for finding the maximum weighted independent set in a k-trapezoid graph is super-polynomial. Therefore, for large values of k, we reduce the latter problem to a restricted version of LCS. Given k sequences, each consisting of a permutation of the elements in a set Σ, the problem is to find the longest sequence that is a subsequence of all input sequences.

Lemma 5. *Restricted LCS can be solved in $O(kn^2)$ time, where $n = |\Sigma|$.*

Proof. We first compute and store the ranks of all elements in all sequences in an $n \times k$ table, which takes $O(kn)$ time. For any $a, b \in \Sigma$, we can then determine if

a occurs before b in all sequences, if b occurs before a in all sequences, or neither. Build a directed graph $G = (\Sigma, E)$ where $(a, b) \in E$ if and only if a occurs before b in all sequences. G is clearly acyclic and can be constructed in $O(kn^2)$ time. A topological sort on G yields the longest directed path and this corresponds to a longest common subsequence. This can be done within the given time bound. $\quad\square$

In our reduction, each point corresponds to an element in Σ and each input tree to a sequence of elements. The reduction must take into account two difficulties: our points are weighted and two points may have the same coordinate for some input tree. The latter is handled by representing each input tree by *two* sequences. If a set of points share a coordinate for a tree, we put them in arbitrary order with respect to each other in the first sequence and reverse this order in the second sequence to ensure that no pair of such points can occur together in the solution. The weights of the points are accounted for by replacing an element a of weight w in all the $2k$ sequences by w consecutive elements $a_1, ..., a_w$, uniformly ordered throughout the sequences. Since the weights correspond to disjoint sets of leaves, the length of the sequences will be at most n.

The construction of the set of the remaining leaves as well as the construction of the set of points $p(b)$ and $q(b)$ can be carried out in total time $O(nk)$, the latter by using lexicographic sort (see for instance [1]). The weights of the points q can be determined in total time $O(n^2)$. The two level application of Lemma 4 or Lemma 5 in order to determine maximum weight independent sets of points q and p takes $O(n \min\{kn, \log^{k-1} n\})$ time. Hence, we obtain the following lemma.

Lemma 6. *If, for each pair a_m, a_q which follows a_l and precedes a_r in the leaf ordering of all trees in $\mathcal{T}$, the cardinalities of solutions to $OMHT_{a_m, a_q}$ are already stored then $OMHT_{a_l, a_r}$ can be solved in $O(n \min\{kn, n + \log^{k-1} n\})$ time.*

There are $O(n^2)$ subproblems $OMHT_{a_l, a_r}$. For any a_m, a_q in Lemma 6, the quadruple a_l, a_m, a_q, a_r is in particular a subsequence of the leaf ordering in the first tree. Hence, it suffices to solve the subproblems in order of non-decreasing distance between a_l and a_r in the leaf ordering of the first tree to solve them with dynamic programming by Lemma 6. We thus obtain our next main result.

Theorem 2. *OMHT can be solved in $O(n^3 \min\{nk, \; n + \log^{k-1} n\})$ time.*

5   Polynomial-Time Solutions for UOMIT and OMIT

Our polynomial-time algorithms for UOMHT and OMHT can easily be adapted to UOMIT and OMIT, respectively. In order to solve the corresponding subproblems $UOMIT_{a_l, a_r}$ and $OMIT_{a_l, a_r}$, we rely on the following lemma.

Lemma 7. *(Smolenskii [14]) Two labeled trees are isomorphic if and only if the distance between any two leaves with corresponding labels is the same.*

By Lemma 7, we can require the subtrees induced by the leaves a_l and a_r to be pairwise isomorphic. In particular, the paths connecting a_l with a_r must be of equal length. Also, any relevant remaining leaf must have the same distances to a_l and a_r in all input trees. This immediately implies that such a leaf has the same lowest ancestor on the path connecting a_l with a_r in all the input trees. Hence, any pair of such leaves with different lowest ancestors on the aforementioned path can occur in a feasible solution to $UOMIT_{a_l,a_r}$ or $OMIT_{a_l,a_r}$, respectively. Consequently, the corresponding weighted graphs have no edges and there is no need to use special algorithms for computing maximum weighted independent sets. The time analysis simplifies and we obtain the following theorems.

Theorem 3. $UOMIT$ can be solved in $O(kn^3)$ time.

Theorem 4. $OMIT$ can be solved in $O((k+n)n^3)$ time.

References

1. A. Aho, J. Hopcroft, and J. Ullman. *The Design and Analysis of Computer Algorithms*. Addison-Wesley, 1974.
2. A. Amir and D. Keselman. Maximum agreement subtree in a set of evolutionary trees: Metrics and efficient algorithms. *SIAM J. Computing*, 26(6):1656–1669, 1997.
3. H. Bodlaender, R. Downey, M. Fellows, and T. Wareham. The parameterized complexity of sequence alignment and consensus. *Theor. Comput. Sci.*, 147:31–54, 1995.
4. P. Bonizzoni, G. Della Vedova, and G. Mauri. Approximating the maximum isomorphic agreement subtree is hard. *International Journal of Foundations of Computer Science*, 11(4):579–590, 2000.
5. D. Bryant. *Building Trees, Hunting for Trees, and Comparing Trees: Theory and Methods in Phylogenetic Analysis*. PhD thesis, University of Canterbury, 1997.
6. M. Farach, T. Przytycka, and M. Thorup. On the agreement of many trees. *Information Processing Letters*, 55:297–301, 1995.
7. M. Fellows, M. Hallett, and U. Stege. Analogs & duals of the MAST problem for sequences & trees. *Journal of Algorithms*, 49(1):192–216, 2003.
8. S. Felsner, R. Müller, and L. Wernisch. Trapezoid graphs and generalizations, geometry and algorithms. *Discrete Applied Mathematics*, 74:13–32, 1997.
9. L. Gąsieniec, J. Jansson, A. Lingas, and A. Östlin. Inferring ordered trees from local constraints. In proceedings of CATS'98, vol. 20(3) of *Australian Computer Science Communications*, pages 67–76. Springer-Verlag, 1998.
10. L. Gąsieniec, J. Jansson, A. Lingas, and A. Östlin. On the complexity of constructing evolutionary trees. *Journal of Combinatorial Optimization*, 3:183–197, 1999.
11. J. Hein, T. Jiang, L. Wang, and K. Zhang. On the complexity of comparing evolutionary trees. *Discrete Applied Mathematics*, 71:153–169, 1996.
12. M.-Y. Kao, T.-W. Lam, W.-K. Sung, and H.-F. Ting. An even faster and more unifying algorithm for comparing trees via unbalanced bipartite matchings. *Journal of Algorithms*, 40(2):212–233, 2001.
13. D. Maier. The complexity of some problems on subsequences and supersequences. *Journal of the ACM*, 25(2):322–336, 1978.
14. E. A. Smolenskii. Jurnal Vicisl. Mat. i Matem. Fiz, 2:371–372, 1962.
15. M. Steel and T. Warnow. Kaikoura tree theorems: Computing the maximum agreement subtree. *Information Processing Letters*, 48:77–82, 1993.
16. W.-K. Sung. *Fast Labeled Tree Comparison via Better Matching Algorithms*. PhD thesis, University of Hong Kong, 1998.
17. V. G. Timkovsky. Complexity of common subsequence and supersequence problems and related problems. *Cybernetics*, 25:1–13, 1990.

Small Phylogeny Problem: Character Evolution Trees

Arvind Gupta[1,*], Ján Maňuch[1,**], Ladislav Stacho[2,***], and Chenchen Zhu[3,†]

[1] School of Computing, Simon Fraser University, Canada
{arvind,jmanuch}@sfu.ca
[2] Department of Mathematics, Simon Fraser University, Canada
lstacho@sfu.ca
[3] Microsoft, Redmond, USA
chenchen@microsoft.com

Abstract. Phylogenetics is a science of determining connections between groups of organisms in terms of ancestor/descendent relationships, usually expressed by phylogenetic trees, also called "trees of life", cladograms, or dendograms. In parsimony approach to reconstruct the phylogenetic trees, the goal is to find the most parsimonious tree, i.e., the tree requiring the smallest number/score of evolutionary steps. For all reasonable measures this problem is NP-hard. Assuming the structure of the tree is given, we are left with, in some cases tractable, problem of "small phylogeny": how to assign characters to the internal nodes representing extinct species. We propose a new approach together with the corresponding parsimony criteria for working with nonlinear transformation series of states of a character: a character evolution trees. We use tools of structural graph theory to reconcile a character tree with a phylogenetic tree. For this purpose, we introduce two new scoring metrics: the *bag cost*, analogous to unweighted parsimony, and the *arc cost*, analogous to weighted parsimony. We will provide several linear time algorithms solving small phylogeny problem while minimizing the above scoring functions.

1 Introduction

Phylogenetics – discovering patterns of evolution, i.e., ancestral and familial relationships between species, is receiving increasing attention amongst biologists, geologists, ecologists, and, most recently computer scientists. The *large phylogeny problem* is to reconstruct a phylogenetic tree based on characters of

* Research supported in part by NSERC (Natural Science and Engineering Research Council of Canada) grant.
** Research supported in part by PIMS (Pacific Institute for Mathematical Sciences.
*** Research supported in part by NSERC (Natural Science and Engineering Research Council of Canada) grant, and VEGA grant No. 2/3164/23 (Slovak grant agency).
† A part of the work was obtained while enrolled in MSc program in School of Computing, Simon Fraser University, Canada.

S.C. Sahinalp et al. (Eds.): CPM 2004, LNCS 3109, pp. 230–243, 2004.
© Springer-Verlag Berlin Heidelberg 2004

extant species (represented as leaves of the tree), that is to define the internal structure of the tree and to assign states of characters to the internal nodes representing hypothetical extinct organisms. The *small phylogeny problem* assumes that the internal structure of the tree is given, and only tries to deduce characteristics of the extinct species. Here, the principle of *parsimony* is usually applied: the goal is to find the most parsimonious tree, i.e., the tree requiring the smallest number/score of evolutionary steps such as the loss of one character, or the modification or gain of another.

The first reconstructions of phylogenies were based on the study of fossil records. New techniques include constructing the best fit for a set of characters from matrices of characters, maximum likelihood constructions, and pair-wise distance constructions which assume a certain mutation rate [FM67,SN87,Fel81]. This virtual explosion of techniques and algorithms has led to the publication of many new phylogenies which can often be contradictory. Statistical approaches have been developed to assess their quality and closeness of their fit to the given data [SI89,TTN94]. Recently, constructing phylogenetic trees using molecular data, called gene trees, has achieved considerable prominence. However, as pointed out in [PN88,Wu91], the processes of gene duplication, loss, and lineage sorting, result in incongruence between the genes trees and usual phylogenetic trees based on character data, called species trees. Therefore, it was proposed in [Doy92] to treat the gene trees as another character trees.

The problem of constructing a phylogenetic tree from the character matrix, the *large* phylogeny problem, is NP-complete even when each character is binary (can take on only two values) [FG82,DS86]. For the small phylogeny problem, there are polynomial time algorithms both for the case of uniform cost of each state change [Fit71] and non-uniform cost [San75].

In this work, we further investigate the small phylogeny problem where partial information of the evolutionary order of a multistate character is also given. In particular, we consider the case that such evolutionary order is represented as a rooted tree, called a *character evolution tree*. In what follows, we will review the history of character evolution trees and explain motivations of our work.

1.1 History of Character Evolution Trees

A character phylogeny: a character transformation series [Hen66] or character state tree [Far70] of a multistate character is a hypothesis that specifies which states of the character evolve directly into which other states. To determine the character transformation series, both the character state polarity and character state order need to be known. The character state order only describes which states are intermediate, but does not specify evolutionary direction. However, the character state polarity explains which state is plesiomorphic or ancestral. The character state polarity can be determined by using the outgroup comparison, parsimony analysis [Far82,Fit71,Mic82], fossil and stratigraphic data or ontogenetic criteria. To determine the character state order, various methods have been utilized. One direction is to impose a rule on how the character evolved [MW90]. The other direction is to maximize congruence among characters such as non-additive analysis [Fit71] or transformation series analysis which runs in an iterative procedure [Mic82].

In many cases, a character state tree needs to be encoded into multiple binary characters using additive binary coding [Far70,CS65] since many program packages require linear variables (e.g., Hennig86, NTSYS, PAUP, PHYLIP). This approach enables to use these algorithms, however it also brings several potential difficulties, such as creation of artificial homoplasy, obscuring relationships between species due to an arbitrary division of multiple states into two or more binary characters, and the ignorance of synapomorphic evidence, as pointed out in [Lip92,HHS97,OD87,PM90]. Therefore having a method of comparing a character evolution tree directly with a phylogenetic tree of species without being coded into binary characters is highly desirable.

1.2   Our Approach

The problem considered in this work can be summarized as follows: We are given a *character evolution tree* representing the evolution of some character (recall, this can be a gene tree built using the molecular data as argued in [Doy92]). The vertices of the character evolution tree represent states of this character. We are also given a set of species each taking on one state of the character. The task is to find a parsimonious phylogenetic tree consistent with the character tree. If the internal structure of the phylogenetic tree is not given, then for one character, it is trivial to construct a phylogenetic tree congruent with the character tree. The problem for a set of characters without any state order ("perfect phylogeny") is NP-complete [FG82] which suggests that this problem is difficult for a set of character trees. Instead we consider the small phylogeny problem in which the internal structure of the phylogenetic tree is known. Since a transformation series is tested against a phylogenetic tree constructed from other characters, the small phylogeny problem also models Lipscomb's problem [Lip92] of testing transformation series.

Our techniques are based on finding graph minor embeddings of labeled trees. Graph minors are generalizations of isomorphisms in which a vertex of the source graph is mapped to a connected component of the target graph preserving the adjacency relation of the source graph. Tree minors are the basis of the seminal work of Robertson and Seymour who used them to prove Wagner's conjecture [RS86] and the flavor of their techniques is carried forward here. We define three generalizations of graph minors, *rooted tree minor*, *relax-minor* and *pseudo-minor* which reflect structures arising in phylogenetic trees.

We will investigate the *small parsimony* problem under two different optimality criteria. In the first: the *bag cost*, the subgraph of the phylogenetic tree induced by a particular state has as few connected components as possible. It also reflects the non-congruence of scattering introduced in [ML91] (multiple occurrence of the same state in non-adjacent species) because less components implies less scattering. In the second: the *arc cost*, we impose the cost for each state transition (represented as the arc cost in the phylogenetic tree) and look for trees that minimize the sum of the costs over all arcs. Similarly, it reflects the non-congruence of hierarchical discordance introduced in [ML91] (incorrect polarity or state order) with less arc costs implying less occurrences of hierarchical discordance. In both cases we find linear time algorithms for these problems. Finally, we show that certain variations of these problems (even when the internal structure of the phylogenetic tree is known) are NP-hard.

The most of the proofs is omitted due to space limitations.

2 Preliminaries

2.1 Basic Definitions

Let (T, r) be a rooted tree with $r \in T$ as a root. The distance of two vertices is the length of the (shortest) path connecting them. We say that a vertex is at level i if its distance from root is i. Note that every edge of (T, r) is connecting vertices on consecutive levels. Hence, we can easily assign orientations to the edges as follows: each edge goes from a vertex u on level j to a vertex v on level $j + 1$, for some j. We say that u is the parent of v and that v is a child of u. Let $A(T)$ be the set of all oriented edges $\langle u, v \rangle$ (also called *arcs*). If there is an oriented path from u to v, we say that u is an *ancestor* of v, and that v is an *descendent* of u, and write $u \prec v$. We say that two states u and v are *incomparable*, denoted by $u \approx v$ if u is neither an ancestor, nor the descendant of v.

Definition 1. *The* least common ancestor $\mathrm{LCA}(v_1, \ldots, v_k)$ *of* $v_1, \ldots, v_k \in T$ *is a vertex* u *such that*

- $u \prec v_j$, *for* $j = 1, \ldots, k$, *and*
- *for any other vertex* u' *such that* $u' \prec v_j$, *for* $j = 1, \ldots, k$, *we have* $u' \prec u$.

A node with the out-degree zero, is called a *leaf*. Let $L(T)$ be the set of all leaves of a tree (T, r).

2.2 Small Parsimony with Character Evolution

A *character evolution tree* (H, h) is a rooted tree with vertices representing the states of the character and oriented edges representing possible evolution between states. A *phylogenetic tree* (G, g) is a rooted tree representing evolution of species. Leaves of (G, g) represent extant species, while the internal vertices represent hypothetical ancestors.

In practice, the states of the character of most of the extant species are known. Hence, another input to the small parsimony problem is a partial function $p : L(G) \to H$, called a *leaf labeling*. If p is a function then p is called a *complete leaf labeling*. Note that a complete leaf labeling assign labels to all extant species of the phylogenetic tree. We say that a function $l : G \to H$ is p-*constrained* if for every $u \in L(G)$, either $l(u)$ is undefined, or $p(u) = l(u)$.

The *small parsimony problem* is to assign states to the hypothetical and unlabeled extant species, i.e., to find a p-constrained function l, so that if a species v is a child of a species u in the phylogenetic tree than the character state $l(v)$ is either equivalent to, or a child of the character state $l(v)$ in the character evolution tree. The goal is do find such an assignment which realizes the most of the evolution steps and minimizes the number of scatterings. (In this work, by scattering, we mean occurrence of two nodes with the same state in the phylogenetic tree which are separated by a node with a different state on the path between these two nodes.) To formalize this concept we need a few definitions.

Definition 2 (Realization of evolution step). *Let (G, g) be a phylogenetic tree, (H, h) a character evolution tree, and $l : G \rightarrow H$ a labeling function. If for an arc $\langle a, b \rangle \in A(H)$, there exists an arc $\langle u, v \rangle \in A(G)$ such that $l(u) = a$ and $l(v) = b$, we say $\langle a, b \rangle$ is realized by l on $\langle u, v \rangle$. Furthermore let $r_l(a, b)$ denote the number of arcs in $A(G)$ that realize $\langle a, b \rangle$ by l.*

The arcs of the character evolution tree (H, h) represent the evolution steps. Hence, $r_l(a, b)$ counts how many times the evolution from a state a to state b has happened. To avoid hierarchical discordance (in this work, skipping a state of the character tree), this number has to be at least one, and the value $r_l(a, b) > 1$ indicates the existence of scattering in the phylogenetic tree.

Definition 3 (Bag-set). *Let (G, g) be a phylogenetic tree, (H, h) a character evolution tree, and $l : G \rightarrow H$ a labeling function. For every $v \in H$, consider the subgraph of (G, g) induced by vertices in $l^{-1}(v)$. Let B_v^l be the set of components of this subgraph, called the* bag-set *of v induced by l. Any particular component of B_v^l is referred to as a* bag *of v. The number of components of B_v^l is denoted by $c(B_v^l)$.*

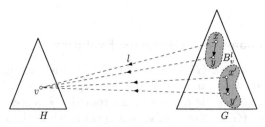

Fig. 1. The character tree (H, h) on the left and the phylogenetic tree (G, g) on the right. Dash lines illustrate the function l. The bag-set B_v^l consists of two components (shadowed areas), i.e. $c(B_v^l) = 2$.

An example of a bag-set B_v^l containing two bags is shown in Figure 1. Note that if $c(B_v^l) = 0$ then the state v does not occur in the phylogenetic tree at all (hierarchical discordance). On other hand, if $c(B_v^l) > 1$, the state v has evolved from its ancestor state several times, i.e., we are witnessing scattering.

3   Reconciling the Character Evolution and Phylogenetic Trees without Incongruences

In this section we will consider the reconstruction of the states of extinct species under requirement that the resulting phylogenetic tree has to be fully congruent with the character evolution tree.

Definition 4 (Rooted-tree minor). *Let (G, g) be a phylogenetic tree, (H, h) a character evolution tree, and $p : L(G) \rightarrow H$ a leaf labeling. We say that H is a rooted-tree minor of G with respect to p, denoted by $(H, h) \leq_{rm} (G, g, p)$, if there exists a p-constrained functions $l : G \rightarrow H$ satisfying the following two conditions:*

(1) for each character state $a \in H$, we have $c(B_a^l) = 1$, and
(2) for each evolution step $\langle a, b \rangle \in A(H)$, we have $r_l(a, b) \geq 1$.

Note that necessarily $l(g) = h$. Let $M(H, G, p)$ be the set of all p-constrained functions $l : G \rightarrow H$ satisfying the above conditions.

Observation 1. The conditions (1) and (2) are equivalent to a single condition:

(3) for each evolution step $\langle a, b \rangle \in A(H)$, we have $r_l(a, b) = 1$.

Note that the character evolution tree is a rooted-tree minor of the phylogenetic tree if and only if every state is present in the phylogenetic tree and this does not contain any occurrence of scattering. Formally, we are interested in the following problem.

Problem 1 (Rooted-tree minor problem). Given two rooted trees (H, h) and (G, g) with a leaf labeling $p : L(G) \rightarrow H$. Decide whether H is a rooted-tree minor of G with respect to p.

3.1   Complexity of the Rooted-Tree Minor Problem

In this section we will consider two versions of the rooted-tree minor problem: (a) without a leaf labeling, (b) with a complete leaf labeling. We will show that in the first case, we deal with an NP-complete problem, while the second can be decided in linear time. First, let us prove that the rooted-tree minor problem is NP-complete. The proof is based on the following result proved in [MT92].

Theorem 1 (Tree minor problem). *Given two trees H and G, it is NP-complete to decide whether H is a minor of G.*

Theorem 2 (Rooted-tree minor problem). *Given two rooted trees (H, h) and (G, g). It is NP-complete to decide whether (H, h) is a rooted-tree minor of (G, g).*

Proof. We show that the tree minor problem can be reduced to the rooted-tree minor problem.

Let H and G be an instance of unrooted tree minor problem. We construct new rooted trees (H', α) and (G', γ) as follows:

Let $H' = H \cup \{\alpha, \beta_2, \ldots, \beta_n\}$ where $n = |G|$. Pick an arbitrary vertex $u \in H$ and attached it together with nodes $\beta_2, \ldots, \beta_n$ to the new root α. The resulting tree (H', α) is depicted in Figure 2(a).

For every vertex v_i of $G = \{v_1, \ldots, v_n\}$, create a new copy of the tree G and root it in v_i. Let us call this new rooted-tree (G_i, v_i). Let $G' = G_1 \cup \cdots \cup G_n \cup \{\gamma\}$. Attach the roots of trees $G_1, \ldots, G_n$ to the new root γ. The resulting tree (G', γ) is depicted in Figure 2(b).

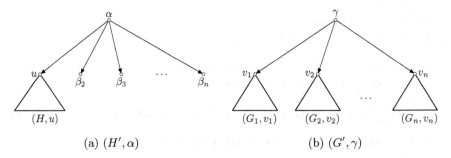

(a) (H', α) (b) (G', γ)

Fig. 2. The construction if rooted trees (H', α) and (G', γ) from the unrooted trees H and G.

Now it suffices to prove the following claim.

Claim. H is a minor of G if and only if (H', α) is a rooted-tree minor of (G', γ).

Proof. First, assume that H is a minor of G, and let $l : G \to H$ be the function satisfying conditions (1) and (2) of Definition 4. Without loss of generality assume that v_1 is in the bag-set of $u \in H$, i.e., $l(v_1) = u$. Then (H, u) is a rooted-tree minor of (G_i, v_i). By setting the bag-set of α to $\{\gamma\}$, and the bag-sets of $\beta_2, \ldots, \beta_n$ to $G_2, \ldots, G_n$, respectively, we can conclude that (H', α) is a rooted-tree minor of (G', γ).

Second, assume (H', α) is a rooted-tree minor of (G', β), and let $l : G' \to H'$ be the corresponding function. Consider the bag-set B_1 of α and the bag-set B_2 of u. Since $\gamma \in B_1$ and the vertices α and γ have degree n, B_2 must be contained in some (G_i, v_i). Then (H, u) is a rooted-tree minor of (G_i, v_i), and therefore, H is a minor of G.

Finally, Claim 3.1 together with Theorem 1 yield NP-completeness of the rooted-tree minor problem.

Assuming that a leaf labeling of a phylogenetic tree is complete, it is possible to decide the rooted-tree minor problem in linear time. The next proof informally describes a linear time algorithm.

Definition 5. *A path $P = (u, \ldots, v)$ in a rooted tree G is called a* single branch path *if every inner vertex of the path has only one child, and u (v) either has more than one child, or is the root of G (is a leaf). Moreover, for any labeling function l, let $l(P) = (l(u), \ldots, l(v))$ be the corresponding path to P in H.*

Theorem 3 (Rooted-tree minor problem with complete leaf labeling).
Given two rooted trees (H, h) and (G, g) with a complete leaf labeling $p : G \to H$. The problem whether (H, h) is a rooted-tree minor of (G, g) with respect to p can decided in polynomial time.

Proof. Here is a description of linear time algorithm:

The algorithm tries to build a labeling $l \in M(H, G, p)$ keeping the track of $r_l(a, b)$ values for all $\langle a, b \rangle \in A(H)$. The algorithm can reject the input (which means that (H, h) is not a rooted-tree minor of (G, g) with respect to p) at any step if it finds it is impossible to complete the construction of the labeling function.

Step 1 Check whether $L(H) \subseteq p(L(G))$. If not, reject the input.

Step 2 Set $l(u) := p(u)$ for every $u \in L(G)$, and $r_l(a, b) := 0$ for every arc $\langle a, b \rangle \in A(H)$.

Step 3 Traverse G in *post-order*. For every internal vertex $u \in G$, assign $l(u) :=$ LCA$(\{l(v); \langle u, v \rangle \in A(G)\})$. Note that computation of LCA can be done in the constant time after a linear time preprocessing on H, cf. [HT84]. For each child v of u such that $l(u) \neq l(v)$ and v is either a leaf or a internal vertex with more than one child, check whether $\langle l(u), l(v) \rangle \in A(H)$. If so, increase the value $r_l(l(u), l(v))$ by one. Otherwise, reject the input.

Step 4 Fix the labels of inner vertices of all single branch paths. Let $P = (v_k, \ldots, v_1)$ be a single branch path. Note that $l(v_i) = l(v_1)$, for all $i = 2, \ldots, k - 1$. If P satisfies any of the following three conditions:

1. $l(v_k) \neq l(v_1)$ and $\langle l(v_k), l(v_1) \rangle \in A(H)$;
2. $v_k \neq g$ and $l(v_1) = l(v_k)$;
3. $v_k = g$ and $l(v_1) = l(v_k) = h$;

no changes are needed. Otherwise, consider the corresponding path $l(P) = (w_m, \ldots, w_1)$ in H, i.e., $w_m = l(v_k)$ and $w_1 = l(v_1)$. If v_k is a root of G, set v_m to h and adjust $l(P)$. If $l(P)$ is not a single branch path, or if $l(P)$ is longer than P, it is not possible to realize every arc of $l(P)$ on P, and the algorithm rejects the input. Otherwise, update the labels of v_i ($i = 2, \ldots, k$) as follows: $l(v_k) := w_m$, $l(v_{k-1}) := w_{m-1}$, $l(v_{k-2}) := w_{m-2}$, $\ldots$, $l(v_{k-m+1}) := w_1$. The labels of vertices $v_1, \ldots, v_{k-m}$ on P remain unchanged. In mean time, we keep updating values of $r_l(a, b)$ for every arc $\langle a, b \rangle$ on the path $l(P)$.

Step 5 If there exists any $\langle a, b \rangle \in A(H)$ such that $r_l(a, b) > 1$, then the algorithm rejects the input. Otherwise, it accepts the input: "$(H, h) \leq_{rm} (G, g, p)$".

The proof of correctness of the algorithm is omitted due to the space limitation. Note that if the (H, h) is a rooted-tree minor of (G, g) then the above algorithm constructs the corresponding labeling l.

4   Reconciling the Character Evolution and Phylogenetic Trees with Incongruences

Gene duplications resulting in paralogous genes can create a situation when a certain state of a character occurs (is developed from its ancestor state) in several places of the phylogenetic tree. In such a situation, it becomes impossible to map the character evolution tree to the tree of species as a rooted-tree minor. Hence, it is necessary to allow incongruences between the character and species trees and use the parsimony principle: find a labeling of the internal nodes of species tree minimizing the number of incongruences. In what follows, we will categorize all possible incongruences, define two parsimony criteria dealing with certain types of incongruences, and finally modify the concept of the rooted-tree minor in two different ways.

4.1   Incongruences

Definition 6. *The following are five types of* incongruences *between the evolutionary order of species given by phylogenetic tree* (G, g) *with labeling l and the order of states given by character evolution tree* (H, h):

- *An* inversion *occurs if for some evolution step in the phylogenetic tree* $\langle u, v \rangle \in A(G)$, $l(v) \prec l(u)$.
- *A* transitivity, *also called a* hierarchical discordance, *occurs if for some evolution step in the phylogenetic tree* $\langle u, v \rangle \in A(G)$, *the state* $a = l(u)$ *is a non-direct ancestor of the state* $b = l(v)$ *($a \prec b$ but $\langle a, b \rangle \notin A(H)$).*
- *An* addition *occurs if for some evolution step in the phylogenetic tree* $\langle u, v \rangle \in A(G)$, $l(u) \approx l(v)$.
- *A* separation, *also called* scattering, *occurs if there exist three vertices* u, v, w *in G such that* $v \prec u$ *and* $v \prec w$, *and* $l(u) = l(w) \neq l(v)$ *and* $u \approx w$.
- *A* negligence *occurs if for some* $\langle a, b \rangle \in A(H)$ *there is no* $\langle u, v \rangle \in A(G)$ *with* $l(u) = a$ *and* $l(v) = b$.

All five incongruences are illustrated in Figure 3.

Note that the concept of the rooted-tree minor (cf. Definition 4) does not allow any of the above incongruences.

Theorem 4. *Given two rooted trees* (H, h) *and* (G, g) *with a leaf labeling p. None of the five incongruences will occur for any labeling* $l \in M(H, G, p)$.

4.2   Parsimony Criteria for the Rooted-Tree Minor Problem

The two standard parsimony criteria for measuring the quality of the labeling l correspond to the unweighted and the weighted cost. The unweighted parsimony assumes a constant cost for every state change, while the weighted parsimony treats the different state changes differently by taking the cost of each state change into consideration. In our approach, we define two metrics to reflect

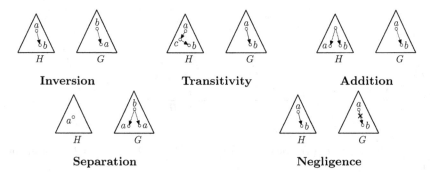

Fig. 3. Illustration of five types of incongruences. Solid lines represent arcs, dashed lines paths of length at least one.

these two criteria, the bag cost for unweighted and the arc cost for weighted (let d be the weight function defined on the set of arcs of the character tree) parsimony.

Definition 7 (Arc and bag costs). *Given two rooted trees (H, h) and (G, g) with a labeling $l : G \rightarrow H$, the* arc cost *and the* bag cost *of l are defined as follows:*

$$\operatorname{arccost}(H, G, l) := \sum_{\langle u,v \rangle \in A(G)} d(l(u), l(v)), \qquad \operatorname{bagcost}(H, G, l) := \sum_{v \in H} c(B_v^l).$$

The bag cost expresses the number of state changes: the number of state changes is the bag cost minus one. This number corresponds to the number of scatterings. The arc cost weights each state change by the distance between the two states. If the canonical weight function (weight of each arc in H is 1 and the weight of any other pair of states is 0) is used, the arc cost corresponds to the number of hierarchical discordances occurring between the phylogenetic and character evolution trees.

4.3   Relaxations of the Rooted-Tree Minor Allowing Incongruences

In the section we define two relaxations of the rooted-tree minor. Each of them allows three types of incongruences listed in the previous section.

Definition 8 (Relax-minor). *Given two rooted trees (H, h) and (G, g) with a leaf labeling p, we say that H is a* relax-minor *of G with respect to p if there exists a p-constrained labeling function $l : G \rightarrow H$ satisfying the following two conditions:*

(1) for each arc $\langle a, b \rangle \in A(H)$, $r_l(a, b) \geq 1$; and
(2) if for some $u, v \in G$ $u \prec v$, then $l(v) \not\prec l(u)$ in H.

Let $R(H, G, p)$ be the set of all such labeling functions.

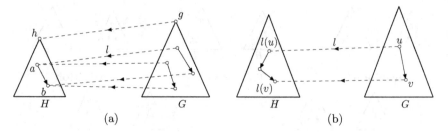

Fig. 4. An illustration of (a) a relax-minor with $|l^{-1}(a)| = 2, |l^{-1}(b)| = 2$; (b) a smooth function.

The main idea of a relax-minor is depicted in Figure 4(a).

Definition 9 (Smooth labeling function). *Let* (H, h) *and* (G, g) *be directed trees. A labeling function* $l : G \to H$ *is called* smooth *if for every arc* $\langle u, v \rangle \in A(G)$, *there is a directed path from* $l(u)$ *to* $l(v)$ *in* H; *see Figure 4(b). Note that a single vertex is considered as a directed path of length 0.*

Definition 10 (Pseudo-minor). *Given two rooted trees* (H, h) *and* (G, g) *with a leaf labeling* p, *we say that* H *is a* pseudo-minor *of* G *if there exists a smooth* p-constrained labeling function $l : G \to H$. *Let* $Q(H, G, p)$ *be the set of all such labeling functions.*

The following table summarizes the potentiality of the five incongruences in rooted-tree minor, relax-minor, and pseudo-minor respectively.

Table 1. Properties of *rooted-tree minor*, *relax-minor* and *pseudo-minor*.

	rooted-tree minor	relax-minor	pseudo-minor
inversion	N	N	N
transitivity	N	Y	Y
addition	N	Y	N
separation	N	Y	Y
negligence	N	N	Y

Note that all three incongruences allowed by a pseudo-minor can naturally occur when reconciling the species and character evolution trees: (a) *transitivity* indicates omission of one or several intermediate extinct species from the phylogenetic tree; (b) *separation* can be cause by gene duplication and horizontal gene transfer as discussed above; and (c) *negligence* indicates incomplete data: either omission of intermediate species, or missing extant species with particular state of the considered character. On the other hand, *addition* allowed by a relax-minor is more likely caused by inconsistency in the structure of phylogenetic tree, or more severe incompatibilities in labeling of this tree. Therefore, it seems that the concept of pseudo-minor is more useful from the practical point of view.

We will measure the quality of relax-minor and pseudo-minor mappings in terms of their bag cost and arc cost, respectively. For this purpose, we define the following three problems.

Problem 2 (Minimum relax-minor bag cost). Given a character tree (H, h) and a phylogenetic tree (G, g) with a leaf labeling p, find a labeling function $l \in R(H, G, p)$ minimizing the bag cost of l.

Problem 3 (Minimum pseudo-minor bag cost). Given a character tree (H, h) and a phylogenetic tree (G, g) with a leaf labeling p, find a labeling function $l \in Q(H, G, p)$ minimizing the bag cost of l.

Problem 4 (Minimum pseudo-minor arc cost). Given a character tree (H, h) and a phylogenetic tree (G, g) with a leaf labeling p, find a labeling function $l \in Q(H, G, p)$ minimizing the arc cost of l.

Note that since the relax-minor allows additions, it is not always possible to compute the arc cost.

4.4   Complexities of Relax-Minor and Pseudo-minor Problems

We showed that Problem 2 is NP-hard, while Problems 3 and 4 can be solved in linear time. Due to space limitation, we skip the proof of NP-hardness of Problem 2, and give only sketches of the linear time algorithms for Problems 3 and 4 without proofs of correctness.

Theorem 5. *It is NP-hard to solve Problem 2.*

Theorem 6. *There is a linear time algorithm solving Problem 3.*

Proof. Description of the linear time algorithm:
Set $l(u) := p(u)$ and $x(u) := 1$ for every $u \in L(G)$ for which $p(u)$ is defined. For all other $u \in G$, set $x(u) := 0$. The algorithm works in two stages.

In the first stage, the tree (G, g) is traversed in *post-order*. For each internal vertex $u \in G$, let $l(u) := \mathrm{LCA}(\{l(v); \langle u, v \rangle \in A(G)\})$. This requires a linear time preprocessing on H as described in the previous algorithm. If there exists a child v of u such that $l(u) = l(v)$ and $x(v) = 1$, then set $x(u) := 1$. Otherwise, the value of $x(u)$ stays unchanged (0).

In the second stage, the tree (G, g) is traversed in *pre-order*. For each vertex $u \in G - \{g\}$, let v be the parent of u. If $x(u) = 0$, then set $l(u) := l(v)$.

Finally, the number of bags, which is initially set to $|G|$, is calculated by subtracting one for every arc $\langle u, v \rangle \in A(G)$ with $l(u) = l(v)$.

Theorem 7. *There is a linear time algorithm solving Problem 4.*

Proof. Description of the linear time algorithm:
Set $l(u) := p(u)$ for every leaf $u \in L(G)$. Traverse the tree (G, g) in *post-order* assigning values $l(u) := \text{LCA}(\{l(v); \langle u, v \rangle \in A(G)\})$ for every internal vertex $u \in G$. As above, the computation of LCA in the tree H can be done in constant time after a linear time preprocessing on H.

The preprocessing algorithm can easily be modified so that computations of $d(a, b)$ take constant time for any given $a, b \in H$ with $a \prec b$. The cost of each arc $\langle u, v \rangle$ is calculated as the distance $d(l(u), l(v))$. The total arc cost of l is the sum of the costs over all arcs in $A(G)$.

Note that for any character tree (H, h) and phylogeny tree (G, g), both the above algorithm and Sankoff's algorithm [SC83] output the labeling with the minimum arc cost since any character tree H can be transformed into a cost matrix M by letting $M_{ij} := d(i, j)$ for $i, j \in H$. However, our algorithm has better performance, since it runs in time $O(|G| \cdot \log_2 \delta + |H|)$, compared to Sankoff's algorithm which runs in time $O(|G| \cdot \delta \cdot |H|)$, where δ is the maximal degree of nodes of the phylogenetic tree (G, g).

References

[CS65]   J. H. Camin and R. R. Sokal. A method for deducing branching sequences in phylogeny. *Evolution*, 19:311–326, 1965.

[Doy92] J. J. Doyle. Gene trees and species trees: Molecular systematics as one-character taxonomy. *Systematic Botany*, 17:144–163, 1992.

[DS86]   W.I.E. Day and D. Sankoff. Computational complexity of inferring phylogenies by compatibility. *Systematic Zoology*, 35:224–229, 1986.

[Far70] J. S. Farris. Methods for computing Wagner trees. *Systematic Zoology*, 19:83–92, 1970.

[Far82] J. S. Farris. Outgroups and parsimony. *Zoology*, 31:314–320, 1982.

[Fel81] J. Felsenstein. Evolutionary trees from DNA sequences: a maximum likelihood approach. *Journal of Molecular Evolution*, 17:368–376, 1981.

[FG82]   L. R. Foulds and R. L. Graham. The Steiner problem in phylogeny is NP-complete. *Advances In Applied mathematics*, 3:43–49, 1982.

[Fit71] W. M. Fitch. Toward defining the course of evolution: Minimum change for a specific tree topology. *Systematic Zoology*, 20:406–416, 1971.

[FM67]   W. M. Fitch and E. Margoliash. Construction of phylogenetic trees. *Science*, 155:279–284, 1967.

[Hen66] W. Hennig. *Phylogenetic Systematics*. University of Illinois Press., 1966.

[HHS97] J. A. Hawkins, C. E. Hughes, and R. W. Scotland. Primary homology assessment, characters and character states. *Cladistics.*, 13:275–283, 1997.

[HT84]   D. Harel and R. Tarjan. Fast algorithms for finding nearest common ancestors. *SIAM Journal on Computing*, 13:338–355, 1984.

[Lip92] D. L. Lipscomb. Parsimony, homology and the analysis of multistate characters. *Cladistics.*, 8:45–65, 1992.

[Mic82] M. F. Mickevich. Transformation series analysis. *Systematic Zoology*, 31:461–478, 1982.

[ML91]   M. F. Mickevich and D. L. Lipscomb. Parsimony and the choice between different transformations for the same character set. *Cladistics.*, 7:111–139, 1991.

[MT92]   J. Matousek and R. Thomas. On the complexity of finding iso- and other morphisms for partial k-trees. *Journal of Algorithms*, 108:343–364, 1992.

[MW90]   M. F. Mickevich and S. Weller. Evolutionary character analysis: Tracing character change on a cladogram. *Cladistics*, 6:137–170, 1990.

[OD87]   R. T. O'Grady and G.B. Deets. Coding mulitistate characters, with special reference to the use of parasites as characters of their hosts. *Systematic Zoology*, 36:268–279, 1987.

[PM90]   M. Pogue and M. F. Michevich. Character definitons and character state delineations: the bete noire of phylogenetics. *Cladistics.*, 6:365–369, 1990.

[PN88]   P. Pamilo and M. Nei. Relationships between gene trees and species trees. *Mo. Biol. Evol.*, 5:568–583, 1988.

[RS86]   N. Robertson and P. D. Seymour. Graph minors II. Algorithmic aspects of tree-width. *Journal of Algorithms*, 7:309–322, 1986.

[San75]   D. D. Sankoff. Minimal mutation trees of sequences. *SIAM Journal on Applied Mathematics*, 28:35–42, 1975.

[SC83]   D. Sankoff and R. Cedergren. Simultaneous comparisons of three or more sequences related by a tree. In D. Sankoff and J. Kruskal, editors, *Time Warp, String Edits, and Macromolecules: the Theory and Practice of Sequence Comparison*, pages 253–264. Addison Wesley, Reading Mass., 1983.

[SI89]   N. Saitou and T. Imanishi. Relative efficiencies of the Fitch-Margoliash, maximum parsimony, maximum likelihood, minimum-evolution, and neighbor-joining methods of phylogenetic tree construction in obtaining the correct tree. *Journal of Molecular Evolution*, 6:514–525, 1989.

[SN87]   N. Saitou and M. Nei. The neighbor-joining method: a new method for reconstructing phylogenetic trees. *Molecular Biology and Evolution*, 4:406–425, 1987.

[TTN94]   Y. Tateno, N. Takezaki, and M. Nei. Relative efficiencies of the maximum-likelihood, neighbor-joining and maximum-parsimony methods when substitution rate varies with site. *Journal of Molecular Evolution*, 11:261–277, 1994.

[Wu91]   C.-I. Wu. Inferences of species phylogeny in relation to segregation of acient polymorphisms. *Genetics*, 127:429–435, 1991.

The Protein Sequence Design Problem in Canonical Model on 2D and 3D Lattices

Piotr Berman[1], Bhaskar DasGupta[2], Dhruv Mubayi[3], Robert Sloan[2], György Turán[3], and Yi Zhang[2]

[1] Department of Computer Science and Engineering, Pennsylvania State University, University Park, PA 16802
berman@cse.psu.edu
[2] Department of Computer Science, University of Illinois at Chicago, Chicago, IL 60607-7053
{dasgupta,sloan,yzhang3}@cs.uic.edu
[3] Department of Mathematics, Statistics & Computer Science, University of Illinois at Chicago, Chicago, IL 60607-7045
mubayi@math.uic.edu, gyt@uic.edu

Abstract. In this paper we investigate the **protein sequence design (PSD)** problem (also known as the **inverse protein folding** problem) under the **Canonical model**[1] on **2D and 3D lattices** [12, 25]. The Canonical model is specified by **(i)** a *geometric representation* of a target protein structure with amino acid residues via its *contact graph*, **(ii)** a *binary folding code* in which the amino acids are classified as *hydrophobic* (H) or *polar* (P), **(iii)** an *energy function* Φ defined in terms of the target structure that should *favor* sequences with a *dense hydrophobic core* and *penalize* those with *many solvent-exposed hydrophobic residues* (in the Canonical model, the energy function Φ gives an H-H residue contact in the contact graph a value of -1 and all other contacts a value of 0), and **(iv)** to prevent the solution from being a biologically meaningless all H sequence, the number of H residues in the sequence S is limited by fixing an upper bound λ on the ratio between H and P amino acids. The sequence S is designed by specifying which residues are H and which ones are P in a way that realizes the *global minima* of the energy function Φ. In this paper, we prove the following results:

(1) An earlier proof of NP-completeness of finding the global energy minima for the PSD problem on 3D lattices in [12] was based on the NP-completeness of the same problem on 2D lattices. However, the reduction was not correct and we show that the problem of finding the global energy minima for the PSD problem for 2D lattices can be solved *efficiently* in *polynomial time*. But, we show that the problem of finding the global energy minima for the PSD problem on 3D lattices is indeed NP-complete by a providing a different reduction from the problem of finding the largest clique on graphs.

---

[1] The Canonical model is neither the same nor a subset of the Grand Canonical (GC) model in [19, 24]; see Section 1.3 for more details.

S.C. Sahinalp et al. (Eds.): CPM 2004, LNCS 3109, pp. 244–253, 2004.
© Springer-Verlag Berlin Heidelberg 2004

(2) Even though the problem of finding the global energy minima on 3D lattices is NP-complete, we show that an *arbitrarily* close approximation to the global energy minima can indeed be found efficiently by taking *appropriate combinations* of optimal global energy minima of substrings of the sequence S by providing a polynomial-time approximation scheme (PTAS). Our algorithmic technique to design such a PTAS for finding the global energy minima involves using the *shifted slice-and-dice approach* in [6, 17, 18]. This result improves the previous best polynomial-time approximation algorithm for finding the global energy minima in [12] with a performance ratio of $\frac{1}{2}$.

1   Introduction

In protein structure studies the single most important research problem is to understand how protein sequences fold into their native 3D structures, *e.g.*, see [3, 5, 7, 9, 12–16, 21, 22, 26, 27]. This problem can be investigated at two *complementary* levels. At a *lower* level, one wishes to determine how an individual protein sequence folds. The problem of using sequence input to generate 3D structure output is referred to as the *ab initio protein structure prediction* problem and has been shown to be NP-hard [3, 5, 7]. At a *higher* level, one wants to analyze the *protein landscapes*, *i.e.*, the relationship between the space of all protein sequences and the space of native 3D structures. A formal framework for analyzing protein landscapes is established by a model that relates a set S of protein sequences to a set P of protein structures. Typically this is given by a real-valued *energy* function $\Phi : S \times P \to \mathbb{R}$ that models the "fit" of a sequence $s \in S$ to a structure $p \in P$ according to the principles of statistical mechanics. A functional relationship between sequences and structures is obtained by *minimizing* Φ with respect to the structures, *i.e.*, a structure q *fits* a sequence s if $\Phi(s, q) = min_{p \in P}\Phi(s, p)$. Typically the values of Φ are assumed to model notions of free energy and the minimization is supposed to provide approximations to the *most probable structure* obtained from thermodynamical considerations.

The exact nature of Φ depends on the particular model but, for any given specification, there is natural interest in the fine-scale structure of Φ. For example, one might ask whether a certain kind of protein structure is more likely to be the native structure of a diverse collection of sequences (thus making structure prediction from sequences difficult). One approach to investigating the structure of Φ is to solve what is called the *protein sequence design* (PSD) or the *inverse protein folding* problem: given a target 2D or 3D structure as input, return a *fittest* sequence with respect to Φ. Three criteria have been proposed for evaluation of the fitness of the protein sequence with respect to the target structure: (a) the sequence should fold to the target structure, (b) there should be *no degeneracy* in the ground state of the sequence and (c) there should be a *large gap* between the energy of the sequence in the target structure and the energy of the sequence in any other structure. Some researchers [27] have proposed weakening condition (b) by requiring that the degeneracy of the sequence be no greater than the degeneracy of any other sequence that also folds to the target structure. The

PSD problem has been investigated in a number of studies [4, 8, 10, 12, 19, 23–25, 27]. The computational complexity of PSD in its full generality as described above is unknown but conjectured to be NP-hard; the currently best known algorithms are by exhaustive search or Monte Carlo simulations.

One possible mode of handling the PSD problem is by defining a *heuristic sequence design* (HSD) problem where a simplified pair-wise interaction function is used to compute the landscape energy function Φ. The implicit assumption is that a sequence that satisfies the HSD problem also solves PSD. Several quantitative models have been proposed for the HSD problem in the literature [8, 24, 25]. This paper is concerned with the Canonical model of Shakhnovich and Gutin [25]. This model is specified by **(1)** a *geometric representation* of a target protein structure with n amino acid residues via its *contact graph*, **(2)** a *binary folding code* in which the amino acids are classified as *hydrophobic* (H) or *polar* (P) [9, 20], and **(3)** an *energy function* Φ defined in terms of the target structure that should *favor* sequences with a *dense hydrophobic core* and *penalize* those with *many solvent-exposed hydrophobic residues*. To design a sequence S, we must specify which residues are H and which ones are P. Thus, S is a sequence of n symbols each of which is either H or P. In the Canonical model, the energy function Φ gives a H-H residue contact in the contact graph a value of -1 and all other contacts a value of 0. To prevent the solution from being a biologically meaningless all H sequence, the number of H residues in S is limited by fixing an upper bound λ of the ratio between H and P amino acids. The Canonical model gives rise to the following *special case* of the *densest subgraph problem* on K vertices (denoted by the PSDC$_2$ and the PSDC$_3$ Problems):

Definition 1.
(a) *A d-dimensional lattice is a graph $G(n, d) = (V(n, d), E(n, d))$ with $V(n, d) = \times_{i=1}^{d}\{-n, -n+1, \ldots, n-1, n\}$ for some positive integer n and*
$E(n, d) = \{\{(i_1, \cdots, i_d), (j_1, \cdots, j_d)\} : \sum_{k=1}^{d} |i_k - j_k| = 1\}$ *($X \times Y$ denote the Cartesian product of two sets X and Y).*

(b) *A 2D sequence (resp. 3D sequence) $S = (V, E)$ is a graph that is a simple path in $G(n, 2)$ (resp. $G(n, 3)$) for some n; the contact graph of such a 2D sequence (resp. 3D sequence) S is a graph $\bar{G} = (\bar{V}, \bar{E})$ where $\bar{E}$ consists of all edges $\{u, v\} \in E(n, 2)$ (resp. $\{u, v\} \in E(n, 2)$) such that $u, v \in V$ and $\{u, v\} \notin E$ and $\bar{V}$ is the set of end points of the edges in $\bar{E}$.*

Problem 1 (DS Problem). The Densest Subgraph (DS) problem has a graph $G = (V, E)$ and a positive integer K as inputs, and the goal is to find a $V' \subseteq V$ with $|V'| \leq K$ that *maximizes* $|\{(u, v) \in E : u, v \in V'\}|$.

Problem 2 (PSDC$_2$/PSDC$_3$ Problems). The PSD problem for the Canonical model on a 2D (resp. 3D) lattice, denoted by PSDC$_2$ (resp. PSDC$_3$), is an instance of the DS problem when the input graph G is the contact graph realized by a 2D (resp. 3D) sequence.

References [1, 2] consider the DS problem for general graphs. Hart [12] considers both PSDC$_2$ and PSDC$_3$ problems, provides approximation algorithm for

PSDC$_3$ with an approximation ratio of $\frac{1}{2}$ and an *almost* optimal algorithm for PSDC$_2$. The following property of the contact graph of a 2D/3D sequence is easy to observe [12]:

> the contact graph G for a 2D sequence (resp. 3D sequence) is a graph that is a subgraph of the 2D lattice (respectively, 3D lattice) with at most two vertices of degree 3 (resp. 5) and all other vertices of degree at most 2 (resp. 4).

1.1   Our Results

Throughout the rest of the paper, G is the given input graph in our problems, K is the maximum number of residues that can be hydrophobic and $V(H)$ (resp. $E(H)$) is the vertex set (resp. edge set) of any graph H. Our results are:

(I) We show that the problem of finding the global energy minima for the PSD problem for 2D lattices can be solved in polynomial time (see Section 2).

(II) We show that the problem of finding the global energy minima for the PSD problem on 3D lattices is NP-complete by showing that the PSDC$_3$ decision problem is NP-complete via a reduction from the problem of finding the largest clique on graphs (see Section 3.1). An earlier proof of NP-completeness of this problem in [12] was based on an incorrect proof of NP-completeness of the same problem on 2D lattices.

(III) Even though the problem of finding the global energy minima on 3D lattices is NP-complete, we show that an arbitrarily close approximation to the global energy minima can indeed be found efficiently by taking appropriate combinations of optimal global energy minima of substrings of the sequence S by providing a polynomial-time approximation scheme (PTAS) for the PSDC$_3$ problem (see Section 3.2). This result improves the previous best polynomial-time approximation algorithm for finding the global energy minima in [12] which had a performance ratio of $\frac{1}{2}$.

1.2   Summary of Algorithmic Techniques Used

- The polynomial-time algorithm in Result **(I)** uses the polynomial-time Generalized Knapsack problem, the special topology of the input contact graph as mentioned at the end of the introduction and the fact that the range of Φ are small integers.
- The NP-completeness reduction in Result **(II)** uses the NP-completeness reduction in [11] from the maximum clique problem to the densest subgraph problem on general graphs. The challenging and tedious parts in our reduction is to make sure that the reduction works for the special topology of our input contact graph and that such a contact graph can in fact be realized by a 3D sequence.
- The PTAS in Result **(III)** is designed using the *shifted slice-and-dice approach* in [6, 17, 18].

1.3 Difference between the Canonical and the Grand Canonical Model

To avoid possible confusion due to similar names, we would like to point out that the Canonical model considered in this paper is neither the same nor a subset of the Grant Canonical (GC) model for the protein sequence design problem [19, 24]. The GC model is defined by a different choice of the energy function Φ. In particular, let S_H to denote the set of numbers i such that the ith position in S is equal to H. Then, Φ is defined by the equation $\Phi(S) = \alpha \sum_{i,j \in S_H, i<j-2} g(d_{ij}) + \beta \sum_{i \in S_H} s_i$, where $\alpha < 0$, $\beta > 0$, s_i is the area of the solvent-accessible contact surface for the residue (in Å), d_{ij} is the distance between the residues i and j (in Å) and $g = \begin{cases} 1/[1 + \exp(d_{ij} - 6.5)] & \text{when } d_{ij} \leq 6.5 \\ 0 & \text{when } d_{ij} > 6.5 \end{cases}$ is a *sigmoidal* function. The scaling parameters α and β have default values -2 and $\frac{1}{3}$, respectively.

1.4 Basic Definitions and Notations

For two graphs G_1 and G_2, $G_1 \cup G_2$ denotes the graph with $V(G_1 \cup G_2) = V(G_1) \cup V(G_2)$ and $E(G_1 \cup G_2) = E(G_1) \cup E(G_2)$. H_S is the subgraph of H induced by the vertex set S, i.e., $V(H_S) = S$ and $E(H_S) = \{(x,y) \in E(H) \mid x, y \in S\}$. $n_0(H), n_1(H)$ and $n_2(H)$ denote the number of vertices in the connected components of a graph H with zero, one or two cycles, respectively. $H \backslash S$ denotes the graph obtained from a graph H by removing the vertices in S and all the edges incident to these vertices in S. For a vertex (x, y, z) of the 3D lattice, x, y and z are the 1st, 2nd and 3rd coordinate, respectively. $[i, j]$ and $[i, j)$ denote the set of integers $\{i, i+1, i+2, \ldots, j\}$ and $\{i, i+1, i+2, \ldots, j-1\}$, respectively. $OPT(G, K)$ denotes the number of edges in an optimal solution to the $PSDC_2$ or $PSDC_3$ problem. A δ-approximate solution (or simply a δ-approximation) of a maximization problem is a solution with an objective value no smaller than δ times the value of the optimum; an algorithm of *performance* or *approximation ratio* δ produces an δ-approximate solution. A *polynomial-time approximation scheme* (PTAS) for a maximization problem is an algorithm that, for any given *constant* $\varepsilon > 0$, runs in polynomial time and produces an $(1 - \varepsilon)$-approximate solution.

2 The PSDC$_2$ Problem

In [12] Hart provided a proof of NP-completeness of $PSDC_2$. Unfortunately, the proof was not correct because the reduction from the Knapsack problem was pseudo-polynomial time and Knapsack problem is not strongly NP-complete. We show in the following lemma that $PSDC_2$ can indeed be solved in polynomial time. Due to space limitations, we omit the proof of the following lemma.

Lemma 1. *There exists an $O(K|V(G)|)$ time algorithm that solves $PSDC_2$.*

3   The PSDC$_3$ Problem

In the first subsection, we show that the PSDC$_3$ problem is NP-complete even though the PSDC$_2$ problem is not. In the second subsection, we show how to design a PTAS for the PSDC$_3$ problem using the shifted slice-and-dice technique.

3.1   NP-Completeness Result for PSDC$_3$

Theorem 1. *The PSDC$_3$ problem is NP-complete.*

Proof. It is trivial to see that PSDC$_3$ is in NP. To show NP-hardness, we provide a reduction from the CLIQUE problem on graphs whose goal is to decide, for a given graph G and an integer k, if there is a complete subgraph (clique) of G of k vertices. Let us denote by 3DS problem the DS problem on graphs with a maximum degree of 3. We will use a minor modification of a reduction of Feige and Seltser [11] from the CLIQUE problem to the the 3DS problem along with additional arguments. Consider an instance (G, k) of the CLIQUE problem where $V(G) = (v_1, \ldots, v_n)$ with $|V(G)| = n$. We can assume without loss of generality that n is an *exact* power of 2, n is sufficiently large and the vertex v_n has zero degree[2]. Let $t_1 \ll t_2 \ll t_3 \ll t_4 \ll t_5 \ll t_6$ be six sufficiently large polynomials in n; for example, $t_1 = n^{20}$ and $t_i = t_{i-1}^2$ for $i \in [2, 6]$ suffices. From G, we construct an instance graph H of the 3DS problem using a minor modification of the construction in Section 3 of Feige and Seltser [11] as follows:

- Replace each vertex v_i by a simple cycle of "cycle" edges

$$C^i = \{v_1^i, v_2^i\}, \{v_2^i, v_3^i\}, \ldots, \{v_{2nt_4-1}^i, v_{2nt_4}^i\}, \{v_{2nt_4}^i, v_1^i\} \in E(H)$$

  on the $2nt_4$ new "cycle" vertices $v_1^i, v_2^i, \ldots, v_{2nt_4}^i \in V(H)$.
- Replace each edge $\{v_i, v_j\} \in E(G)$ with $i < j$ by a simple path of "path" edges

$$P^{ij} = \{\{v_{(n+j)t_4}^i, u_1^{ij}\}, \{u_1^{ij}, u_2^{ij}\} \ldots, \{u_{kt_5-1}^{ij}, u_{kt_5}^{ij}\}, \{u_{kt_5}^{ij}, v_{(n+i)t_4}^j\}\} \subseteq E(H)$$

  of $kt_5 + 2 > 2nkt_4$ vertices between $v_{(n+j)t_4}^i$ and $v_{(n+i)t_4}^j$ where $u_1^{ij}, u_2^{ij}, \ldots, u_{kt_5}^{ij} \in V(H)$ are the new "path" vertices.
- Finally, we add a set of s additional separate connected components $Q_1, Q_2, \ldots, Q_s$, which will be specified later, such that all vertices in $\cup_{i=1}^s Q_i$ are of degree *at most* 2, no Q_i is an odd cycle and $\cup_{i=1}^s |V(Q_i)|$ is a polynomial in n.

Let $K = 2nkt_4 + \binom{k}{2}kt_5$ and $m = 2nkt_4 + \binom{k}{2}(kt_5 + 1)$. The same proof in Feige and Seltser [11] works to show that, for *any* selection of $Q_1, \ldots, Q_s$, there exists a subgraph with K vertices and at least m edges in H if and only if G has a clique of k vertices. Thus, to complete our reduction, we need to show the following:

---

[2] The degree assumption for v_n helps us to design the sequence $\mathcal{S}$ whose contact map will correspond to the graph H for the 3DS problem that we generate from an instance of the CLIQUE problem.

Step 1 (embedding H in the 3D lattice) H can be embedded in the 3D lattice.

Step 2 (realizing H as a contact graph) For some choice of $Q_1, Q_2, \ldots, Q_s$ H is the contact graph of a 3D sequence $\mathcal{S}$.

Details of both these steps are omitted due to space limitations.

Corollary 2 *3DS is NP-complete even if G is a subgraph of the 3D lattice.*

3.2   An Approximation Scheme via Shifted Slice-and-Dice

All the graphs discussed in this section are subgraphs of the 3D lattice. For notational convenience and simplifications we assume, without loss of generality, that our input graph G satisfies $V(G) \subseteq \times_{i=1}^3 [0, n_i)$ for some n_1, n_2, n_3 with $|V(G)| \geq \max\{n_1, n_2, n_3\}$. We classify an edge $\{(i_1, i_2, i_3), (j_1, j_2, j_3)\} \in E(G)$ as *horizontal, vertical* or *lateral* if $i_1 \neq j_1$, $i_2 \neq j_2$ or $i_3 \neq j_3$, respectively. Let E_-, $E_|$ and $E_/$ be the set of horizontal, vertical and lateral edges in an optimal solution.

Theorem 3. *For every $\varepsilon > 0$, there is an $O\left(\frac{K}{\varepsilon^3} 2^{1/\varepsilon^3} |V(G)|\right)$ time algorithm that returns a solution of the $PSDC_3$ problem with at least $(1 - \varepsilon)OPT(G, K)$ edges.*

Proof. We use the shifted slice-and-dice technique of [6, 17, 18]. For convenience, we use the following notations:

- $\nu_j = \left\lfloor \frac{n_j - 1}{\ell} \right\rfloor$ for $j \in [1, 3]$,
- $\kappa_1 = [i\ell + \alpha, \min\{(i+1)\ell, n_1\} + \alpha)$ $\kappa_2 = [j\ell + \alpha, \min\{(j+1)\ell, n_2\} + \alpha)$ and $\kappa_3 = [k\ell + \alpha, \min\{(k+1)\ell, n_3\} + \alpha)$ for some specified values i, j, k and number α.

We first need the following definition.

Definition 2. *For a given positive integer (partition length) $\ell > 0$ and three positive integers (shifts) $0 \leq \alpha, \beta, \gamma < \ell$, an (α, β, γ)-shifted ℓ-partition of G, denoted by $\Pi_\ell^{\alpha, \beta, \gamma}[G]$ is the subgraph of G in which $V(\Pi_\ell^{\alpha, \beta, \gamma}[G]) = V(G)$ and $E(\Pi_\ell^{\alpha, \beta, \gamma}[G])$ is exactly*

$$E(G) \cap$$
$$\left(\bigcup_{i=0}^{\nu_1} \bigcup_{j=0}^{\nu_2} \bigcup_{k=0}^{\nu_3} \{ \{(x, y, z), (x', y', z')\} \mid x, x' \in \kappa_1 \ \& \ y, y' \in \kappa_2 \ \& \ z, z' \in \kappa_3 \} \right)$$

See Figure 1 for a simple illustration of the above definition.

Let $\ell = \lceil 1/\varepsilon \rceil$. It is trivial to compute the $\Pi_\ell^{\alpha, \beta, \gamma}[G]$'s for all $0 \leq \alpha, \beta, \gamma < \ell$ in $O(\ell^3 |V(G)|)$ time. For each $\Pi_\ell^{\alpha, \beta, \gamma}[G]$, $OPT(\Pi_\ell^{\alpha, \beta, \gamma}[G], K)$ can be calculated in $O(K 2^{\ell^3} |V(G)|)$ time since:

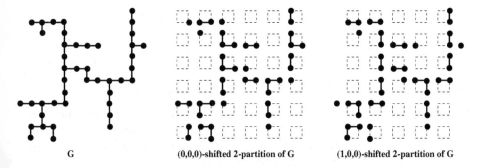

Fig. 1. Illustration of Definition 2 for a G embeddable in the 2D lattice (*i.e.*, $n_3 = 2$).

- For each $i \in [0, \nu_1]$, $j \in [0, \nu_2]$ and $k \in [\nu_3]$, the subgraph $G_{i,j,k,\alpha,\beta,\gamma}$ of $\Pi_\ell^{\alpha,\beta,\gamma}[G]$ induced by the set of vertices $V(G_{i,j,k,\alpha,\beta,\gamma}) = V(G) \cap \{x, y, z \mid x \in \kappa_1 \ \& \ y \in \kappa_2 \ \& \ z \in \kappa_3\}$ is not connected by any edge of $\Pi_\ell^{\alpha,\beta,\gamma}[G]$ to any remaining vertex of $\Pi_\ell^{\alpha,\beta,\gamma}[G]$. Thus, we can compute $\mathrm{OPT}(G_{i,j,k,\alpha,\beta,\gamma}, \mu)$ for all $1 \le \mu \le K$ by exhaustive enumeration in $O(K2^{\ell^3})$ time. Since there are at most $|V(G)|$ $G_{i,j,k,\alpha,\beta,\gamma}$'s that are not empty, the total time for this step is $O(K2^{\ell^3}|V(G)|)$.
- We now use the dynamic programming algorithm for the General Knapsack (GK) problem. For each $i \in [0, \nu_1]$, $j \in [0, \nu_2]$ and $k \in [0, \nu_3]$, we have a set of K objects $\mathcal{A}_{i,j,k} = \{a_{i,j,k}^1, a_{i,j,k}^2, \ldots, a_{i,j,k}^K\}$ with $s(a_{i,j,k}^\mu) = \mu$ and $v(a_{i,j,k}^\mu) = \mathrm{OPT}(G_{i,j,k,\alpha,\beta,\gamma}, \mu)$ for $\mu \in [1, K]$, and moreover we set $\mathbf{b} = K$. We can solve this instance of the GK problem to determine in $O(K|V(G)|)$ time a subset of indices $\{(i_1, j_1, k_1), (i_2, j_2, k_2), \ldots, (i_t, j_t, k_t)\}$ such that $\sum_{p=1}^t |V(G_{i_p, j_p, k_p, \alpha, \beta, \gamma})| \le K$ and $\sum_{p=1}^t |E(G_{i_p, j_p, k_p, \alpha, \beta, \gamma})|$ is maximized. Obviously, $\mathrm{OPT}(\Pi_\ell^{\alpha,\beta,\gamma}[G], K) = \sum_{p=1}^t |E(G_{i_p, j_p, k_p, \alpha, \beta, \gamma})|$.

Our algorithm then outputs $\max_{\alpha,\beta,\gamma} \mathrm{OPT}(\Pi_\ell^{\alpha,\beta,\gamma}[G], K)$ as the approximate solution. The total time taken by the algorithm is therefore $O(K2^{\ell^3}\ell^3|V(G)|) = O(K|V(G)|)$ since $\varepsilon > 0$ is a constant. We now show that $\max_{\alpha,\beta,\gamma} \mathrm{OPT}(\Pi_\ell^{\alpha,\beta,\gamma}[G], K) \ge \left(1 - \frac{1}{\ell}\right)\mathrm{OPT}(G, K) \ge (1 - \varepsilon)\mathrm{OPT}(G, K)$. For each $0 \le \alpha, \beta, \gamma < \ell$, let $E_-(\alpha, \beta, \gamma) = E_- - E(\Pi_\ell^{\alpha,\beta,\gamma}[G])$, $E_|(\alpha, \beta, \gamma) = E_| - E(\Pi_\ell^{\alpha,\beta,\gamma}[G])$ and $E_/(\alpha, \beta, \gamma) = E_/ - E(\Pi_\ell^{\alpha,\beta,\gamma}[G])$. Now we observe the following:

- The sets $E_-(\alpha, \beta, \gamma)$, $E_|(\alpha, \beta, \gamma)$ and $E_/(\alpha, \beta, \gamma)$ are mutually disjoint.
- For any $e \in E_-$ (respectively, $e \in E_|$, $e \in E_/$), $|\{E_-(\alpha, \beta, \gamma) \mid e \in E_-(\alpha, \beta, \gamma)\}| \le \ell^2$ (respectively, $|\{E_|(\alpha, \beta, \gamma) \mid e \in E_|(\alpha, \beta, \gamma)\}| \le \ell^2$, $|\{E_/(\alpha, \beta, \gamma) \mid e \in E_/(\alpha, \beta, \gamma)\}| \le \ell^2$). We prove the case for $e \in E_-$ only; the other cases are similar. Suppose that $e \in E_-(\alpha, \beta, \gamma)$ for some α, β and γ. Then, $e \notin E_-(\alpha', \beta', \gamma')$ if $\alpha' \ne \alpha$.

- Thus, $\sum_{\alpha=0}^{\ell-1} \sum_{\beta=0}^{\ell-1} \sum_{\gamma=0}^{\ell-1} \mathrm{OPT}(\Pi_\ell^{\alpha,\beta,\gamma}[G], K)$ is at least

$$\ell^3 \mathrm{OPT}(G, K) - \sum_{\alpha=0}^{\ell-1} \sum_{\beta=0}^{\ell-1} \sum_{\gamma=0}^{\ell-1} (E_-(\alpha, \beta, \gamma) + E_|(\alpha, \beta, \gamma) + E_/(\alpha, \beta, \gamma)$$
$$\geq \ell^3 \mathrm{OPT}(G, K) - \ell^2(|E_-| + |E_|| + |E_/|) \geq \ell^3 \mathrm{OPT}(G, K) - \ell^2 \mathrm{OPT}(G, K)$$

Hence, $\max_{\alpha,\beta,\gamma} \mathrm{OPT}(\Pi_\ell^{\alpha,\beta,\gamma}[G], K) \geq \mathrm{OPT}(G, K) - \frac{1}{\ell} \mathrm{OPT}(G, K)$.

Remark 1. The PTAS can be generalized in an obvious manner when the given graph is embeddable in a d-dimensional lattice for $d > 3$; however the running time grows exponentially with d. We do not describe the generalization here since it has no applications to the PSD problem.

Acknowledgements

Berman was supported by NSF grant CCR-0208821, DasGupta was supported in part by NSF grants CCR-0296041, CCR-0206795 and CCR-0208749, Mubayi was supported by NSF grant DMS-9970325, Sloan and Turán were supported by NSF grant CCR-0100336 and Zhang was supported by NSF grant CCR-0208749.

References

1. Y. Asahiro, K. Iwama,H. Tamaki and T. Tokuyama. *Greedily Finding a Dense Subgraph*, Journal of Algorithms 34,203-221,2000.
2. Y. Asahiro, R. Hassin and K. Iwama. *Complexity of finding dense subgraphs*, Discrete Applied Mathematics 121, 15-26,2002.
3. J. Atkins and W. E. Hart. *On the intractability of protein folding with a finite alphabet of amino acids*, Algorithmica, 25(2-3):279–294, 1999.
4. J. Banavar, M. Cieplak, A. Maritan, G. Nadig, F. Seno, and S. Vishveshwara. *Structure-based design of model proteins*, Proteins: Structure, Function, and Genetics, 31:10–20, 1998.
5. B. Berger and T. Leighton. *Protein folding in the hydrophobic-hydrophilic (HP) model is NP-complete*, Journal of Computational Biology, 5(1):27–40, 1998.
6. P. Berman, B. DasGupta and S. Muthukrishnan. *Approximation Algorithms For* MAX-MIN *Tiling*, Journal of Algorithms, 47 (2), 122-134, July 2003.
7. P. Crescenzi, D. Goldman, C. Papadimitriou, A. Piccolboni, and M. Yannakakis. *On the complexity of protein folding*, Journal of Computational Biology, 423–466, 1998.
8. J. M. Deutsch and T. Kurosky. *New algorithm for protein design*, Physical Review Letters, 76:323–326, 1996.
9. K. A. Dill, S. Bromberg, K. Yue, K. M. Fiebig, D. P. Yee, P. D. Thomas, and H. S. Chan. *Principles of protein folding — A perspective from simple exact models*, Protein Science, 4:561–602, 1995.
10. K. E. Drexler. *Molecular engineering: An approach to the development of general capabilities for molecular manipulation*, Proceedings of the National Academy of Sciences of the U.S.A., 78:5275–5278, 1981.
11. U. Feige and M. Seltser. *On the densest k-subgraph problems.* Technical Report # CS97-16, Faculty of Mathematics and Computer Science, Weizmann Institute of Science, Israel
    (available online at http://citeseer.nj.nec.com/feige97densest.html).

12. W. E. Hart. *On the computational complexity of sequence design problems*, Proceedings of the 1st Annual International Conference on Computational Molecular Biology, 128–136, 1997.

13. W. E. Hart and S. Istrail. *Fast protein folding in the hydrophobic-hydrophilic model within three-eighths of optimal*, Journal of Computational Biology, 3(1):53–96, 1996.

14. W. E. Hart and S. Istrail. *Invariant patterns in crystal lattices: Implications for protein folding algorithms (extended abstract)*, Lecture Notes in Computer Science 1075: Proceedings of the 7th Annual Symposium on Combinatorial Pattern Matching, 288–303, 1996.

15. W. E. Hart and S. Istrail. *Lattice and off-lattice side chain models of protein folding: Linear time structure prediction better than 86% of optimal*, Journal of Computational Biology, 4(3):241–260, 1997.

16. V. Heun. *Approximate protein folding in the HP side chain model on extended cubic lattices*, Lecture Notes in Computer Science 1643: Proceedings of the 7th Annual European Symposium on Algorithms, 212–223, 1999.

17. D. Hochbaum. *Approximation Algorithms for NP-hard problems*, PWS Publishing Company, 1997.

18. D. S. Hochbaum and W. Mass. *Approximation schemes for covering and packing problems in image processing and VLSI*, Journal of ACM, 32(1):130–136, 1985.

19. J. Kleinberg. *Efficient Algorithms for Protein Sequence Design and the Analysis of Certain Evolutionary Fitness Landscapes.*, Proceedings of the 3rd Annual International Conference on Computational Molecular Biology, 226-237, 1999.

20. K. F. Lau and K. A. Dill. *A lattice statistical mechanics model of the conformational and sequence spaces of proteins*, Macromolecules, 22:3986–3997, 1989.

21. G. Mauri, G. Pavesi, and A. Piccolboni. *Approximation algorithms for protein folding prediction*, Proceedings of the 10th Annual ACM-SIAM Symposium on Discrete Algorithms, 945–946, 1999.

22. K. M. Merz and S. M. L. Grand, editors. *The Protein Folding Problem and Tertiary Structure Prediction*, Birkhauser, Boston, MA, 1994.

23. J. Ponder and F. M. Richards. *Tertiary templates for proteins*, Journal of Molecular Biology, 193:63–89, 1987.

24. S. J. Sun, R. Brem, H. S. Chan, and K. A. Dill. *Designing amino acid sequences to fold with good hydrophobic cores*, Protein Engineering, 8(12):1205–1213, Dec. 1995.

25. E. I. Shakhnovich and A. M. Gutin. *Engineering of stable and fast-folding sequences of model proteins*, Proc. Natl.Acad.Sci., 90:7195-7199, 1993.

26. T. F. Smith, L. L. Conte, J. Bienkowska, B. Rogers, C. Gaitatzes, and R. H. Lathrop. *The threading approach to the inverse protein folding problem*, Proceedings of the 1st Annual International Conference on Computational Molecular Biology, 287–292, 1997.

27. K. Yue and K. A. Dill. *Inverse protein folding problem: Designing polymer sequences*, Proceedings of the National Academy of Sciences of the U.S.A., 89:4163–4167, 1992.

A Computational Model
for RNA Multiple Structural Alignment

Eugene Davydov and Serafim Batzoglou

Dept. of Computer Science, Stanford University,
Stanford CA 94305, USA
{edavydov,serafim}@cs.stanford.edu

Abstract. This paper addresses the problem of aligning multiple sequences of non-coding RNA genes. We approach this problem with the biologically motivated paradigm that scoring of ncRNA alignments should be based primarily on secondary structure rather than nucleotide conservation. We introduce a novel graph theoretic model (NLG) for analyzing algorithms based on this approach, prove that the RNA multiple alignment problem is NP-Complete in this model, and present a polynomial time algorithm that approximates the optimal structure of size S within a factor of $O(\log^2 S)$.

1   Introduction

Noncoding RNA (*ncRNA*) genes are among the biologically active features in genomic DNA. They are polymers of four nucleotides: A (adenine), C (cytosine), G (guanine), and U (uracil). Unlike regular genes, ncRNAs are not translated into protein, but rather fold directly into secondary and tertiary structures, which can have a variety of structural, catalytic, and regulatory functions [5].

The structural stability and function of ncRNA genes are largely determined by the formation of stable secondary structures through complementarity of nucleotides, whereby G-C, A-U, and G-U form hydrogen bonds that are energetically favored. This secondary structure can be predicted from the nucleotide sequence as one minimizing (some approximation of) the free energy [18,19], which is largely determined by the formation of the hydrogen bonds. In ncRNAs, such bonds almost always occur in a nested fashion, which allows the optimal structure to be computed in time $O(n^3)$ in the length of the input sequence using a dynamic programming approach [13,11]. Algorithms that do not assume a nested structure are even more computationally expensive [15]. However, the stability of ncRNA secondary structures is not sufficiently different from the predicted stability of random genomic fragments to yield a discernible statistical signal [16], limiting the application of current ncRNA detection methods to the simplest and best understood structures, such as tRNAs [12].

One of the most promising ways of detecting ncRNA genes and predicting reliable secondary structures for them is comparative sequence analysis. During the course of genome evolution, mutations that occur in functional regions of the genome tend to be deleterious, and therefore unlikely to fix, while mutations that occur in non-functional regions tend to be neutral and accummulate. As a result, functional regions of the genome tend to exhibit significant sequence similarity across related genomes, whereas regions that are not functional are usually much less conserved. This difference in the rate of sequence conservation is used as a powerful signal for detecting protein-coding genes [1, 2] and regulatory sites [14, 10], and could be applied to ncRNA genes. However, their function is largely determined by their secondary structure, which in turn is determined by nucleotide complementarity: RNA genes across different species are similar in the pattern of nucleotide complementarity rather than in the genomic sequence. As a result, conventional sequence alignment methods are not able to properly align ncRNAs [6].

One biologically meaningful approach to ncRNA multiple alignment is finding the largest secondary structure common to all the sequences, lining up the nucleotides forming this structure, and then aligning corresponding leftover pieces as one would align genomic sequences which have no evolutionary pressure favoring complementary substitution. However, this approach has never been applied in practice because the task of finding the largest common secondary structure among several sequences is computationally challenging: the straightforward extention of the dynamic programming algorithm using stochastic context-free grammars (SCFGs) has a running time of $O(n^{3k})$, where k is the number of sequences being aligned, which is prohibitive even for two sequences of moderate length [9].

The problem of aligning multiple DNA sequences has been proven to be NP-Complete for certain scoring schemes and metrics [17, 3] . However, when analyzing the computational complexity of ncRNA multiple alignment, it's more relevant to focus on the complexity of finding the largest common secondary structure, because for most biologically meaningful ncRNAs the remaining pieces should be relatively short and easy to align.

In this paper we introduce a novel theoretical framework for analyzing the problem of ncRNA multiple structural alignment. We present the **Nested Linear Graph (NLG)** model and formulate the problem of computing the largest common secondary structure in this model in terms of finding the largest common nested subgraph. We then prove this problem to be NP-Complete, and present a polynomial-time algorithm which approximates the optimal solution within a factor of $O(\log^2 S)$, where S is the size of the optimal solution. We conclude with a discussion of the NLG model in general and our algorithm and results in particular.

2   A Graph Theoretic Formulation

A *linear graph* is a graph whose vertices, V, are points on some line L. Genomic sequences naturally yield themselves to linear graph representations, because each of their nucleotides can correspond to a point, and the sequence can correspond to the line. For modeling ncRNA folding and secondary structure, we form the linear graph with edges connecting pairs of vertices that represent complementary nucleotide pairs (A-U, C-G, and G-U). A typical linear graph induced by an RNA sequence is shown in fig. 1.

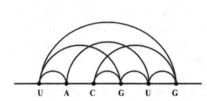

Fig. 1. A linear graph representation of the RNA sequence $UACGUG$. The nucleotides are represented by points on a line in the same order as in the sequence. Each edge is represented by an arc to emphasize that it does not pass through the nodes between its two endpoints. Edges are drawn between nodes representing complementary nucleotide pairs A-U, C-G, and G-U.

Two edges $\widehat{ab}$ and $\widehat{cd}$ of a linear graph intersect if exactly one of c and d lies on the line segment $\overline{ab}$ (and vice versa). A linear graph is *nested* if no two edges of the graph intersect each other. For a linear graph derived from an RNA sequence, a nested subgraph represents a plausible fold of that sequence. Thus, in the NLG model, the problem of finding the largest secondary structure of an ncRNA is precisely the problem of finding the largest nested subgraph in the linear graph derived from the sequence. For multiple ncRNA alignment, where we seek the largest common secondary structure, the appropriate NLG formulation is finding the **largest common nested linear subgraph (MAX-NLS)** among the linear graphs induced by the sequences (see fig. 2). We now formulate this problem precisely and formally analyze its computational complexity.

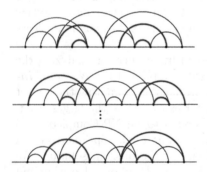

Fig. 2. The MAX-NLS of several linear graphs; its edges have been emphasized in bold to distinguish them from the edges of the original linear graphs. Note that the MAX-NLS is not necessarily unique, but its size is.

3   Complexity Analysis of MAX-NLS

Let $G_1, \ldots, G_m$ be the linear graphs derived from ncRNA sequences $S_1, \ldots, S_m$ respectively. The MAX-NLS of these graphs is the largest nested graph G_c such that G_c is a subgraph of G_i for all $i = 1, \ldots, m$. For any problem instance $I = \{G_1, \ldots, G_m\}$, we write MAX-NLS(I) to indicate this G_c. To represent the order (number of edges) in this graph, we use the notation $|\text{MAX-NLS(I)}|$.

Note that the MAX-NLS problem represents a slight generalization of the RNA multiple alignment problem, in that we do not constrain the linear graphs to be derived from RNA strings by connecting *every* pair of nodes corresponding to complementary nucleotides with an edge. We motivate this relaxation in the discussion section of the paper.

The MAX-NLS is an *optimization problem*, because our objective is to maximize the order of the common nested subgraph of $\{G_1, \ldots, G_m\}$. We now formulate the corresponding *decision problem*, where our objective is to answer a boolean query.

Definition 1. *The NLS decision problem (D-NLS) is to determine, given an input $G_1, \ldots, G_m$ and a positive integer k (where $1 < k < \min_i |G_i|$), whether there exists a common nested linear subgraph of $G_1, \ldots, G_m$ with order $\geq k$.*

Theorem 1. *D-NLS is NP-Complete.*

Proof (of Theorem 1). We proceed by demonstrating a polynomial reduction to 3-SAT, a well-known NP-Complete problem [4].

Definition 2. *Let $x_1, \ldots, x_k$ be boolean variables. Let $\psi_1, \ldots, \psi_n$ be logical clauses, with each clause ψ_i being a disjunction of 3 literals, where each literal is either a variable x_j or its negation, $\neg x_j$. The 3-SAT problem is to determine, given this as input, whether there exists an assignment for the k variables which satisfies all n clauses simultaneously.*

To establish the reduction we need to demonstrate that the existence of a polytime algorithm for D-NLS yields a polytime algorithm for 3-SAT. As such, we show that given any input instance I_{3-SAT} and a polytime algorithm A for D-NLS, we can construct, in polynomial time and space, an instance I_{D-NLS} such that computing $A(I_{D-NLS})$ will allow us to answer whether the instance I_{3-SAT} is satisfiable. However, to simplify the description of this construction, we must define the notion of a *c-thick edge* (see fig. 3).

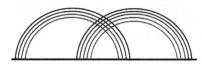

Fig. 3. A 4-thick edge intersecting a 5-thick edge. Any edge not shown must intersect either all edges in either stack, or none at all.

Definition 3. *In a linear graph, an edge $\widehat{ab}$ contains an edge $\widehat{cd}$ if and only if both c and d lie strictly between a and b on the line. $\widehat{ab}$ directly contains $\widehat{cd}$ whenever $\widehat{ab}$ contains $\widehat{cd}$ and there is no other edge e that contains $\widehat{cd}$ but is itself contained by $\widehat{ab}$.*

A c-thick "edge" in a linear graph is a set of c edges $e_1, \ldots, e_c$ with the properties that:

(i) for all i, j such that $i > j$, e_i contains e_j
(ii) for any other edge e', either e' intersects all e_i, or it intersects none of them

We can now describe the construction of I_{D-NLS} from I_{3-SAT}, as depicted in fig. 4. Given the set of variables $x_1, \ldots, x_k$ and the clauses $\psi_1, \ldots, \psi_n$, we construct $k + 1$ linear graphs: one corresponding to each boolean variable in I_{3-SAT}, and an extra graph x' whose purpose will be clarified shortly. Each of the n graphs consists of two intersecting c_1-thick edges, each of which contains a sequence of k similar groups of edges, where each group corresponds to a particular clause ψ_i. Such a group is depicted in detail at the bottom of fig. 4.

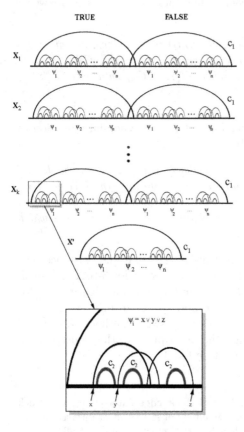

Fig. 4. Constructing an instance of D-NLS from an instance of 3-SAT. Each variable x_j gives rise to a linear graph x_j, which consists of two intersecting c_1-thick edges, each of which contains n edge groups corresponding to clauses ψ_i. Every clause group consists of 3 c_2-thick edges in sequence, as well as up to 3 mutually intersecting selection edges, which are present if ψ_i does not depend on x_j, or the truth value induced upon x_j by the label of the c_1 edge makes ψ_i TRUE. Finally, I_{D-NLS} contains an extra linear graph x', consisting of only one c_1 edge, which contains the standard collection of $3nc_2$ edges, as well as all possible selection edges. The goal of the x' graph is to force an alignment where every other graph x_j, x' aligns to either the TRUE or FALSE portion of x_j, thus corresponding to a truth assignment to all variables of the 3-SAT problem.

The edge group varies slightly depending on which x_j and ψ_i it corresponds to. The portion common to all such groups consists of three c_2-thick edges none of which intersects or contains the other. Beyond these, each group has up to three of the following set of mutually intersecting edges: an edge that contains the first and second c_2-thick edges, an edge that contains only the second, and an edge that contains the third, as illustrated in fig. 4. An edge is missing from the group only if the corresponding literal in the clause ψ_i is not in agreement with the truth assignment induced by the c_1 edge to x_j. To be more precise, let $\psi_i = \eta_a x_a \vee \eta_b x_b \vee \eta_c x_c$, where $a \leq b \leq c$, and the corresponding η can either be the identity or negation $\neg$. The edge corresponding to $\eta_a x_a$ is absent if and only if $a = j$ and $\eta_a x_a$ is false under the truth assignment induced by the c_1 edge of x_j. If $a = b$ and $\eta_a = \neg\eta_b$, the edge is present; since if a clause contains the disjunction $x_j \vee \neg x_j$, it is automatically satisfied and the edge should exist.

The $k + 1^{st}$ graph consists of only one c_1 edge, and n clause groups each of which contains all 3 selector edges in addition to the $3c_2$ edges. The basic premise of this construction is that if (and only if) there is a satisfying assignment, we will be able to match the x' graph to the corresponding c_1 edge in each of the k graphs, and align the n clause groups within. Only because the assignment is satisfying will we be able to align one additional selector edge from *every* clause, giving us the largest possible common subgraph.

Lemma 1. *Let $c_2 = n+1$ and let $c_1 = 3n^2+4n+1$. Under the scheme described above, the $k+1$ linear graphs have a common nested subgraph of order $6n^2+8n+1$ if and only if $\psi_1, \ldots, \psi_k$ are satisfiable.*

Proof (of Lemma 1).

Suppose the clauses are satisfiable, that is, there exists some assignment to $x_1, \ldots, x_k$ which satisfies them all. We align the c_1 edge of the x' graph with the c_1 edge of graph j that corresponds to the value of x_j in this truth assignment. We then align the c_2 edges to each other. Now consider a particular selector edge in some clause ψ_i. Because of the way we aligned the c_1 edges, if this edge is absent in any of the half-graphs we selected, it is because its corresponding literal is false in that clause given the truth assignment. However, since we assumed our assignment is satisfying, every clause must have a literal that evaluates to TRUE. The corresponding selector edge must be present in every graph.

We can choose at most one selector edge per clause, since they all intersect each other. Because we can choose one from every clause, we have a total of $c_1 + 3nc_2 + n = 6n^2 + 8n + 1$.

Now suppose we indeed have a common nested subgraph of order $6n^2+8n+1$. As there are a total of $6nc_2$ c_2 edges and up to $2n$ selector edges that may be chosen simultaneously, we could only have $6n^2 + 8n$ edges without choosing a c_1 edge. Thus, we must align a c_1 edge, in which case we might as well align the whole stack of them. That leaves $3n^2 + 4n$ edges to be included. Note that each c_2 stack contributes more than the selector edges could simultaneously, so we must choose all $3n$ c_2 stacks for a total of $3n^2 + 3n$ edges. This leaves n edges to be accounted for, all of which must be selector edges, one from each clause.

Note that the c_1 alignment we choose induces a truth assignment to our variables. As we just showed, the order of our alignment implies not only that the c_1 and c_2 edges are aligned, but also that under this truth assignment, every clause has a selector edge that is present in every graph's chosen c_1 half. In particular, that edge is present in the graph corresponding to its literal, meaning that under this induced truth assignment, the clause is satisfied because the literal is TRUE. Since this applies to all the clauses, $\psi_1, \ldots, \psi_k$ are all satisfied.
□

The time required for this construction is $O(kn^2)$; thus, we have demonstrated a polynomial reduction from D-NLS to 3-SAT, and D-NLS is NP-Complete.
□

4 Approximating MAX-NLS with MAX-FLS

In view of Theorem 1 there is little hope for a tractable exact algorithm for MAX-NLS. Therefore, we present a polynomial time approximation algorithm that guarantees optimality within a factor of $O(\log^2 S)$, where S is the size of the optimal solution. The polynomial time is achieved by restricting attention to a subclass of nested linear graphs and finding the optimal member of this restricted subclass. The main tradeoff here is the choice of the restriction: if the subclass is too narrow, our approximation will be poor; otherwise, finding the optimal member of the subclass may still be NP-Complete.

The restriction that yields our algorithm is best presented as a composition of two restrictions. First, we first consider the subclass of NLGs that are *flat*.

Definition 4. *A branching edge in a nested linear graph is an edge e that contains two edges e_1 and e_2, neither of which contains the other. A nested linear graph is flat if it contains no branching edges. The flat order of a nested linear graph is the order of its largest flat subgraph.*

The optimization problem corresponding to this restriction is that of finding the largest common flat nested linear graph (**MAX-FLS**). We now show that this restriction yields a solution that is suboptimal by a factor of at most $O(\log S)$.

Theorem 2. *Every nested linear graph G with flat order F_G satisfies $|G| \leq F_G \log(F_G)$.*

Proof (of Theorem 2).

We begin by introducing the tree representation of nested linear graphs in order to relate the main notions of our argument to familiar data structures. The basic transformation is mapping each NLG edge to a node in the tree, as shown in fig. 5. We first add an edge containing the entire NLG, for the sake of uniformity. We then construct the tree by mapping each edge e_i to a tree node n_i. A node n_i is a parent of another node n_j whenever its corresponding edge e_i directly contains e_j (see Definition 3 for the notion of direct containment). While this transformation is rather elementary, it affords us insights into the notion of flat order. Noting that the notion of a branching edge in an NLG corresponds precisely to a branching node in the tree, we observe the following:

(i) When viewed as a subtree, the path from the root to any leaf contains no branching nodes and is therefore flat. Thus, the flat order F_T satisfies $F_T \geq h(T)$, where $h(T)$ is the node height of T (number of nodes in the longest root-leaf path).

(ii) Consider any disjoint subtrees of T satisfying the property that nodes in different subtrees cannot be descendants or ancestors of each other in T. The union of their flat subtrees will also be flat, as no branching nodes can be introduced by taking the union of flat constituents that have no ancestor relationships amongst one another. Consequently, for any split node in the tree, the sum of the flat orders of its subtrees is $\leq F_T$.

Fig. 5. A tree representation of a nested linear graph. Each node in the tree corresponds to an edge in the graph. Node i is an ancestor of node j in the tree if and only if the corresponding edge i contains j in the graph. For unity of representation, an edge containing all other edges in the NLG is added so that the result of the transformation is a tree rather than a forest. This edge and the corresponding root vertex are represented with dashed lines in the diagram.

We now examine an arbitrary tree T_n with flat order n. We show that $|T_n| \leq n \log(n) + 1$. We establish the general result by strong induction: assuming the formula holds for every $n' < n$, we show that it holds for n. We enumerate the required base cases in figure 7.

Each tree can be represented as an initial trunk of length $\ell \geq 0$, followed by a split into some number of subtrees. Among these we then consider the subtree with the largest flat order. If its flat order is $> n/2$, we recursively divide that subtree into a trunk, a splitting node, and the subtrees at that node. We continue this process until no subtree has flat order $> n/2$, as shown in fig. 6. Note that there can only be one subtree with flat order $> n/2$, so we will never have to subdivide more than one subtree at each level.

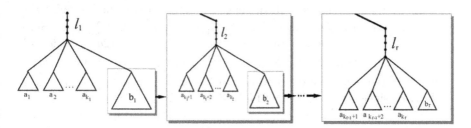

Fig. 6. An upper bound on a tree's size N_T as a function of its flat order F_T. Every tree is representable as an ℓ-edge ($\ell \geq 0$) chain from its root to the first node where a split into subtrees $T_1, \ldots, T_{k_1}, T_{b_1}$ occurs. These subtrees (labeled with their flat order) are arranged from left to right according to increasing flat order. We continue splitting the largest subtree, b_i, recursively, until we have no subtrees with flat order $> F_T/2$. This allows us to prove that $N_T = O(F_T \log(F_T))$.

We can now write the formula for the number of nodes in T_n. From the diagram,

$$|T_n| = \sum_{i=1}^{k_r} |T_{a_i}| + |T_{b_r}| + \sum_{j=1}^{r} \ell_j.$$

By the inductive assumption $|T_{a_i}| \leq a_i \log(a_i) + 1$ and $|T_{b_r}| \leq b_r \log(b_r) + 1$, so

$$|T_n| \leq \sum_{i=1}^{k_r} (a_i \log(a_i) + 1) + (b_r \log(b_r) + 1) + \sum_{j=1}^{r} \ell_j.$$

By construction, all a_i and b_r are $\leq n/2$. Furthermore, since $n \geq \sum_{i=1}^{k_r} a_i + b_r$, at most 3 of $\{a_1, \ldots, a_{k+r}, b_r\}$ may be $> n/4$. When $a_i \leq n/4$, $a_i \log(a_i) + 1 \leq a_i \log(a_i) + a_i \leq a_i \log(2a_i) \leq a_i \log(n/2)$, similarly for b_r. Thus,

$$|T_n| \leq \sum_{i=1}^{k_r} a_i \log(n/2) + b_r \log(n/2) + 3 + \sum_{j=1}^{r} \ell_j.$$

To prove that this implies $|T_n| \leq n \log(n) + 1$, we now consider 3 cases:

(1) $h(b_r) \geq 2$

Then, according to observation (i), $n \geq \sum_{j=1}^{r} \ell_j + h(b_r) \geq \sum_{j=1}^{r} \ell_j + 2$, therefore,

$$|T_n| \leq \log(n/2)(b_r + \sum_{i=1}^{k_r} a_i) + 1 + n \leq n \log(n/2) + n + 1 = n \log(n) + 1.$$

(2) $h(b_r) = 0$

Then T_{b_r} has no nodes, and since by construction it is the largest subtree in its level, it must be that the splitting node at the bottom of trunk ℓ_r has no children. This means that either the entire tree is a single trunk, in which case $|T_n| = n \leq n \log(n) + 1$, or that $\ell_r > n/2$, since we had to subdivide $T_{b_{r-1}}$. In this case, we have $|T_n| \leq \sum_{i=1}^{k_r-1} (a_i \log(a_i) + 1) + \sum_{j=1}^{r} \ell_j$. Since $a_i \leq n/2$ by construction, we have $a_i \log(a_i) + 1 \leq a_i \log(2a_i) \leq a_i \log(n)$, and therefore $|T_n| \leq \log(n) \sum_{i=1}^{k_r-1} a_i + \sum_{j=1}^{r} \ell_j$, which transforms to $|T_n| < (n/2) \log(n) + n$ since observation (ii) implies $\sum_{i=1}^{k_r-1} a_i \leq n - \ell_r < n/2$. Finally, since $n \leq (n/2) \log(n)$ for $n \geq 4$, we have

$$|T_n| < n \log(n).$$

(3) $h(b_r) = 1$

In this case T_{b_r} consists of a single node, so $b_r = 1$. We may now write $|T_n| \leq \log(n/2) \sum_{i=1}^{k_r} a_i + 3 + 1 \log(1) + \sum_{j=1}^{r} \ell_j$, since at most 3 elements of $\{a_i\}$ may be $> n/4$. Noting that $1 = b_r \log(2) \leq b_r \log(n/2)$ as long as $n \geq 4$, we have $|T_n| \leq \log(n/2)(b_r + \sum_{i=1}^{k_r} a_i) + 1 + \sum_{j=1}^{r} \ell_j$. Applying the results of observations (i) and (ii), we have the familiar inequalities $b_r + \sum_{i=1}^{k_r} a_i \leq n$ and $1 + \sum_{j=1}^{r} \ell_j \leq n$, yielding

$$|T_n| \leq n \log(n/2) + n + 1 = n \log(n) + 1.$$

The assumption $n \geq 4$ can be eliminated by noting that the largest trees with flat order < 4 still obey the equation. These trees are shown in fig. 7.

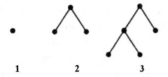

Fig. 7. The largest possible trees with flat order 1, 2, and 3, respectively.

Thus, for an arbitrary tree T with flat order F_T, $|T| \leq F_T \log(F_T) + 1 = O(F_T \log(F_T))$, which is precisely the statement of the theorem for nested flat graphs. □

It is noteworthy to observe that this bound is asymptotically tight. Consider the family of trees T_i defined recursively as:

- $T_0 = $ a single node.
- $T_{i+1} = $ a trunk of length 2^i nodes, which splits into two subtrees T_i, as shown in fig. 8.

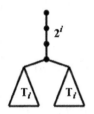

Fig. 8. A family of trees T_i with flat order F_i that attain the asymptotic upper bound of $O(\log(F_i))$ on the ratio $|T_i|/F_i$. These particular tree depicted in the diagram is T_{i+1}.

By induction, it is clear that both the height and the flat order of T_i are equal to 2^i. The number of nodes is defined by the recurrence $|T_{i+1}| = 2|T_i| + 2^i$, the solution to which is $|T_i| = 2^{i-1}(i+2)$. Thus, for any tree T of this family,

$$|T| = (1/2)F_T(2 + \log(F_T)) = \Theta(F_T \log(F_T)).$$

5   Approximating MAX-FLS with MAX-LLS

We now further restrict the subclass of NLGs to examine by introducing the notion of **level** flat graphs, and the corresponding optimization problem **MAX-LLS**. First, however, we prove a useful property of flat linear graphs.

Theorem 3. *Any flat nested linear graph G can be written as a union of $k \geq 0$ disjoint subsets, $G = \bigcup_{i=1}^{k} C_i$, where each C_i is a column of edges, i.e. a $|C_i|$-thick edge.*

Proof (of Theorem 3).
    Consider any edge $e \in G$, and let E be the set of edges that either contain or are contained by e. Because G is flat, E must form a column: if two distinct edges in E both contain e, they must contain each other or intersect; if they are both contained by e, they must contain each other, otherwise e is a branching

edge. Now note that by exactly the same reasoning, there can be no edge $e' \in E$ that contains or is contained by an edge $g \in G - E$, since g and e cannot contain or be contained by one another: if e' contains them both it must be a branching edge, if e' is contained by both then they must intersect, and if e' contains one and is contained by the other, then one must contain the other.

Thus, E is completely disjoint with respect to containment from $G - E$. Thus, we can let $C_1 = E$, and continue subdividing $G - E$ in this manner to obtain $C_2, \ldots, C_k$. In the end, each C_i is a column separate from one another, and $G = \bigcup_{i=1}^{k} C_i$. □

Definition 5. *Consider any flat nested linear graph* $G = \bigcup_{i=1}^{k} C_i$, *where each* C_i *is a column.* G *is level if* $|C_1| = \ldots = |C_i|$.

The MAX-LLS optimization problem is therefore to find the largest level flat subgraph in a set of linear graphs. We now show that this further restriction yields an approximation within a factor of $O(\log |G_F|)$ of the optimal solution G_F to MAX-FLS.

Theorem 4. *For any flat nested linear graph* G_F, *its largest level subgraph* G_L *with order* $L = |G_L|$ *satisfies* $|G_F| = O(L \log L)$.

Proof (of Theorem 4).
We first define two properties of linear graphs that are particularly important for level graphs.

Definition 6. *The length* ℓ_G *of a linear graph* G *is the order of the largest subgraph of* G *that consists solely of edges that do not intersect or contain one another, i.e. a flat graph where* $|C_i| = 1$ *for all* i. *The height* h_G *of* G *is the order of the largest subgraph of* G *that consists solely of one column, i.e. a flat graph consisting of one* h_G-*thick edge.*

These definitions are applicable to any linear graphs, but for level graphs they induce a compact representation since each level graph corresponds to an ordered pair (h, ℓ), as shown in fig. 9.

Fig. 9. Level graphs [a] $(h, \ell) = (2, 7)$ and [b] $(h, \ell) = (4, 2)$. These particular graphs represent points on the level signature of the flat graph shown in fig. 10.

We now consider an arbitrary flat graph G_F with height h_G and length ℓ_G. For each $h = 1, \ldots, h_G$, we let $F(h)$ be the largest value such that the level graph $(h, F(h))$ is a subgraph, noting that $1 \leq F(h) \leq \ell_G$ (see fig. 10). The discrete function F is thus uniquely defined for any flat graph G_F. We call this function the **level signature** of a flat graph. Note that the level signature is unique for any flat graph, although two distinct flat graphs may produce the same level signature simply because of different order of the columns.

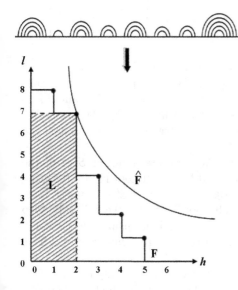

Fig. 10. Representing the possible level subgraphs of a flat graph G with a discrete nonincreasing function F, its level signature. Each point $(h, F(h))$ corresponds to a level graph with $F(h)$ columns of height h that is the largest level subgraph of G of height h. The shaded area represents L, the order of the largest level subgraph of G. The hyperbola $\hat{F}$ has the equation $h\ell = L$ and lies above all other points of F.

Each point $(h, F(h))$ corresponds to a level subgraph of G_F, as depicted in fig. 9. The order of this subgraph is $hF(h)$, therefore, the largest level subgraph of G_F corresponds to the point with the largest $hF(h)$, say (h^*, ℓ^*). Thus, $L = |G_L| = h^*\ell^*$.

Let $\hat{F}$ be the hyperbola passing through (h^*, ℓ^*) with the equation $h\ell = L$. By definition of (h^*, ℓ^*), all points on F must lie below this hyperbola. Note that the area under F given by $\sum_{h=1}^{h_G} F(h)$ gives the order of the original flat graph G_F, because $F(h)$ counts the number of columns containing an edge at height h. We now rewrite the sum as $|G_F| = \ell_G + \sum_{h=2}^{h_G} F(h)$, noticing that the area represented by the sum is a subset of the area under $\hat{F}$ from $h = 1$ to $h = h_G$. Thus,

$$|G_F| \leq \ell_G + \int_1^{h_G} \hat{F}(h)dh.$$

Since $(1, \ell_G)$ and $(h_G, 1)$ are both points of the level signature, $\ell_G \leq L$ and $h_G \leq L$. Evaluating the integral, we have $\int_1^{h_G} \hat{F}(h)dh = \int_1^{h_G} L/h\,dh = L \log h_G$. Thus, $|G_F| \leq L + L \log L = O(L \log L)$. □

6   A Polytime Algorithm for MAX-LLS

To briefly summarize the results of theorems 2 and 4, a nested linear graph G_N with order S has a flat subgraph G_F with order F satisfying $S = O(F \log F)$. Rewriting, we have $F = \Omega(S/\log F) = \Omega(S/\log S)$ since $F \leq S$. The flat graph G_F in turn has a level subgraph G_L with order L satisfying $F = O(L \log L)$, which can be similarly rewritten as $L = \Omega(F/\log S)$ (since $L \leq S$). Combining these equations yields $L = \Omega(S/\log^2 S)$.

Since the largest level flat subgraph of G_N has order $L = \Omega(S/\log^2 S)$, and the optimal common level subgraph MAX-LLS has by definition order $\geq L$, we have thus shown that MAX-LLS approximates MAX-NLS within a factor of at most $O(\log^2 S)$. We now present an algorithm to compute MAX-LLS in polynomial time.

The main idea of the algorithm is to efficiently search the space of level subgraphs for the one with the largest order. For an input instance I consisting of k linear graphs $G_1, \ldots, G_k$, let $\ell_I = \min_{i=1}^{k} length(G_i)$, and $h_I = \min_{i=1}^{k} height(G_i)$; these will be computed in the course of the algorithm. We now demonstrate how to find, for any $h \leq h_{G_i}$, the largest level (h, ℓ) which is a subgraph of G_i.

For any edge $e = \widehat{x_i x_j}$ where $x_i < x_j$, we compute a subset $S(e)$ of the edges containing e. Each edge in $S(e)$ is indexed by its left coordinate. Iterating through all edges of G_i, we only an add edge $e' = \widehat{x_{i'} x_{j'}}$ if $i' < i$ and $j' > j$. If $S(e)$ already contains an edge e_c with left coordinate $x_{i'}$, we will only keep whichever of e' and e_c has a smaller right coordinate. This ensures that we only keep the smallest edge containing e for each left coordinate. Thus, $S(e)$ will have size $O(n)$ for every edge.

Using $S(e)$ we can compute the height of every edge in the graph (the height of an edge e is the height of the tallest column where that e is the top edge). We simply iterate the following procedure until convergence:

1. $h_{old}(e) \leftarrow 1$ for every $e \in G_i$
2. **until convergence do**
3. $h_{new}(e) \leftarrow h_{old}(e)$ **foreach** e
4.   **foreach** $e \in G_i$ **do**
5.   **foreach** $e' \in S(e)$ **do**
6. $h_{new}(e') \leftarrow 1 + h_{old}(e)$ **if** $h_{new}(e') < 1 + h_{old}$
7. $h_{old}(e) \leftarrow h_{new}(e)$ **foreach** e

This variation of the Floyd-Warshall algorithm [7] computes all-pairs longest paths in a directed acyclic graph. The nodes of this DAG are the edges of our linear graph, and the edge $e \rightarrow e'$ is present in the DAG if and only if $e' \in S(e)$. After $h_{G_i} + 1$ iterations the algorithm converges, and all edges are labeled with their height. We call this procedure *vertical labeling*.

Similarly, we compute $R(e)$, a subset of edges that lie to the right of e. We only add an edge if its left coordinate $x_{i'} > x_j$, and we only keep one such edge per left coodinate, the one with the smaller right coordinate, ensuring $|R(e)| = O(n)$ for any edge e. Using the same iteration scheme as in phase 1, but with $R(e)$ instead of $S(e)$, we obtain a labeling of edges according to the length of the largest flat sequence of non-intersecting edges ending at the given edge. The largest label in the graph will have value ℓ_{G_i}, the length of the graph.

We generalize this approach to produce the largest level subgraph of height h. Starting with the labeling of the edges by height obtained during the vertical labeling phase, we compute $R_h(e)$, which is the same as $R(e)$ in the subset of G_i that has height $\geq h$. In other words, we disregard all edges of height $< h$, and label the remaining edges with what is essentially "length at height h" (see fig. 11). The largest level subgraph of G_i with height h will be $(h, F(h))$, where

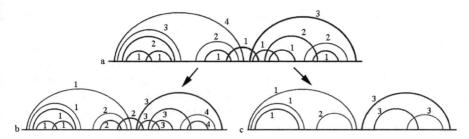

Fig. 11. The algorithm for finding the MAX-LLS of a linear graph. First, all edges in the graph are marked with their height (see [a]), using the vertical labeling iteration scheme. Next, for each h, all edges of height $< h$ are ignored, and the remaining edges are marked using horizontal iteration. For $h = 1$ and $h = 2$ the results are shown in [b] and [c] respectively.

$F(h)$ is the largest label in the graph obtained in this manner. We call this *horizontal labeling*.

Using this procedure, we can now find MAX-LLS for an instance I as follows:

1. Label the edges of each graph G_i according to height using the iteration in the vertical direction.
2. Let $h_I = min_{G \in I} h_G$.
3. For $h = 1, \ldots, h_I$ and each $G \in I$, compute the length $F_G(h)$ of the largest level subgraph of G with height h, using the iteration in the horizational direction. For each h, let $\ell_h = min_{G \in I} F_G(h)$. The level graph (h, ℓ_h) is the largest level common subgraph for the instance I of height h.
4. While iterating from $h = 1$ to $h = h_I$, keep track of the largest level subgraph (h, ℓ_h) produced in the previous step. Return this subgraph.

Suppose the k linear graphs in the input I each have size $\leq n$. Each iteration of both horizontal and vertical labeling takes $O(ne)$, as $O(e)$ edges must be updated, each update taking $O(n)$ time due to the size of $S(e)$ or $R_h(e)$. Vertical labeling takes overall time $O(hne)$, as it takes $O(h)$ iterations to converge. Horizontal labeling takes $O(\ell_h)$ iterations to converge, and must be performed for every h. Both types of labeling must be done for each of the k linear graphs. Thus, the overall running time, dominated by horizontal iteration, is $O(kh\ell ne) = O(kn^5)$.

7   Conclusions

We have introduced a novel computational model for RNA multiple structural alignment, by representing each RNA sequence as a linear graph and the multiple alignment as a common nested subgraph. We noted earlier that the MAX-NLS problem represents a relaxation of RNA multiple structural alignment, because a linear graph derived from an RNA sequence by connecting *all* complementary nucleotide pairs has certain constraints dictating which edges must exist.

There are sound biological and computational reasons to adopt the more general NLG model. At times the complementarity of nucleotides is not sufficient for the formation of a stable hydrogen bond. For instance, adjacent complementary nucleotides are rarely paired in real structures, because geometric constraints prevent them from achieving an orientation that allows participation in hydrogen bonding. It is therefore common to explicitly prevent the structure from pairing such nucleotides (or more generally, nucleotides that are less than some fixed number of bases apart) by modifying the algorithm used to compute it. In the NLG model, this can be accomplished simply by not adding such edges to the linear graphs constructed from each sequence. In general, the NLG model is flexible enough to allow easy incorporation of biological insights that modify the space of permissible pairings. Insights that reduce this space are particularly valuable because by decreasing the number of edges in the resulting linear graphs, the running time of our approximation algorithm improves accordingly. In addition, heuristic approaches to prune certain edges, which are deemed unlikely to be included in the final structure, could be combined with our algorithm in order to reduce running time further. Such enhancements are likely to be incorporated into any practical algorithm that finds biologically meaningful structures.

The approximation quality, while bounded by $O(\log^2 S)$ in the worst case, will vary depending on the class of ncRNAs being aligned. When mapped back to the RNA sequence, a level graph consists of ℓ groups of stems, each consisting of h complementary pairs. Thus, for ncRNA families whose secondary structure fits this pattern well, such as tRNAs, our algorithm will perform more accurately.

Compared to the elaborate free energy functions used by several structure-prediction programs [18, 19], the NLG model uses a fairly rough approximation. The main advantage of the NLG model is the ability to incorporate multiple sequence information without having a fixed alignment. The approximation algorithm we presented could be used to obtain a rough alignment and structure, which could then be refined using heuristic methods with more elaborate scoring models. Such a hybrid would combine theoretical bounds on approximation quality derived in the NLG framework with the benefits of heuristic approaches.

Acknowledgements

The authors were supported in part by NSF grant EF-0312459. ED was also supported by NIH-NIGMS training grant GM063495. Special thanks to Marina Sirota for assistance with manuscript preparation and editing.

References

1. Bafna V, Huson DH. The conserved exon method for gene finding. *Proceedings of the Fifth International Conference on Intelligent Systems for Molecular Biology* 3-12, 2000.
2. Batzoglou S, Pachter L, Mesirov JP, Berger B, Lander ES. Human and mouse gene structure: Comparative analysis and application to exon prediction. *Genome Research* **10**:950-958, 2000.

3. Bonizzoni P, Vedova GD. The complexity of multiple sequence alignment with SP-score that is a metric. *Theoretical Computer Science* **259**:63-79, 2001.

4. Cook SA. The complexity of theorem-proving procedures. *Proceedings of the Third ACM Symposium on Theory of Computing* 151-158, 1971.

5. Eddy SR. Noncoding RNA genes and the modern RNA world. *Nature Review Genetics* **2**:919-929, 2001.

6. Eddy SR. Computational genomics of noncoding RNA genes. *Cell* **109**:137-140, 2002.

7. Floyd RW. Algorithm 97: Shortest path. *Comm. ACM* **5**:345, 1962.

8. Hardison RC, Oeltjen J, Miller W. Long Human-Mouse Sequence Alignments Reveal Novel Regulatory Elements: A Reason to Sequence the Mouse Genome. *Genome Research* **7**:959-966, 1997.

9. Holmes I, Rubin GM. Pairwise RNA Structure Comparison with Stoachastic Context-Free Grammars. *Pacific Symposium on Biocomputing* **7**:175-186.

10. Jareborg N, Birney E, Durbin R. Comparative analysis of noncoding regions of 77 orthologous mouse and human gene pairs. *Genome Research* **9**:815-824, 1999.

11. Kasami T. An efficient recognition and syntax algorithm for context-free languages. Technical Report AF-CRL-65-758, Air Force Cambridge Research Laboratory, Bedford, MA, 1965.

12. Lowe TM, Eddy SR. tRNAscan-SE: a Program For Improved Detection of Transfer RNA genes in Genomic Sequence. *Nucleic Acids Research* **25**:955-964, 1997.

13. Nussinov R, Pieczenik G, Griggs JR, Kleitman DJ. Algorithms for loop matching. *SIAM Journal of Applied Mathematics* **35**:68-82, 1978

14. Pennacchio L, Rubin E. Genomic strategies to identify mammalian regulatory sequences. *Nature Reviews* **2**:100-109, 2001.

15. Rivas E, Eddy SR. A dynamic programming algorithm for RNA structure prediction including pseudoknots. *Journal of Molecular Biology* **285**:2053-2068, 1999.

16. Rivas E, Eddy SR. Secondary structure alone is generally not statistically significant for the detection of noncoding RNAs. *Bioinformatics* **16**:573-583, 2000.

17. Wang L, Jiang T. On the complexity of multiple sequence alignment. *Journal of Computational Biology* **1**:337-348, 1994.

18. Zuker M, Stiegler P. Optimal computer folding of large RNA sequences using thermodynamics and auxiliary information. *Nucleic Acids Research* **9**:133-148, 1981.

19. Zuker M. Computer Prediction of RNA structure. *Methods in Enzymology* **180**:262-288, 1989.

Computational Design
of New and Recombinant Selenoproteins

Rolf Backofen and Anke Busch

Friedrich-Schiller-Universitaet Jena
Institute of Computer Science, Chair for Bioinformatics
Ernst-Abbe-Platz 1-2, D-07743 Jena, Germany
{backofen,busch}@inf.uni-jena.de

Abstract. Selenoproteins contain the 21th amino acid Selenocysteine, which is encoded by the STOP-codon UGA. For its insertion it requires a specific mRNA sequence downstream the UGA-codon that forms a hairpin-like structure (called Sec insertion sequence (SECIS)). Selenoproteins have gained much interest recently since they are very important for human health.

In contrast, very little is known about selenoproteins. For example, there is only one solved crystal structure available. One reason for this is that one is not able to produce enough amount of selenoproteins by using recombinant expression in a system like *E. coli*. The reason is that the insertion mechanisms are different between *E. coli* and eukaryotes. Thus, one has to redesign the human/mammalian selenoprotein for the expression in *E. coli*. In this paper, we introduce an polynomial-time algorithm for solving the computational problem involved in this design, and we present results for known selenoproteins.

Keywords: Selenocysteine, SECIS, Protein Design

1 Introduction

Selenocysteine (Sec) is a rare amino acid, which was discovered as the 21st amino acid [6] about a decade ago. Proteins containing selenocysteine are called *selenoproteins*. Selenocysteine is encoded by the UGA-codon, which is usually a STOP-codon. It has been shown [6] that, in the case of selenocysteine, termination of translation is inhibited in the presence of a specific mRNA sequence in the 3'-region after the UGA-codon that forms a hairpin-like structure (called **Sec** insertion sequence (SECIS), Figure 1).

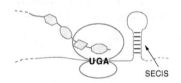

Fig. 1. Translation of mRNA requires a SECIS-element in case of selenocysteine.

Selenoproteins have gained much interest recently since selenoproteins are of fundamental importance to human health. They are an essential component of several major metabolic pathways, including such important ones as the antioxidant defence systems, the thyroid hormone metabolism, and the immune func-

S.C. Sahinalp et al. (Eds.): CPM 2004, LNCS 3109, pp. 270–284, 2004.
© Springer-Verlag Berlin Heidelberg 2004

tion (for overviews see e.g. [7]). Just recently, the complete mammalian seleno-proteome was determined [11]. For this reason, there is an enormous interest in the catalytic properties of selenoproteins, especially since selenocysteine is more reactive than its counterpart cysteine. Results of the research have shown that a selenocysteine containing protein has greatly enhanced enzymatic activites compared to the cysteine homologues [8]. The general way of investigating properties of proteins is to provide 1.) large amounts of pure proteins, and 2.) generating variants by site directed mutagenesis. In both cases, one has to use recombinant protein expression systems with *E.coli* being the simplest one to handle. However, there are differences between the mechanisms for inserting selenocysteine in eukaryotes and bacteria. In eukaryotes, the SECIS-element is located in the 3' UTR of the mRNA with a distance from the UGA-codon that varies from 500 to 5300 nucleotides [13]. In bacteria, the situation is quite different. We consider the case of *E.coli*, where the mechanism of selenocysteine insertion is well understood and all corresponding factors are identified [15]. The SECIS-element is located immediately downstream the UGA-codon [17], which implies that the SECIS-element is in the coding part of the protein. A displacement of the SECIS-element by more than one codon or a displacement not preserving the reading-frame results in a drastic reduction of selenocysteine insertion efficiency [9].

Therefore, for selenoproteins, recombinant expression is complicated and fails very often [16]. To our knowledge, there is only one solved structure of a selenoprotein: glutathione peroxidase (PDB 1GP1). Besides, Sandalova et al. [14] determined the three-dimensional structure of a Cys-mutant of the rat selenoprotein thioredoxin reductase (TrxR) which is expected to fold similar to the Sec-variant. Furthermore, there are only two cases of successive heterologous expression [8,1], which both required a careful, hand-crafted design of the nucleotide sequence. Thus, we have the following implications. First, an eukaroytic selenoprotein cannot directly be expressed in the *E.coli* system. Second, the expression of an eukaroytic selenoprotein requires the design of an appropriate SECIS-element directly after the UGA-position. Third, this design always changes the protein sequence. Therefore, one has to make a compromise between changes in the protein sequence and the efficiency of selenocysteine insertion (i.e. the quality of the SECIS-element).

The corresponding bioinformatic problem is to search for similar proteins under sequential and structural constraints imposed on the mRNA by the SECIS-elements. A first positive solution to this problem was given in [2]. This algorithm considers a fixed SECIS-element without any insertions/deletions. In [3], it was shown that the problem becomes NP-hard if more complicated structures (i.e. pseudoknots) are considered and an approximation algorithm was given for this case.

In this paper, we consider the extension of [2] where we allow insertions and deletions in the amino acid sequence. The reason is that, albeit the SECIS-element is fixed in the mRNA, it is possible to place it on different parts of the amino acid sequence via insertions and deletions. Furthermore, we can gain information about possible insertions/deletions from well known alignments of the corresponding protein sequence (e.g. from seed alignments of the Pfam-database [5]) and from additional biological knowledge that restricts insertions/deletions to certain positions. As we will explain in the next section, in this problem insertions and deletions are more complicated than in the usual alignment problems, since an insertion or deletion changes the mapping between the mRNA and the protein sequence. In addition, we introduce a second optimization criteria: optional bonds. The reason for this extension is that in the SECIS-element, only some part of the structure is fixed. Other elements (like the lower part of the hairpin stem) are not really required, albeit they improve the quality of the SECIS-elements. This is captured by the concept of optional bonds.

2   The Computational Problem

As input, we have the consensus sequence of the SECIS-element $S = S_1...S_{3n}$ and the original amino acid sequence $A = A_1...A_n$ from the 3'-region of the position where we wish to insert selenocysteine. The process of inserting a SECIS-element poses the problem of finding an appropriate mRNA sequence $N = N_1...N_{3n}$. This mRNA sequence must contain a SECIS-element (sequence and structure) at the right position. This implies that we do not have to measure only the similarity on the mRNA sequence but also to guarantee that the SECIS structure can be formed. Thus, the corresponding nucleotides in N must be complementary bases. In addition, we also need to require that the encoded amino acid sequence $A' = A'_1...A'_n$ has maximum similarity with A. Thus, we have the following picture for the similarity without insertions and deletions:

$$
\begin{array}{ccccc}
A = & A_1 & ... & A_i & ... & A_n \\
 & \sim & & \sim & & \sim \\
A' = & A'_1 & ... & A'_i & ... & A'_n \\
N = & \overbrace{N_1 N_2 N_3} & ... & \overbrace{N_{3i-2} N_{3i-1} N_{3i}} & ... & \overbrace{N_{3n-2} N_{3n-1} N_{3n}} \\
 & \sim \; \sim \; \sim & & \sim \quad \sim \quad \sim & & \sim \quad \sim \quad \sim \\
S = & S_1 S_2 S_3 & ... & S_{3i-2} S_{3i-1} S_{3i} & ... & S_{3n-2} S_{3n-1} S_{3n}
\end{array}
$$

$N = N_1...N_{3n}$ is the mRNA sequence to be searched for. $\sim$ indicates the similarity on both the amino acid $(A \sim A')$ and the nucleotide level $(N \sim S)$. This similarity will be measured by functions $F_{A_i}^{S_{3i-2} S_{3i-1} S_{3i}}(N_{3i-2} N_{3i-1} N_{3i})$. These functions also consider conditions on the nucleotide and the amino acid level: Some positions of the SECIS-consensus sequence S as well as some positions of the amino acid sequence A are highly conserved and ensure the biological function of the SECIS-element and the protein. They must not be changed. Therefore, we restrict our problem by nucleotide and amino acid conditions which penalize changes at conserved positions. Of course, a problem arises if the nucleotide and amino acid conditions are contradictory. In this paper we show how such contractictions can be solved. We will consider two extensions to the original problem:

1.) Insertions and Deletions. In the original problem there is a direct mapping from nucleotide positions to amino acid positions, where the codon

$N_{3i-2}N_{3i-1}N_{3i}$ on the nucleotide level correspond to the i-th position on the amino acid level. This changes if one considers insertions and deletions additionally. Since the SECIS-element is relatively fixed, we consider only insertions and deletions on the amino acid level.

Now consider fixed $N_{3i-2}N_{3i-1}N_{3i}$ corresponding to the i-th amino acid. Suppose that we have no insertions and deletions so far. An *insertion* at this position means that we insert a new amino acid A_i', which does not have a counterpart in A. This implies that 1.) only the similarity between $N_{3i-2}N_{3i-1}N_{3i}$ and the corresponding part of the SECIS-consensus $S_{3i-2}S_{3i-1}S_{3i}$ is measured, 2.) a gap cost for the insertion is added, and 3.) the mapping from the nucleotide sequence to the original amino acid sequence is changed. Therefore, we have to compare $N_{3j-2}N_{3j-1}N_{3j}$ with the $j-1$-th amino acid (for $j > i$). Furthermore, this means that we can have at most one *insertion* per codon position in the SECIS-element.

In comparison, a *deletion* implies that we compare the codon $N_{3i-2}N_{3i-1}N_{3i}$ with A_{i+1} (thus skipping A_i), and there could be several deletions per SECIS-codon site. Since only few insertions and deletions are allowed, and since we are restricted by one insertion per SECIS-codon site by the nature of the problem, we confine ourselves to one deletion per SECIS-codon site.

By the given restriction, it is easier to skip the alignment notation and to represent insertions and deletions directly by a vector t with $t_i \in \{-1, 0, 1\}$ that shows which operation is applied on the i-th SECIS-codon site. Here, -1 indicates a deletion, $+1$ an insertion and 0 a substition. Furthermore, these values determine the offset that has to be added in order to find the mapping between A_i' and the corresponding position in the original amino acid sequence. An example is given in Figure 2a. Here, we have $t_1 = 0$, which implies that we have a substitution and compare A_1' with A_1. For position $i = 2$, we have a deletion indicated by $t_2 = -1$. Hence, we have to compare A_2' with A_3. Thus, position j of the original amino acid has to be compared with the amino acid A_i' in the newly generated amino acid sequence. In this case j can be calculated by $j = i - \sum_{k \leq i} t_k$. A further consequence of the restriction is that each deletion must be followed by a substitution. Figure 2a shows a correct deletion followed by a substitution at SECIS-position 2 whereas Figure 2b exemplifies the incorrect case of two successive deletions at SECIS-positions 2 and 3[1].

Therefore, we have to consider a modified similarity function, where the changes in the index are taken into account. Thus, we introduce $f_i(L_i, a_i, t_i)$, where

L_i represents the *codon* corresponding to the nucleotides $N_{3i-2}N_{3i-1}N_{3i}$

t_i as defined above

a_i is the difference between the number of insertions and the number of deletions up to position i, i.e. $a_i = \sum_{j=1}^{i} t_j$, which reflects the relative displacement of the old and the new amino acid sequence to each other.

---

[1] To allow for successive deletions (as indicated in Figure 2b), one must allow two or more deletions per SECIS-codon site within the *basic case* (see equation (3)).

a) $A:$	A_1	A_2	A_3	A_4	$-$	$-$	A_5
$A':$	A'_1	$-$	A'_2	A'_3	A'_4	A'_5	A'_6
$N:$	$N_1N_2N_3$	$---$	$N_4N_5N_6$	$N_7N_8N_9$	$N_{10}N_{11}N_{12}$	$N_{13}N_{14}N_{15}$	$N_{16}N_{17}N_{18}$
$S:$	$S_1S_2S_3$	$---$	$S_4S_5S_6$	$S_7S_8S_9$	$S_{10}S_{11}S_{12}$	$S_{13}S_{14}S_{15}$	$S_{16}S_{17}S_{18}$
$t:$	0	-1		0	$+1$	$+1$	0
SECpos	1	2		3	4	5	6
b) $A:$	A_1	A_2	A_3	A_4	$-$	$-$	A_5
$A':$	A'_1	$-$	$-$	A'_2	A'_3	A'_4	A'_5
$N:$	$N_1N_2N_3$	$---$	$---$	$N_4N_5N_6$	$N_7N_8N_9$	$N_{10}N_{11}N_{12}$	$N_{13}N_{14}N_{15}$
$S:$	$S_1S_2S_3$	$S_4S_5S_6$	$---$	$S_7S_8S_9$	$S_{10}S_{11}S_{12}$	$S_{13}S_{14}S_{15}$	$S_{16}S_{17}S_{18}$
$t:$	0	-1	-1	$+1$	$+1$	0	
SECpos	1	2	3	4	5	6	

Fig. 2. Allowed and disallowed deletion patterns. a) Possible solution, each deletion must be followed by a substitution, b) Not allowed solution, a deletion not followed by a substitution.

Three different cases of the new similarity function have to be taken into account:

$$f_i(L_i, a_i, 0) = F^{S_{3i-2}S_{3i-1}S_{3i}}_{A_{i-a_i}}(N_{3i-2}N_{3i-1}N_{3i})$$

$$f_i(L_i, a_i, +1) = IP + F^{S_{3i-2}S_{3i-1}S_{3i}}(N_{3i-2}N_{3i-1}N_{3i})$$

$$f_i(L_i, a_i, -1) = \begin{cases} f_i(L_i, a_i, 0) + DP & \text{if no amino acid cond. at } i - (a_i - t_i) \\ -\infty & \text{otherwise} \end{cases}$$

If an insertion is done at position i, the similarity is calculated by a combination of an *insertion-penalty* IP and the similarity on the nucleotide level. Then there is no amino acid restriction because inserting each amino acid should be allowed. If the amino acid at position i is deleted, one has to distinguish between two cases: (a) if there is no amino acid condition at position $i - a_i - 1$ (original amino acid position to be deleted), the similarity is calculated the normal way with an additional *deletion-penalty* DP, and (b) if an amino acid condition exists at position $i - a_i - 1$, deleting this amino acid is not allowed and the similarity function has to be -∞.

2.) Optional Bonds. For the second extension, we now declare some bonds as being optional. In contrast to the mandatory bonds, these optional ones are not fixed. It would be of advantage if they form but they are not necessary to ensure the function of the SECIS-element.

Due to these two improvements (allowing insertions/deletions and declaring some optional bonds) there are two different and possibly competing values to be maximized: the similarity and the number of realized optional bonds.

3   The Formal Description of the Problem

An undirected graph is said to be *outer-planar* if, when drawn, it satisfies the following conditions: 1.) all the vertices lie on a line, 2.) all the edges lie on one side of the line, and 3.) two edges that intersect in the drawing do so only at their end points. The solution to the problems we address in this paper are a vector whose values are from $\{-1, 0, 1\}$ representing deletions/insertions and a string whose letters are from $\Sigma_{RNA} = \{A, C, G, U\}$. The elements of this set are referred to as *nucleotides*, an element of $(\Sigma_{RNA})^3$ is called *codon*. We allow standard Watson-Crick bonds (*A-U*, *C-G*) and in some cases, we consider G and U as complements to each other, too. The complement set of a variable is denoted by the superscript C.

Input: Table 1 and Figure 3 show the input to our problem. We have an egde-labeled graph $G = (V, E, Lab)$ on $3n$ vertices with $V = \{v_1, ..., v_{3n}\}$. For every edge $\{v_k, v_l\} \in E$, the label $Lab(v_k, v_l)$ is taken from the set $\{-2, -1, 1, 2, 3, 4\}$. Note that the la-

Table 1. Edge labels.

$Lab(v_k, v_l)$	Condition	Meaning
+1	$N_k \in N_l^C$	bond
+2	$N_k \in \Sigma_{RNA}$	optional bond
+3	$N_k \in N_l^{C(G-U)}$	G-U bond
+4	$N_k \in \Sigma_{RNA}$	opt. G-U bond
-1	$N_k \notin N_l^C$	prohibited bond
-2	$N_k \in \Sigma_{RNA}$	opt. prohib. bond

beling of the egdes is chosen arbitrarily. We decided to label prohibited bonds with negative values and favoured bonds with positive ones. $Lab(v_k, v_l)$ determines the bonding between nucleotides at positions k and l in the mRNA according to Table 1. The only label that requires a bit of explanation is $Lab(v_k, v_l) = -2$ (optionally prohibited bond). This implies that it would be advantageous if there was no bond. Prohibited bonds are necessary since the SECIS-element needs some bulged nucleotides. In the following we assume that $\{v_k, v_l\} \in E(G)$ and $Lab(v_k, v_l) = -1$ imply that v_k and v_l are not part of a bond. Figure 3 shows a small example.

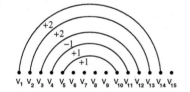

N7 N8
N6   N9
N5—N10
N4—N11
N3   N12
N2—N13
N1—N14
N15

N7 N8
N6   N9
N5—N10
N4—N11
N3   N12
N2—N13
N1   N14
N15

N7 N8
N6   N9
N5—N10
N4—N11
N3   N12
N2   N13
N1—N14
N15

N7 N8
N6   N9
N5—N10
N4—N11
N3   N12
N2   N13
N1
N14
N15

Fig. 3. Graph example and possible structures.

The number of all possible optional bonds, optional G-U bonds and optionally prohibited bonds is denoted as $maxOp$. Furthermore n functions $f_1, ..., f_n$ and conditions on the nucleotide and amino acid level are also part of the input. The set of functions f_i is named f and the set of conditions is denoted as C. f_i is associated with $\{v_{3i-2}, v_{3i-1}, v_{3i}\}$, $1 \leq i \leq n$, see above.

Output and Kinds of Bonds: Find vectors $t = (t_1, ..., t_n) \in \{-1, 0, 1\}^n$ and $N = (N_1, .., N_{3n}) \in \{A, C, G, U\}^{3n}$, so that N_k is assigned to v_k, t represents the succession of insertions and deletions and

1. all egdes $\{v_k, v_l\} \in E(G)$ are labeled according to Table 1
2. depending on the optimization criteria, $\sum_{i=1}^{n} f_i(N_{3i-2}N_{3i-1}N_{3i}, a_i, t_i) = \sum_{i=1}^{n} f_i(L_i, a_i, t_i)$ (referred to as the *similarity* of the assignment) is maximized (a) over all assignments or (b) among all assignments having the maximal number of realized optional bonds.

We denote the problems by $MRSODI_S(G, f, C)$ (2a) and $MRSODI_B(G, f, C)$ (2b). MRSODI cuts *mRNA Structure Optimization allowing Deletions and Insertions* short. In the following we refer to condition 1 as the *complementary condition*.

Additional Notations: G is referred to as the *structure graph* on the mRNA level. Accordingly, we need such a graph on the amino acid level as well. Given an mRNA structure graph G, we define the *implied graph* G^{impl} as a graph on the vertices $V(G^{impl}) = \{u_1, ..., u_n\}$ with

$$E(G^{impl}) = \left\{ \{u_i, u_j\} \middle| \begin{array}{l} \exists r \in \{3i-2, 3i-1, 3i\} : \\ \exists s \in \{3j-2, 3j-1, 3j\} : \{v_r, v_s\} \in E(G) \end{array} \right\}.$$

Note that in the implied graph every node has at most degree 3 if the nodes of the input graph G have at most degree 1. (This holds in our problem since we have at most one prohibited bond per node.) Hence, up to 3 edges can emerge from a node in the implied graph. We follow the convention that, independent of the subscript or superscript, u denotes a vertex from $V(G^{impl})$ and v denotes a vertex from $V(G)$. Given $J \subseteq \{1...n\}$, U_J denotes $\{u_j | j \in J\}$. $E(G^{impl})|_J$ is the set of egdes of the induced subgraph of G^{impl} on U_J and $E(G)||_J$ denotes the edge set of the induced subgraph of G on $\{v_{3j-2}, v_{3j-1}, v_{3j} | j \in J\}$.

A codon sequence $L_1...L_n$ is said to *satisfy* $E(G)$ iff the corresponding nucleotides $\{N_{3i-2}N_{3i-1}N_{3i}\}_{i=1}^n$ satisfies the complementary conditions for G. L_i and L_j are said to be *valid for* $\{u_i, u_j\}$ w.r.t. $E(G)$ if the corresponding nucleotides $N_{3i-2}N_{3i-1}N_{3i}$ and $N_{3j-2}N_{3j-1}N_{3j}$ satisfy the complementary conditions imposed by $E(G)$.

4   Algorithm for SECIS-Like Structures Allowing Insertions, Deletions and Optional Bonds

We present a polynomial time recursive algorithm that solves $MRSODI_S$ and $MRSODI_B$ when G^{impl} is outer-planar *and* every node in G has at most degree one. The hairpin shape of the SECIS-element is captured by the outer-planarity of G^{impl}. Our algorithm is based on a recurrence relation that we introduce in this section.

We fix $u_1, ..., u_n$ as the ordering of the vertices from left to right on a line in an outer-planar embedding of G^{impl}. Let f_i be the function associated with u_i. Having fixed an embedding of the graph on a line, we do not distinguish between a vertex and its index. That is, the interval $[i...i+k]$ denotes the set of vertices $\{u_i, ..., u_{i+k}\}$.

We now define the notion of compatibility between codons L_i and L_j assigned to vertices u_i and u_j, respectively. The compatibility with respect to the complementary conditions in $E(G)$ is denoted by $\equiv_{E(G)}$. Thus, for $1 \le i, j \le n$, we define $(i, L_i) \equiv_{E(G)} (j, L_j)$ by

$$
(i, L_i) \equiv_{E(G)} (j, L_j) =
\begin{cases}
\text{true} & \text{if } \{u_i, u_j\} \notin E(G^{impl}) \\
\text{true} & \text{if } \{u_i, u_j\} \in E(G^{impl}) \text{ and} \\
& L_i \text{ and } L_j \text{ are } valid \text{ for } \{u_i, u_j\} \text{ w.r.t. } E(G) \\
\text{false} & \text{otherwise}
\end{cases}
$$

Some additional variables have to be defined: The number of realized optional bonds, optional G-U bonds and optionally prohibited bonds in $E(G)\|_{[i...i+k]}$ (in short: $ob(L_i...L_{i+k})$) is named m, where $m \leq maxOp$. The difference of insertions and deletions up to position i (at all pos. $\leq i$) is denoted as l ('left'), whereas s ('inside') corresponds to the difference of insertions and deletions at and between positions $i+1$ and $i+k$ ($i+1 \leq$ pos. $\leq i+k$). Furthermore, s^I is the number of insertions at and between positions $i+1$ and $i+k$, accordingly s^D is the number of deletions. As mentioned above only one insertion or deletion per position is allowed.

Therefore, it is necessary that $|l| \leq i$ and $|s| \leq k$. If these two inequations are fulfilled, (l, s) are said to be *VALID* for $\{i, k\}$. Let $SPLIT_k^s$ be a set of tupels (s^I, s^D) which fulfil

$$
s^I - s^D = s, \qquad s^I + s^D \leq k, \quad \text{and} \quad 0 \leq s^I, s^D. \tag{1}
$$

Let $V(a, b, L)$ be the set of vectors with length L having a coordinates with value $+1$, b with value -1 and $L - (a + b)$ coordinates being 0. It is necessary that $a + b \leq L$.

We now define the central function which will be implemented by dynamic programming. $w_{i+k}^i(L_i, L_{i+k}, m, l, s)$ gives the maximal similarity of the designed sequence to the original one between positions i and $i + k$ where L_i is at position i and L_{i+k} at position $i + k$ of the new sequence. The number of realized optional and optionally prohibited bonds in the interval $[i...i + k]$ is m. l and s are defined as given above. l is used to find the relative position of the original sequence to the new one resulting from possible insertions and deletions before and at i which is important to the similarity function[2]. Formally this means that $w_{i+k}^i(L_i, L_{i+k}, m, l, s)$ equals

$$
\max_{\substack{L_{i+1}...L_{i+k-1} \\ t \in V(s^I, s^D, k) \\ (s^I, s^D) \in SPLIT_k^s}}
\left\{
\sum_{i<j\leq i+k} f_j\left(L_j, l + \sum_{a=1}^{j-i} t_a, t_{j-i}\right)
\left|
\begin{array}{l}
L_i...L_{i+k} \text{ satisfies} \\
E(G)\|_{[i...i+k]} \text{ and} \\
ob(L_i...L_{i+k}) = m
\end{array}
\right.
\right\} \tag{2}
$$

If the set over which the maxima are taken is empty, the value of the function is considered as $-\infty$. This function forms the central part of our algorithm because at first w treats both objective functions separately (maximizing the similarity and maximizing the number of realized optional bonds and optionally prohibited bonds) and one can combine them afterwards, as follows, to get the value of interest:

---

[2] **Notice:** In contrast to position $i + k$, the similarity at position i is not included here.

1. $MRSODI_S$: Maximize the similarity overall and among all assignment having this maximal similarity, choose the one with the highest number of realized optional and optionally prohibited bonds. The value of interest is $\max_{(L_1,L_n,m)} \max_{|l|\leq 1,|s|\leq n-1} \{w_n^1(L_1,L_n,m,l,s) + f_1(L_1,l,l)\}$.

2. $MRSODI_B$: Maximize the number of realized optional bonds and optionally prohibited bonds first. Therefore, the value of interest is $\max_{(L_1,L_n)} \max_{|l|\leq 1,|s|\leq n-1} \{w_n^1(L_1,L_n,m_{max},l,s) + f_1(L_1,l,l)\}$,   where $m_{max} \leq maxOp$ is the maximal value which can be adopted by m without violating any conditions.

The *basic case* for w is the following: ($m = 0, ..., maxOp$, $|l| \leq i$, $|s| \leq 1$)

$$w_{i+1}^i(L_i, L_{i+1}, m, l, s) = \begin{cases} f_{i+1}(L_{i+1}, l+s, s) & \text{if } (i, L_i) \equiv_{E(G)} (i+1, L_{i+1}), \\ & \text{and } (l, s) \text{ are VALID for } \{i, 1\}, \\ & \text{and } ob(L_iL_{i+1}) = m \\ -\infty & \text{otherwise} \end{cases}$$

(3)

It shows that at most one insertion or deletion per position is allowed. This is an algorithmic reason for inequations $|l| \leq i$ and $|s| \leq k$ mentioned above. Nevertheless, it is not possible to prevent two or more insertions one after another straightforward. The insertion-penalty has to be chosen appropriately.

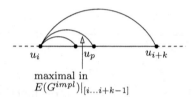

u_i u_p u_{i+k}

maximal in
$E(G^{impl})|_{[i...i+k-1]}$

Fig. 4. Definition of $next(i, i+k)$.

We solve the problem on an interval $[i...i + k]$ by splitting it into two parts and solving the resulting subproblems. If there is no edge between i and any vertex in $[i...i + k - 1]$, the interval is split into $[i...i + 1]$ and $[i + 1...i + k]$. Otherwise, we choose the farthest vertex p in $[i...i + k - 1]$ which is adjacent to i. The edge $\{i, p\}$ is called the *maximal* edge in $E(G^{impl})|_{[i...i+k-1]}$. Then, the interval is split into $[i...p]$ and $[p...i + k]$ (see Figure 4). Hence, we define $next(i, i + k)$ for $k \geq 2$ by

$$next(i, i+k) = \begin{cases} i+r & \text{if } \{i, i+r\} \text{ is maximal in } E(G^{impl})|_{[i...i+k-1]} \\ i+1 & \text{otherwise.} \end{cases}$$

Theorem 1 (Recurrence). *Let* (G, f, C) *be an instance of* $MRSODI_S$ *or* $MRSODI_B$. *For* $k \geq 2$, $1 \leq i \leq n - k$, (l, s) *are VALID for* $\{k\}$, $0 \leq m \leq maxOp$, *let* $p = next(i, i + k)$. *Then* $w_{i+k}^i(L_i, L_{i+k}, m, l, s)$ *equals*

$$\begin{cases} -\infty & \text{if } (i, L_i) \not\equiv_{E(G)} (i+k, L_{i+k}) \\ \max\limits_{\substack{L_p \\ m_1+m_2=m_{L_{i+k}}^{L_i} \\ m_1,m_2 \geq 0}} \max\limits_{\substack{s_1+s_2=s; \\ |s_1| \leq p-i; \\ |s_2| \leq i+k-p}} \begin{pmatrix} w_p^i(L_i, L_p, m_1, l, s_1) + \\ w_{i+k}^p(L_p, L_{i+k}, m_2, l+s_1, s_2) \end{pmatrix} & \\ & \text{if } (i, L_i) \equiv_{E(G)} (i+k, L_{i+k}) \end{cases}$$

(4)

where $m_{L_{i+k}}^{L_i} = m - opt_{L_{i+k}}^{L_i}$ and $opt_{L_{i+k}}^{L_i}$ is the number of realized optional bonds only between codons L_i and L_{i+k}.

A proof of the Theorem is given in the appendix B.

Remark 1. The similarity at the left boundary of the interval $[i...i + k]$ is not included in $w_{i+k}^i(L_i, L_{i+k}, m, l, s)$. It suffices to keep the codon on this position to combine the subproblems properly.

4.1 Complexity of the Algorithm

We present the properties of the algorithm at level $[i...i + k]$ of the recursion. Recall that our goal is to compute $w_{i+k}^i(L_i, L_{i+k}, m, l, s)$ for all choices of L_i, L_{i+k}, m, l, s, corresponding assignments of codons and insertions and deletions. Therefore, $O(n^2)$ (to be exact: $64^2 * maxOp * (2i + 1) * (2k + 1)$) different cases have to be considered. Let $p = next(i, i + k)$. Since each vertex in G^{impl} has, at most, degree 3, p can be found in not more than 3 steps. If $w_p^i(L_i, L_p, m_1, l, s_1)$ and $w_{i+k}^p(L_p, L_{i+k}, m_{L_{i+k}}^{L_i} - m_1, l + s_1, s - s_1)$ are known for all choices of L_p, m_1 and s_1, we can compute $w_{i+k}^i(L_i, L_{i+k}, m, l, s)$. This computation takes at most linear time because only s_1 depends on n. Therefore, a subproblem can be solved in $O(n^3)$.

Now we have to check how many subproblems will be generated. Note that the subproblems are determined by the result of $next(.,.)$. Since G^{impl} is outer-planar, the application of $next(.,.)$ to some specific subproblem will always return a new position (a position that has not been considered in any other subproblem). Hence, we get only $n - 2$ subproblems which can be split again and take $O(n^3)$ time each. (Notice that the basic-case-subproblems

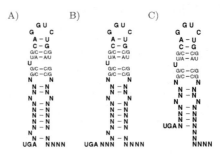

Fig. 5. SECIS-consensus sequence and structure: A) standard, B) an additional codon between UGA and the actual SECIS-element, C) standard, lacking the first codon.

need only $O(n)$ time.) Thus, we can compute our goal in $O(n^4)$. Analogous to a fixed parameter tractability, this can be reduced to $O(n^2)$ by restricting the number of insertions and deletions to a constant number.

4.2 Results

We have implemented the algorithm in C++ and used it to search for mammalian selenoproteins in which a typical SECIS-element of *E.coli* (*FdhF*) can be designed directly after the UGA-position with only a few changes within the amino acid sequence. According to [10, 12] we chose SECIS-consensus sequences and structures as shown in Figure 5.

A) mouse MsrB	... U I F S S S L K F V P K G K E ...	
new_RecSec	... U I F S S L P G L V P K G K E ...	Sim = **43**
RecSec [4]	... U I F S T V A G L H P K G K E ...	Sim = 35
new_mRNA_seq	... UGAAUUUUCUCUUCGCUACCAGGUCUGGUGCCAAAAGGAAAAGAA	
new_mRNA_struc	... . (((((. ((. ((. ((((....)))))) .)) .)))))	opt_bonds = 9
B) mouse MsrB	... U I F S S S L K - F V P K G K E ...	
new_RecSec	... U I F S S S L P G L V P K G K E ...	Sim = 57 + DP
new_mRNA_seq	... UGAAUAUUUUCCUCUUCGCUACCAGGUCUGGUGCCAAAAGGAAAAGAA	
new_mRNA_struc	... . (. (((((. ((. ((. ((((....)))))) .)) .)))))	opt_bonds = 10
C) mouse MsrB	... U I F S S S L K F V P K G K E ...	
new_RecSec	... U I F S L P - G L V P K G K E ...	Sim = 41 + IP
new_mRNA_seq	... UGAAUCUUUUCGCUACCAGGUCUGGUGCCAAAAGGUAAAGAA	
new_mRNA_struc	... ... (((((((. ((. ((((....)))))) .)))))))	opt_bonds = 7

Fig. 6. Results for MsrB, designing (A) the standard SECIS-element shown in Fig. 5A, (B) the SECIS-element of Fig. 5B, (C) the standard SECIS-element missing the first codon ($IP = -10$, $DP = -5$, changed positions are underlined) where RecSec is the sequence designed in [4], new_RecSec is the one designed by our algorithm.

The mammalian methionine sulfoxide reductase B (MsrB) has already been studied in [4]. We gained an even better sequence with a higher similarity to the original amino acid sequence (BLOSUM62-score = 43) than the one described in [4] which yields an BLOSUM62-score of 35. Our results are shown in Figure 6. Using our program, we have found modifications of several human selenoproteins that have high similarity to both the original amino acid sequence and the SECIS-element. Based on these modifications the coding mRNA is able to form the required hairpin structure. For more results see appendix A.

A   Results

		C) Q92813	... P P F T S Q L P A F R K L ...
A) P49895	... P S F M F K - F D Q F K R L ...	Q92813'	... P P F L A G L Q A F R K L ...
P49895'	... P S F M V P G L D Q F K R L ...		
		P07203	... G T T V R D Y T Q M N E L ...
Q9NZV5	... G S G R T L R E T V L E S S P ...	P07203'	... G T T V A G L L Q M N E L ...
Q9NZV5'	... G _ G R T L P G L V L E S S P ...		
		P22352	... G L T G Q Y I E L N A L Q ...
P49908₅₉	... Y L C I I E A S K L E D L R V ...	P22352'	... G L T L P G L E L N A L Q ...
P49908₅₉'	... Y L C I V - A G L L E D L R V ...		
		Q9NZJ3	... G Y R R V - F E E Y M R V ...
		Q9NZJ3'	... G Y R L P G L E E Y M R V ...
B) Q9NZV5	... G S G R T L R E T V L E S S P ...		
Q9NZV5'	... G S G R T L P G L V L E S S P ...	P59797	... S Y S L R Y I L L K K S L ...
		P59797'	... S Y S L A G L L L K K S L ...
P49908₅₉	... Y L C I I E A S K L E D L R V ...		
P49908₅₉'	... Y L C I I V A G L L E D L R V ...	O15532	... G Y K S K Y L Q L K K K L ...
		O15532'	... G Y K L A G L Q L K K K L ...
P49908₃₁₈	... Q C K E N L P S L C S C Q G L ...		
P49908₃₁₈'	... Q C K E N L P G L V A C Q G L ...	Q99611	... G C K V P Q E A L L K L L ...
		Q99611'	... G C K V P G L V L L K L L ...

Fig. 7. Results for human selenoproteins, designing (A) the standard SECIS-element shown in Figure 5A, (B) the SECIS-element of Figure 5B, (C) the SECIS-element missing the first codon when $IP = -10$ and $DP = -5$ (changed positions are underlined).

B   Proof of the Recurrence (Theorem 1)

Proof. In the first case, when $(i, L_i) \neq_{E(G)} (i + k, L_{i+k})$ the result is true as we follow the convention that the maximum over an empty set is $-\infty$. Thus, we just have to consider the second case where $(i, L_i) \equiv_{E(G)} (i+k, L_{i+k})$. Let $k \geq 2$ and the other conditions of the theorem be fulfilled. We wish to evaluate,

$$\max_{\substack{L_p \\ m_1 + m_2 = m_{L_{i+k}}^{L_i} \\ m_1, m_2 \geq 0}} \max_{\substack{s_1 + s_2 = s \\ |s_1| \leq p - i \\ |s_2| \leq i + k - p}} \left(\begin{array}{c} w_p^i(L_i, L_p, m_1, l, s_1) \\ + w_{i+k}^p(L_p, L_{i+k}, m_2, l + s_1, s_2) \end{array} \right). \tag{5}$$

where $w_p^i(L_i, L_p, m_1, l, s_1)$ and $w_{i+k}^p(L_p, L_{i+k}, m_2, l + s_1, s_2)$ are defined as given in equation (2).

Now we have to show that equation (5) equals

$$\max_{\substack{L_{i+1}...L_{i+k-1} \\ t \in V(s^I, s^D, k) \\ (s^I, s^D) \in SPLIT_k^s}} \left\{ \sum_{i<j\leq i+k} f_j\left(L_j, l + \sum_{a=1}^{j-i} t_a, t_{j-i}\right) \left| \begin{array}{c} L_i...L_{i+k} \text{ satisfies} \\ E(G)||_{[i...i+k]} \text{ and} \\ ob(L_i...L_{i+k}) = m \end{array} \right. \right\} \tag{6}$$

There are no edges between vertices in the interval $[i + 1...p - 1]$ and vertices in $[p + 1...i + k - 1]$ because we have assumed as input an outer-planar drawing of G^{impl}. Firstly, we have to prove the compatibility of the codons to the graph in correspondence with the number of realized optional bonds, optional G-U bonds and optionally prohibited bonds and secondly, we have to show that all possible vectors $\in \{-1, 0, 1\}^n$ (representing insertions and deletions) on all possible assignments of s_1^I, s_2^I, s_1^D and s_2^D are considered in both quantities.

Let $L'_{i+1}...L'_{p-1}$ be some codon sequence such that $L_i L'_{i+1}...L'_{p-1} L_p$ satisfies $E(G)||_{[i...p]}$ and $ob(L_i...L_p) = m_1$, and $L''_{p+1}...L''_{i+k-1}$ be a codon sequence such that $L_p L''_{p+1}...L''_{i+k-1} L_{i+k}$ satisfies $E(G)||_{[p...i+k]}$ and $ob(L_p...L_{i+k}) = m_2$. Now we have to show that $L_i L'_{i+1}...L'_{p-1} L_p L''_{p+1}...L''_{i+k-1} L_{i+k}$ satisfies $E(G)||_{[i...i+k]}$ and $ob(L_i...L_{i+k}) = m$. L_i and L_{i+k} are compatible since we have assumed $(i, L_i) \equiv_{E(G)} (i + k, L_{i+k})$. The number of realized optional bonds, optional G-U bonds and optionally prohibited bonds only between L_i and L_{i+k} is $opt_{L_{i+k}}^{L_i}$ (fixed for fixed L_i and L_{i+k}). The compatibility within the regions $[i...p]$ and $[p...i + k]$ holds by assumption. The compatibility of vertex pair (u_r, u_s) with $r \in [i...p - 1]$ and $s \in [p + 1...i + k]$ holds since (i, p) is a maximal edge and there are no edges in $E(G^{impl})$ (and corresponding edges and optional edges in $E(G)||_{[i...i+k]}$) between r and s. Therefore, $ob(L_i...L_{i+k})$ can be identified by adding $ob(L_i...L_p)$, $ob(L_p...L_{i+k})$ and $opt_{L_{i+k}}^{L_i}$:

$$\begin{aligned} ob(L_i...L_{i+k}) &= ob(L_i...L_p) + ob(L_p...L_{i+k}) + opt_{L_{i+k}}^{L_i} \\ &= m_1 + m_2 + opt_{L_{i+k}}^{L_i} \\ &= m_{L_{i+k}}^{L_i} + opt_{L_{i+k}}^{L_i} = m. \end{aligned}$$

The maximum over all possible values of m_1 (and $m_2 = m - opt_{L_{i+k}}^{L_i} - m_1$) ensures that every assignment of realized optional bonds etc. is considered.

Now we have to show, that all possible vectors $t \in V = V(s^I, s^D, k)$ for all possible s^I and s^D are included in the combination of all possible vectors

$t^1 \in V^1 = V^1(s_1^I, s_1^D, p - i)$ and $t^2 \in V^2 = V^2(s_2^I, s_2^D, i + k - p)$ on all possible assignments of s_1^I, s_2^I, s_1^D and s_2^D and vice versa, i.e. that

$$\bigcup_{\substack{(s^I, s^D) \\ \in SPLIT_k^s}} V = \bigcup_{\substack{s_1 + s_2 = s \\ |s_1| \leq p - i \\ |s_2| \leq i + k - p}} \left(\bigcup_{\substack{(s_1^I, s_1^D) \\ \in SPLIT_{p-i}^{s_1}}} V^1 \times \bigcup_{\substack{(s_2^I, s_2^D) \\ \in SPLIT_{i+k-p}^{s_2}}} V^2 \right)$$

The left set (let us name it L, where the right set is denoted as R) includes all vectors with valid numbers of -1 (s^D) and $+1$ (s^I). We have to show, that all vectors included in L are also included in R and vice versa.

Case $L \supseteq R$: Let $s_1 + s_2 = s$ be a valid splitting of s. Let $t^1 = t_1^1 t_2^1 ... t_{p-i}^1$ be a vector $\in V^1(s_1^I, s_1^D, p - i)$ so that $(s_1^I, s_1^D) \in SPLIT_{p-i}^{s_1}$, and $t^2 = t_1^2 t_2^2 ... t_{i+k-p}^2$ be a vector $\in V^2(s_2^I, s_2^D, i + k - p)$ so that $(s_2^I, s_2^D) \in SPLIT_{i+k-p}^{s_2}$. We have to show that $t^1 t^2 = t_1^1 t_2^1 ... t_{p-i}^1 t_1^2 t_2^2 ... t_{i+k-p}^2$ is a vector $\in \bigcup_{\substack{(s^I, s^D) \\ \in SPLIT_k^s}} V(s^I, s^D, k)$.

The vectors t^1 and t^2 have the following properties

t^1: s_1^I many $+1$ t^2: s_2^I many $+1$

               s_1^D many -1 s_2^D many -1

$(p - i) - (s_1^I + s_1^D)$ many 0 $(i + k - p) - (s_2^I + s_2^D)$ many 0

Thus, it is true that $t^1 t^2$ consists of

$$s_1^I + s_2^I \quad \text{many} \quad +1$$
$$s_1^D + s_2^D \quad \text{many} \quad -1$$
$$k - (s_1^I + s_2^I) - (s_1^D + s_2^D) \quad \text{many} \quad 0$$

It follows from this that $t^1 t^2 \in V(s_1^I + s_2^I, s_1^D + s_2^D, k) = V(s^I, s^D, k)$ with $(s_1^I + s_2^I, s_1^D + s_2^D) = (s^I, s^D) \in SPLIT_k^s$ since:

- $(s_1^I + s_2^I) - (s_1^D + s_2^D) = (s_1^I - s_1^D) + (s_2^I - s_2^D) = s_1 + s_2 = s$

- $(s_1^I + s_2^I) + (s_1^D + s_2^D) = \underbrace{(s_1^I + s_1^D)}_{\leq p-i} + \underbrace{(s_2^I + s_2^D)}_{\leq i+k-p} \leq k$

- $s_1^I, s_2^I \geq 0 \Rightarrow s^I \geq 0$ and $s_1^D, s_2^D \geq 0 \Rightarrow s^D \geq 0$

Up to now we have shown that $L \supseteq R$.

Case $L \subseteq R$: We still have to prove, that the relation holds in the other direction as well (that $L \subseteq R$).

Let $t = t_1 t_2 ... t_k$ be a vector $\in V(s^I, s^D, k)$ so that $(s^I, s^D) \in SPLIT_k^s$. Then t can be split into $t^1 = t_1 t_2 ... t_{p-i}$ and $t^2 = t_{p-i+1} t_{p-i+2} ... t_k$. The number of insertions ($+1$) and deletions (-1) in t^1 is $s_1^I \geq 0$ and $s_1^D \geq 0$ and in t^2 it is $s_2^I \geq 0$ and $s_2^D \geq 0$. Obviously, it is true that $s_1^I + s_2^I = s^I$

and $s_1^D + s_2^D = s^D$. Furthermore, it is true that $t^1 \in V^1(s_1^I, s_1^D, p - i)$ with $(s_1^I, s_1^D) \in SPLIT_{p-i}^{s_1 = s_1^I - s_1^D}$ since $s_1^I + s_1^D \le p - i$ because t^1 has only $p - i$ positions. Analogously, $t^2 \in V^2(s_2^I, s_2^D, i + k - p)$ with $(s_2^I, s_2^D) \in SPLIT_{i+k-p}^{s_2 = s_2^I - s_2^D}$ since $s_2^I + s_2^D \le i + k - p$. It can also be confirmed that the splitting of t results in vectors t^1 and t^2 which have valid values s_1 and s_2 concerning to the right side (R) since

$$- \; s_1 + s_2 = (s_1^I - s_1^D) + (s_2^I - s_2^D) = (s_1^I + s_2^I) - (s_1^D + s_2^D) = s^I - s^D = s$$

$$- \; |s_1| = |s_1^I - s_1^D| \le p - i \text{ because } 0 \le s_1^I, s_1^D \le p - i \text{ since } t^1 \text{ has only } p - i$$
positions

$$- \; |s_2| = |s_2^I - s_2^D| \le i + k - p \text{ because } 0 \le s_2^I, s_2^D \le i + k - p \text{ since } t^2 \text{ has}$$
only $i + k - p$ positions

Now we have also shown that $L \subseteq R$. From this and $L \supseteq R$ it follows that $L = R$. Therefore, each possible vector t of insertions and deletions is considered.

The last point to be mentioned is the correct calculation of the similarity. The numbers of insertions and deletions in the interval $[i + 1...p]$ result in a further relative displacement (besides l) of the original to the new amino acid sequence. This displacement is the only effect the intervals $[i + 1...p]$ and $[p + 1...i + k]$ have on each other, which has to be considered during the calculation of the similarity within $[p + 1...i + k]$. This is achieved through the similarity function $f_x = f_x(L_x, \sum_{j < x} t_j, t_x)$ taking into account all insertions and deletions made before x. Since the similarity function is additive, the similarity can be calculated separately in both intervals.

Therefore, the two quantities above are equal because the maximum for the problem in the region $[i...i + k]$ for fixed L_i, L_p, L_{i+k}, m_1, l and s_1 can be obtained by finding $w_p^i(L_i, L_p, m_1, l, s_1)$ (maximum in the region $[i...p]$) and $w_{i+k}^p(L_p, L_{i+k}, m_{L_{i+k}}^{L_i}, l + s_1, s - s_1)$ (maximum in the region $[p...i + k]$). To find the maximum in $[i...i + k]$ with only L_i and L_{i+k} fixed, we have to maximize over all possible assignments of L_p, m_1 and s_1.   □

References

1. E. S. Arner, H. Sarioglu, F. Lottspeich, A. Holmgren, and A. Bock. High-level expression in escherichia coli of selenocysteine-containing rat thioredoxin reductase utilizing gene fusions with engineered bacterial-type SECIS elements and co-expression with the selA, selB and selC genes. *Journal of Molecular Biology*, 292(5):1003–16, 1999.
2. Rolf Backofen, N. S. Narayanaswamy, and Firas Swidan. Protein similarity search under mRNA structural constraints: application to targeted selenocysteine insertion. *In Silico Biology*, 2(3):275–90, 2002.
3. Rolf Backofen, N.S. Narayanaswamy, and Firas Swidan. On the complexity of protein similarity search under mrna structure constraints. In H. Alt and A. Ferreira, editors, *Proc. of 19th International Symposium on Theoretical Aspects of Computer*

Science (STACS2002),, volume 2285 of *Lecture Notes in Computer Science*, pages 274–286, Berlin, 2002. Springer Verlag.

4. Shoshana Bar-Noy and Jackob Moskovitz. Mouse methionine sulfoxide reductase b: effect of selenocysteine incorporation on its activity and expression of the seleno-containing enzyme in bacterial and mammalian cells. *Biochem Biophys Res Commun*, 297(4):956–61, 2002.

5. A. Bateman, E. Birney, L. Cerruti, R. Durbin, L. Etwiller, S. R. Eddy, S. Griffiths-Jones, K. L. Howe, M. Marshall, and E. L. Sonnhammer. The Pfam protein families database. *Nucleic Acids Res*, 30(1):276–80, 2002.

6. A. Böck, K. Forchhammer, J. Heider, and C. Baron. Selenoprotein synthesis: an expansion of the genetic code. *Trends Biochem Sci*, 16(12):463–467, 1991.

7. K. M. Brown and J. R. Arthur. Selenium, selenoproteins and human health: a review. *Public Health Nutr*, 4(2B):593–9, 2001.

8. Stephane Hazebrouck, Luc Camoin, Zehava Faltin, Arthur Donny Strosberg, and Yuval Eshdat. Substituting selenocysteine for catalytic cysteine 41 enhances enzymatic activity of plant phospholipid hydroperoxide glutathione peroxidase expressed in *escherichia coli*. *Journal of Biological Chemistry*, 275(37):28715–28721, 2000.

9. J. Heider, C. Baron, and A. Böck. Coding from a distance: dissection of the mRNA determinants required for the incorporation of selenocysteine into protein. *EMBO J*, 11(10):3759–66, 1992.

10. A. Hüttenhofer and A. Böck. RNA structures involved in selenoprotein synthesis. In R. W. Simons and M. Grunberg-Manago, editors, *RNA Structure and Function*, pages 603–639. Cold Spring Habor Laboratory Press, Cold Spring Habor, 1998.

11. Gregory V. Kryukov, Sergi Castellano, Sergey V. Novoselov, Alexey V. Lobanov, Omid Zehtab, Roderic Guigo, and Vadim N. Gladyshev. Characterization of mammalian selenoproteomes. *Science*, 300(5624):1439–43, 2003.

12. Z. Liu, M. Reches, I. Groisman, and H. Engelberg-Kulka. The nature of the minimal 'selenocysteine insertion sequence' (SECIS) in Escherichia coli. *Nucleic Acids Research*, 26(4):896–902, 1998.

13. Susan C. Low and Marla J. Berry. Knowing when not to stop: selenocysteine incorporation in eukaryotes. *Trends in Biochemical Sciences*, 21(6):203–208, 1996.

14. T. Sandalova, L. Zhong, Y. Lindqvist, A. Holmgren, and G. Schneider. Three-dimensional structure of a mammalian thioredoxin reductase: implications for mechanism and evolution of a selenocysteine-dependent enzyme. *Proc. Natl. Acad. Sci. USA*, 98(17):9533–8, 2001.

15. G. Sawers, J. Heider, E. Zehelein, and A. Böck. Expression and operon structure of the sel genes of escherichia coli and identification of a third selenium-containing formate dehydrogenase isoenzyme. *J Bacteriol.*, 173(16):4983–93, 1991.

16. R. M. Tujebajeva, J. W. Harney, and M. J. Berry. Selenoprotein P expression, purification, and immunochemical characterization. *Journal of Biological Chemistry*, 275(9):6288–94, 2000.

17. F. Zinoni, J. Heider, and A. Böck. Features of the formate dehydrogenase mRNA necessary for decoding of the UGA codon as selenocysteine. *Proc. Natl. Acad. Sci. USA*, 87(12):4660–4, 1990.

A Combinatorial Shape Matching Algorithm for Rigid Protein Docking

Vicky Choi[1],* and Navin Goyal[2],**

[1] Department of Computer Science, Duke University
vchoi@cs.duke.edu
[2] Department of Computer Science, Rutgers University
ngoyal@cs.rutgers.edu

Abstract. The protein docking problem is to predict the structure of protein-protein complexes from the structures of individual proteins. It is believed that shape complementarity plays a dominant role in protein docking. Recently, it has been shown empirically by Bespamayt-nikh et al [4] that the shape complementarity (measured by a score function) is sufficient for the bound protein docking problem, in which proteins are taken directly from the known protein-protein complex and reassembled, treating each protein as a rigid body. In this paper, we study the shape complementarity measured by their score function from a theoretical point of view. We give a combinatorial characterization of the docked configuration achieved by the maximum score. This leads to a simple polynomial time algorithm to find such a configuration. The arrangement of spheres inspired by the combinatorial characterization plays an essential role in an efficient local search heuristic of Choi et al [7] for rigid protein docking. We also show that our general idea can be used to give simple algorithms for some point pattern matching problems in any dimension.

1 Introduction

Protein-protein docking problem is one of the current challenges in computational structural biology. In this problem, we are given two protein molecules which are known to dock with each other, the problem is to find the configuration of the docked protein-protein complex. See Figure 1 for an illustration. A recent survey can be found in Mendez et al [14]. The general problem is far from being solved as the physical chemistry of protein-protein interactions is not well understood and the conformational space is high dimensional because proteins undergo conformational changes upon binding. There are no known computationally feasible methods to perform conformational searches during docking. Hence researchers often start with *bound protein docking* problem in which proteins are taken directly from the known protein-protein complex and

---

* Supported by NSF under grant CCR-00-86013 and a BGT Postdoc Program from Duke University.
** Supported in part by NSF grant CCR-9988526 and in part by NSF Career 0315147.

S.C. Sahinalp et al. (Eds.): CPM 2004, LNCS 3109, pp. 285–296, 2004.

Fig. 1. Given two proteins shown on left, the protein docking problem is to predict the docked configuration as shown on right.

reassembled. In this case the proteins are treated as rigid bodies, i.e., we assume that they do not undergo changes upon interaction as in the real (unbound) case, limiting the dimensionality of search space to six – three for translations and three for rotations. The protein docking problem with the rigidity assumption of proteins (not necessarily taken directly from the protein-protein complex) is known as *rigid protein docking* problem. Thus bound docking is a special case of rigid docking. Efficient rigid docking algorithms are useful because most of the conformational change on docking is believed to be small (or at least this has been and continues to be the assumption of most current research) [15].

For the bound protein docking case, it is believed that geometric complementarity alone is sufficient to solve the problem [15]. This is formalized using the notion of score, i.e., the shape complementarity is measured by certain score functions and the docked configuration is assumed to correspond to the one maximizing the score function. Many different score functions have been used in the literature, e.g. Fast Fourier Transform-based [12], Geometric Hashing-based [9], with varying degrees of success in that the near correctly docked configuration is generally among some number of top score configurations (and not necessarily the highest score configuration). Recently, in Bespamaytnikh et al [4], it was shown empirically that the highest score configuration was always the near correctly docked configuration. Their score function (to be defined below; a similar score function was used by Chen et al [6] which was also shown empirically to be quite successful) approximates van der Waals force. This is consistent with the belief that the van der Waals (vdW) forces lead to surface complementarity [5]. In this paper we study this score function from a theoretical point of view. We give a combinatorial characterization of the docked configurations achieved by maximum score. This also leads to a simple polynomial-time algorithm to find such a configuration. The algorithm is not practical due to high running time. However, the arrangement of spheres inspired by the combinatorial characterization plays an essential role in an efficient local search heuristic for rigid protein docking [7].

The rest of this paper is organized as follows. Section 2 contains definitions, in section 3 we give our combinatorial characterization of docking configurations, in section 4 we use this characterization to obtain a docking algorithm, sections 5 and 6 contain some applications, we conclude with a brief discussion in section 7.

2 Problem Statement

A protein molecule consists of a set of atoms, each of which is represented by a ball (a solid sphere) in R^3. Let $\mathcal{A} = \{\alpha_1, \ldots, \alpha_n\}$, where $\alpha_i = (a_i, r_i)$ specifies the ith atom of protein $\mathcal{A}$ with center $a_i \in R^3$ and radius r_i (which is Van der Waals radius in Å). Denote the corresponding sphere by $\partial(a_i, r_i)$. Similarly, protein $\mathcal{B} = \{\beta_1, \ldots, \beta_m\}$ where $\beta_j = (b_j, s_j)$ specifies the jth atom of protein $\mathcal{B}$. Our proteins are rigid: So the distance between any two atoms in a protein remains constant. There are five typical atom types found in a Protein Data Bank (PDB) file and is shown in the following table:

Atom Type	C	N	O	P	S
Radius in Angstrom	1.548	1.400	1.348	1.880	1.808

Fix the configuration of $\mathcal{A}$, we want to move $\mathcal{B}$ (using rigid motion) towards $\mathcal{A}$ such that the transformed $\mathcal{B}$ best complements $\mathcal{A}$, where the complementarity is measured by the score defined below. For $\alpha = (a, r) \in \mathcal{A}$, $\beta = (b, s) \in \mathcal{B}$,

$$score(\alpha, \beta) = \begin{cases} 1 & \text{if } r + s \le ||a - b|| \le r + s + \lambda, \\ 0 & \text{otherwise.} \end{cases}$$

$$bump(\alpha, \beta) = \begin{cases} 1 & \text{if } r + s > ||a - b||, \\ 0 & \text{otherwise.} \end{cases}$$

where $|| \cdot ||$ is the Euclidean distance and constant $\lambda = 1.5$ (Å). Define $score(\mathcal{A}, \mathcal{B}) = \sum_{i=1}^{n} \sum_{j=1}^{m} score(\alpha_i, \beta_j)$, $bump(\mathcal{A}, \mathcal{B}) = \sum_{i=1}^{n} \sum_{j=1}^{m} bump(\alpha_i, \beta_j)$. That is, $score(\mathcal{A}, \mathcal{B})$ counts the number of atom pairs within a distance cutoff λ. $bump(\mathcal{A}, \mathcal{B})$ counts the number of atom pairs colliding. In [4], it was shown empirically by exhaustive search, which consists of a sampling of rigid motion space and evaluating each rigid motion using the score function, that the configuration corresponding to the maximum score and bump $\le \Theta$ (constant $\Theta = 5$) is always near the correct docked configuration (measured by root-mean-square-distance). So our rigid docking problem for the above specific score function is: Fix the configuration of protein $\mathcal{A}$ in R^3, the goal is to find a rigid motion p of protein $\mathcal{B}$ such that $score(\mathcal{A}, p(\mathcal{B}))$ is maximized and $bump(\mathcal{A}, p(\mathcal{B})) \le \Theta$, where constant $\Theta = 5$.

It is easy to see that the above setting makes sense for R^d in general – by treating $\mathcal{A}$ and $\mathcal{B}$ as sets of balls in R^d, and our characterization also work in this more general setting.

3 Combinatorial Characterization

Fix the configuration of $\mathcal{A}$ in R^d (we are mostly interested in $d = 2, 3$), let $\mathcal{C}$ be the space of all rigid motions of $\mathcal{B}$. For example, for $d = 3$, $\mathcal{C} = R^3 \times SO(3)$ (R^3 for translations and $SO(3)$ for rotations). Each point (rigid motion) $p \in \mathcal{C}$ corresponds to a configuration of $\mathcal{B}$, denoted by $p(\mathcal{B})$.

If a rigid body is allowed to move in some restricted way, then the dimension of the space of the motion of that body is known as its *degree of freedom* (DOF). So, for example, a rigid body in R^3 has 6 DOFs. In general, there are $d(d+1)/2$ DOFs for a rigid body in R^d. In the Robot Motion Planning community, $\mathcal{C}$ is called the configuration space of $\mathcal{B}$ and each point $p \in \mathcal{C}$ is a configuration of $\mathcal{B}$, while $p(\mathcal{B})$ is called the placement of $\mathcal{B}$ (and denoted by $\mathcal{B}(p)$). See, e.g., [13] and references therein for background on these notions including DOFs.

We define an equivalence relation on $\mathcal{C}$: $p \in \mathcal{C}$ is equivalent to $q \in \mathcal{C}$ iff $score(\alpha_i, p(\beta_j)) = score(\alpha_i, q(\beta_j))$ and $bump(\alpha_i, p(\beta_j)) = bump(\alpha_i, q(\beta_j))$ for all $\alpha_i \in \mathcal{A}$, $\beta_j \in \mathcal{B}$. Call the decomposition of $\mathcal{C}$ by the equivalence classes (called *cells*) the *arrangement* $\Pi(\mathcal{A}, \mathcal{B})$ of $\mathcal{C}$. Note that our cells are not necessarily connected. By the definition of a cell, the score and bump of all configurations in a cell are the same. Hence we can talk about the score and bump of a cell.

For ease of exposition, we assume henceforth that the balls in $\mathcal{B}$ all have the same radius s (generalization to include the case where radii may be different easily follows). Observe that for two balls $(a, r) \in \mathcal{A}$ and $(b, s) \in \mathcal{B}$, $score((a, r), (b, s)) = 1 \iff r + s \leq ||a - b|| \leq r_i + s + \lambda \iff b$ lies on the annulus formed by spheres $\partial(a, r + s)$, $\partial(a, r + s + \lambda)$ as shown in Figure 2. Let $G_s(\mathcal{A}) = \{\partial(a_i, r_i + s), \partial(a_i, r_i + s + \lambda) : \forall (a_i, r_i) \in \mathcal{A}\}$ the set of two enlarged spheres corresponding to each ball in $\mathcal{A}$.

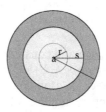

Fig. 2. The three circles in the figure have radii $r, r+s, r+s+\lambda$. $score((a, r), (b, s)) = 1 \iff r + s \leq ||a - b|| \leq r + s + \lambda \iff b$ lies on the shaded region.

The assumption above that the balls in $\mathcal{B}$ all have the same radius affords a somewhat more intuitive picture of the arrangement $\Pi(\mathcal{A}, \mathcal{B})$: A cell in $\Pi(\mathcal{A}, \mathcal{B})$ is defined by assigning to each center b_i of balls in $\mathcal{B}$ a cell γ_i in the arrangement of spheres in $G_s(\mathcal{A})$ (this arrangement should not be confused with $\Pi(\mathcal{A}, \mathcal{B})$) and consists of $p \in \mathcal{C}$ such that $p(b_i) \in \gamma_i$ for all i. This is the way we will think about $\Pi(\mathcal{A}, \mathcal{B})$.

We have to deal with one technical issue. The boundary of a cell can be complicated in the sense that it may be open at some places and closed at others. Our goal will be to find a representative point for each cell, but the procedure that finds this point guarantees only that the point lies in the closure of the cell. This causes the problem that a point may belong to more than one such closures and cannot be used to uniquely characterize a cell. This problem can be resolved by making two copies of each sphere in $G_s(\mathcal{A})$, slightly enlarging one of them and shrinking the other, and appropriately defining the cells. Under this new definition cells are closed sets and their boundaries are disjoint. We omit the tedious details. To simplify matters, in this paper we will just deal with the closures of cells.

We need some more definitions. For given $\mathcal{A}$ and $\mathcal{B}$ and $p \in \mathcal{C}$, a *contact-constraint* on configuration $p(\mathcal{B})$ is specified by the center b of a ball $(b, s) \in \mathcal{B}$, a sphere ω of $G_s(\mathcal{A})$ and the condition that $p(b)$ lie on ω. Also, we slightly abuse the definitions by saying a contact-constraint is satisfied by a rigid motion p if the corresponding configuration $p(\mathcal{B})$ satisfies the contact-constraint.

We assume throughout that $\mathcal{A}$ and $\mathcal{B}$ are in *general position*. We say that $\mathcal{A}$ and $\mathcal{B}$ are in general position, if no set of more than $d(d+1)/2$ contact-constraints is satisfiable by any rigid motion in $\mathcal{C}$. This is a natural assumption, as we expect one constraint to reduce DOF by one, and so if we have more constraints than available DOF, then we can no longer satisfy all of them. This assumption can be ascertained by applying small perturbation to the positions of the centers of balls.

Theorem 1. *Let $\mathcal{A}$ and $\mathcal{B}$ be two sets of balls in R^d in general position, and let $\Pi(\mathcal{A}, \mathcal{B})$ be defined as above. Then for any $F \in \Pi(\mathcal{A}, \mathcal{B})$, there is a set S of $k \leq d(d + 1)/2$ contact-constraints such that the rigid motions corresponding to the configurations satisfying them form $2^{O(d^3)}$ connected components, and at least one of the components is contained in the closure of F.*

Proof. Fix a cell $F \in \Pi(\mathcal{A}, \mathcal{B})$. Let p be a point in F such that $p(\mathcal{B})$ satisfies the maximal number of contact-constraints. Denote by S the set of contact-constraints; by Center_S the set of centers of balls in $\mathcal{B}$ involved in contact-constraints in S; and $\mathsf{Spheres}_S$ the set of spheres in $G_s(\mathcal{A})$ involved in S. By our general position assumption we have $|S| \leq d(d + 1)/2$ and thus $|\mathsf{Center}_S| \leq d(d + 1)/2$ and $|\mathsf{Sphere}_S| \leq d(d + 1)/2$.

Each contact-constraint in S amounts to an equation of degree 2 of a sphere in Sphere_S in the variables corresponding to the coordinates of some center in Center_S. Configurations satisfying the contact-constraints in S can be found by solving a system of equations including contact-constraints equation and centers distance equations (because $\mathcal{B}$ is rigid). Denote by $\mathsf{Solution}_S$ the set of solutions to the system of these equations.

Now the theorem of Oleinik-Petrovsky/Thom/Milnor (see, e.g. [3]) from real algebraic geometry asserts that the number of connected components of the set of solutions to a system of equations of degree $\leq D$ in v variables is bounded above by $D(2D - 1)^{v-1}$. In our case, $D = 2$ and $v \leq d \cdot d(d + 1)/2$ (the number of centers whose coordinates appear in the equations is $\leq d(d+1)/2$, and each has d coordinates, hence the bound on the number of variables). So the number of connected components in Solution_S is $\leq 2^{O(d^3)}$.

Let the set of rigid motions such that the corresponding configurations satisfy the contact-constraints in S be Motion_S. We claim that the number of connected components in Motion_S is the same as that in Solution_S. Elements in Solution_S provide values of the coordinates of the centers in Center_S, and from these we can obtain the set of rigid motions whose corresponding configurations satisfy the constraints. Given the coordinates of the centers in Center_S we may have two cases: (1) These are sufficient to determine a unique rigid motion. (2) We still have some degree of freedom left and we can find a set of rigid motions on each of the corresponding configuration the coordinates of the centers take on the given values; call this set of rigid motions the *image* of the given element in Center_S. It is clear that in the first case, the number of connected components in Motion_S is the same as that in Solution_S. For the second case, note that given some (valid) values of coordinates of centers in Center_S, its image is connected. E.g., if we fix two points of a rigid body in R^3 then the remaining motion is just the rotation around the line connecting the two fixed points; these rigid motions form a connected subset. Therefore, the image of any connected component of Solution_S in Motion_S is also connected. So the number of connected components in Solution_S is the same as that in Motion_S.

Denote the component of Motion_S containing p by P. We claim that P is contained in the closure of F. Suppose not, i.e. there exists $p' \in P \setminus cl(F)$ and there is a path connecting p and p' within P. Consider a point q where this path crosses a component of F. The reason the path gets out of the closure of F is that some center b in $\mathcal{B}$ gets out of its cell in the arrangement of $G_s(\mathcal{A})$ by crossing some sphere ω. The contact-constraint that b lie on ω is not in S, because it is not always satisfied in P, in particular on some points of the above path. So at point q, which is in the closure of F, we have contact-constraints in S and one extra contact-constraint satisfied. But this contradicts the maximality assumption. Hence P must be contained in the closure of F.   □

The theorem above says that any cell in $\Pi(\mathcal{A}, \mathcal{B})$ can be almost uniquely characterized by a set of contact-constraints in the sense that at least one of the small number of connected components formed by the configurations satisfying these contact-constraints is in the given cell. While the above proof may appear somewhat abstract, the principle behind it is quite intuitive, which is illustrated below by an example in R^2.

Example in R^2. As mentioned above, for each cell $F \in \Pi(\mathcal{A}, \mathcal{B})$, we want to find a point which has a simple description. This point will be found by satisfying as many contacts between centers of the balls in $\mathcal{B}$ and circles in $G_s(\mathcal{A})$ as possible while staying inside $cl(F)$, so that these (almost) uniquely identify the cell. Let $p \in F$ and suppose that $p(\mathcal{B})$ does not satisfy any contact-constraints. Choose a direction to translate $p(\mathcal{B})$ so that the center b_1 of some ball of $\mathcal{B}$ hits a circle ω_1 from $G_s(\mathcal{A})$. This is always possible (e.g. the direction of the closest ball pairs). This translation corresponds to a new point p' in $\mathcal{C}$

and it is not difficult to see that $p' \in cl(F)$. Next rotate $p'(\mathcal{B})$ about $p'(b_1)$, let $p'' \in \mathcal{C}$ be the corresponding point in the configuration space (if any) such that the center b_2 of some ball in $\mathcal{B}$ hits a circle w_2 from $G_s(\mathcal{A})$. (Note: in this example, we only consider rotating $p'(\mathcal{B})$ around $p'(b_1)$ and we might not find a second-contact, but it is possible that the second-contact does exist by a rigid motion which moves $p'(b_1)$ to a different position on w_1, and then rotates $p'(\mathcal{B})$. Also, we only require $(b_1, w_1) \neq (b_2, w_2)$, and it is possible that $b_1 = b_2$ (in this case the constraint is that b_1 ($= b_2$) is at an intersection point of w_1 and w_2), and similarly it is possible $w_1 = w_2$ with distinct b_1, b_2.) Since $\mathcal{B}$ has 3 DOFs in R^2, and each condition of the above kind reduces one DOF (by a general position assumption), hence $\mathcal{B}$ has one more DOF to move while preserving the 2 contact-constraints. One of the following three events happens in the course of this rigid motion, each giving a set of contact-constraints. (1) One of the two points b_1 and b_2 hits a vertex (an intersection point of two circles) in $G_s(\mathcal{A})$. (2) A new point b_3 in $\mathcal{B}$ hits a circle in $G_s(\mathcal{A})$. (3) None of the above. In the first two cases we only have a constant number of solutions, while in the last case we get a curve in the rigid motion space.

For general d, let $h = \frac{1}{2}d(d+1)$. Each cell in $\Pi(\mathcal{A}, \mathcal{B})$ is characterized by a system of contact-constraints as in Theorem 1. Each such system of contact-constraints is specified by at most h spheres in $\mathcal{A}$ and h spheres in $\mathcal{B}$. There are a total $O(m^h n^h)$ possible systems of contact-constraints. From Theorem 1, each cell contains at least one connected component of some system of contact-constraints. Since each such system has at most $2^{O(d^3)}$ components, we have

Corollary 1. *In* R^d, $\Pi(\mathcal{A}, \mathcal{B})$ *has at most* $O(2^{O(d^3)} m^h n^h)$ *cells, where* $h = \frac{1}{2}d(d+1)$.

4   The Algorithm

In this section we describe our algorithm for finding a maximum score configuration with bump less than a given threshold. We content ourselves with a high-level description of the algorithm, as in its present form it is mainly of theoretical interest.

Basic idea is simple: Enumerate all systems of contact-constraints which can arise in Theorem 1. For each system of (at most $d(d+1)/2$) contact-constraints, it corresponds to a set of equations of degree 2. Then we generate a point (the coordinates of the centers) in each connected component of the solutions to the system of equations. The generation can be done by one of the many available algorithms in real algebraic geometry, see for example Chapter 11.6 of [3], which takes time, in our situation, $2^{O(d^3)}$. We then recover a rigid motion for each solution, which corresponds to a point in each connected component of the configurations satisfying these contact-constraints. This can be done in time polynomial in d. Finally, compute the score and bump for each configuration. Having examined all systems, take the one with the maximum score and bump below the threshold. By theorem 1, this procedure will generate at least one point in each cell in $\Pi(\mathcal{A}, \mathcal{B})$, and thus will find the desired optimal solution.

To analyze the running time, note that the number of constraint systems that we examine is $O(m^h n^h)$, with $h = \frac{1}{2}d(d+1)$, and for each such system we spend $2^{O(d^3)}$ time generating at most $2^{O(d^3)}$ points. For each of the corresponding configurations $p(\mathcal{B})$, we compute $score(\mathcal{A}, p(\mathcal{B}))$ and $bump(\mathcal{A}, p(\mathcal{B}))$. Doing this in the obvious way takes time $O(mn)$. Hence, the total time taken is $O(2^{O(d^3)} m^{h+1} n^{h+1})$.

In the following we will improve on this algorithm in R^3 by making use of a property satisfied by protein molecules. The following density property of the spheres in a protein (and also in $G_s(\mathcal{A})$) was given by [10]. It says that the atoms in a protein overlap in a somewhat sparse fashion and do not crowd together. For the sake of completeness, we include the easy proof; bound here is better than in [10].

Theorem 2. Let $\mathcal{A} = \{\alpha_i = (a_i, r_i) : i = 1, \ldots, n\}$ be a set of n spheres in R^3. Suppose there is a $\delta > 0$ such that $\|a_i - a_j\| > \delta$, $1 \leq i, j \leq n$ $(i \neq j)$. Then for any $1 \leq i \leq n$, $|\{\alpha_j \in \mathcal{A} : \alpha_i \cap \alpha_j \neq \emptyset\}| \leq (4\frac{r_{max}}{\delta} + 1)^3$, where $r_{max} = \max_{j=1..n} r_j$.

Proof. Since for any $1 \leq i, j \leq n$ $(i \neq j)$, $\|a_i - a_j\| > \delta$, we have $(a_i, \frac{\delta}{2}) \cap (a_j, \frac{\delta}{2}) = \emptyset$. Consider $\alpha_i \in \mathcal{A}$, for any $\alpha_j \in \mathcal{A}$ with $\alpha_i \cap \alpha_j \neq \emptyset$, $a_j \in (a_i, r_i + r_{max})$. Thus, $(a_j, \frac{\delta}{2}) \subset (a_i, r_i + r_{max} + \frac{\delta}{2})$. By volume argument, $|\{\alpha_j \in \mathcal{A} : \alpha_i \cap \alpha_j \neq \emptyset\}| \leq (\frac{a_i \cdot r + r_{max} + \frac{\delta}{2}}{\frac{\delta}{2}})^3 \leq (4\frac{r_{max}}{\delta} + 1)^3$. $\square$

For the set of spheres in a protein (or $G_s(\mathcal{A})$), $\frac{r_{max}}{\delta}$ is a constant. That is, each sphere with radius at most r_{max} intersects only constantly many spheres in the set. With this property, we can then use data structure of [10] to compute the score and bump of each corresponding configuration in $O(m)$ time (by preprocessing a data structure using $O(n)$ space in $O(n)$ expected time for the $2n$ spheres of $G_s(\mathcal{A})$), and output a configuration with the maximum score and bump $\leq \Theta$. It is possible that there are more than one such configurations. For the application to rigid protein docking this is not problematic in the light of empirical results of Bespamaytnikh et al [4], which show that such configurations are all very close to each other.

Therefore, we have

Theorem 3. Let $\mathcal{A}, \mathcal{B}$ be two proteins (in general position) in R^3 and the score function defined as above. The docked configuration achieved by the maximum score and bump $\leq \Theta$ can be computed in $O(m^6 n^6 \min\{m, n\})$ time.

This algorithm is not practical due to the high running time. However, the arrangement of the spheres in $G_s(\mathcal{A})$ used in the algorithm has important applications described in the next section.

5   Applications of the Arrangement of Spheres

In this section, we describe two immediate applications of the arrangement of spheres in $G_s(\mathcal{A})$. It provides a useful representation of the complement of a protein; and second, it gives a simple way to identify surface atoms. Recall that the arrangement of spheres in $G_s(\mathcal{A})$ decomposes the space into regions (might be disconnected) with score and bump constant in a region. By the above theorem on the density property of spheres in $G_s(\mathcal{A})$, each sphere intersects with only constantly many other spheres and hence there are only linearly many vertices in the arrangement of spheres in $G_s(\mathcal{A})$. Again, by the same data structure as in [10], we can compute the vertex set of the arrangement in $O(n \log n)$ time implemented by binary search tree or $O(n)$ expected time using randomized perfect hashing, both with $O(n)$ space. Also, we can compute the score and bump of each vertex, and associate each vertex with the atoms which contribute to the score, in extra $O(1)$ time. In particular, the bump-free vertices (i.e. with bump $= 0$) play an important role in an efficient local search algorithm [7] in that they describe discrete positions of the complement of the protein measured by the score function. See Figure 3 for an example of these vertices.

Fig. 3. The bump-free vertices of protein A are shown by dots.

Another application of the arrangement of spheres is an easy and useful way to identify "surface" atoms from "interior" atoms. The set of surface atoms is just the set of atoms associated with the bump-free vertices. It is possible that some of these atoms actually lie inside the molecule, and the bump-free vertices associated to them are bump-free because of holes in the molecule. For applications, we are generally not interested in such surface atoms. To identify the set of surface atoms which are not inside the molecule we define a graph on the bump-free vertices such that two vertices are adjacent if they are close. Now the vertices corresponding to the outer surface form a connected component in this graph and this can be easily identified. See Figure 4 for an example.

Our surface atom definition is tailored for use in rigid docking. To solve rigid docking problem one can restrict attention to the configurations for which surface atoms contribute to the interaction; the configurations for which interior atoms contribute can be safely ignored. While the formal molecular surfaces such as

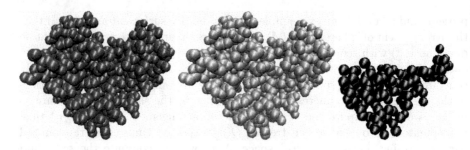

Fig. 4. The original protein (left) = surface atoms only (middle) + interior atoms (right).

solvent accessible surface ([10]), which involves computing the surface area, can also be used to identify the surface atoms but they are much complicated to compute [10].

6 Application of Our Framework to Point Pattern Matching

Recently, Agarwal et al [1] considered using partial shape matching under Hausdorff distance to measure the shape complementarity under translation. See the references there for related literature on point pattern matching. An observation somewhat similar to ours has been made by Alt et al [2] in solving a congruence problem in R^2. In the general d-dimensional version of this problem we are given two sets $\mathcal{A}$ and $\mathcal{B}$ in R^d of n points each, and a number δ. The goal is to decide if there exists a one-one matching between the points in $\mathcal{A}$ and $\mathcal{B}$ and a rigid motion p of $\mathcal{B}$ so that the maximum of the distance between the matched pairs in $\mathcal{A}$ and $p(\mathcal{B})$is at most δ. The algorithm given in Alt et al [2] for $d = 2$, which considered enumerates sets of constraints of size 2, has $O(n^8)$ time. They are not able to generalize their algorithm to higher dimensions. Here we show how to use our framework and results of [8] to get an algorithm with time $\widetilde{O}(n^{7.5})$ in R^2; an $O(n^{13+5/6+\epsilon})$ time algorithm in R^3 (for any positive ϵ); and an $O(n^{d(d+1)+2.5})$ time algorithm in R^d for $d > 3$.

Similarly to the algorithm in the Section 4, the present algorithm has two parts: (1) Enumeration, and (2) testing. Algorithm enumerates a set of constraints, and for each such set it checks if the configurations satisfying these constraints provide a positive answer. Enumeration part is very similar to our previous algorithm; the testing part is done using bipartite bottleneck matching.

We first describe the enumeration part which works for general d. We define the notion of a constraint similar to the earlier one. Assume that $\mathcal{A}$ is fixed. A constraint on the configuration of $\mathcal{B}$ specifies a sphere with center in $\mathcal{A}$ and radius δ, and a point in $\mathcal{B}$, and requires that this point lie on the given sphere. Denote by $G(\mathcal{A})$ the arrangement of R^d induced by spheres of radius δ with centers in $\mathcal{A}$. We can decompose the configuration space of $\mathcal{B}$ into equivalence classes: Two configurations $p(\mathcal{B})$ and $p'(\mathcal{B})$ are equivalent if for all points $b \in \mathcal{B}$, $p(b)$ and $p'(b)$ lie in the same cell of $G(\mathcal{A})$. If the answer to the problem is yes then there is a one-one correspondence between $\mathcal{A}$ and $\mathcal{B}$, and there is nonempty equivalence class such that the configurations $p(\mathcal{B})$ in it satisfy: $\forall b \in \mathcal{B}$, $p(b)$ is in the sphere of radius δ around the point of $\mathcal{A}$ matched to b. Denote this class by M. By the arguments similar to those in the proof of Theorem 1, there is a set of at most $d(d+1)/2$ constraints with the following property. At least one connected component of the configurations satisfying these constraints is contained in M, and there are at most $2^{O(d^3)}$ connected components. The enumeration part is now clear: We consider all possible sets of $\leq d(d+1)/2$ constraints; for each such set we generate a point in each connected component of the configurations satisfying the constraints in the set.

The testing part checks if a given configuration $p(\mathcal{B})$ solves the problem. We can do this using the bipartite bottleneck matching: Given two sets of equal cardinality, the problem is to find a one-one matching between them such that the maximum distance (cost) between the matched points is minimized. Clearly, for our problem, if we have a configuration $p(\mathcal{B})$ which is in M, then the bottleneck matching between $\mathcal{A}$ and $p(\mathcal{B})$ will have cost at most δ. And also, if the answer to our decision problem is no, then for all configurations $p(\mathcal{B})$, bottleneck matching between $\mathcal{A}$ and $p(\mathcal{B})$ will have cost more than δ. In general, we can construct a bipartite graph with points in $\mathcal{A}$ on one side, and points in $p(\mathcal{B})$ on the other side, and two points on different sides are connected if their distance is $\leq \delta$. Then check if the bipartite graph has a perfect matching. This can be done by the algorithm of Hopcroft and Karp [11] in time $O(v^{2.5})$ where v is the number of vertices. There is a bottleneck matching between $\mathcal{A}$ and $p(\mathcal{B})$ with cost at most δ iff the bipartite graph has a perfect matching. This gives an algorithm of $O(n^{d(d+1)+2.5})$ time for general d. For $d = 2, 3$, the bipartite bottleneck matching problem can be solved more efficiently in $\widetilde{O}(n^{1.5})$ time in R^2, and $O(n^{11/6+\epsilon})$ time in R^3 (for any positive ϵ) by Efrat et al [8]. Thus, we have an algorithm of $\widetilde{O}(n^{7.5})$ for $d = 2$ and $O(n^{13+5/6+\epsilon})$ for $d = 3$.

7   Discussion

We have given a simple algorithm for rigid protein docking under the score function of [4]. This algorithm is not practical due to high running time and numerical issues. A natural question is if it can be substantially improved. Note that we did not use the protein density property in the analysis of the arrangement. We believe that by using this property and by pruning the search space, e.g. by excluding the configurations with large bumps from the enumeration, we can get faster algorithms.

Acknowledgment

We thank Saugata Basu for answering our question regarding real algebraic geometry. We also thank Pankaj K. Agarwal, Herbert Edelsbrunner and Raimund Seidel for comments.

References

1. P. Agarwal, S. Har-Peled, M. Sharir and Y. Wang. Hausdorff distance under translation for points, disks and balls. *Proc. 19th ACM Symp. on Computational Geometry* (2003), 282-291.
2. H. Alt, K. Mehlhorn, H. Wagener, E. Welzl. Congruence, similarity, and symmetries of geometric objects. *Discrete Comput. Geom.* 3 (1988), no. 3, 237-256.
3. S. Basu, R. Pollack, M.-F. Roy. Algorithms in Real Algebraic Geometry. *Springer* 2003.
4. S. Bespamaytnikh, V. Choi, H. Edelsbrunner and J. Rudolph. Accurate Bound Protein Docking by Shape Complementarity Alone. *Technical Report, Department of Computer Science, Duke University, 2003.*
5. C.J. Camacho and S. Vajda , Protein docking along smooth association pathways. *Proc Natl Acad Sci USA* 98 (2001), pp. 10636-10641.
6. R. Chen, Z. Weng. A novel shape complementarity scoring function for protein-protein docking. *Proteins: Structure, Function, and Genetics* 51:3, 2003, 397-408.
7. V. Choi, P. K. Agarwal J. Rudolph and H. Edelsbrunner. Local Search Heuristic for Rigid Protein Docking. *To be submitted.*
8. A. Efrat, A. Itai and M. J. Katz. Geometry helps in bottleneck matching and related problems. *Algorithmica* 31 (2001), no. 1, 1-28.
9. D. Fischer, S.L. Lin, H.J. Wolfson and R. Nussinov. A suite of molecular docking processes. *J Mol Biol* 248 (1995), pp. 459-477.
10. D. Halperin and M.H. Overmars. Spheres, molecules, and hidden surface removal. *Computational Geometry: Theory and Applications* 11 (2), 1998, 83-102.
11. J. Hopcroft and R.M.Karp. An $n^{5/2}$ algorithm for maximum matchings in bipartite graphs. *SIAM J. Computing*, 2 (1973), 225-231.
12. E. Katchalski-Katzir, I. Shariv, M. Eisenstein, A.A. Friesem, C. Aflalo and I.A. Vakser. Molecular surface recognition: determination of geometric fit between proteins and their ligands by correlation techniques. *Proc Natl Acad Sci USA* 89 (1992), pp. 2195-2199.
13. J.-C. Latombe. Robot Motion Planning. Kluwer Academic, Boston, 1991.
14. R. Mendez, R. Leplae, L. D. Maria, S. J. Wodak. Assessment of blind predictions of protein-protein interactions: Current status of docking methods. *Proteins: Structure, Function, and Genetics* 52:1, 2003, 51-67.
15. G.R. Smith and M. JE. Sternberg. Predictions of Protein-Protein Interactions by Docking Methods. *Current Opinion in Structural Biology.* **12**:28-35, 2002.

Multi-seed Lossless Filtration

(Extended Abstract)

Gregory Kucherov[1], Laurent Noé[1], and Mikhail Roytberg[2]

[1] INRIA/LORIA, 615, rue du Jardin Botanique,
B.P. 101, 54602, Villers-lès-Nancy, France
{Gregory.Kucherov,Laurent.Noe}@loria.fr
[2] Institute of Mathematical Problems in Biology,
Pushchino, Moscow, Region, 142290, Russia
roytberg@impb.psn.ru

Abstract. We study a method of seed-based lossless filtration for approximate string matching and related applications. The method is based on a simultaneous use of several spaced seeds rather than a single seed as studied by Burkhardt and Karkkainen [1]. We present algorithms to compute several important parameters of seed families, study their combinatorial properties, and describe several techniques to construct efficient families. We also report a large-scale application of the proposed technique to the problem of oligonucleotide selection for an EST sequence database.

1 Introduction

Filtering is a widely-used technique in various string processing applications. Applied to the approximate string matching problem [2], it can be summarized by the following two-stage scheme: to find approximate occurrences (matches) of a given string in a text, one first quickly discards (filters out) those text areas where matches cannot occur, and then checks out the remaining parts of the text for actual matches. The filtering is done according to small patterns of a specified form that the searched string is assumed to share, in the exact way, with its approximate occurrences. A similar filtration scheme is used by heuristic local alignment algorithms ([3–6], to mention a few): they first identify potential similarity regions that share some patterns and then actually check whether those regions represent a significant similarity by computing a corresponding alignment.

Two types of filtering should be distinguished – *lossless* and *lossy*. A lossless filtration guarantees to detect *all* text fragments under interest, while a lossy filtration may miss some of them, but still tries to detect the majority of them. Local alignment algorithms usually use a lossy filtration. On the other hand, the lossless filtration has been studied in the context of approximate string matching problem [7, 1]. In this paper, we focus on the lossless filtration.

In the case of lossy filtration, its efficiency is measured by two parameters, usually called *selectivity* and *sensitivity*. The sensitivity measures the part of text fragments of interest that are missed by the filter (false negatives), and the

S.C. Sahinalp et al. (Eds.): CPM 2004, LNCS 3109, pp. 297–310, 2004.
© Springer-Verlag Berlin Heidelberg 2004

selectivity indicates what part of detected fragments don't actually represent a solution (false positives). In the case of lossless filtration, only the selectivity parameter makes sense and is therefore the main characteristic of the filtration efficiency.

The choice of patterns that must be contained in the searched text fragments is a key ingredient of the filtration algorithm. *Gapped seeds* (spaced seeds, gapped q-grams) have been recently shown to significantly improve the filtration efficiency over the "traditional" technique of contiguous seeds. In the framework of lossy filtration for sequence alignment, the use of designed gapped seeds has been introduced by the PATTERNHUNTER method [4] and then used by some other algorithms (e.g. [5, 6]). In [8, 9], spaced seeds have been shown to improve indexing schemes for similarity search in sequence databases. The estimation of the sensitivity of spaced seeds (as well as of some extended seed models) has been subject of several recent studies [10–14]. In the framework of lossless filtration for approximate pattern matching, gapped seeds were studied in [1] (see also [7]) and have been also shown to increase the filtration efficiency considerably.

In this paper, we study an extension of the single-seed filtration technique [1]. The extension is based on using *seed families* rather than individual seeds. In Section 3, we present dynamic programming algorithms to compute several important parameters of seed families. In Section 4, we first study several combinatorial properties of families of seeds, and, in particular, seeds having a periodic structure. These results are used to obtain a method for constructing efficient seed families. We also outline a heuristic genetic programming algorithm for constructing seed families. Finally, in Section 5, we present several seed families we computed, and we report a large-scale experimental application of the method to the practical problem of oligonucleotide design.

2   Multiple Seed Filtering

A *seed* Q (called also *spaced seed* or *gapped q-gram*) is a list $\{p_1, p_2, \ldots, p_w\}$ of positive integers, called *matching positions*, such that $p_1 < p_2 < \ldots < p_w$. By convention, we always assume $p_1 = 0$. The *span* of a seed Q, denoted $s(Q)$, is the quantity $p_w + 1$. The number w of positions is called the *weight* of the seed and denoted $w(Q)$. Often we will use a more visual representation of seeds, adopted in [1], as words of length $s(Q)$ over the two-letter alphabet $\{\#, -\}$, where # occurs at all matching positions and - at all positions in between. For example, seed $\{0, 1, 2, 4, 6, 9, 10, 11\}$ of weight 8 and span 12 is represented by word ###-#-#--###. The character - is called a *joker*. Note that, unless otherwise stated, the seed has the character # at its first and last positions.

Intuitively, a seed specifies the set of patterns that, if shared by two sequences, indicate a possible similarity between them. Two sequences are similar if the Hamming distance between them is smaller than a certain threshold. For example, sequences CACTCGT and CACACTT are similar within Hamming distance 2 and this similarity is detected by the seed ##-# at position 2. We are interested in seeds that detect *all* similarities of a given length with a given Hamming distance.

Formally, a *gapless similarity* (hereafter simply similarity) of two sequences of length m is a binary word $w \in \{0,1\}^m$ interpreted as a sequence of matches (1's) and mismatches (0's) of individual characters from the alphabet of input sequences. A seed $Q = \{p_1, p_2, \ldots, p_w\}$ *matches* a similarity w at position i, $1 \leq i \leq m - p_w + 1$, iff $\forall j \in [1..w]$, $w[i + p_j] = 1$. In this case, we also say that seed Q *has an occurrence* in similarity w at position i. A seed Q is said to *detect* a similarity w if Q has at least one occurrence in w.

Given a similarity length m and a number of mismatches k, consider all similarities of length m containing k 0's and $(m - k)$ 1's. These similarities are called (m, k)-similarities. A seed Q *solves the detection problem* (m, k) (for short, the (m, k)-problem) iff for all $\binom{m}{k}$ (m, k)-similarities w, Q detects w. For example, one can check that seed `#-##--#-##` solves the $(15, 2)$-problem.

Note that the weight of the seed is directly related to the *selectivity* of the corresponding filtration procedure. A larger weight improves the selectivity, as less similarities will pass through the filter. On the other hand, a smaller weight reduces the filtration efficiency. Therefore, the goal is to solve an (m, k)-problem by a seed with the largest possible weight.

Solving (m, k)-problems by a single seed has been studied by Burkhardt and Kärkkäinen [1]. An extension we propose here is to use a *family of seeds*, instead of a single seed, to solve the (m, k)-problem. Formally, a finite family of seeds $F = < Q_l >_{l=1}^{L}$ *solves the* (m, k)-*problem* iff for all $\binom{m}{k}$ (m, k)-similarities w, there exists a seed $Q_l \in F$ that detects w.

Note that the seeds of the family are used in the complementary (or disjunctive) fashion, i.e. a similarity is detected if it is detected by *one of the seeds*. This differs from the conjunctive approach of [7] where a similarity should be detected by two seeds *simultaneously*.

The following example motivates the use of multiple seeds. In [1], it has been shown that a seed solving the $(25, 2)$-problem has the maximal weight 12. The only such seed (up to reversal) is `###-#--###-#--###-#`. However, the problem can be solved by the family composed of the following two seeds of weight 14: `#####-##---#####-##` and `#-##---#####-##---####`.

Clearly, using these two seeds increases the selectivity of the search, as only similarities having 14 or more matching characters pass the filter vs 12 matching characters in the case of single seed. On uniform Bernoulli sequences, this results in the decrease of the number of candidate similarities by the factor of $|A|^2/2$, where A is the input alphabet. This illustrates the advantage of the multiple seed approach: it allows to increase the selectivity while preserving a lossless search. The price to pay for this gain in selectivity is a double work spent on identifying the seed occurrences. In the case of large sequences, however, this is largely compensated by the decrease in the number of false positives caused by the increase of the seed weight.

3   Computing Properties of Seed Families

Burkhardt and Kärkkäinen [1] proposed a dynamic programming algorithm to compute the *optimal threshold* of a given seed – the minimal number of its occurrences over all possible (m, k)-similarities. In this section, we describe an extension of this algorithm for seed families and, on the other hand, describe dynamic programming algorithms for computing two other important parameters of seed families that we will use in a latter section.

Consider an (m, k)-problem and a family of seeds $F =< Q_l >_{l=1}^{L}$. We need the following notation.

- $s_{max} = max\{s(Q_l)\}_{l=1}^{L}$, $s_{min} = min\{s(Q_l)\}_{l=1}^{L}$,
- for a binary word w and a seed Q_l, $suff(Q_l, w) = 1$ if Q_l matches w at position $(|w| - s(Q_l) + 1)$ (i.e. matches a suffix of w), otherwise $suff(Q_l, w) = 0$,
- $last(w) = 1$ if the last character of w is 1, otherwise $last(w) = 0$,
- $zeros(w)$ is the number of 0's in w.

3.1   Optimal Threshold

Given an (m, k)-problem, a family of seeds $F =< Q_l >_{l=1}^{L}$ has the *optimal threshold* $T_F(m, k)$ if every (m, k)-similarity has at least $T_F(m, k)$ occurrences of seeds of F and this is the maximal number with this property. Note that overlapping occurrences of a seed as well as occurrences of different seeds at the same position are counted separately. As an example, the singleton family {###-##} has threshold 2 for the $(15, 2)$-problem.

Clearly, F solves an (m, k)-problem if and only if $T_F(m, k) > 0$. If $T_F(m, k) > 1$, then one can strengthen the detection criterion by requiring several seed occurrences for a similarity to be detected. This shows the importance of the optimal threshold parameter.

We now describe a dynamic programming algorithm for computing the optimal threshold $T_F(m, k)$. For a binary word w, consider the quantity $T_F(m, k, w)$ defined as the minimal number of occurrences of seeds of F in all (m, k)-similarities which have the suffix w. By definition, $T_F(m, k) = T_F(m, k, \varepsilon)$. Assume that we precomputed values $T_F(j, w) = T_F(s_{max}, j, w)$, for all $j \leq max\{k, s_{max}\}$, $|w| = s_{max}$. The algorithm is based on the following recurrence relations on $T_F(i, j, w)$, for $i \geq s_{max}$.

$$T_F(i, j, w[1..n]) = \begin{cases} T_F(j, w), & \text{if } i = s_{max}, \\ T_F(i-1, j-1, w[1..n-1]), & \text{if } w[n] = 0, \\ T_F(i-1, j, w[1..n-1]) + [\sum_{l=1}^{L} suff(Q_l, w)], & \text{if } n = s_{max}, \\ min\{T_F(i, j, 1.w), T_F(i, j, 0.w)\}, & \text{if } zeros(w) < j, \\ T_F(i, j, 1.w), & \text{if } zeros(w) = j. \end{cases}$$

The first relation is an initial condition of the recurrence. The second one is based on the fact that if the last symbol of w is 0, then no seed can match a suffix of w (as the last position of a seed is always a matching position). The third relation reduces the size of the problem by counting the number of suffix seed occurrences. The fourth one splits the counting into two cases, by considering two possible characters occurring on the left of w. If w already contains j 0's, then only 1 can occur on the left of w, as stated by the last relation.

A dynamic programming implementation of the above recurrence allows to compute $T_F(m, k, \varepsilon)$ in a bottom-up fashion, starting from initial values $T_F(j, w)$ and applying the above relations in the order in which they are written. A straightforward dynamic programming implementation requires $O(m \cdot k \cdot 2^{(s_{max}+1)})$ time and space. However, the space complexity can be immediately improved: if values of i are processed successively, then only $O(k \cdot 2^{(s_{max}+1)})$ space is needed. Furthermore, for each i and j, it is not necessary to consider all $2^{(s_{max}+1)}$ different strings w, but only those which contain up to j 0's. The number of those w is $g(j, s_{max}) = \sum_{i=0}^{j} \binom{s_{max}}{i}$. For each i, j ranges from 0 to k. Therefore, for each i, we need to store $f(k, s_{max}) = \sum_{j=0}^{k} g(j, s_{max}) = \sum_{i=0}^{k} \binom{s_{max}}{i} \cdot (k - i + 1)$ values. This yields the same space complexity as for computing the optimal threshold for one seed [1].

The quantity $\sum_{l=1}^{L} suff(Q_l, w)$ can be precomputed for all considered words w in time $O(L \cdot g(k, s_{max}))$ and space $O(g(k, s_{max}))$, under the assumption that checking an individual match is done in constant time. This leads to the overall time complexity $O(m \cdot f(k, s_{max}) + L \cdot g(k, s_{max}))$ with the leading suff $m \cdot f(k, s_{max})$ (as L is usually small compared to m and $g(k, s_{max})$ is smaller than $f(k, s_{max})$).

3.2   Number of Undetected Similarities

We now describe a dynamic programming algorithm that computes another characteristic of a seed family, that will be used later in Section 4.4. Consider an (m, k)-problem. Given a seed family $F = < Q_l >_{l=1}^{L}$, we are interested in the number $U_F(m, k)$ of (m, k)-similarities that are not detected by F. For a binary word w, define $U_F(m, k, w)$ to be the number of undetected (m, k)-similarities that have the suffix w.

Similar to [10], let $X(F)$ be the set of binary words w such that (i) $|w| \leq s_{max}$, (ii) for any $Q_l \in F$, $suff(Q_l, 1^{s_{max}-|w|}w) = 0$, and (iii) no proper suffix of w verifies (ii). Note that word 0 belongs to $X(F)$, as the last position of every seed is a matching position.

The following recurrence relations allow to compute $U_F(i, j, w)$ for $i \leq m$, $j \leq k$, and $|w| \leq s_{max}$.

$$U_F(i, j, w[1..n]) = \begin{cases} \binom{i-|w|}{j-zeros(w)}, & \text{if } i < s_{min}, \\ 0, & \text{if } \exists l \in [1..L], \ suff(Q_l, w) = 1, \\ U_F(i-1, j-last(w), w[1..n-1]), & \text{if } w \in X(F), \\ U_F(i, j, 1.w) + U(i, j, 0.w), & \text{if } zeros(w) < j, \\ U_F(i, j, 1.w), & \text{if } zeros(w) = j. \end{cases}$$

The first condition says that if $i < s_{min}$, then no word of length i will be detected, hence the binomial formula. The second condition is straightforward. The third relation follows from the definition of $X(F)$ and allows to reduce the size of the problem. The last two conditions are similar to those from the previous section.

The set $X(F)$ can be precomputed in time $O(L \cdot g(k, s_{max}))$ and the worst-case time complexity of the whole algorithm remains $O(m \cdot f(k, s_{max}) + L \cdot g(k, s_{max}))$.

3.3   Contribution of a Seed

Using a similar dynamic technique, one can compute, for a given seed of the family, the number of (m, k)-similarities that are detected only by this seed and not by the others. Together with the number of undetected similarities, this parameter will be used later in Section 4.4.

Given an (m, k)-problem and a family $F =< Q_l >_{l=1}^{L}$, we define $S_F(m, k, l)$ to be the number of (m, k)-similarities detected by the seed Q_l exclusively (through one or several occurrences), and $S_F(m, k, w, l)$ to be the number of those similarities ending with the suffix w. A dynamic programming algorithm similar to the one described in the previous sections can be applied to compute $S_F(m, k, l)$. Below we give only the main recurrence relations for $S_F(m, k, w, l)$ and leave out initial conditions.

$$S_F(i, j, w[1..n], l) = \begin{cases} \sum_{x \in \{0,1\}} S_F(i-1, j-1+x, x.w[1..n-1], l) & \text{if } \mathit{suff}(Q_l, w) = 1 \text{ and} \\ \quad + U_F(i-1, j-1+x, x.w[1..n-1]), & \forall l' \neq l, \ \mathit{suff}(Q_{l'}, w) = 0, \\ \sum_{x \in \{0,1\}} S_F(i-1, j-1+x, x.w[1..n-1], l) & \text{if } \forall l', \ \mathit{suff}(Q_{l'}, w) = 0. \end{cases}$$

The first relation allows to reduce the problem when Q_l matches a suffix of w, but not the other seeds of the family. The second one applies if no seed matches a suffix of w. The complexity of computing $S_F(m, k, l)$ for a given l is the same as the complexity of dynamic programming algorithms from the previous sections.

4   Seed Design

In the previous Section we showed how to compute various useful characteristics of a given family of seeds. A much more difficult task is to find an efficient seed family that solves a given (m, k)-problem. Note that there exists a trivial solution where the family consists of all $\binom{m}{k}$ position combinations, but this is in general unacceptable in practice because of a huge number of seeds. Our goal is to find families of reasonable size (typically, with the number of seeds smaller than ten), with a good filtration efficiency.

In this section, we present several results that contribute to this goal. In Section 4.1, we start with the case of single seed with a fixed number of jokers and show, in particular, that for one joker, there exists one best seed in a sense that will be defined. We then show in Section 4.2 that a solution for a larger problem can be obtained from a smaller one by a regular expansion operation. In Section 4.3, we focus on seeds that have a periodic structure and show how those seeds can be constructed by iterating some smaller seeds. We then show a way to build efficient families of periodic seeds. Finally, in Section 4.4, we briefly describe a heuristic approach to constructing efficient seed families that we used in the experimental part of this work presented in Section 5.

4.1   Single Seeds with a Fixed Number of Jokers

Assume that we fixed a class of seeds under interest (e.g. seeds of a given minimal weight). One possible way to define the seed design problem is to fix the similarity length m and find a seed that solves the (m, k)-problem with the largest possible value of k. A complementary definition is to fix k and minimize m provided that the (m, k)-problem is still solved. In this section, we adopt the second definition and present an optimal solution for one particular case.

For a seed Q and a number of mismatches k, define the k-critical value for Q as the minimal value m such, that Q solves the (m, k)-problem. For a class of seeds $\mathcal{C}$ and a value k, a seed is k-optimal in $\mathcal{C}$ if Q has the minimal k-critical value among all seeds of $\mathcal{C}$.

One interesting class of seeds $\mathcal{C}$ is obtained by putting an upper bound on the possible number of jokers in the seed, i.e. on the number $(s(Q) - w(Q))$. We have found a general solution of the seed design problem for the class $\mathcal{C}_1(n)$ consisting of seeds of weight n with only one joker.

Theorem 1. *Let n be an integer and $r = \lfloor n/3 \rfloor$. For every $k \geq 2$, seed $Q(n) =$ #$^{n-r}$-#^{r} is k-optimal among the seeds of $\mathcal{C}_1(n)$.*

To illustrate Theorem 1, seed `####-##` is optimal among all seeds of weight 6 with one joker. This means that this seed solves the $(m, 2)$-problem for all $m \geq 16$ and this is the smallest possible bound over all seeds of this class. Similarly, this seed solves the $(m, 3)$-problem for all $m \geq 20$, which is the best possible bound, etc.

4.2   Regular Expansion and Contraction of Seeds

We now show that seeds solving larger problems can be obtained from seeds solving smaller problems, and vice versa, using a regular expansion and regular contraction operations.

Given a seed Q , its *i-regular expansion* $i \otimes Q$ is obtained by multiplying each matching position by i. This is equivalent to inserting $i - 1$ jokers between every two successive positions along the seed. For example, if $Q = \{0, 2, 3, 5\}$ (or `#-##-#`), then the 2-regular expansion of Q is $2 \otimes Q = \{0, 4, 6, 10\}$ (or `#---#-#---#`). Given a family F, its *i-regular expansion* $i \otimes F$ is the family obtained by applying the i-regular expansion on each seed of F.

Lemma 1. *If a family F solves the (m, k)-problem, then the $(im, (i+1)k - 1)$-problem is solved both by family F and by its i-regular expansion $F_i = i \otimes F$.*

Proof. Consider an $(im, (i+1)k - 1)$-similarity w. By the pigeon hole principle, it contains at least one substring of length m with k mismatches or less, and therefore F solves the $(im, (i+1)k - 1)$-problem. On the other hand, consider i disjoint subsequences of w each one consisting of m positions equal modulo i. Again, by the pigeon hole principle, at least one of them contains k mismatches or less, and therefore the $(im, (i+1)k - 1)$-problem is solved by $i \otimes F$.

The following lemma is the inverse of Lemma 1, it states that if seeds solving a bigger problem have a regular structure, then a solution for a smaller problem can be obtained by the regular contraction operation, inverse to the regular expansion.

Lemma 2. *If a family $F_i = i \otimes F$ solves the (im, k)-problem, then F solves both the (im, k)-problem and the $(m, \lfloor k/i \rfloor)$-problem.*

Proof. Similar to Lemma 1.

Example 1. To illustrate the two lemmas above, we give the following example pointed out in [1]. The following two seeds are the only seeds of weight 12 that solve the $(50, 5)$-problem: `#-#-#---#-----#-#-#---#-----#-#-#---#` and `###-#--###-#--###-#`. The first one is the 2-regular expansion of the second. The second one is the only seed of weight 12 that solves the $(25, 2)$-problem.

The regular expansion allows, in some cases, to obtain an efficient solution for a larger problem by reducing it to a smaller problem for which an optimal or a near-optimal solution is known.

4.3   Periodic Seeds

In this section, we study seeds with a periodic structure that can be obtained by iterating a smaller seed. Such seeds often turn out to be among maximally weighted seeds solving a given (m, k)-problem. Interestingly, this contrasts with the lossy framework where optimal seeds usually have a "random" irregular structure.

Consider two seeds Q_1, Q_2 represented as words over $\{$`#`,`-`$\}$. We denote $[Q_1, Q_2]^i$ the seed defined as $(Q_1 Q_2)^i Q_1$. For example, $[$`###-#`$, $`--`$]^2 = $`###-#--`
`###-#--###-#`.

We also need a modification of the (m, k)-problem, where (m, k)-similarities are considered modulo a cyclic permutation. We say that a seed family F solves a *cyclic (m, k)-problem*, if for every (m, k)-similarity w, F detects one of cyclic permutations of w. Trivially, if F solves an (m, k)-problem, it also solves the cyclic (m, k)-problem. To distinguish from a cyclic problem, we call sometimes an (m, k)-problem a *linear* problem.

We first restrict ourselves to the single-seed case. The following lemma demonstrates that iterating smaller seeds solving a cyclic problem allows to obtain a solution for bigger problems, for the same number of mismatches.

Lemma 3. *If a seed Q solves a cyclic (m, k)-problem, then for every $i \geq 0$, the seed $Q_i = [Q, -^{(m-s(Q))}]^i$ solves the linear $(m \cdot (i+1) + s(Q) - 1, k)$-problem. If $i \neq 0$, the inverse holds too.*

Example 2. Observe that the seed `###-#` solves the cyclic $(7, 2)$-problem. From Lemma 3, this implies that for every $i \geq 0$, the $(11 + 7i, 2)$-problem is solved by the seed $[$`###-#`$, $`--`$]^i$ of span $5 + 7i$. Moreover, for $i = 1, 2, 3$, this seed is optimal (maximally weighted) over all seeds solving the problem.

By a similar argument based on Lemma 3, the periodic seed $[$`#####-##`$, $`---`$]^i$ solves the $(18 + 11i, 2)$-problem. Note that its weight grows as $\frac{7}{11}m$ compared to $\frac{4}{7}m$ for the seed from the previous paragraph. However, this is not an asymptotically optimal bound, as we will see later.

The $(18 + 11i, 3)$-problem is solved by the seed $(\texttt{\#\#\#-\#--\#}, \texttt{---})^i$, as seed $\texttt{\#\#\#-\#--\#}$ solves the cyclic $(11, 3)$-problem. For $i = 1, 2$, the former is a maximally weighted seed among all solving the $(18 + 11i, 3)$-problem.

One question raised by these examples is whether there exists a general periodic seed which is asymptotically optimal, i.e. has a maximal asymptotic weight. The following theorem establishes a tight asymptotic bound on the weight of an optimal seed, for a fixed number of mismatches. It gives a negative answer to this question, as it shows that the maximal weight grows faster than any linear fraction of the problem size.

Theorem 2. *Consider a fixed k. Let $w(m)$ be the maximal weight of a seed solving the cyclic (m, k)-problem. Then $(m - w(m)) = \Theta(m^{\frac{k-1}{k}})$.*

The following simple lemma is also useful for constructing efficient seeds.

Lemma 4. *Assume that a family F solves an (m, k)-problem. Let F' be the family obtained from F by cutting out l characters from the left and r characters from the right of each seed of F. Then F' solves the $(m - r - l, k)$-problem.*

Example 3. The $(9 + 7i, 2)$-problem is solved by the seed $[\texttt{\#\#\#}, \texttt{-\#--}]^i$ which is optimal for $i = 1, 2, 3$. Using Lemma 4, this seed can be immediately obtained from the seed $[\texttt{\#\#\#-\#}, \texttt{--}]^i$ from Example 2, solving the $(11 + 7i, 2)$-problem.

We now apply the above results for the single seed case to the case of multiple seeds.

For a seed Q considered as a word over $\{\texttt{\#},\texttt{-}\}$, we denote by $Q_{[i]}$ its cyclic shift to the left by i characters. For example, if $Q = \texttt{\#\#\#\#-\#-\#\#--}$, then $Q_{[5]} = \texttt{\#-\#\#--\#\#\#\#-}$. The following lemma gives a way to construct seed families solving bigger problems from an individual seed solving a smaller cyclic problem.

Lemma 5. *Assume that a seed Q solves a cyclic (m, k) problem and assume that $s(Q) = m$ (otherwise we pad Q on the right with $(m - s(Q))$ jokers). Fix some $i > 1$. For some $L > 0$, consider a list of L integers $0 \le j_1 < \cdots < j_L < m$, and define a family of seeds $F = \langle \|(Q_{[j_l]})^i\| \rangle_{l=1}^{L}$, where $\|(Q_{[j_l]})^i\|$ stands for the seed obtained from $(Q_{[j_l]})^i$ by deleting the joker characters at the left and right edges. Define $\delta(l) = ((j_{l-1} - j_l) \mod m)$ (or, alternatively, $\delta(l) = ((j_l - j_{l-1}) \mod m)$) for all l, $1 \le l \le L$. Let $m' = \max\{s(\|(Q_{[j_l]})^i\|) + \delta(l)\}_{l=1}^{L} - 1$. Then F solves the (m', k)-problem.*

We illustrate Lemma 5 with two examples that follow.

Example 4. Let $m = 11$, $k = 2$. Consider the seed $Q = \texttt{\#\#\#\#-\#-\#\#--}$ solving the cyclic $(11, 2)$-problem. Choose $i = 2$, $L = 2$, $j_1 = 0$, $j_2 = 5$. This gives two seeds $Q_1 = \|(Q_{[0]})^2\| = \texttt{\#\#\#\#-\#-\#\#--\#\#\#\#-\#-\#\#}$ and $Q_2 = \|(Q_{[5]})^2\| = \texttt{\#-\#\#--\#\#\#\#-\#-\#\#--\#\#\#\#}$ of span 20 and 21 respectively, $\delta(1) = 6$ and $\delta(2) = 5$. $\max\{20 + 6, 21 + 5\} - 1 = 25$. Therefore, family $F = \{Q_1, Q_2\}$ solves the $(25, 2)$-problem.

Example 5. Let $m = 11$, $k = 3$. The seed $Q = $ `###-#--#---` solving the cyclic $(11, 3)$-problem. Choose $i = 2$, $L = 2$, $j_1 = 0$, $j_2 = 4$. The two seeds are $Q_1 = \|(Q_{[0]})^2\| = $ `###-#--#---###-#--#` (span 19) and $Q_2 = \|(Q_{[4]})^2\| = $ `#--#---###-#--#---###` (span 21), with $\delta(1) = 7$ and $\delta(2) = 4$. $\max\{19 + 7, 21 + 4\} - 1 = 25$. Therefore, family $F = \{Q_1, Q_2\}$ solves the $(25, 3)$-problem.

4.4 Heuristic Seed Design

Results of Sections 4.1-4.3 allow to construct efficient seed families in certain cases, but still do not allow to perform a systematic seed design. In this section, we briefly outline a heuristic genetic programming algorithm for designing seed families. The algorithm was used in the experimental part of this work, that we present in the next section. Note that this algorithm uses dynamic programming algorithms of Section 3.

The algorithm tries to iteratively improve the characteristics of a *population* of seed families. A sample of family candidates are processed, then some of them are *mutated* and *crossed over* according to the set of (m, k)-similarities they do not detect.

The first step of each iteration is based on screening current families against sets of *difficult similarities*, which are similarities that have been detected by fewer families. These sets are permanently reordered and updated according to the number of families that don't detect those similarities.

For those families that pass through the screening filter, the number of undetected similarities is computed by the dynamic programming algorithm of Section 3.2. The family is kept if it produces a smaller number than the families currently known. To detect seeds to be improved inside a family, we compute the contribution of each seed by the dynamic programming algorithm of Section 3.3. The seeds with the least contribution are then modified.

The entire heuristic procedure does not guarantee finding optimal seeds families but often allows to compute efficient or even optimal solutions in a reasonable time. For example, in ten runs of the algorithm we found 3 of the 6 possible families of two seeds of weight 14 solving the $(25, 2)$-problem. The whole computation took less than 1 hour, compared to a week of computation needed to exhaustively test all seed pairs.

5   Experiments

We describe two groups of experiments we have made. The first one concerns the design of efficient seed families, and the second one applies a multi-seed lossless filtration to the identification of unique oligos in a large set of EST sequences.

Seed Design Experiments

We considered several (m, k)-problems. For each problem, and for a fixed number of seeds in the family, we computed families solving the problem and realizing as large seed weight as possible (under a natural assumption that all seeds in a family have the same weight). We also kept track of the ways (periodic seeds, genetic programming heuristics, exhaustive search) in which those families can be computed.

Tables 1 and 2 summarize some results obtained for the $(25, 2)$-problem and the $(25, 3)$-problem respectively. Families of periodic seeds (that can be found using Lemma 5) are marked with ^{p}, those that are found using a genetic algorithm are marked with ^{g}, and those which are obtained by an exhaustive search are marked with ^{e}. Only in this latter case, the families are guaranteed to be optimal. Families of periodic seeds are shifted according to their construction (see Lemma 5).

Moreover, to compare the selectivity of different families solving a given (m, k)-problem, we estimated the probability δ for at least one of the seeds of the family to match at a given position of a uniform Bernoulli four-letter sequence.

Note that the simple fact of passing from a single seed to a two-seed family results in a considerable gain in efficiency: in both examples shown in the tables there a change of about one order magnitude in the selectivity estimator δ.

Table 1. Seed families for $(25, 2)$-problem

size	weight	family seeds	δ
1	$12^{e,p,g}$	`###-#--###-#--###-#`	$5.96 \cdot 10^{-8}$
2	$14^{e,p,g}$	`####-#-##--####-#-##`	$7.47 \cdot 10^{-9}$
		`#-##--####-#-##--####`	
3	15^p	`#--##-#-######--##-#-##`	$2.80 \cdot 10^{-9}$
		`#-######--##-#-#####`	
		`####--##-#-######--##`	
4	16^p	`###-##-#-###--#######`	$9.42 \cdot 10^{-10}$
		`##-#-###--#######-##-#`	
		`###--#######-##-#-###`	
		`#######-##-#-###--###`	
6	17^p	`##-#-##--#######-####-#`	$3.51 \cdot 10^{-10}$
		`#-##--#######-####-#-##`	
		`#######-####-#-##--###`	
		`###-####-#-##--#######`	
		`####-#-##--#######-###`	
		`##--#######-####-#-##--#`	

Oligo Design Using Multi-seed Filtering

An important practical application of lossless filtration is the design of reliable oligonucleotides for DNA micro-array experiments. Oligonucleotides (oligos) are small DNA sequences of fixed size (usually ranging from 10 to 50) designed to hybridize only with a specific region of the genome sequence. In micro-array experiments, oligos are expected to match ESTs that stem from a given gene and not to match those of other genes. The problem of oligo design can then be formulated as the search for strings of a fixed length that occur in a given sequence but do not occur, within a specified distance, in other sequences of a

Table 2. Seed families for $(25, 3)$-problem

size	weight	family seeds	δ
1	$8^{\ e, p, g}$	`###-#-----###-#`	$1.53 \cdot 10^{-5}$
2	10^p	`####-#-##--#---##`	$1.91 \cdot 10^{-6}$
		`##--#---####-#-##`	
3	11^p	`#---####-#-##--#---##`	$7.16 \cdot 10^{-7}$
		`###-#-##--#--####`	
		`##--#--####-#-##--#`	
4	12^p	`#---####-#-##--#---###`	$2.39 \cdot 10^{-7}$
		`###-#-##--#---####-#`	
		`#-##--#---####-#-##--#`	
		`##--#---####-#-##--#---#`	

given (possibly very large) sample. Different approaches to oligo design apply different distance measures and different algorithmic techniques [15–18]. The experiments we briefly present here demonstrate that the multi-seed filtering provides an efficient solution to the oligo design problem.

family size	weight	δ
1	7^e	$6.10 \cdot 10^{-5}$
2	8^e	$3.05 \cdot 10^{-5}$
3	9^g	$1.14 \cdot 10^{-5}$
4	10^g	$3.81 \cdot 10^{-6}$
6	11^g	$1.43 \cdot 10^{-6}$
10	12^g	$5.97 \cdot 10^{-7}$

```
{ ####---#--------#---#--#### ,
  ###--#--##--------#-#### ,
  ####----#--#--##-### ,
  ###-#-#---##--#### ,
  ###-##-##--#-#-## ,
  ####-##-#-#### }
```

Fig. 1. Computed seed families for the considered $(32, 5)$-problem and the chosen family (6 seeds of weight 11)

We adopt the formalization of the oligo design problem as the problem of identifying in a given sequence (or a sequence database) all substrings of length m that have no occurrences elsewhere in the sequence within the Hamming distance k. The parameters m and k were set to 32 and 5 respectively. For the $(32, 5)$-problem, different seed families were designed and their selectivity was estimated. Those are summarized in the table in Figure 1, using the same conventions as in Tables 1 and 2 above. The family composed of 6 seeds of weight 11 was selected for the filtration experiment (shown in Figure 1).

The filtering has been applied to a database of rice EST sequences composed of 100015 sequences for a total length of 42.845.242 Mb[1]. Substrings matching other substrings with 5 substitution errors or less were computed. The computation took slightly more than one hour on a Pentium 4 3GHz computer. Before applying the filtering using the family for the $(32, 5)$-problem, we made a rough pre-filtering using one spaced seed of weight 16 to detect, with a high selectivity, almost identical regions. As a result of the whole computation, 87% percent of the database were removed by the filtering procedure, leaving the remaining part as oligo candidates.

---

[1] source : `http://bioserver.myongji.ac.kr/ricemac.html`, The Korea Rice Genome Database

6   Conclusion

In this paper, we studied a lossless filtration method based on multi-seed families and demonstrated that it represents an improvement compared to the single-seed approach considered in [1]. We showed how some important characteristics of seed families can be computed using the dynamic programming. We presented several combinatorial results that allow one to construct efficient families composed of seeds with a periodic structure. Finally, we described experimental results providing evidence of the applicability and efficiency of the whole method.

The results of Sections 4.1-4.3 establish several combinatorial properties of seed families, but many more of them remain to be elucidated. We conjecture that the structure of optimal or near-optimal seed families can be studied using the combinatorial design theory, but this relation remains to be clearly established. Another direction is to consider different distance measures, especially the Levenstein distance, or at least to allow some restricted insertion/deletion errors. The method proposed in [19] does not seem to be easily generalized to multi-seed families, and a further work is required to improve lossless filtering in this case.

After this work was completed, it has been brought to our attention that in the context of lossy filtering for local alignement algorithms, the use of optimized multi-seed families has been recently proposed in PATTERNHUNTER II software [20], and the design of those families has been also studied in paper [21].

Acknowledgements

G. Kucherov and L. Noé have been supported by the French *Action Spécifique "Algorithmes et Séquences"* of CNRS. A part of this work has been done during a stay of M. Roytberg at LORIA, Nancy, supported by INRIA. M. Roytberg has been supported by the Russian Foundation for Basic Research (project nos. 03-04-49469, 02-07-90412) and by grants from the RF Ministry for Industry, Science, and Technology (20/2002, 5/2003) and NWO.

References

1. Burkhardt, S., Kärkkäinen, J.: Better filtering with gapped q-grams. Fundamenta Informaticae **56** (2003) 51–70 Preliminary version in Combinatorial Pattern Matching 2001.
2. Navarro, G., Raffinot, M.: Flexible Pattern Matching in Strings – Practical on-line search algorithms for texts and biological sequences. Cambridge University Press (2002) ISBN 0-521-81307-7. 280 pages.
3. Altschul, S., Madden, T., Schäffer, A., Zhang, J., Zhang, Z., Miller, W., Lipman, D.: Gapped BLAST and PSI-BLAST: a new generation of protein database search programs. Nucleic Acids Research **25** (1997) 3389–3402
4. Ma, B., Tromp, J., Li, M.: PatternHunter: Faster and more sensitive homology search. Bioinformatics **18** (2002) 440–445

5. Schwartz, S., Kent, J., Smit, A., Zhang, Z., Baertsch, R., Hardison, R., Haussler, D., Miller, W.: Human–mouse alignments with BLASTZ. Genome Research **13** (2003) 103–107
6. Noe, L., Kucherov, G.: YASS: Similarity search in DNA sequences. Research Report RR-4852, INRIA (2003) `http://www.inria.fr/rrrt/rr-4852.html`.
7. Pevzner, P., Waterman, M.: Multiple filtration and approximate pattern matching. Algorithmica **13** (1995) 135–154
8. Califano, A., Rigoutsos, I.: Flash: A fast look-up algorithm for string homology. In: Proceedings of the 1st International Conference on Intelligent Systems for Molecular Biology. (1993) 56–64
9. Buhler, J.: Provably sensitive indexing strategies for biosequence similarity search. In: Proceedings of the 6th Annual International Conference on Computational Molecular Biology (RECOMB02), Washington, DC (USA), ACM Press (2002) 90–99
10. Keich, U., Li, M., Ma, B., Tromp, J.: On spaced seeds for similarity search. Discrete Applied Mathematics (2004) to appear.
11. Buhler, J., Keich, U., Sun, Y.: Designing seeds for similarity search in genomic DNA. In: Proceedings of the 7th Annual International Conference on Computational Molecular Biology (RECOMB03), Berlin (Germany), ACM Press (2003) 67–75
12. Brejova, B., Brown, D., Vinar, T.: Vector seeds: an extension to spaced seeds allows substantial improvements in sensitivity and specificity. In Benson, G., Page, R., eds.: Proceedings of the 3rd International Workshop in Algorithms in Bioinformatics (WABI), Budapest (Hungary). Volume 2812 of Lecture Notes in Computer Science., Springer (2003) 39–54
13. Kucherov, G., Noe, L., Ponty, Y.: Estimating seed sensitivity on homogeneous alignments. In: Proceedings of the IEEE 4th Symposium on Bioinformatics and Bioengineering (BIBE2004), May 19-21, 2004, Taichung (Taiwan), IEEE Computer Society Press (2004) to appear.
14. Choi, K., Zhang, L.: Sensitivity analysis and efficient method for identifying optimal spaced seeds. Journal of Computer and System Sciences (2003) to appear.
15. Li, F., Stormo, G.: Selection of optimal DNA oligos for gene expression arrays. Bioinformatics **17** (2001) 1067–1076
16. Kaderali, L., Schliep, A.: Selecting signature oligonucleotides to identify organisms using DNA arrays. Bioinformatics **18** (2002) 1340–1349
17. Rahmann, S.: Fast large scale oligonucleotide selection using the longest common factor approach. Journal of Bioinformatics and Computational Biology **1** (2003) 343–361
18. Zheng, J., Close, T., Jiang, T., Lonardi, S.: Efficient selection of unique and popular oligos for large est databases. In: LNCS 2676, Proceedings of Symposium on Combinatorial Pattern Matching (CPM'03), Springer (2003) 273–283
19. Burkhardt, S., Karkkainen, J.: One-gapped q-gram filters for Levenshtein Distance. In: Proceedings of the 13th Symposium on Combinatorial Pattern Matching (CPM'02). Volume 2373., Springer (2002) 225–234
20. Li, M., Ma, B., Kisman, D., Tromp, J.: PatternHunter II: Highly sensitive and fast homology search. Journal of Bioinformatics and Computational Biology (2004) Earlier version in GIW 2003 (International Conference on Genome Informatics).
21. Sun, Y., Buhler, J.: Designing multiple simultaneous seeds for DNA similarity search. In: Proceedings of the 8th Annual International Conference on Research in Computational Molecular Biology (RECOMB 2004), ACM Press (2004)

New Results for the 2-Interval Pattern Problem

Guillaume Blin[1], Guillaume Fertin[1], and Stéphane Vialette[2]

[1] LINA, FRE CNRS 2729
Université de Nantes, 2 rue de la Houssinière
BP 92208 44322 Nantes Cedex 3, France
{blin,fertin}@lina.univ-nantes.fr
[2] LRI, UMR CNRS 8623
Faculté des Sciences d'Orsay, Université Paris-Sud
Bât 490, 91405 Orsay Cedex, France
vialette@lri.fr

Abstract. We present new results concerning the problem of finding a constrained pattern in a set of 2-intervals. Given a set of n 2-intervals $\mathcal{D}$ and a model R describing if two disjoint 2-intervals can be in precedence order ($<$), be allowed to nest ($\sqsubset$) and/or be allowed to cross ($\between$), the problem asks to find a maximum cardinality subset $\mathcal{D}' \subseteq \mathcal{D}$ such that any two 2-intervals in $\mathcal{D}'$ agree with R. We improve the time complexity of the best known algorithm for $R = \{\sqsubset\}$ by giving an optimal $O(n \log n)$ time algorithm. Also, we give a graph-like relaxation for $R = \{\sqsubset, \between\}$ that is solvable in $O(n^2 \sqrt{n})$ time. Finally, we prove that the problem is **NP**-complete for $R = \{<, \between\}$, and in addition to that, we give a fixed-parameter tractability result based on the crossing structure of $\mathcal{D}$.

1 Introduction

The general problem of establishing a general representation of structured patterns, *i.e.*, *macroscopic describers* of RNA secondary structures, was considered in [Via02,Via04]. The approach was to set up a *geometric* description of helices by means of a natural generalization of intervals, namely a 2-*interval*. A 2-interval is the disjoint union of two intervals on the line. The geometric properties of 2-intervals provide a possible guide for understanding the computational complexity of finding structured patterns in RNA sequences. Using a model to represent non sequential information allows us for varying restrictions on the complexity of the pattern structure. Indeed, two disjoint 2-intervals, *i.e.*, two 2-intervals that do not intersect in any point, can be in precedence order ($<$), be allowed to nest ($\sqsubset$) and/or be allowed to cross ($\between$). Furthermore, the set of 2-intervals and the pattern can have different restrictions. These different combinations of restrictions alter the computational complexity of the problems, and need to be examined separately. This examination produces efficient algorithms for more restrictive structured patterns, and hardness results for those less restrictive.

There are basically two lines of research our results refer to. The first one is that of arc annotated sequences and the other one is that of protein topologies. In the context of arc annotated sequences, the Arc-Preserving Subsequence (APS) and Longest Arc-Preserving Common Subsequence

S.C. Sahinalp et al. (Eds.): CPM 2004, LNCS 3109, pp. 311–322, 2004.
© Springer-Verlag Berlin Heidelberg 2004

(LAPCS) problems are useful in representing the structural information of RNA and protein sequences [Eva99,JLMZ00,GGN02,AGGN02]. The basic idea is to provide a measure for similarity, not only on the sequence level, but also on the structural level. Moreover, a similar problem to compare the three-dimensional structure of proteins is the CONTACT MAP OVERLAP problem described by Goldman *et al* [GIP99]. Viksna and Gilbert described algorithms for pattern matching and pattern learning in TOPS diagram (formal description of protein topologies) [VD01].

Our results are also related to the independent set problem in different extensions of 2-interval graphs. A graph G is a t-interval graph if there is an intersection model whose objects consist of collections of t intervals, $t \geq 1$, such that G is the intersection graph of this model [TH79,GW79]. From this definition, it is clear that every interval graph is a 1-interval graph. Of particular interest is the class of 2-interval graphs. For example, line graphs, trees and circular-arc graphs are 2-interval graphs. However, West and Shmoys [WS84] have shown that the recognition problem for t-interval graphs is **NP**-complete for every $t \geq 2$ (this has to be compared with linear time recognition of 1-interval graphs). In the context of sequence similarity, [JMT92] contains an application of graphs having interval number at most two. In [BYHN+02], the authors considered the problem of scheduling jobs that are given as groups of non-intersecting segments on the real line. Of particular importance, they showed that the maximum weighted independent set for t-interval graphs ($t \geq 2$) is **APX**-hard even for highly restricted instances Also, they gave a $2t$-approximation algorithm for general instances based on a fractional version of the Local Ratio Technique.

The problem of finding the longest 2-interval pattern in a set of 2-intervals $\mathcal{D}$ with respect to a given abstract model, the so-called 2-INTERVAL PATTERN problem, has been introduced by Vialette [Via02,Via04]. Vialette divides the problem in different classes based on the structure of the model and gives for most of them either **NP**-completeness results or polynomial time algorithms. In the present paper, we focus on three classes: the model $\{\sqsubset\}$ over an unlimited support, the model $\{\sqsubset, \emptyset\}$ over a disjoint support and the model $\{<, \emptyset\}$ over a unitary support. We give precise results for these three classes. Those three classes are of importance since each one is a straightforward extension of the PATTERN MATCHING OVER SET OF 2-INTERVALS problem introduced in [Via04] and hence is strongly related, in the context of molecular biology, to pattern matching over RNA secondary structures. The results given in the present paper almost complete the table proposed by Vialette [Via04] (see Table 1) and provide an important step towards a better understanding of the precise complexity of 2-interval pattern matching problems.

The remainder of the paper is organized as follows. In Section 2 we briefly review the terminology introduced in [Via04]. In Section 3, we improve the time complexity of the best known algorithm for model $R = \{\sqsubset\}$ over an unlimited support. In Section 4, we give a graph-like relaxation for model $\{\sqsubset, \emptyset\}$ that is solvable in polynomial time. In Section 5, we prove that the 2-interval pattern problem for model $R = \{<, \emptyset\}$ is **NP**-complete even when restricted to unitary support thereby answering an open problem posed in [Via04]. In addition to that latter result, we give in Section 6 a fixed-parameter tractability result based on the crossing structure of $\mathcal{D}$.

2   Preliminaries

An interval and a 2-interval represent respectively a sequence of contiguous bases and pairings between two intervals, *i.e.*, *stems*, in RNA secondary structures. Thus, 2-intervals can be seen as *macroscopic describers* of RNA structures.

Formally, a 2-*interval* is the disjoint union of two intervals on a line. We denote it by $D = (I_1, J_1)$ where I_1 and J_1 are intervals such that $I_1 < J_1$ (here $<$ is the strict precedence order between intervals); in that case we write also $\mathsf{Left}(D) = I_1$ and $\mathsf{Right}(D) = J_1$. If $[x : y]$ and $[x' : y']$ are two intervals such that $[x : y] < [x' : y']$, we will sometimes write $D = ([x : y], [x' : y'])$ to emphasize on the precise definition of the 2-interval D. Let $D_1 = (I_1, J_1)$ and $D_2 = (I_2, J_2)$ be two 2-intervals. They are called *disjoint* if $(I_1 \cup J_1) \cap (I_2 \cup J_2) = \emptyset$ (*i.e.*, involved intervals do not intersect). The *covering interval* of a 2-interval D, written $\mathsf{Cover}(D)$, is the least interval covering both $\mathsf{Left}(D)$ and $\mathsf{Right}(D)$.

Of particular interest is the relation between two disjoint 2-intervals $D_1 = (I_1, J_1)$ and $D_2 = (I_2, J_2)$. We will write $D_1 < D_2$ if $I_1 < J_1 < I_2 < J_2$, $D_1 \sqsubset D_2$ if $I_2 < I_1 < J_1 < J_2$ and $D_1 \between D_2$ if $I_1 < I_2 < J_1 < J_2$. Two 2-intervals D_1 and D_2 are τ-comparable for some $\tau \in \{<, \sqsubset, \between\}$ if $D_1 \tau D_2$ or $D_2 \tau D_1$. Let $\mathcal{D}$ be a set of 2-intervals and $R \subseteq \{<, \sqsubset, \between\}$ be non-empty. The set $\mathcal{D}$ is R-*comparable* if any two distinct 2-intervals of $\mathcal{D}$ are τ-comparable for some $\tau \in R$. Throughout the paper, the non-empty subset R is called a *model*. Clearly, if a set of 2-intervals $\mathcal{D}$ is R-comparable then $\mathcal{D}$ is a set of disjoint 2-intervals. The *support* of a set of 2-intervals $\mathcal{D}$, written $\mathsf{Support}(\mathcal{D})$, is the set of all *simple* intervals involved in $\mathcal{D}$, *i.e.*, $\mathsf{Support}(\mathcal{D}) = \bigcup_{D \in \mathcal{D}} (\mathsf{Left}(D) \cup \mathsf{Right}(D))$. The *leftmost* (resp. *rightmost*) element of a set of disjoint 2-intervals $\mathcal{D}$ is the 2-interval $D_i \in \mathcal{D}$ such that $\mathsf{Left}(D_i) < \mathsf{Left}(D_j)$ (resp. $\mathsf{Right}(D_j) < \mathsf{Right}(D_i)$) for all $D_j \in \mathcal{D} - D_i$. Observe that it could be the case that D_i is both the leftmost and rightmost element of $\mathcal{D}$ (this is indeed the case if $|\mathcal{D}| = 1$ or if $D_j \sqsubset D_i$ for all $D_j \in \mathcal{D} - D_i$). Some parameters can be defined. The *width* of $\mathcal{D}$, written $\mathsf{Width}(\mathcal{D})$, is the size of a maximum cardinality $\{<\}$-comparable subset of $\mathcal{D}$, the *height* of $\mathcal{D}$, written $\mathsf{Height}(\mathcal{D})$, is the size of a maximum cardinality $\{\sqsubset\}$-comparable subset of $\mathcal{D}$ and the *depth* of $\mathcal{D}$, written $\mathsf{Depth}(\mathcal{D})$, is the size of a maximum cardinality $\{\between\}$-comparable subset of $\mathcal{D}$. Observe that these three parameters can be computed in polynomial time [Via04]. Finally, the *forward crossing number* of $\mathcal{D}$, written $\mathsf{FCrossing}(\mathcal{D})$, is defined by $\mathsf{FCrossing}(\mathcal{D}) = \max_{D_i \in \mathcal{D}} |\{D_j : D_i \between D_j\}|$. Clearly, $\mathsf{Depth}(\mathcal{D}) \leq \mathsf{FCrossing}(\mathcal{D})$.

In [Via04], Vialette proposed two restrictions on the support:

1. all the intervals of the support are of the same size;
2. all the intervals of the support are disjoint, *i.e.*, if two intervals $I, I' \in \mathsf{Support}(\mathcal{D})$ overlap then $I = I'$.

Using restrictions on the support allows us for varying restrictions on the complexity of the 2-interval set structure, and hence on the complexity of the problems. These two restrictions involve three levels of complexity:

- UNLIMITED: no restrictions
- UNITARY: restriction 1
- DISJOINT: restrictions 1 and 2

Given a set of 2-intervals $\mathcal{D}$, a model $R \subseteq \{<, \sqsubset, \lozenge\}$ and a positive integer k, the 2-INTERVAL PATTERN problem consists in finding a subset $\mathcal{D}' \subseteq \mathcal{D}$ of cardinality greater than or equal to k such that $\mathcal{D}'$ is R-comparable. For the sake of brevity, the 2-INTERVAL PATTERN problem with respect to a model R over an unlimited (resp. unitary, disjoint) support is abbreviated in 2-IP-UNL-R (resp. 2-IP-UNI-R, 2-IP-DIS-R).

Vialette proved in [Via04] that 2-IP-UNI-$\{<, \sqsubset, \lozenge\}$ and 2-IP-UNI-$\{\sqsubset, \lozenge\}$ are **NP**-complete. Moreover, he gave polynomial algorithms for the problem with respect to the models $\{<\}$, $\{\sqsubset\}$, $\{\lozenge\}$ and $\{<, \sqsubset\}$ (cf. Table 1).

In this article, we answer three open problems and we improve the complexity of another one as shown in Table 1. Moreover, we show that 2-IP-UNI-$\{<, \lozenge\}$ is fixed parameter tractable when parameterized by the forward crossing number of $\mathcal{D}$.

Table 1. 2-INTERVAL PATTERN problem complexity where $n = |\mathcal{D}|$. When not specified, the complexity comes from [Via04]. ⋆ contributions of the present paper. • improvement of the existing complexity (which was $O(n^2)$ in [Via04]).

2-INTERVAL PATTERN PROBLEM			
	SUPPORT		
MODEL	UNLIMITED	UNITARY	DISJOINT
$\{<, \sqsubset, \lozenge\}$	**NP**-complete		$O(n\sqrt{n})$ [MV80]
$\{\sqsubset, \lozenge\}$	**NP**-complete		$O(n^2\sqrt{n})$ ⋆
$\{<, \sqsubset\}$	$O(n^2)$		
$\{<, \lozenge\}$	**NP**-complete ⋆		?
$\{<\}$	$O(n \log n)$		
$\{\sqsubset\}$	$O(n \log n)$ ⋆ •		
$\{\lozenge\}$	$O(n^2 \log n)$		

3   Improving the Complexity of 2-IP-UNL-$\{\sqsubset\}$

The problem of finding the largest $\{\sqsubset\}$-comparable subset in a set of 2-intervals was considered in [Via04]. Observing that this problem is equivalent to finding a largest clique in a comparability graph (a linear time solvable problem [Gol80]), an $O(n^2)$ time algorithm was thus proposed. We improve that result by giving an optimal $O(n \log n)$ time algorithm for finding a largest $\{\sqsubset\}$-comparable subset in a set of 2-intervals.

The inefficiency of the algorithm proposed in [Via04] lies in the effective construction of a comparability graph. We show that this construction can be avoided by considering trapezoids in place of 2-intervals. Recall that a *trapezoid graph* is the intersection graph of a finite set of trapezoids between two parallel lines [DGP88] (it is easily seen that trapezoid graphs generalize both interval graphs and permutation graphs). Analogously to 2-intervals, we will denote by $T = ([x : y], [x' : y'])$ the trapezoid with upper interval $[x : y]$ and lower interval $[x' : y']$.

Proposition 1. 2-IP-UNL-$\{\sqsubset\}$ *is solvable in* $O(n \log n)$ *time.*

Proof. Let $\mathcal{D} = \{D_1, D_2, \ldots, D_n\}$ be a collection of 2-intervals of the real line. Construct a collection of trapezoids $\mathcal{T} = \{T_1, T_2, \ldots, T_n\}$ between two parallel lines as follows. For each 2-interval $D_i = ([x : y], [x' : y']) \in \mathcal{D}$, we add the trapezoid $T_i = ([x : y], [-y' : -x'])$ to $\mathcal{T}$.

Claim. For all $1 \leq i \leq j \leq n$, the 2-intervals D_i and D_j are $\{\sqsubset\}$-comparable if and only if the trapezoids T_i and T_j are non-intersecting.

Proof (of Claim 3). Let $D_i = ([x_i : y_i], [x'_i : y'_i])$ and $D_j = ([x_j : y_j], [x'_j : y'_j])$ be two 2-intervals of $\mathcal{D}$ and $T_i = ([x_i : y_i], [-y'_i : -x'_i])$ and $T_j = ([x_j : y_j], [-y'_j : -x'_j])$ be the two corresponding trapezoids in $\mathcal{T}$. Suppose that D_i and D_j are $\{\sqsubset\}$-comparable. Without loss of generality, we may assume $D_j \sqsubset D_i$. Thus, we have $y_i < x_j$ and $y'_j < x'_i$. It follows immediately that $-x'_i < -y'_j$, and hence the two trapezoids T_i and T_j are non-intersecting. The proof of the converse is identical. $\square$

Clearly, the collection $\mathcal{T}$ can be constructed in $O(n)$ time. Based on a geometric representation of trapezoid graphs by boxes in the plane, Felsner *et al.* [FMW97] have designed a $O(n \log n)$ algorithm for finding a maximum cardinality subcollection of non-intersecting trapezoids in a collection of trapezoids, and the proposition follows. $\square$

Based on Fredman's bound for the number of comparisons needed to compute maximum increasing subsequences in permutation [Fre75], Felsner *et al.* [FMW97] argued that their $O(n \log n)$ time algorithm for finding a maximum cardinality subcollection of non-intersecting trapezoids in a collection of trapezoids is optimal. Then it follows from Proposition 1 that our $O(n \log n)$ time algorithm for finding a maximum cardinality $\{\sqsubset\}$-comparable subset in a set of 2-intervals is optimal as-well.

4   A Polynomial Time Algorithm for 2-IP-DIS-$\{\sqsubset, \between\}$

In this section, we give a $O(n^2 \sqrt{n})$ time algorithm for the 2-IP-DIS-$\{\sqsubset, \between\}$ problem, where n is the cardinality of the set of 2-intervals $\mathcal{D}$. Recall that given a set of 2-intervals $\mathcal{D}$ over a disjoint support, the problem asks to find the size of a maximum cardinality $\{\sqsubset, \between\}$-comparable subset $\mathcal{D}' \subseteq \mathcal{D}$. Observe that the

2-IP-Dis-$\{\sqsubset, \lozenge\}$ problem has an interesting formulation in terms of constrained matchings in general graphs: Given a graph G together with a linear ordering π of its vertices, the 2-IP-Dis-$\{\sqsubset, \lozenge\}$ problem is equivalent to finding a maximum cardinality matching $\mathcal{M}$ in G with the property that for any two distinct edges $\{u, v\}$ and $\{u', v'\}$ of $\mathcal{M}$ neither $\max\{\pi(u), \pi(v)\} < \min\{\pi(u'), \pi(v')\}$ nor $\max\{\pi(u'), \pi(v')\} < \min\{\pi(u), \pi(v)\}$ occur.

Roughly speaking, our algorithm is based on a three-step procedure. First, the interval graph of all the covering intervals of 2-intervals in $\mathcal{D}$ is constructed. Next, all the maximal cliques of that graph are efficiently computed. Finally, for each maximal clique we construct a new graph and find a solution using a maximum cardinality matching algorithm. The size of a best solution found in the third step is thus returned. Clearly, the efficiency of our algorithm relies upon an efficient algorithm for finding all the maximal cliques in the intersection of the covering intervals. We now proceed with the details of our algorithm.

Let $\mathcal{D} = \{D_i : 1 \leq i \leq n\}$ be a set of 2-intervals. Consider the set $\mathcal{C}_\mathcal{D}$ composed of all the covering intervals of the 2-intervals in $\mathcal{D}$, $i.e.$, $\mathcal{C}_\mathcal{D} = \{\mathsf{Cover}(D) : D \in \mathcal{D}\}$. Now, let $\Omega(\mathcal{C}_\mathcal{D})$ be the interval graph associated with $\mathcal{C}_\mathcal{D}$. The graph $\Omega(\mathcal{C}_\mathcal{D})$ has a vertex v_i for each interval $\mathsf{Cover}(D_i)$ in $\mathcal{C}_\mathcal{D}$ and two vertices v_i and v_j of $\Omega(\mathcal{C}_\mathcal{D})$ are joined by an edge if the two associated intervals $\mathsf{Cover}(D_i)$ and $\mathsf{Cover}(D_j)$ intersect. Most in the interest in the interval graph $\Omega(\mathcal{C}_\mathcal{D})$ stems from the following lemma.

Lemma 1. *Let $\mathcal{D}$ be a set of 2-intervals and $\mathcal{D}'$ be a $\{\sqsubset, \lozenge\}$-comparable subset of $\mathcal{D}$. Then, $\{v_i : D_i \in \mathcal{D}'\}$ induces a complete graph in $\Omega(\mathcal{C}_\mathcal{D})$.*

Proof. Let D_i and D_j be two distinct 2-intervals of $\mathcal{D}'$. Since D_i and D_j are $\{\sqsubset, \lozenge\}$-comparable then it follows that either intervals $\mathsf{Cover}(D_i)$ and $\mathsf{Cover}(D_j)$ overlap or one interval is included in the other. In both cases, intervals $\mathsf{Cover}(D_i)$ and $\mathsf{Cover}(D_j)$ intersect and hence vertices v_i and v_j are joined by an edge in $\Omega(\mathcal{C}_\mathcal{D})$. Therefore $\{v_i : D_i \in \mathcal{D}'\}$ induces a complete graph in $\Omega(\mathcal{C}_\mathcal{D})$. $\quad\square$

Observe that the converse is false since the intersection of two 2-intervals in $\mathcal{D}$ results in an edge in $\Omega(\mathcal{C}_\mathcal{D})$, and hence two 2-intervals associated to two distinct vertices in the maximal clique C may not be $\{\sqsubset, \lozenge\}$-comparable. However, thanks to Lemma 1 we now only need to focus on maximal cliques of $\Omega(\mathcal{C}_\mathcal{D})$. Several problems that are **NP**-complete on general graphs have polynomial time algorithms for interval graphs. The problem of finding all the maximal cliques of a graph is one such example. Indeed, an interval graph $G = (V, E)$ is a chordal graph and as such has at most $|V|$ maximal cliques [FG65]. Furthermore, all the maximal cliques of a chordal graph can be found in $O(n + m)$ time, where $n = |V|$ and $m = |E|$, by a modification of Maximum Cardinality Search (MCS) [TY84,BP93].

Let C be a maximal clique of $\Omega(\mathcal{C}_\mathcal{D})$. As observed above, any two 2-intervals associated to two distinct vertices in the maximal clique C may not be $\{\sqsubset, \lozenge\}$-comparable. Let $\mathcal{D}' \subseteq \mathcal{D}$ be the set of all 2-intervals associated to vertices in the maximal clique C. Based on C, consider the graph $G_C = (V_C, E_C)$ defined by $V_C = \mathsf{Support}(\mathcal{D}')$ and $E_C = \{\{I, J\} : D = (I, J) \in \mathcal{D}'\}$. In other words, the

---

Max $\{\sqsubset, \lozenge\}$-Comparable 2-Interval Pattern

Input: A set of 2-intervals $\mathcal{D}$ with disjoint support
Output: The size of a maximum cardinality $\{\sqsubset, \lozenge\}$-comparable subset of $\mathcal{D}$

1. Construct the interval graph $\Omega(\mathcal{C}_\mathcal{D})$
2. Compute all maximal cliques in $\Omega(\mathcal{C}_\mathcal{D})$
3. For each maximal clique C in $\Omega(\mathcal{C}_\mathcal{D})$
   3.1. Construct the graph G_C
   3.2. Compute a maximal matching $\mathcal{M}$ in G_C
   3.3. Store the cardinality of $\mathcal{M}$ in $m(C)$
4. Return $\max\{m(C) : C$ is a maximal clique of $\Omega(\mathcal{C}_\mathcal{D})\}$

Fig. 1. Algorithm Max $\{\sqsubset, \lozenge\}$-Comparable 2-Interval Pattern.

set of vertices of G_C is the support of $\mathcal{D}'$ and the edges of G_C is the 2-interval subset $\mathcal{D}'$ itself viewed as a set of subsets of size 2. Note that the construction of G_C is possible only because $\mathcal{D}'$ has disjoint support. The following lemma is an immediate consequence of the definition of G_C and Lemma 1.

Lemma 2. *Let C be a clique in $\Omega(\mathcal{C}_\mathcal{D})$ and $G_C = (V_C, E_C)$ be the graph constructed as detailed above. Then, $\{(I_{i_1}, J_{i_1}), (I_{i_2}, J_{i_2}), \ldots, (I_{i_k}, J_{i_k})\}$ is a $\{\sqsubset, \lozenge\}$-comparable subset if and only if $\{\{I_{i_1}, J_{i_1}\}, \{I_{i_2}, J_{i_2}\}, \ldots, \{I_{i_k}, J_{i_k}\}\}$ is a matching in G_C.*

Proposition 2. *The 2-IP-DIS-$\{\sqsubset, \lozenge\}$ problem is solvable in $O(n^2\sqrt{n})$ time, where n is the number of 2-intervals in $\mathcal{D}$.*

Proof. Consider the algorithm given in Figure 1. Correctness of this algorithm follows from Lemmas 1 and 2. What is left is to prove the time complexity. Clearly, the interval graph $\Omega(\mathcal{C}_\mathcal{D})$ can be constructed in $O(n^2)$ time. All the maximal cliques of $\Omega(\mathcal{C}_\mathcal{D})$ can be found in $O(n+m)$ time, where m is the number of edges in $\Omega(\mathcal{C}_\mathcal{D})$ [TY84,BP93]. Summing up, the first two steps can be done in $O(n^2)$ time since $m < n^2$. We now turn to the time complexity of the loop (in fact the dominant term of our analysis). For each maximal clique C of $\Omega(\mathcal{C}_\mathcal{D})$, the graph G_C can be constructed in $O(n)$ time since $|C| \le n$. We now consider the computation of a maximal matching in G_C. Micali and Vazirani [MV80] (see also [Vaz94]) gave an $O(\sqrt{|V|}|E|)$ time algorithm for finding a maximal matching in a graph $G = (V, E)$. But G_C has at most n edges (as each edge corresponds to a 2-interval) and hence has at most $2n$ vertices. Then it follows that a maximum matching $\mathcal{M}$ in G_C can be found in $O(n\sqrt{n})$ time. Since $\Omega(\mathcal{C}_\mathcal{D})$ is an interval graph with n vertices, it has at most n maximal cliques [FG65], we conclude that the algorithm as a whole runs in $O(n^2\sqrt{n})$ time. $\qquad\square$

5 2-IP-UNI-$\{<, \lozenge\}$ Is NP-complete

Theorem 1 below completes the analysis of 2-IP-UNI-R and 2-IP-UNL-R for any model $R \subseteq \{<, \sqsubset, \lozenge\}$ (see Table 1).

Theorem 1. *The* 2-IP-UNI-$\{<, \emptyset\}$ *problem is* **NP**-*complete.*

Proof. The proof is by reduction from the EXACT 3-CNF SAT problem. Due to space considerations, the rather technical proof is deferred to the full version of this paper.

6   A Fixed-Parameter Algorithm for 2-IP-UNI-$\{<, \emptyset\}$

According to Theorem 1, finding the largest $\{<, \emptyset\}$-comparable subset in a set of 2-intervals on a unitary support is an **NP**-complete problem. In this section, we give an exact algorithm for that problem with strong emphasis on the crossing structure of the set of 2-intervals. More precisely, we consider the time complexity of the problem with respect to the *forward crossing number* of the input. Indeed, in the context of 2-intervals, one may reasonably expect the forward crossing number to be small compared to the number of 2-intervals. Therefore, a natural direction seems to be the question for the fixed-parameter tractability with respect to parameter $\mathsf{FCrossing}(\mathcal{D})$. In response to that question, we show that the problem can be solved for any support by means of dynamic programming in $O(n \cdot \mathsf{FCrossing}(\mathcal{D}) \cdot 2^{\mathsf{FCrossing}(\mathcal{D})}(\log(n) + \mathsf{FCrossing}(\mathcal{D})))$ time where n is the number of 2-intervals in $\mathcal{D}$, and hence is fixed-parameter tractable with respect to parameter $\mathsf{FCrossing}(\mathcal{D})$.

For any $D_i \in \mathcal{D}$, let $T(D_i)$ denote the size of the largest $\{<, \emptyset\}$-comparable subset $\mathcal{D}' \subseteq \mathcal{D}$ of which the 2-interval D_i is the rightmost element. Furthermore, for any $D_i, D_j \in \mathcal{D}$ such that $D_j \emptyset D_i$, let $T(D_j \mid D_i)$ denotes the size of the largest $\{<, \emptyset\}$-comparable subset $\mathcal{D}' \subseteq \mathcal{D}$ such that (1) the 2-interval D_j is the rightmost element of $\mathcal{D}'$ and (2) the 2-interval D_i is not part of the subset $\mathcal{D}'$ but can safely be added to $\mathcal{D}'$ to obtain a new $\{<, \emptyset\}$-comparable subset of size $|\mathcal{D}'| + 1$.

Clearly, a maximum cardinality $\{<, \emptyset\}$-comparable subset $\mathcal{D}' \subseteq \mathcal{D}$ of which the 2-interval D_i is the rightmost element can be obtained either (1) by adding D_i to a maximum cardinality $\{<, \emptyset\}$-comparable subset $\mathcal{D}'' \subseteq \mathcal{D}$ whose rightmost 2-interval D_j precedes the 2-interval D_i, *i.e.*, $D_j < D_i$, or (2) by adding D_i to a maximum cardinality $\{<, \emptyset\}$-comparable subset $\mathcal{D}'' \subseteq \mathcal{D}$ whose rightmost 2-interval D_j crosses the 2-interval D_i, *i.e.*, $D_j \emptyset D_i$, and such that D_i crosses or precedes any 2-interval of $\mathcal{D}''$. Here is another way of stating these observations:

$$\forall D_i \in \mathcal{D}, \qquad T(D_i) = 1 + \max \begin{cases} \max \{T(D_j) : D_j < D_i\} \\ \max \{T(D_j \mid D_i) : D_j \emptyset D_i\} \end{cases} \tag{1}$$

What is left is thus to compute $T(D_j \mid D_i)$. To this aim, we extend the notation $T(D_j \mid D_i)$ as follows: for any $\{\emptyset\}$-comparable subset $\{D_{i_1}, D_{i_2}, \ldots, D_{i_k}\} \subseteq \mathcal{D}$, $k \geq 1$, satisfying $\mathsf{Right}(D_{i_1}) < \mathsf{Right}(D_{i_2}) < \ldots < \mathsf{Right}(D_{i_k})$, we let $T(D_{i_1} \mid D_{i_2}, \ldots, D_{i_k})$ stand for the size of a largest $\{<, \emptyset\}$-comparable subset $\mathcal{D}' \subseteq \mathcal{D}$ such that (1) the 2-interval D_{i_1} is the rightmost element of $\mathcal{D}'$ and (2) the 2-intervals $\{D_{i_2}, D_{i_3}, \ldots, D_{i_k}\}$ are not part of the subset $\mathcal{D}'$ but can safely be added to $\mathcal{D}'$ to obtain a new $\{<, \emptyset\}$-comparable subset of size $T(D_{i_1} \mid$

$D_{i_2}, \ldots, D_{i_k}) + k - 1$. A straightforward extension of the calculation (1) yields the following recurrence relation for computing the entry $T(D_{i_1} \mid D_{i_2}, \ldots, D_{i_k})$ of the dynamic programming table:

$$T(D_{i_1} \mid D_{i_2}, \ldots, D_{i_k}) = 1+$$
$$\max \begin{cases} \max \{T(D_j) \mid D_j \text{ satisfies condition (1)}\} \\ \max \{T(D_j \mid D_{i_1}) \mid D_j \text{ satisfies condition (2)}\} \\ \max \{T(D_j \mid D_{i_1}, D_{i_2}) \mid D_j \text{ satisfies condition (3)}\} \\ \vdots \\ \max \{T(D_j \mid D_{i_1}, D_{i_2}, \ldots, D_{i_k}) \mid D_j \text{ satisfies condition } (k+1)\} \end{cases} \quad (2)$$

where condition (i), $1 \le i \le k + 1$, is defined as follows:

$$\text{condition } (i) \quad \begin{cases} D_j \between D_{i_r} & \text{for all } 0 < r < i \quad \text{(crossing conditions)} \\ D_j < D_{i_s} & \text{for all } i \le s < k+1 \quad \text{(precedence conditions)} \end{cases}$$

It follows from the above recurrence relation that entries of the form $T(D_i \mid *)$ depend only on entries of the form $T(D_j \mid *)$ where $D_j < D_i$ or $D_j \between D_i$. From a computational point of view, this implies that the calculation of entries of the form $T(D_i \mid *)$ depends only on of the calculation of entries of the form $T(D_j \mid *)$ where $\mathsf{Right}(D_j) < \mathsf{Right}(D_i)$. The following easy lemma gives an upper-bound on the size of the dynamic programming table T with respect to the forward crossing number of $\mathcal{D}$.

Lemma 3. *The number of distinct entries of the dynamic programming table T is upper-bounded by $|\mathcal{D}| \cdot 2^{\mathsf{FCrossing}(\mathcal{D})}$.*

Proof. For any 2-interval $D_i \in \mathcal{D}$, the number of distinct $\{\between\}$-comparable subsets of which D_i is the leftmost element is upper-bounded by $2^{\mathsf{FCrossing}(\mathcal{D})}$, and hence there exist at most $2^{\mathsf{FCrossing}(\mathcal{D})}$ distinct entries of the form $T(D_i \mid *)$ in the dynamic programming table T. □

The overall algorithm for finding the size of the largest $\{<, \between\}$-comparable subset in a set of 2-intervals is given in Figure 2. Using a suitable data structure for efficiently searching 2-intervals, we have the following result (proof deferred to the full version of this paper).

Proposition 3. *Algorithm* Max $\{<, \between\}$-Comparable 2-Interval Pattern *returns the size of a maximum cardinality $\{<, \between\}$-comparable subset of a set of 2-intervals $\mathcal{D}$ in $O(n^2 \cdot \mathsf{FCrossing}(\mathcal{D}) \cdot 2^{\mathsf{FCrossing}(\mathcal{D})}(\log(n) + \mathsf{FCrossing}(\mathcal{D})))$ time where n is the number of 2-intervals in $\mathcal{D}$.*

Corollary 1. *The* 2-IP-Uni-$\{\sqsubset, \between\}$ *problem is fixed-parameter tractable with respect to parameter $\mathsf{FCrossing}(\mathcal{D})$.*

It remains open, however, whether the 2-IP-Uni-$\{\sqsubset, \between\}$ problem is fixed-parameter tractable with respect to parameter $\mathsf{Depth}(\mathcal{D})$ (recall indeed that $\mathsf{FCrossing}(\mathcal{D}) \ge \mathsf{Depth}(\mathcal{D})$).

Max $\{<, \lozenge\}$-Comparable 2-Interval Pattern

Input: A set $\mathcal{D}$ of n 2-intervals.
Output: The maximum size of a $\{<, \lozenge\}$-comparable pattern in $\mathcal{D}$.

1. Sort the set $\mathcal{D}$ according to their right interval. For the sake of clarity, let us assume that the ordered 2-intervals set is now given by $\mathcal{D} = \{D_1, D_2, \ldots, D_n\}$, i.e., $\mathsf{Right}(D_i) < \mathsf{Right}(D_j)$ implies $i < j$. All ordered subsets considered in the following of the algorithm are to be understood as ordered with respect to that order.
2. For i from 1 to n
   2.1. Fill the entry $T(D_i)$.
   2.2. For all ordered non-empty set $\{D_{i_1}, D_{i_2}, \ldots, D_{i_q}\} \subseteq \mathcal{D}$ such that $\{D_i\} \cup \{D_{i_1}, D_{i_2}, \ldots, D_{i_q}\}$ is an ordered subset of $\{\lozenge\}$-comparable 2-intervals with $\mathsf{Right}(D_i) < \mathsf{Right}(D_{i_1}) < \ldots < \mathsf{Right}(D_{i_q})$, fill the entry $T(D_i \mid D_{i_1}, D_{i_2}, \ldots, D_{i_q})$ according to the recurrence relation (2).
3. Return the largest entry $T(D_i)$

Fig. 2. Algorithm Max $\{<, \lozenge\}$-Comparable 2-Interval Pattern.

7 Conclusion

In the context of structured pattern matching, we considered the problem of finding an occurrence of a given structured pattern in a set of 2-intervals and solved three open problems of [Via04]. We gave an optimal $O(n \log n)$ algorithm for model $R = \{\sqsubset\}$ thereby improving the complexity of the best known algorithm. Also, we described a $O(n^2 \sqrt{n})$ time algorithm for model $R = \{\sqsubset, \lozenge\}$ over a disjoint support. Finally, we proved that the problem is **NP**-complete for model $R = \{<, \lozenge\}$ over a unitary support, and in addition to that, we gave a fixed parameter-tractability result based on the crossing structure of the set of 2-intervals. These results almost complete the table of complexity classes for the 2-interval pattern problem proposed by Vialette [Via04] (see Table 1).

An interesting question would be to answer the last remaining open problem in that area, that is to determine whether there exists a polynomial time algorithm for 2-IP-Dis-$\{<, \lozenge\}$, i.e., finding the largest $\{<, \lozenge\}$-comparable subset of a set of 2-intervals over a disjoint support[1]. In the light of Table 1, we conjecture that problem to be polynomial time solvable.

---

[1] The 2-IP-Dis-$\{<, \lozenge\}$ problem has an immediate formulation in terms of constrained matchings in general graphs: Given a graph G together with a linear ordering π of the vertices of G, the 2-IP-Dis-$\{<, \lozenge\}$ problem is equivalent to finding a maximum cardinality matching $\mathcal{M}$ in G with the property that for any two distinct edges $\{u, v\}$ and $\{u', v'\}$ of $\mathcal{M}$ neither $\min\{\pi(u), \pi(v)\} < \min\{\pi(u'), \pi(v')\}$ and $\max\{\pi(u'), \pi(v')\} < \max\{\pi(u), \pi(v)\}$ nor $\min\{\pi(u'), \pi(v')\} < \min\{\pi(u), \pi(v)\}$ and $\max\{\pi(u), \pi(v)\} < \max\{\pi(u'), \pi(v')\}$ occur.

References

[AGGN02] J. Alber, J. Gramm, J. Guo, and R. Niedermeier, *Towards optimally solving the longest common subsequence problem for sequences with nested arc annotations in linear time*, Proceedings of the 13th Annual Symposium on Combinatorial Pattern Matching (CPM 2002), Lecture Notes in Computer Science, vol. 2373, Springer-Verlag, 2002, pp. 99–114.

[BP93] J.R.S. Blair and B. Peyton, *An introduction to chordal graphs and clique trees*, Graph Theory and Sparse Matrix Computation **56** (1993), 1–29.

[BYHN+02] R. Bar-Yehuda, M.M. Halldorsson, J. Naor, H. Shachnai, and I. Shapira, *Scheduling split intervals*, Proceedings of the 13th Annual ACM-SIAM Symposium on Discrete Algorithms, 2002, pp. 732–741.

[DGP88] I. Dagan, M.C. Golumbic, and R.Y. Pinter, *Trapezoid graphs and their coloring*, Discrete Applied Mathematics **21** (1988), 35–46.

[Eva99] P. Evans, *Finding common subsequences with arcs and pseudoknots*, Proceedings of the 10th Annual Symposium Combinatorial Pattern Matching (CPM 1999), Lecture Notes in Computer Science, vol. 1645, Springer-Verlag, 1999, pp. 270–280.

[FG65] D.R. Fulkerson and O.A. Gross, *Incidence matrices and interval graphs*, Pacific Journal of Math. **15** (1965), 835–855.

[FMW97] S. Felsner, R. Müller, and L. Wernisch, *Trapezoid graphs and generalizations: Geometry and algorithms*, Discrete Applied Math. **74** (1997), 13–32.

[Fre75] M.L. Fredman, *On computing the length of longest increasing subsequences*, Disrete Mathematics **11** (1975), 29–35.

[GGN02] J. Gramm, J. Guo, and R. Niedermeier, *Pattern matching for arc-annotated sequences*, Proceedings of the the 22nd Conference on Foundations of Software Technology and Theoretical Computer Science (FSTTCS 2002), Lecture Notes in Computer Science, vol. 2556, 2002, pp. 182–193.

[GIP99] D. Goldman, S. Istrail, and C.H. Papadimitriou, *Algorithmic aspects of protein structure similarity*, Proceedings of the 40th Annual Symposium of Foundations of Computer Science (FOCS99), 1999, pp. 512–522.

[Gol80] M.C. Golumbic, *Algorithmic graph theory and perfect graphs*, Academic Press, New York, 1980.

[GW79] J.R. Griggs and D.B. West, *Extremal values of the interval number of a graph, I*, SIAM J. Alg. Discrete Methods **1** (1979), 1–7.

[JLMZ00] T. Jiang, G.-H. Lin, B. Ma, and K. Zhang, *The longest common subsequence problem for arc-annotated sequences*, In Proc. 11th Annual Symposium on Combinatorial Pattern Matching (CPM 2000), Lecture Notes in Computer Science, vol. 1848, Springer-Verlag, 2000, pp. 154–165.

[JMT92] D. Joseph, J. Meidanis, and P. Tiwari, *Determining DNA sequence similarity using maximum independent set algorithms for interval graphs*, Proceedings of the Third Scandinavian Workshop on Algorithm Theory (SWAT 92), Lecture Notes in Computer Science, Springer-Verkag, 1992, pp. 326–337.

[MV80] S. Micali and V.V. Vazirani, *An $O(\sqrt{|V|}|E|)$ algorithm for finding maximum matching in general graphs*, Proceedings of the 21st Annual Symposium on Foundation of Computer Science, IEEE, 1980, pp. 17–27.

[TH79]   W.T. Trotter and F. Harary, *On double and multiple interval graphs*, J. Graph Theory **3** (1979), 205–211.

[TY84]   R.E. Tarjan and M Yannakakis, *Simple linear-time algorithms to test chordality of graphs, test acyclicity of hypergraphs, and selectively reduce acyclic hypergraphs*, SIAM J. Comput. **13** (1984), 566–579.

[Vaz94]   V.V. Vazirani, *A theory of alternating paths and blossoms for proving correctness of the $O(\sqrt{|V|}|E|)$ maximum matching algorithm*, Combinatorica **14** (1994), no. 1, 71–109.

[VD01]   J. Viksna and D.Gilbert, *Pattern matching and pattern discovery algorithms for protein topologies*, Lecture Notes in Computer Science, vol. 2149, Springer, 2001, pp. 98–111.

[Via02]   S. Vialette, *Pattern matching over 2-intervals sets*, In Proc. 13th Annual Symposium Combinatorial Pattern Matching (CPM 2002), Lecture Notes in Computer Science, vol. 2373, Springer-Verlag, 2002, pp. 53–63.

[Via04]   _____, *On the computational complexity of 2-interval pattern matching*, Theoretical Computer Science **312** (2004), no. 2-3, 223–249.

[WS84]   D.B. West and D.B. Shmoys, *Recognizing graphs with fixed interval number is NP-complete*, Discrete Applied Mathematics **8** (1984), 295–305.

A Linear-Time Algorithm
for Computing Translocation Distance
between Signed Genomes⋆

Guojun Li[1,2], Xingqin Qi[2], Xiaoli Wang[2], and Binhai Zhu[3]

[1] Institute of Software, Chinese Academy of Sciences, Beijing 100080, P.R. China
[2] School of Mathematics and Systems Science, Shandong University,
Jinan 250100, P.R. China
gjli@sdu.edu.cn
[3] Department of Computer Science, Montana State University,
Bozeman, MT 59717-3880, USA
bhz@cs.montana.edu

Abstract. The study of evolution based on rearrangements leads to a rearrangement distance problem, i.e., computing the minimum number of rearrangement events required to transform one geonome to another. In this paper we study the translocation distance problem, modeling the evolution of genomes by translocations. We present a linear-time algorithm for computing the translocation distance between signed genomes in this paper, improving a previous $O(n^3)$ bound by Hannenhalli in 1996.

1 Introduction

With the advent of large-scale DNA physical mapping and sequencing, investigation of genome rearrangements becomes increasingly important in evolutionary molecular biology. The most common rearrangement events in mammalian evolution are *translocations*, which exchange genetic material between different chromosomes, and *reversals*, which rearrange genetic material within a chromosome. Multichromosomal genomes frequently evolve by translocations. A computational approach to evolutionary studies based on rearrangements was pioneered by Sankoff [1, 2].

We adopt the following abstraction for our purpose. A chromosome can be represented as a sequence of genes, each is represented by an integer. A genome is a set of chromosomes. A translocation is said to act on chromosomes X and Y when the chromosomes are cleaved in (X_1, X_2) and (Y_1, Y_2) respectively, and the segments of the chromosomes are swapped, thus transforming chromosomes X and Y into two new chromosomes. We study *reciprocal* translocation only, where each of the four segments is non-empty. A translocation on a pair of chromosomes of genome A transforms A into another genome. Given two genomes A and B, *translocation distance* between A and B, $d(A, B)$, is the minimum number of

---

⋆ This research is partially supported by NSFC under Grants 10271065 and 60373025.

S.C. Sahinalp et al. (Eds.): CPM 2004, LNCS 3109, pp. 323–332, 2004.

translocations required to transform A into B. We refer to any sequence of translocations transforming A to B as *evolution* of A into B.

The importance of computing the rearrangement distance, under most of the rearrangement events, has motivated researchers to develop efficient algorithms for various types of rearrangements. Besides translocations, rearrangement with reversals have also been well-studied [5–8]. Bader, Moret and Yan [4] gave a linear-time algorithm for computing reversal distance between signed permutations. Hannenhalli [3] gave a polynomial algorithm for computing the translocation distance (including the actual list of translocations) between two signed genomes, whose complexity is $O(n^3)$.

In this paper we present a linear-time algorithm for computing the translocation distance between signed genomes, i.e., a shortest sequence of translocations transforming one genome into another.

In our model, every *gene* is represented by a positive integer and an associated sign "+" or "−" reflecting the direction of the gene. A *chromosome* is a sequence of genes, and a *genome* is defined as a set of chromosomes. Given two genomes, $A = \{(a_{11}, a_{12}, \ldots, a_{1m_1}), (a_{21}, a_{22}, \ldots, a_{2m_2}), \ldots, (a_{N1}, a_{N2}, \ldots, a_{Nm_N})\}$, and $B = \{(b_{11}, b_{12}, \ldots, b_{1n_1}), (b_{21}, b_{22}, \ldots, b_{2n_2}), \ldots, (b_{N1}, b_{N2}, \ldots, b_{Nn_N})\}$, we assume that they contain the same set of genes and that every gene appears in each genome exactly once. For an arbitrary sequence $S = s_1 s_2 \cdots s_k$ of genes, we will denote the reverse ordering of S by $-S$, i.e., $-S = -s_k - s_{k-1} \cdots - s_1$. A chromosome Y is said to be *identical* to a chromosome X iff either $Y = X$ or $Y = -X$. Genomes A and B are said to be *identical* ($A = B$) iff the sets of chromosomes corresponding to A and B are the same.

A translocation exchanges genes between two chromosomes. Given two chromosomes $\pi = (\pi_1 \cdots \pi_{i-1} \pi_i \cdots \pi_n)$ and $\sigma = (\sigma_1 \cdots \sigma_{j-1} \sigma_j \cdots \sigma_m)$. A translocation is a *prefix-prefix* translocation if the prefix of one chromosome is swapped with the prefix of the other chromosome. Formally, a prefix-prefix translocation $\rho(\pi, \sigma, i, j)$ transforms π and σ into two new chromosomes $(\pi_1 \cdots \pi_{i-1} \sigma_j \cdots \sigma_m)$ and $(\sigma_1 \cdots \sigma_{j-1} \pi_i \cdots \pi_n)$. A translocation is a *prefix-suffix* translocation if the prefix of one chromosome is swapped with the suffix of the other chromosome. Formally, a prefix-suffix translocation $\rho_s(\pi, \sigma, i, j)$ transforms π and σ into two new chromosomes $(\pi_1 \cdots \pi_{i-1} - \sigma_{j-1} \cdots - \sigma_1)$ and $(-\pi_n \cdots - \pi_i \sigma_j \cdots \sigma_m)$ (Figure 1(a)).

A translocation on a pair of chromosomes of genome A transforms genome A into another genome. In this paper, we assume that a target genome B is fixed. For each chromosome $\pi = (\pi_1 \cdots \pi_{i-1} \pi_i \cdots \pi_n)$, genes $+\pi_1$ and $-\pi_n$ are called *nodal* genes of π. In this paper, we restrict our discussion to the case when both prefix-prefix and prefix-suffix reciprocal translocations are allowed, so any translocation acting on two chromosomes of one genome will not change the set of nodal genes. Therefore, if genome A can be transformed into genome B by a sequence of reciprocal translocations, then the set of nodal genes is the same for A and B.

In the following, we revisit an important graph – *cycle graph*, which is the basis of our analysis of translocation distance. In a chromosome $X = (x_1, \cdots, x_k)$, replace every positive integer $+x_i$ by ordered pair (x_i^t, x_i^h) of vertices and replace every negative integer $-x_i$ by ordered pair (x_i^h, x_i^t) of vertices. Then each chromosome is transformed into an unsigned sequence. We call the first and the last vertices of each sequence *tails*. Note that the ordered pair of each nodal gene contains a tail. We say that vertices u and v are *neighbors* if they are adjacent in the ordered list constructed in aforementioned manner. For gene x, vertices x^t and x^h are always neighbors and for simplicity, we exclude them from the definition of "neighbors" in the following discussion.

We construct the bi-colored *cycle graph* $G_A = G_{AB}(V, E)$ of a genome A (with respect to a fixed target genome B) as follows. The vertex set V contains the pair of vertices x^t and x^h for every gene x in A, i.e., $V = \{u : u$ is either x^t or x^h, x is a gene in $A\}$. Edges of G are colored either gray or black. Vertices u and v are connected by a *black* edge iff they are neighbors in A. Vertices u and v are connected by a *gray* edge iff they are neighbors in the target genome B. Each tail is an isolated vertex in G_A. The number of black edges (gray edges) in G_A is $n - N$, where n is the number of genes in A and N is the number of chromosomes. Clearly, each vertex except for tails in G_A is adjacent to exactly one black edge and one gray edge. Hence the graph can be uniquely decomposed into a number of disjoint cycles. We denote the number of cycles in G_A as c_A. Obviously, A is identical to the target genome iff the number of cycles is maximized, i.e., iff $c_A = n - N$. We call a cycle with l gray (black) edges as l-*cycle*. A translocation on genome A will act on two black edges on two different chromosomes, i.e., cut off these black edges to produce a new genome.

A *segment* is defined as an interval $I = x_i x_{i+1} \cdots x_j$ within a chromosome $X = (x_1 \cdots x_m)$ in A. Let V_I be the set of vertices induced by the genes in I. We refer to the left vertex corresponding to x_i and the right vertex corresponding to x_j as $LEFT(I)$ and $RIGHT(I)$, respectively. Define $IN(I) = V_I \setminus \{LEFT(I), RIGHT(I)\}$. An edge (u, v) in G_A is said to be *inside* the interval I if $u, v \in IN(I)$.

A *subpermutation* (SP) is an interval of genes $x_i x_{i+1} \cdots x_j$ within a chromosome X in A such that there exists a segment $x_i, permutation(x_{i+1}, \ldots, x_{j-1}), x_j$ within Y (or $-Y$), where Y is a chromosome in the target genome B, and $permutation(x_{i+1}, \ldots, x_{j-1}) \neq (x_{i+1}, \ldots, x_{j-1})$. Equivalently, SP is an interval I within some chromosome of A such that (1) there exists no edge (u, v) such that $v \notin IN(I)$ and $u \in IN(I)$ and (2) there is at least one k-cycle, $k > 1$, involving edges inside I.

If $I = x_{i+1} \cdots x_j$ is a SP within a chromosome X in genome A, then $LEFT(I)$ and $RIGHT(I)$ are isolated vertices in the subgraph $G[I]$ of G_A induced by V_I, we call $G[I] \setminus \{LEFT(I), RIGHT(I)\}$ a *co-SP* corresponding to I in G_A.

A *minimal subpermutation* (MSP) is a SP not containing any other SP. We use s_A to denote the number of the MSPs in A. Similarly, corresponding to every MSP, we define a *co-MSP* in G_A: I is an MSP iff $G'_I = G[I] \setminus \{LEFT(I), RIGHT(I)\}$ is a *co-MSP* in G_A. Hence, the number of MSPs in A

is the same as the number of *co-MSPs* in G_A. In the following we can use the properties of *co-MSPs* to describe *MSPs*.

We introduce another definition that is also important to the translocation distance. Genome A has an *even-isolation* if (i) all the *MSPs* of A reside in a single chromosome, (ii) s_A is even and (iii) all the *MSPs* are contained within a single *SP*.

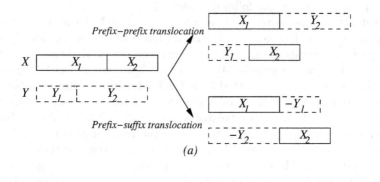

(a)

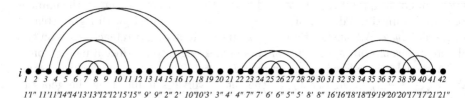

(b)

Fig. 1. The signed genomes A, B and their various representations, in which $A = \{(1, 11, -14, 13, -12, 15), (9, -2, -10, 3), (4, -7, 6, -5, 8), (16, 18, 19, 20, -17, 21)\}$, $B = \{(1, 2, 3), (4, 5, 6, 7, 8), (9, 10, 11, 12, 13, 14, 15), (16, 17, 18, 19, 20, 21)\}$. For the ease of labeling in the figure, $x^t = x'$ and $x^h = x''$.

Figure 1(b) illustrates these concepts. The set of nodal genes of A is $\{1, -15, 9, -3, 4, -8, 16, -21\}$, the same as B's. The tail set of G_A is $\{1^t, 15^h, 9^t, 3^h, 4^t, 8^h, 16^t, 21^h\}$. There are nine cycles in G_A. There are three *MSPs* in A: $\{-14, 13, -12\}$, $\{-7, 6, -5\}$ and $\{16, 18, 19, 20, -17, 21\}$. There are three *co-MSPs* in G_A corresponding to these *MSPs*.

Let S_1 and S_2 be two *MSPs* within chromosome X of A such that S_1 is to the left of S_2. A gray edge (u, v) *separates* S_1 and S_2 if the vertex v belongs to a chromosome different from X and the vertex u is in between the vertices $RIGHT(S_1)$ and $LEFT(S_2)$ in X. Obviously, if there exist *MSPs* S_1 and S_2 with a separating gray edge in A, then genome A has no *even-isolation*.

2 Preliminary Results

Bader *et al.* [4] gave a linear-time algorithm, which requires the computation of connected components of an interleaving graph (to be defined), for computing reversal (inversion) distance between signed permutations. We restate the idea of the linear-time algorithm for computing connected components here, as it will be useful in what follows. We begin with some definitions from [4].

Given a signed permutation (chromosome) π of n genes, we transform it into an unsigned permutation $\pi_1 \pi_2 \ldots \pi_{2n}$ in the same way as described in Section 1, then extend it to the set of $\{\alpha, \pi_1, \ldots, \pi_{2n}, \delta\}$ by setting $\pi_0 = \alpha$, $\pi_{2n+1} = \delta$, where α and δ are integers not appeared in $\{\pi_1, \ldots, \pi_{2n}\}$. Let $\pi = \pi_0 \pi_2 \ldots \pi_{2n} \pi_{2n+1}$. Now given two permutations π and σ, we want to compute the reversal distance from π to σ. Assume that the two permutations have been turned in this manner into unsigned permutations, then we can obtain the cycle graph G_π of π. The *interleaving graph* H_π of a signed permutation π is a graph whose vertex set is the cycles in G_A and there exists an edge between two vertices iff the two cycles represented by the two vertices are interleaving in G_A (i.e., whose corresponding intervals overlap). Notice that this graph could have quadratic size, so explicitly computing it is too costly. We denote the cycle including the vertex π_i in G_π by $C[i]$. The *extent* of cycle C is the interval $[C.B, C.E]$, where $C.B = \min\{i | \pi_i \in C\}$ and $C.E = \max\{i | \pi_i \in C\}$. The extent of a set of cycles $\{C_1, \ldots, C_k\}$ is $[B, E]$, with $B = \min\{C_i.B : 1 \le i \le k\}$ and $E = \max\{C_i.E : 1 \le i \le k\}$. For example, the extent of cycle $\{11^h\ 12^t\ 15^t\ 14^h\}$ in Figure 1(b) is $[4, 11]$.

We summarize the result in [4] as follows.

Theorem 1. [4] *In linear time, one can construct a forest in which each tree is composed exactly of those vertices that form a connected component in H_π.*

Hannenhalli [3] gave a polynomial-time algorithm and a duality theorem for computing the translocation distance.

Theorem 2. [3] *For an arbitrary genome A and target genome B, $d(A, B) = n - N - c_A + s_A + o_A + 2i_A$.*

In Theorem 2, n is the number of genes in A, N is the number of the chromosomes in A, and c_A is the number of cycles in the cycle graph G_A. s_A is the number of *MSP*s in A. If s_A is even, $o_A = 0$, else $o_A = 1$. If A has an even-isolation, then $i_A = 1$, else $i_A = 0$.

Let ψ_A be a parameter ψ associated with genome A (or G_A). For a transloca-tion ρ on A, we denote the increase in ψ as $\Delta(\psi)$, i.e., $\Delta(\psi) = \psi_{A.\rho} - \psi_A$. Given a translocation ρ on genome A, we say ρ is *proper* if $\Delta(c_A) = 1$, ρ is *improper* if $\Delta(c_A) = o$, ρ is *bad* if $\Delta(c_A) = -1$, and ρ is *valid* if $\Delta(c_A - s_A - o_A - 2i_A) = 1$. For each gray edge whose two ends belong to different chromosomes, which is called a *proper gray edge*, there must exist a proper translocation acting on the two black edges adjacent to it.

Theorem 3. [3]*For every genome there exists a valid translocation. In other words, every genome can be transformed into the target genome by a series of valid translocations.*

Hannenhalli's algorithm can not only compute the shortest translocation distance, but also generate the shortest list of translocations in $O(n^3)$ time. If we want to compute the translocation distance faster, we need to know the number of *MSP*s and judge whether there is an even-isolation in genome A in $o(n^3)$ time. Let us put the chromosomes one by one on a line and replace each signed integer by an ordered pair of vertices in the manner mentioned in Section 1 to form a new unsigned permutation π_A. It is clear to see that π_A is a permutation of the vertices of G_A. In the following, we use A to represent π_A for convenience.

Definition 1. *(a) A gray edge in G_A is* inter-chromosomal *(or* proper gray edge*) if it connects vertices in different chromosomes of A, and* intra-chromosomal *otherwise.*

(b) A cycle in G_A is inter-chromosomal *if it contains an inter-chromosomal edge and* intra-chromosomal *otherwise.*

The *interleaving graph H_A* of genome A is defined in the same way as that of a permutation. The corresponding vertex in H_A of an inter-chromosomal cycle in G_A is an *inter-chromosomal* vertex.

Definition 2. *A connected component of H_A is* inter-chromosomal *if it contains an inter-chromosomal vertex and* intra-chromosomal *otherwise.*

It is easy to see that all of *co-MSP*s are intra-chromosomal components. We denote the extent of component U by $[U.B, U.E]$, where
$U.B = \min\{C_i.B | C_i \text{ is a cycle in } U\}$ and
$U.E = \max\{C_i.E | C_i \text{ is a cycle in } U\}$. We define the *extent* of a chromosome to be the interval between its two tails. The *extent* of a set of components $U = \{D_1, D_2, \ldots, D_k\}$ is $[U.B, U.E]$, where $U.B = \min\{D_i.B : 1 \le i \le k\}$ and $U.E = \max\{D_i.E : 1 \le i \le k\}$. For example, the extents of the four chromosomes in A shown in Figure 1 are $[1, 12], [13, 20], [21, 30]$ and $[31, 42]$ respectively. There is one inter-chromosomal component $\{1^h, 11^t, 9^h, 2^h, 2^t, 10^h, 10^t, 3^t\}$, its extent is $[2, 19]$. There are two intra-chromosomal components, $\{11^h, 14^h, 12^t, 15^t\}$ and $\{9^h, 2^h, 10^t, 3^t\}$, which are not *co-MSP*s in G_A, their extents are $[4, 11]$ and $[22, 29]$ respectively.

By the above definitions, we can easily obtain three lemmas as follows.

Lemma 1. *Every 1-cycle C is an isolated vertex in H_A, i.e., a component of H_A, and $C.E = C.B + 1$.*

Lemma 2. *Every co-MSP consists of some intra-chromosomal connected components in the interleaving graph H_A of genome A.*

Lemma 3. *For any intra-chromosomal connected component D in the interleaving graph H_A of genome A, all vertices in the extent of D compose a co-SP in G_A.*

From Lemma 2, we can easily obtain the following lemma.

Lemma 4. *For each co-MSP G'_S in G_A, which corresponds to the MSP S in A, there exists exactly one component C of G'_S involving long cycles (of length > 1), and the others are 1-cycles whose extents are included in the extent of G'_S.*

From Lemma 4, every *co-MSP* in G_A is spanned by the vertices in the extent of its unique component which is not a 1-cycle. We can easily obtain the following result.

Lemma 5. *For each intra-chromosomal component D which is not an isolated vertex in H_A, there exists a co-MSP whose extent is involved in D.*

From Lemma 2, we can denote the extent of *co-MSP* S by $[S.B, S.E]$, where $S.B = \min\{U_i.B|U_i$ is a connected component of G_A in $S\}$ and $S.E = \max\{U_i.E|U_i$ is a connected component of G_A in $S\}$. Clearly, the extent of the *MSP* corresponding to S is $[S.B - 1, S.E + 1]$.

3 Linear Algorithm to Compute the Translocation Distance

Now we are ready to present a linear-time algorithm to compute the translocation distance. Notice that the critical task is to decide the parameter s_A in linear time. Let A and B be the two genomes each containing n genes and π_A be the permutation corresponding to A described in Section 2, here π_A is a permutation of length $2n$. For each $i \in \{1, 2, \ldots, 2n - 1, 2n\}$, we use v_i to denote the vertex corresponding to position i of π_A. We denote the kth chromosome in A by X_k and the kth chromosome in B by Y_k. In the following discussion, we also use X_k to denote the segment in π_A corresponding to X_k, and let $m_0 = 0$. X_k involves m_k genes which are located in $2m_k$ positions in π_A: $2\Sigma_{j=0}^{k-1} m_j + 1, \ldots, 2\Sigma_{j=0}^{k} m_j$. If $v_i \in X_k$ is at position i, then $2\Sigma_{j=0}^{k-1} m_j + 1 \leq i \leq 2\Sigma_{j=0}^{k} m_j$. Clearly, $LEFT(X_k) = 2\Sigma_{j=0}^{k-1} m_j + 1$ and $RIGHT(X_k) = 2\Sigma_{j=0}^{k} m_j$ are the positions of two tails in X_k.

Phase 1.
Each vertex has one neighbor in A and one neighbor in B except for the tails in G_A. Scanning A and B, we can obtain the neighbor a_i in A (A-neighbor) and the neighbor b_i in B (B-neighbor) of the vertex v_i in position i. Then we can label each position i with (v_i, a_i, b_i). Scanning the permutation π_A again, from the above labels we can obtain the gray edges and cycles in G_A. We denote the gray edge in position i by $[e_i.B, e_i.E]$, where $e_i.B$ and $e_i.E$ are the left vertex and right vertex of e_i respectively, and denote the cycle $C[i]$ involving position i by the extent $[C_i.B, C_i.E]$ of $C[i]$. This phase can be completed in $O(n)$ time.

Phase 2.
The chromosomes of A are arranged in line and integrated into a permutation π_A in Phase 1. Then we can invoke the algorithm in [4] to find the connected components of the interleaving graph H_A of genome A in $O(n)$ time. Suppose

that the set of components in π_A is $\{D_1, D_2, \ldots, D_s\}$. Recall that each node in H_A corresponds to a cycle in G_A. For each i, the vertex v_i of position i in π_A is labeled with $[D[i].B, D[i].E]$, where $D[i]$ is the connected component involving position i. The position of each tail of chromosomes in G_A is also memorized. For each chromosome X_k in A, $LEFT(X_k)$ and $RIGHT(X_k)$ will be isolated vertices in G_A and not appear in the interleaving graph H_A, hence the labels for them will be the same as their positions, i.e., $i = D[i].B = D[i].E$. This phase can also be completed in $O(n)$ time.

Phase 3.

We can easily learn whether a component of H_A is inter-chromosomal by testing whether the extent of the component is involved in the extent of some chromosome. Whereafter, we perform Algorithm 1 by scanning permutation π_A to find the co-MSPs from the intra-chromosomal components and then obtain the values of s_A, o_A and i_A based on the discussion in Section 2.

Let $D[i] = C_1 \cup C_2 \cup \cdots \cup C_a \cup C_1' \cup C_2' \cup \cdots \cup C_b'$ be the connected component involving vertex v_i, where $C_1, \cdots, C_a$ are the different maximal segments of $D[i]$ included in X_k, and $C_1', \cdots, C_b'$ are the different maximal segments of $D[i]$ not included in X_k. Obviously, $C_1 \cup C_2 \cup \cdots \cup C_a \neq \phi$. $D[i]$ is inter-chromosomal iff $C_1' \cup C_2' \cup \cdots \cup C_b' \neq \phi$.

In Algorithm 1, the current position i which is not skipped at the previous iterations is the first vertex of some segment C_j of $D[i]$. There are four cases for the connected component $D[i]$ involving vertex v_i : (1) 1-cycle not included in any MSP, (2) an inter-chromosomal component, (3) an intra-chromosomal component whose extent will not span a co-MSP, and (4) an intra-chromosomal component whose extent will span a co-MSP.

In case (1), skip the positions corresponding to the 1-cycle and continue to scan the next position. In case (2), find a segment of $D[i]$ and skip the positions in the segment, then continue to scan the following. In case (3), when $D[i]$ has been determined as an intra-chromosomal component that will not span a co-MSP, it will be labeled as a *checked* component (in order to be distinguished from an intra-chromosomal component which will span a co-MSP in G_A), the vertices in each segment of $D[i]$ will be skipped when they are scanned. In case (4), when we find a co-MSP, skip all vertices in co-MSP (up to now these vertices will be scanned in Algorithm 1) and continue to scan the following. At each iteration i, we will skip some positions in one segment on the right of i, whereafter continue the following iteration, we call these positions in Algorithm 1 as skipped position.

For each k, we introduce a parameter α_k, called *separating number*, for the kth chromosome X_k in A, to denote a property of co-MSPs (MSPs) in X_k: $\alpha_k = 0$ if there exists no MSP in X_k; $\alpha_k = 3$ if there exists some gray edge separating two MSPs in X_k; $\alpha_k = 1$ if there exist some MSPs in X_k, but there exists no proper gray edge either separating these MSPs or with one end whose position is to the right of these MSPs in X_k; and $\alpha_k = 2$ otherwise. We use B_k to denote the set of co-MSPs in X_k. For convenience, we use X_k^i to denote the segment of X_k from position $X_k.B$ to position $i - 1$ and use α_k^i to denote the separating number in X_k^i. Let e_i be the gray edge at position i of X_k, obviously e_i is proper iff either $e_i.B < LEFT(X_k)$ or $e_i.E > RIGHT(X_k)$.

Algorithm 1

Input: permutation π_A.
Output: $B_1, B_2, \ldots, B_N, s_A, \alpha_1, \alpha_2, \ldots, \alpha_N$.

1. scan the permutation, label each position i with $D[i].B$, set up $[D[i].B, D[i].E]$ and $[e_i.B, e_i.E]$.
2. initialize the parameters: $s_A \leftarrow 0$, $R \leftarrow -1$.
3. **for** $k \leftarrow 1$ **to** N **do**
   $L \leftarrow R + 3$, $R \leftarrow R + 2m_k$, $B_k \leftarrow \phi, \alpha_k \leftarrow 0$, $i \leftarrow L$.
   (comment: $L = LEFT(X_k) + 1$ and $R = RIGHT(X_k) - 1$.)
   for $i \leftarrow L$ **to** R **do**
   (a) **if** $i = D[i].B = D[i].E - 1$ **then** skip i and $i + 1$
   (b) **if** $D[i].E > R$ or $D[i].B < L$ **then**
       (comment: $D[i]$ is an inter-chromosomal component)
       i. scan $D[i]$ from i in order to find an segment $\{i, i+1, \ldots, i_1 - 1\} \subseteq D[i]$ and $i_1 \notin D[i]$
       ii. **if** there exists some proper gray edge in $\{e_i, e_{i+1}, \ldots, e_{i_1-1}\}$ and $\alpha_k = 1$ **then** $\alpha_k \leftarrow 2$ **else** $\alpha_k \leftarrow \alpha_k$
       iii. skip the segment $\{i, i + 1, \ldots, i_1 - 1\}$
   (c) **if** $L \leq D[i].B < D[i].E - 1 \leq R$ **then**
       (comment: $D[i]$ is an intra-chromosomal component in X_k)
       i. scan $D[i]$ from i in order to find an segment $\{i, i+1, \ldots, i_1 - 1\} \subseteq D[i]$ and $i_1 \notin D[i]$.
       ii. **if** $i_1 - 1 = D[i].E$ and $D[i]$ has been labeled **then**
           skip all positions on the left of i_1 in $D[i]$
           else if $i_1 - 1 = D[i].E$ and $D[i]$ has not been labeled **then**
           (comment: $D[i]$ is a co-MSP)
           (1) $B_k \leftarrow B_k \cup \{D[i]\}$, $s_A \leftarrow s_A + 1$
           (2) **if** $\alpha_k \in \{0, 2\}$ **then** $\alpha_k \leftarrow \alpha_k + 1$ **else** $\alpha_k \leftarrow \alpha_k$
           (3) skip the segment $\{i, i + 1, \ldots, i_1 - 1\}$
           else if $D[i_1]$ is a 1-cycle **then**
           $D[i] \leftarrow D[i] \cup D[i_1]$ and skip the segment $\{i, i + 1, \ldots, i_1\}$
           else (comment: $D[i]$ is not a co-MSP)
           label $D[i]$ and skip the segment $\{i, i + 1, \ldots, i_1 - 1\}$
4. **return** $B_1, B_2, \ldots, B_N, s_A, \alpha_1, \alpha_2, \ldots, \alpha_N$.

From the definition of *MSP*, *even isolation* and α_k, we can easily obtain the following theorem:

Theorem 4. *If one of the following conditions holds, then* $i_A = 0$, *else* $i_A = 1$:
(1) $s_A = 0$ *;(2)* s_A *is odd; (3)for some k,* $\alpha_k = 3$; *(4) at least two* α_k *is not 0.*

Lemma 6. *For each iteration position i of Algorithm 1, suppose that i belongs to $IN(X_k)$ for some k and j is the successive iteration position. By the end of iteration i, we obtain the following results:*

(1) s_A is the number of the co-MSPs in $X_1 \cup X_2 \cup \cdots \cup X_k^j$;
(2) B_k is the set of co-MSPs in X_k^j;
(3) $\alpha_k = \alpha_k^j$.

By Lemma 6, we can easily obtain the following theorem.

Theorem 5. *By the end of Algorithm 1, we obtain B_k, α_k, $k = 1,\ldots N$, and s_A with the following properties:*

(1) B_k is the set of co-MSPs in X_k;
(2) α_k is the separating number of X_k;
(3) s_A is the number of co-MSPs in G_A, which is also the number of MSPs in A.

It is clear to see that Algorithm 1 can be completed in $O(n)$ time, where $n = \sum_{i=1}^{k} m_k$. Using the output of Algorithm 1, by Theorem 2, Theorem 4 and Theorem 5, we can compute the translocation distance $d(A, B)$ in linear time.

Theorem 6. *The translocation distance between two signed genomes can be computed in $O(n)$.*

Remarks. In this paper we obtain a linear-time algorithm to compute the translocation distance between two signed genomes. To compute the actual shortest list of translocations, the $O(n^3)$ bound by Hannenhalli remains to be improved.

References

1. D. Sankoff. Edit distance for genomes comparison based on non-local operation. *Proc. 3rd Ann. Symp. Combinatorial Pattern Matching,* LNCS 644, pp. 121-135, 1992.
2. D. Sankoff, G. Leduc, N. Antoine, B. Paquin, B. Lang and R. Cedergren. Gene order comparisons for phylogenetic inference: evolution of the mitochondrial genome. *Proc. Nat. Sci. USA,* 89:6575-6579, 1992.
3. S. Hannenhalli. Polynomial-time algorithm for computing translocation distance between genomes. *Discrete Applied Mathematics,* 71: 137-151, 1996.
4. D. Bader, B. Moret and M. Yan. A linear-time algorithm for computing inversion distance between signed permutations with experimental study. *J. Comput. Biol.,* 8:483-491, 2001.
5. V. Bafna and P. Pevzner. Genome rearrangements and sorting by reversals. *SIAM J. Comput.,* 25:272-289, 1996.
6. J. Kececioglu and D. Sankoff. Exact and approximation algorithms for the reversal distance between two permutation. *J. Algorithms* 13(1/2): 180-210, 1995.
7. S. Hannenhalli and P. Pevzner. Transforming men into mice-polynomial algorithm for computing genomic distance problem. *Proc. 36th IEEE Symposium on Foundations of Computer Science,* pp. 581-592, 1995.
8. J. Kececioglu and R. Ravi. Of mice and men: evolutionary distances between genomes under translocation. *Proc. 6th Ann. ACM-SIAM Symp. on Discrete Algorithms,* pp. 604-613, 1995.

Sparse Normalized Local Alignment

Nadav Efraty[1,*] and Gad M. Landau[1,2,**]

[1] Department of Computer Science, Haifa University, Haifa 31905, Israel
{landau,nadave}@cs.haifa.ac.il
[2] Department of Computer and Information Science, Polytechnic University,
Six MetroTech Center, Brooklyn, NY 11201-3840, USA

Abstract. Given two strings, X and Y, both of length $O(n)$ over alphabet Σ, a basic problem (*local alignment*) is to find pairs of similar substrings, one from X and one from Y. For substrings X' and Y' from X and Y, respectively, the metric we use to measure their similarity is *normalized alignment value*: $LCS(X', Y')/(|X'| + |Y'|)$. Given an integer M we consider only those substrings whose LCS length is at least M. We present an algorithm that reports the pairs of substrings with the highest normalized alignment value in $O(n \log |\Sigma| + rM \log \log n)$ time ($r-$ the number of matches between X and Y). We also present an $O(n \log |\Sigma| + rL \log \log n)$ algorithm ($L = LCS(X, Y)$) that reports all substring pairs with a normalized alignment value above a given threshold.

1 Introduction

Sequence comparison is an extensively studied topic. Many textbooks are devoted to the subject [2, 6, 7, 9, 15]. Its applications are numerous and include areas such as file comparison, search for similarity between bio-sequences, information retrieval and XML querying, music retrieval, image comparison and an almost infinite number of other sequence comparison applications.

While for applications such as the comparison of protein sequences the methods of scoring can involve arbitrary scores for symbol pairs and for gaps among unaligned symbols, for uses in other contexts such as text comparison or screening sequences, simple unit score schemes suffice. Two of these, the *Longest Common Subsequence* (*LCS*) and the *edit distance* measures, have been studied extensively, for the unit cost nature of their scoring provides combinatorial leverage not found in the more general framework [3, 10, 11, 13].

As the *LCS* and *edit distance* algorithms evolved, the notion of the sparsity of the essential data in the dynamic programming table became the key to the

* Research supported in part by the Israel Science Foundation grant 282/01, and by the FIRST Foundation of the Israel Academy of Science and Humanities.
** Research supported in part by NSF grant CCR-0104307, by the Israel Science Foundation grant 282/01, by the FIRST Foundation of the Israel Academy of Science and Humanities, and by IBM Faculty Partnership Award.

S.C. Sahinalp et al. (Eds.): CPM 2004, LNCS 3109, pp. 333–346, 2004.
© Springer-Verlag Berlin Heidelberg 2004

acceleration of the algorithms. The evolution of the LCS algorithms can be tracked by examining $[1, 3, 8, 10, 11, 14]$.

While the LCS and *edit distance* algorithms are measures of the global similarity between two strings, in many applications, two strings may not be very similar in their entirety, but may contain regions that are very similar. The task is to find and extract a pair of regions, one from each of the two given strings, that exhibit a strong degree of similarity. This is called the *local similarity* or the *local alignment* problem.

The *local similarity* problem is, in many senses, more challenging than that of the *global similarity*. There are no clear starting and ending points, so any entry has the potential of being the first or the last of an optimal alignment. In addition, one single match is always a perfect local alignment. Thus, a local alignment algorithm might report all the matches as optimal alignments, while longer and more meaningful alignments that are imperfect will not be reported.

One of the most important and commonly used local comparison techniques was introduced by Smith and Waterman [16]. Their algorithm is broadly used in molecular biology, as well as in other fields where local sequence comparison is practiced.

According to a recent paper by Arslan, Eğecioğlu and Pevzner [4], the Smith Waterman algorithm has two weaknesses that make it non optimal as a similarity measure. The first weakness is called the *mosaic effect*. This term describes the algorithm's inability to discard poorly conserved intermediate segments, although it can discard poor prefixes or suffixes of a segment. The second weakness is known as the *shadow effect*. This term describes the tendency of the algorithm to lengthen long alignments with a high score rather than shorter alignments with a lower score and a higher degree of similarity. These effects may be avoided by normalizing the values of the alignments by their lengths.

Definition 1. *The normalized alignment value of two substrings, X' and Y', is $S(X', Y')/(|X'| + |Y'|)$, where S is the global alignment value of X' and Y' according to one of the scoring schemes.*

Arslan et al. [4] suggested a measure designed for the purpose of finding the most similar pair of substrings whose length is significant. The measure is based on the reformulation of the above definition of the normalized alignment's value. Their definition is as follows:

Definition 2. *The normalized alignment value of two substrings is $S(X', Y')/(|X'| + |Y'| + L)$, where X' and Y' are substrings of X and Y, $S(X', Y')$ is the global maximal score of the alignment of X' and Y', and L is a positive number that controls the amount of normalization.*

The ratio between L and $(|X'| + |Y'|)$ determines the influence of L on the value of the normalized sequence alignment under that metric. For short alignments it might lower the normalized sequence value dramatically, while for long alignments the effect on the value should be minor. Using this measure, it is less likely that short alignments will receive high normalized sequence alignment values. The weakness in this, otherwise effective, approach for discarding alignments whose length is insufficient is the reformulation of the original definition of the normalized value (definition 1). By altering the definition, the outcomes will accordingly be different than the expected outcomes of the original problem under the original definition. The time complexity of the measure, suggested by Arslan et al., is $O(n^2 \log n)$, where n is the size of the input strings.

In this paper, we present an algorithm designed for the computation of the local similarity normalized values of substrings of the two input strings whose lengths are not too short to be of significance. The presented algorithm utilizes the LCS metric for the computation of the normalized local alignment value and exploits the sparsity of the essential data in the dynamic programming tables.

Definition 3. *An entry (i, j) in the dynamic programming table of two sequences, $|X| = n$ and $|Y| = n$, is called a match if and only if $X_i = Y_j$. The number of such entries in the table is denoted by r where obviously, $r \leq n^2$ [1].*

The LCS, which is a global measure is made into a local measure of similarity by dividing the LCS value of the two substrings by the sum of their lengths. Substrings that maximize that value are the most similar. Note that the alignment with the highest similarity level must begin and end in a match. Otherwise, there is a better alignment with the same LCS and a lower value of $|X| + |Y|$.

Let X' and Y' be substrings of the input strings X and Y, respectively. A minimal length constraint, denoted herein by M, may be either a minimal length constraint on the sum of the lengths $|X'| + |Y'|$, or a minimal length constraint on the length of the longest common subsequence (LCS) of X' and Y'. We chose to refer to the constraint on the LCS of X' and Y' because it better suits an algorithm that exploits the sparsity of the matches in the dynamic programming tables. The minimal length constraint (M) is enforced in a straightforward fashion, without the need to reformulate the original problem that in the case of normalized LCS is $LCS(X', Y')/(|X'| + |Y'|)$. The value of that minimal constraint is expected to be problem related rather than input related, and it is expected to be on a much smaller scale than the input strings' lengths.

Note that for 100% similarity, we demand that

$$|LCS(X, Y)| = |X| \land |LCS(X, Y)| = |Y|;$$

thus, the normalized value is $\frac{1}{2}$. Any other normalized value represents a similarity level that is twice its value.

Definition 4. $Best(M)_Y^X$ - *the highest normalized valued alignment of any of the substrings pairs, of strings X and Y, with LCS value higher than M.*

Results

Given two strings X and Y of length $O(n)$, and a minimal length constraint M, we will introduce two algorithms that compute the value of $Best(M)_Y^X$. The first algorithm is discussed thoroughly in section 2. This normalized local LCS

algorithm reports substring pairs that achieve the value $Best(M)_Y^X$ and whose common subsequence is longer than M. Alternatively, it may output substring pairs whose similarity is higher than a predetermined value and whose common sequence is longer than M. The time complexity of that algorithm is $O(n \log |\Sigma| + rL \log \log n)$ and its space complexity is $O(rL + nL)$ where $L = LCS(X, Y)$.

The second algorithm, discussed in section 3, is similar to the first in its ability to compute the normalized value of $Best(M)_Y^X$, as well as the substring pairs that achieve that value. The time and space complexity of that algorithm are $O(n \log |\Sigma| + rM \log \log n)$ and $O(rM + nM)$, respectively.

Since we expect M to be much smaller than L, the second algorithm is more efficient than the first one. But, it does not report long substring pairs whose similarity exceeds a predetermined value, if this value is lower than the normalized value of $Best(M)_Y^X$.

Our algorithms avoid the shadow and mosaic effects. The shadow effect is avoided since for any number of matches, the shortest alignment is constructed. Longer alignments would be preferable over shorter alignments only if the longer ones contain more matches, and their normalized value (and not score) is higher. The mosaic effect is avoided since the normalized value of a sufficiently long alignment with a poor intermediate segments would be lower than the normalized values of its prefix and suffix, which are computed separately.

2   The $O(rL\mathrm{loglog}n)$ Normalized Local LCS Algorithm

In this section we discuss our basic algorithm for the computation of $Best(M)_Y^X$ and for the computation of the alignments that exceed a certain similarity level. The discussion begins with the definitions and lemmas that are needed for the understanding of the algorithm. Each of the major stages of the algorithm, as well as the complexity analysis, will be discussed in a separate subsection.

The input is two strings, $|X| = n$ and $|Y| = m$ ($m = O(n)$). As in [1], our algorithm constructs a data structure that substitutes the dynamic programming tables that are used by other local similarity algorithms. Implicitly, many of the properties of the dynamic programming tables are maintained in our sparse representation of it. In the following, a match (i, j) is a match that will be in entry (i, j) in the analogous dynamic programming table.

A *chain* was defined in [5] as a sequence of matches that is strictly increasing in both components, i.e., two matches (i, j) and (i', j') may be part of the same chain if and only if $(i < i' \wedge j < j') \vee (i > i' \wedge j > j')$. Let us present the extended definition of a chain that will be used throughout this work.

Definition 5. *A* $k - Chain_{(i',j')}^{(i,j)}$ *denotes a sequence of k matches that is strictly increasing in both components, whose head is the match (i, j) and whose tail is the match (i', j').*

- $k = LCS(X_{j...j'}, Y_{i...i'})$. $X_{j...j'}$ *and* $Y_{i...i'}$ *are substrings of the input strings X and Y, respectively.*

- Length of $k - Chain_{(i',j')}^{(i,j)}$: The length is the sum of the lengths of $X_{j...j'}$ and $Y_{i...i'}$ (i.e. $j' - j + i' - i$).
- $k - Chain^{(i,j)}$ denotes the best chain of k matches starting from (i, j), i.e., the chain of the shortest possible length that has k matches.
- Normalize value of $k - Chain_{(i',j')}^{(i,j)}$: The normalized value is $\frac{k}{j'-j+i'-i}$.

For each match (i, j), the algorithm constructs $k - Chain^{(i,j)}$ for every possible value of k ($1 \leq k \leq LCS(X, Y)$). The algorithm starts by marking the positions of the matches between the input strings. Later, the matches are processed in decreasing row number order (bottom to top). The processing of each row has two stages.

1. First stage: The algorithm constructs the best $k - Chains$ of any possible value of k, starting from each of the matches in the row. This is done using data structures that were prepared during the processing of previous rows.
2. Second stage: The matches of the processed row and additional information regarding their $k - Chains$ are inserted into the data structures, in order to prepare them for future use during the processing of the succeeding rows.

A major obstacle in the process of constructing $k - Chains$ is that any attempt to construct $(k + 1) - Chain^{(i,j)}$ simply by tying another match to the tail of $k - Chain^{(i,j)}$ (which is the best chain of k matches starting from (i, j)) will not necessarily produce optimal results (see figure 1). We chose to deal with that difficulty as follows: from among all of the $k - Chains$ that start lower than and to the right of (i, j), we choose the one that, when concatenated to (i, j) as its head, creates $(k+1) - Chain^{(i,j)}$. The following lemma proves the correctness of this strategy.

Lemma 1. For any given value of k, and for a match (i, j), $(k+1) - Chain^{(i,j)}$ is a chain that starts from (i, j) and continues with $k - Chain^{(i',j')}$, $i' > i \wedge j' > j$.

Proof. Assume that instead of using $k - Chain^{(i',j')}$ we use another chain of k matches starting from (i', j') which yield a better chain of $k+1$ matches for (i, j). Since the length of the chain from (i, j) to (i', j') remains identical, regardless of the k matches' suffix starting from (i', j'), the difference in the length between two potential chains depends only on the length of the chain of k matches starting from (i', j'). Thus, if $(k + 1) - Chain^{(i,j)}$ passes through (i', j'), but its suffix is different than $k - Chain^{(i',j')}$, it implies that we have constructed a better chain of k matches starting from (i', j'), thereby contradicting the definition of $k - Chain^{(i',j')}$ (definition 5). ∎

The preprocessing stage of the algorithm is similar to the typical preprocessing of the sparse LCS algorithms [1]. Its output is a list of the different symbols of Σ, where each symbol has a list of the indices of its appearances in the input string X. After executing this stage, we can view the matches of each row i by examining the list of symbol $\sigma = Y_i$ ($\sigma \in \Sigma$). The two stages of the algorithm, as well as the procedure that reports $Best(M)_Y^X$ will be discussed in the following subsections.

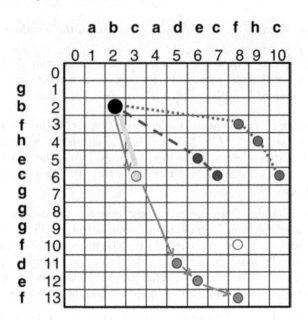

Fig. 1. The dynamic programming table of the strings $abcadecfhc$ (X) and $gbfhecgggfdef$ (Y). The matches are marked as circles. $2 - Chain^{(2,2)}_{(6,3)}$ is, in fact, $2 - Chain^{(2,2)}$ (marked with a solid line). Its length is $6 - 2 + 3 - 2 = 5$ and its normalized value $\frac{2}{5}$. $3 - Chain^{(2,2)}$, $4 - Chain^{(2,2)}$ and $5 - Chain^{(2,2)}$ are marked with dashed lines, dotted lines, and arrowed lines, respectively

2.1 Stage Two – The Creation and Updating of Ranges

The purpose of this stage is to insert the chains that were constructed during the first stage into a data structure that will enable us to narrow the search performed by each of the succeeding matches to a single $k - Chain$. L data structures are maintained for $k - Chains$ of each number of matches k ($1 \leq k \leq L$). Our discussion commences with formal definitions of the intuitive concepts of range and owner.

Definition 6. *Range - A range of a match* (i, j) *is an area of the dynamic programming table that stretches from column* $j - 1$ *and to the left and from row* $i - 1$ *and above, i.e., it is* $(i'...i - 1, j'...j - 1)$ *for each* i' *and* j', $0 \leq i' < i \wedge 0 \leq j' < j$. *Hence each match has* $i \times j$ *such ranges.*

Definition 7. *Mutual range- The range of one match may partially or fully contain a range of another match. The overlap area that is part of the range of both of the matches is called a mutual range.*

Definition 8. *Owner of a range- The match* (i, j) *is the owner of a range if* $k - Chain^{(i,j)}$ *is the suffix of all* $(k + 1) - Chains$ *that start inside the range.*

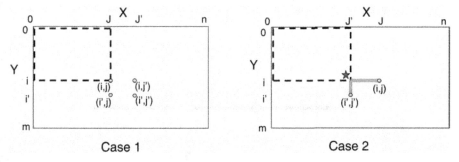

Case 1   Case 2

Fig. 2. The two cases from Lemma 2. In the figure representing case 1, the range that is surrounded by the dashed line is owned by (i, j). In the figure representing case 2, the mutual point is marked with a star and the mutual range is surrounded by a dashed line

L separated lists of ranges and their owners are maintained by the algorithm. The following lemma provides the key to determining the correct ranges and their owners in each of these lists.

Lemma 2. *A mutual range of two matches is owned completely by one of them.*

Proof. The $k - Chain$ that is headed by a match (i, j) may be the suffix of any $k + 1$ matches chain starting from any of the matches in the ranges of (i, j). Note, however, that these chains are not necessarily the $(k+1) - Chain$ of these matches. For all matches that are in a range of a single match (i, j) (i.e., they are not in a mutual range), the only way to construct a $(k+1) - Chain$ is to pass through (i, j). Thus, (i, j) will be the owner of that range. Let us deal with the two different settings of two matches that share a mutual range. These matches will be p $((i, j))$ and q $((i', j'))$.

1. $i \leq i' \wedge j \leq j'$: The mutual range of p and q is $(0...i - 1, 0...j - 1)$. According to their positions, p may use the $k - 1$ suffix of $k - Chain^q$ as part of a possible $k - Chain$ from it. Hence, for each match in the mutual range, a $(k+1) - Chain$ through p is either equal to or better than the chain through q. Thus, p owns the mutual range.
2. $i < i' \wedge j > j'$: The mutual range of p and q is $(0...i - 1, 0...j' - 1)$. Let us define the entry $(i - 1, j' - 1)$ as the *mutual point (MP)* of p and q. MP is the bottommost and rightmost entry of the mutual range, and it is not a match. The length of the chain from any match z in the mutual range to either p or q is equal to the length of the chain from z to MP (which is equal for both p and q) plus the length of the chain from MP to either of the two matches (for match z in coordinates (i'', j''), $i'' < i \wedge j'' < j'$, the length of the chain to p is $(i - i'') + (j - j'')$, and the length of the chain from (i'', j'') to (i, j) that passes through MP is $(i - (i - 1) + j - (j' - 1)) + ((i - 1) - i'' + (j' - 1) - j'') = (i - i'') + (j - j'')$). Since the distances from both matches to MP are predetermined (they are $j - j' + 2$ and $i' - i + 2$ for p and q, respectively), the one whose tail is closer to MP also forms a shorter chain with any match z in the mutual range. ∎

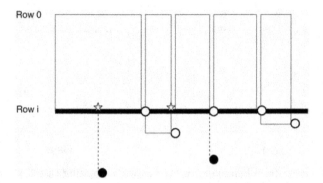

Fig. 3. The LRO^k. Matches that are heads of $k - Chains$ are marked by circles. The white circles are the owners of the ranges that are in LRO^k. Each white circle is the owner of the range to its left. The black circles are owners that were extracted from LRO^k. The stars represent the mutual points, where the boundary of ranges were set according to Lemma 2, case 2

Observations

1. For the given matches p $((i, j))$ and q $((i', j'))$, such that $i < i' \wedge j > j'$, and for the given lengths L^p and L^q of $k - Chain^p$ and $k - Chain^q$, respectively, if $L^p + (j - j') > L^q + (i' - i)$ then the owner of the mutual range is q and the range owned by p is blocked from the left by the range of q. If $L^p + (j - j') < L^q + (i' - i)$, then the owner of the mutual range is p and the range owned by q is blocked from row i and above by the range of p. Since the algorithm processes the matches in decreasing row number order, matches whose row coordinate value is higher than i will not be processed later by the algorithm. Thus, the range owned by q (i.e., $(i...i' - 1, j''...j')$, $j'' < j$) is no longer relevant, and it would not become relevant later. No range above row i would be owned by q, and therefore, it may be extracted from the data structure of the heads of $k - Chains$. In the case of an equality $(L^p + (j - j') = L^q + (i' - i))$, we prefer p over q as the owner of the mutual range because it gives us the opportunity to extract q from the data structure without the loss of important information.

2. For a given group of matches that are the heads of $k - Chains$, the matches whose row number is the lowest (at a given time) must own (at that time) the ranges that stretch between their row and row 0.

The Data Structure: LRO^k denotes the list of ranges and their owners that are the heads of $k - Chains$. Such a list is maintained for each value of k, $1 \leq k \leq L$. Each such list of range owners is ordered by the column. The range of an owner in LRO^k, whose position is (i, j), is $(0...i - 1, j'...j - 1)$, where $j' < j$ is the column of the left neighbor of (i, j) in LRO^k. An example of an LRO^k is given in figure 3. In addition to each owner, we keep the length of the $k - Chain$ starting from it.

The LRO^ks are implemented as Johnson Trees [12]. Explicitly, LRO^k is held in data structures for integers in the range $[0, n]$. These data structures support the operations insert, extract and look for the range that a given match is in.

The algorithm processes the rows in decreasing row number order. Thus, row i is processed only after rows m to $i + 1$ were processed and matches that are the heads of $k - Chains$ were inserted into LRO^k. When the match p $((i, j))$, which is the head of a $k - Chain$, is processed, then according to observation 2 above, it will always be inserted into LRO^k as the range whose right boundary is column $j - 1$. Later, the following update operations are performed in LRO^k:

- Right boundary: If LRO^k has another match q $((i', j'))$ such that $i < i' \wedge j = j'$, then by Lemma 2, the range of q that is above row i is owned completely by p and thus, q is extracted from LRO^k.
- Left boundary: The left neighboring range, whose owner is q $((i', j'),\ i' \geq i \wedge j' < j)$ is examined. If $i' = i$, the left boundary of the range of p is j' (Lemma 2, case 1). If $i' > i$, we use observation 1 to determine the owner of the mutual range of p and q. If q is the owner of the mutual range, it sets the left boundary of the range of p. If p is the owner of the mutual range, q is extracted from the data structure (implicitly, the range of p was extended) and the left neighbor of q is examined in the same fashion.

2.2   Stage One – The Construction of (k+1)-Chains

In this stage, we will compute the $(k+1) - Chains$ of all matches of row i, where $1 \leq k \leq L$ and $1 \leq i \leq m$. The input for this stage is the list of ranges and their owners (LRO^k) that were computed for rows m to $i+1$, as was discussed in the previous subsection.

For a match p, $(k + 1) - Chain^p$ is constructed simply by concatenating p to the match q, which is the owner of the range containing p. Explicitly, q is the match in LRO^k whose column coordinate is the closest to that of p from the right.

The Data Structure: All the matches are ordered according to their positions. Every match has information regarding all the $k - Chain$, $1 \leq k \leq L$, starting from it. For a given match p, the data structure maintains a record where for any given k value, the length of $k - Chain^p$ is recorded, along with a pointer to a match q, such that $(k - 1) - Chain^q$ is the suffix of $k - Chain^p$. Owners of ranges that were extracted from LRO^k are not deleted from that data structure.

2.3   Reporting $Best(M)_Y^X$

After the matches of row 1 have been processed, the data structure wherein every match p has a record with all of the $k - Chain^p$ and their lengths, is completed. Now, the records of all of the matches are examined, the normalized value of any of the $k - Chains$, $k \geq M$, is computed, and the highest valued $k - Chain$, $Best(M)_Y^X$, and its normalized value are computed. $Best(M)_Y^X$ and its corresponding substrings X' and Y' of the input strings X and Y, respectively, may be reported by traversing the pointers of the data structure of matches.

Alternatively, it is possible to report all of the chains and the corresponding substrings whose normalized value is higher than a given normalized value, e.g. 80%. Such sequences may also be reported on the fly during the operation of the algorithm.

2.4   Complexity Analysis

Let us analyze the complexity of each of the stages of the algorithm.

Preprocessing stage: The complexity of the preprocessing stage is $O(n \log |\Sigma|)$, $|\Sigma| \leq m$, and the collective space consumed for the lists of all individual symbols is $O(n)$ [1].

First stage: During the first stage of the processing of each match, attempts are made to construct k chains, $1 \leq k \leq L$, where $L = LCS(X, Y)$ is the highest possible number of matches in any of the chains. Each such attempt requires one query for the nearest neighbors on each of the corresponding LRO^ks. The LRO^ks are implemented as Johnson Trees [12]. The time complexity of each such query is $O(\log \log G)$, where G is the gap between the integer that was the subject of the operation (i.e., the column number of the processed match) and its right and left neighbors in the list. A connected list of the owners of ranges is maintained, hence, in such lists, when a pointer to one of the owners of the ranges is given, its predecessor and successor are reported in $O(1)$ time. The space complexity of such a tree is $O(n)$. Since it is difficult to assess the mean value of G because of the constant changes in LRO^k, we refer to it as n. For all practical purposes, however, the mean value of G is lower than n. Hence, the total complexity of all the iterations of all the r matches is $O(rL \log \log n)$.

Second stage: Each match is inserted and extracted no more than once from each of the LRO^ks. The total time complexity of this entire operation is again $O(rL \log \log n)$.

Reporting $Best(M)^X_Y$: For the retrieval of the highest normalized value and for the construction of the optimal sequence (or the corresponding substrings), the algorithm must examine all the elements in the record of each match with a total time complexity of $O(rL)$.

Henceforth, the time complexity of the algorithm is $O(n \log |\Sigma| + rL \log \log n)$.

The space complexity is $O(rL + nL)$. It is dictated by the size of the data structure for the matches where each match has a record with pointers to no more than L other matches, with one additional length value recorded with each such pointer, and the space needed for L LRO^k data structures that are, in fact, Johnson Trees of $O(n)$ space each.

When $m = o(n)$, the complexity of the algorithm is further reduced to $O(m \log |\Sigma| + n + rL \log \log m)$ time and $O(rL + mL)$ space (the analogous table is comprised of n rows and m columns).

3 The $O(rM\log\log n)$ Normalized Local LCS Algorithm

In this section we present an algorithm for the computation of the normalized value of $Best(M)_Y^X$. Such an algorithm may be ideal for screening input strings that do not reach a desired similarity level. Later, we will show that this algorithm may actually do more than just compute the normalized value of $Best(M)_Y^X$. It may also be used to construct the longest chain that is $Best(M)_Y^X$.

The algorithm that was presented in the previous section is capable of computing $Best(M)_Y^X$ and its corresponding normalized value by constructing the $k - Chains$, $1 \leq k \leq LCS(X, Y)$, starting from each of the matches. In this section we will prove that constructing $k - Chains$ for $k \leq 2M - 1$ is sufficient for the computation of the value of $Best(M)_Y^X$.

Let us start with the definition of a *sub-chain*, that will be followed by the claim that the normalized value of a chain cannot be higher than the normalized value of its best sub-chain.

Definition 9. *sub-chain: A sub-chain of a $k - Chain$ is a path that contains a sequence of $x \leq k$ consecutive matches of the $k - Chain$.*

Note that unlike a $k - Chain$, which always starts and ends with a match, any sub-chain, except the first and the last of a given $k - Chain$, may start and end at any entry of the chain, even if it is not a match. The first sub-chain, which is the prefix of the $k - Chain$, always starts at the head of the $k - Chain$, and the last sub-chain, which is its suffix, always ends at the tail of the $k - Chain$.

Note also that a sub-chain of x matches has a normalized value that is less than or equal to the normalized value of the $x - Chain$ comprised of the same matches, since the sub-chain may have an additional length (at its front and rear).

According to definition 5, the normalized value of a given $k - Chain$ whose length is ℓ is $\frac{k}{\ell}$. Let us split this $k-Chain$ into any number $\leq k$ of non overlapping consecutive sub-chains, such that $k = \sum k_i$ and $\ell = \sum \ell_i$. Hence, $\frac{k}{\ell} = \frac{\sum k_i}{\sum \ell_i}$. The normalized value of each such sub-chain is $\frac{k_i}{\ell_i}$.

Lemma 3. $\frac{k}{\ell} \leq \max(\frac{k_i}{\ell_i})$.

Proof. Let $\frac{k_{i*}}{\ell_{i*}} = \max(\frac{k_i}{\ell_i})$. Thus, for any i, $\frac{k_i}{\ell_i} \leq \frac{k_{i*}}{\ell_{i*}}$. The value of ℓ_i that represents the length of the i's sub-chain must be positive, hence, $\frac{k_i}{\ell_i} \leq \frac{k_{i*}}{\ell_{i*}} \rightarrow k_i \times \ell_{i*} \leq k_{i*} \times \ell_i$. Since it holds for any i, we get $\sum(k_i \times \ell_{i*}) \leq \sum(k_{i*} \times \ell_i)$, and thus, $\ell_{i*} \times \sum k_i \leq k_{i*} \times \sum \ell_i$. Hence, $\frac{k}{\ell} = \frac{\sum k_i}{\sum \ell_i} \leq \frac{k_{i*}}{\ell_{i*}} = \max(\frac{k_i}{\ell_i})$. $\blacksquare$

Note that if $\frac{k}{\ell} = \max(\frac{k_i}{\ell_i})$, then for any sub-chain, $\frac{k_i}{\ell_i} = \frac{k}{\ell}$.

According to Lemma 3, constructing all of the short sub-chains is sufficient to find the value of $Best(M)_Y^X$. Very short sub-chains may have normalized values that are extremely high (e.g., if we consider $1 - Chains$, then each such chain would have a normalized value of $\frac{1}{2}$ which is equal to 100% similarity) but do not reflect significant similarity between the input strings. Thus, in order to compute the value of $Best(M)^X_Y$, it is necessary to construct sub-chains of at least M matches.

Lemma 4. *Constructing all $(2M-1)-Chains$ is sufficient for the computation of the value of $Best(M)^X_Y$.*

Proof. Any $k - Chain$ $(k \geq M)$ can be split into consecutive non overlapping sub-chains of M to $2M - 1$ matches. Chains with less than M matches are not sufficient, and $(2M - 1) - Chains$ can not be split to sub-chains of at least M matches. According to lemma 3, the normalized value of the $k - Chain$ is not better than the normalized value of its best sub-chain. ∎

This concludes our claim that by constructing chains of no more than $2M - 1$ matches, the algorithm can report the value of $Best(M)^X_Y$. Now, let us turn to the claim that the $O(rM \log \log n)$ algorithm may also be used to report the longest chain that is $Best(M)^X_Y$.

When the normalized value of $Best(M)^X_Y$ equals $\frac{1}{2}$ (100% similarity), the $Best(M)^X_Y$ chains and the corresponding substring alignments can be found using the suffix tree of the two input strings. The construction of such a suffix tree is accomplished in $O(n \log(|\Sigma|))$ [17]. In fact, it may be worthwhile to construct a suffix tree and check whether there is a substring of at least M matches that is common to both the input strings even before we turn to the $O(rM \log \log n)$ algorithm for the computation of the normalized value of $Best(M)^X_Y$.

We will prove that when the normalized value of $Best(M)^X_Y$ is lower than $\frac{1}{2}$, the longest $Best(M)^X_Y$ will be a chain of no more than $2M - 1$ matches. This would imply that the $O(rM \log \log n)$ algorithm is also sufficient for the construction of the longest $Best(M)^X_Y$.

Lemma 5. *If the normalized value of $Best(M)^X_Y$ is lower than $\frac{1}{2}$, the longest $Best(M)^X_Y$ is a chain of no more than $2M - 1$ matches.*

Proof. Consider a chain with more than $2M - 1$ matches with normalized value $Best(M)^X_Y$, denoted by LB.

- According to Lemma 4, we may split LB into a number of sub-chains of M matches, followed by a single sub-chain of between M and $2M - 1$ matches.
- According to Lemma 3, the normalized value of each of these sub-chains must be equal to the normalized value of LB.
- According to the definition of a sub-chain (definition 9), if one of the above sub-chains of LB does not start or end with a match, the chain comprised of the same matches has a normalized value that is higher than that of the sub-chain, and thus, higher than the normalized value of LB itself. Hence, all of these sub-chains of LB must start and end with a match.

Let $M - Chain^{(i,j)}_{(i',j')}$ be one of these M matches sub-chains of LB. This sub-chain is, in fact, a chain because it starts and ends at a match. Let the length of $M-Chain^{(i,j)}_{(i',j')}$ be ℓ $(\ell = i'-i+j'-j)$. The normalized value of $M-Chain^{(i,j)}_{(i',j')}$, which is equal to the normalized value of LB, is $\frac{M}{\ell}$. The sub-chain next to $M - Chain_{(i',j')}^{(i,j)}$ must also start at a match. Thus, $(i'+1, j'+1)$, which is the position of the head of the next sub-chain, must be a match, and the length of $(M+1) - Chain_{(i'+1,j'+1)}^{(i,j)}$, which is comprised of the matches of $M - Chain_{(i',j')}^{(i,j)}$ and the match $(i'+1, j'+1)$, is $\ell+2$. Since $\frac{M}{\ell} < \frac{1}{2} \rightarrow \frac{M}{\ell} < \frac{M+1}{\ell+2}$, the normalized value of $(M+1) - Chain_{(i'+1,j'+1)}^{(i,j)}$ is higher than that of $M - Chain_{(i',j')}^{(i,j)}$ alone, and thus, it is also higher than that of LB. Hence, if LB has more than $2M-1$ matches, and if its normalized value is lower than $\frac{1}{2}$, LB must have a sub-chain of at least M matches whose normalized value is higher than the normalized value of LB. Therefore, such LB cannot be $Best(M)_Y^X$. ∎

This concludes our claim that the $O(rM \log \log n)$ algorithm may be used for the construction of the longest $Best(M)_Y^X$.

The $O(rM \log \log n)$ algorithm: The algorithm is identical to the $O(rL \log \log n)$ algorithm from the previous section in all aspects except one; it constructs $k - Chains$ for $1 \le k \le 2M - 1$. Thus, only $2M - 1$ LRO^ks are maintained and updated, and the record of each match in the data structure of matches has at most $2M - 1$ elements listed.

Complexity analysis: In order to construct chains of at most $2M-1$ matches, each match has to issue queries at $2M-1$ LRO^ks. Each match is inserted into and extracted from each LRO^k at most once. Thus, the total time complexity of the algorithm is $O(n \log |\Sigma| + rM \log \log n)$. The space complexity is $O(rM + nM)$. $O(rM)$ is also the time complexity of retrieving $Best(M)_Y^X$.

4   Conclusions and Open Problems

The normalized sequence alignment approach enables us to localize the LCS algorithm, which is global by its nature. This technique enabled us not only to design an algorithm that is both local and sparse, but also to eliminate the mosaic and the shadow effects from which non normalized local similarity algorithms suffer. In addition, the issue of minimal length constraint on the length of the output alignments, which is trivial in the non normalized algorithms, but tends to be problematic for normalized algorithms, is handled simply and without the reformulation of the original normalized alignment problem.

As proved in section 3, the $O(rM \log \log n)$ algorithm is capable of computing the normalized value of $Best(M)_Y^X$ and constructing the longest $Best(M)_Y^X$. Still, for many practical applications, such as local text similarity, the $O(rL \log \log n)$ algorithm that can compute all the substring pairs whose similarities are higher than a predefined value and whose length has no upper bound (except by the length of the input strings) may be the preferred algorithm. Nonetheless, it may be useful to use the $O(rM \log \log n)$ algorithm first to screen out input strings that do not achieve the desired local similarity values.

The modification of the scoring scheme of these algorithms from the LCS metric to other unit cost scorings schemes such as the edit distance remains an open problem.

Acknowledgment

The authors would like to thank Kunsoo Park for introducing the problem to us. We are also grateful to Alberto Apostolico, Klara Kedem, Yuri Rabinovich, Micha Sharir, Alexander Vainshtein and Michal Ziv-Ukelson for fruitful discussions.

References

1. Apostolico, A. String editing and longest common subsequence. in: *Handbook of Formal Languages*, Vol. 2, 361-398, G. Rozenberg and A. Salomaa, editors, Springer Verlag, Berlin, (1997).
2. Apostolico, A., Z. Galil. Pattern matching algorithms. Oxford University Press, 1997.
3. Apostolico, A., C. Guerra. The Longest Common Subsequence Problem Revisited. *Algorithmica*, 2, 315-336, (1987).
4. Arslan, A.N., Ö. Eğecioğlu, P.A. Pevzner. A new approach to sequence comparison: normalized sequence alignment. *Bioinformatics*, 17(4), 327-337, (2001).
5. Claus R. Efficient Computation of All Longest Common Subsequences. *SWAT 2000*, 407-418, (2000).
6. Crochemore M., W. Rytter. Text Algorithms. Oxford University Press, 1994.
7. Crochemore M., W. Rytter. Jewels of Stringology. World Scientific, 2002.
8. Eppstein, D., Z. Galil, R. Giancarlo, G.F. Italiano. Sparse Dynamic Programming I: Linear Cost Functions. *JACM*, 39, 546-567, (1992).
9. Gusfield, D., Algorithms on strings, trees, and sequences. Cambridge University Press (1997).
10. Hirschberg, D.S. Algorithms for the longest common subsequence problem *JACM*, 24(4), 664-675 (1977).
11. Hunt, J.W., T.G. Szymanski. A fast algorithm for computing longest common subsequence. *Communications of the ACM*, 20, 350-353 (1977).
12. Johnson, D.B. A priority queue in which initialization and queue operations take O(loglog D) time. *Math. Syst. Theory*, 15, 295-309 (1982).
13. Levenshtein, V.I., Binary codes capable of correcting, deletions, insertions and reversals. *Soviet Phys. Dokl*, 10, 707-710 (1966)
14. Myers, E.W. Incremental Alignment Algorithms and their Applications. Tech. Rep. 86-22, Dept. of Computer Science, U. of Arizona (1986).
15. Navarro G., M. Raffinot. Flexible pattern matching in strings practical on-line search algorithms for text and biological sequences. Cambridge University Press, 2002.
16. Smith, T.f., M.S. Waterman. The identification of common molecular subsequences. *J. Mol. Biol.*, 147, 195-197 (1981).
17. Ukkonen E., On-line construction of suffix trees. Technical Report No A-1993- 1, Department of Computer Science, University of Helsinki, 1993

Quadratic Time Algorithms for Finding Common Intervals in Two and More Sequences

Thomas Schmidt[1] and Jens Stoye[2]

[1] International NRW Graduate School in Bioinformatics and Genome Research,
Center of Biotechnology, Universität Bielefeld, 33594 Bielefeld, Germany
Thomas.Schmidt@CeBiTec.Uni-Bielefeld.de
[2] Technische Fakultät, Universität Bielefeld, 33594 Bielefeld, Germany
Stoye@TechFak.Uni-Bielefeld.de

Abstract. A popular approach in comparative genomics is to locate groups or clusters of orthologous genes in multiple genomes and to postulate functional association between the genes contained in such clusters. To this end, genomes are often represented as permutations of their genes, and common intervals, i.e. intervals containing the same set of genes, are interpreted as gene clusters. A disadvantage of modelling genomes as permutations is that paralogous copies of the same gene inside one genome can not be modelled.

In this paper we consider a slightly modified model that allows paralogs, simply by representing genomes as sequences rather than permutations of genes. We define common intervals based on this model, and we present a simple algorithm that finds all common intervals of two sequences in $\Theta(n^2)$ time using $\Theta(n^2)$ space. Another, more complicated algorithm runs in $O(n^2)$ time and uses only linear space. We also show how to extend the simple algorithm to more than two genomes, and we present results from the application of our algorithms to real data.

1 Introduction

The availability of completely sequenced genomes for an increasing number of organisms opens up new possibilities for information retrieval by whole genome comparison. The traditional way in genome annotation is establishing orthologous relations to well-characterized genes in other organisms on nucleic-acid or protein level. In the field of high-level genome comparison the attention is directed to gene order and content in related genomes, instead. During the course of evolution, speciation results in the divergence of genomes that initially have the same gene order and content. If there is no selective pressure, successive rearrangements that are common in prokaryotic genomes will eventually lead to a randomized gene order. Therefore the presence of a region of conserved gene order is a source of evidence for some non-random signal that allows, e.g., the prediction of groups of functionally associated genes [13].

S.C. Sahinalp et al. (Eds.): CPM 2004, LNCS 3109, pp. 347–358, 2004.

Usually, two closely related prokaryotes share many *gene clusters*, which are sets of genes in close proximity to each other, but not necessarily contiguous nor in the same order in both genomes [9]. The existence of such gene clusters has been explained in different ways: by functional selection [8], operon formation [3, 7], and other processes in evolution which affect the gene order and content [10]. These papers show that the conservation of gene order is a source of information for many fields in genomic research. Unfortunately, the definition of gene clusters differs as the case arises, and models are based on heuristic algorithms which depend on very specific parameters like the size of gaps between genes. Also all of these approaches lack a statistical analysis to test the significance if an observed gene cluster occurs just by chance. Such an analysis was performed by Durand and Sankoff [5], who present probabilistic models to determine the significance of gene clusters, but leave open the question how to detect these gene clusters in two or more given genomes.

The first rigorous formulation of the concept of a gene cluster was given by Uno and Yagiura [12]. They introduced the notion of common intervals as contiguous regions in each of two permutations containing the same elements, and gave an optimal $O(n + K)$ time algorithm for finding all K common intervals in two permutations of n elements. Heber and Stoye [6] extended this result to common intervals of $k \geq 2$ permutations. But the simplicity of the model makes it unsuitable to be used on real data. Aspects like coding direction, paralogous genes, or the size of interleaving non-coding regions are ignored. On the other hand, model extensions quickly increase the computational complexity of algorithms for detecting gene clusters. As one step of extending the model while still staying within feasible computation time, in this paper we address the integration of paralogous genes, i.e. multiple copies of the same gene in a genome, into the model of common intervals, implying that we work on strings instead of permutations.

In [1], Amir *et al.* developed an algorithm applicable to our problem, using an efficient coding (fingerprints) of the sub-alphabets of substrings. The time complexity of their algorithm is $O(n|\Sigma| \log n \log |\Sigma|)$ where $|\Sigma|$ is the alphabet size. In our application, though, where the number of different genes (the alphabet size) is closely related to the length of the genome (we will always assume that $|\Sigma| \in \Theta(n)$), this becomes $O(n^2 \log^2 n)$. A recent algorithm, presented by Didier in [4], solves our problem using a tree-like data structure in $O(n^2 \log n)$ time, independent of the alphabet size. This algorithm will be further discussed in Section 5, where we show how its running time can be reduced to $O(n^2)$.

The main result of this paper is a worst-case optimal $\Theta(n^2)$ time and space algorithm based on elementary data structures that detects all common intervals of two strings. We also sketch how this algorithm can be extended to find gene clusters in more than two or in a subset of k' out of k genomes. The application of these algorithms on real data presented in Section 6 shows that the incorporation of paralogous genes and regions of internal duplication is a new source of information for research in the field of comparative genomics.

2   Basic Definitions

Given a string S over the finite alphabet of integers $\Sigma := \{1, ..., m\}$, $|S|$ is the length of S, $S[i]$ refers to the ith character of S, and $S[i, j]$ is the substring of S that starts with the ith and ends with the jth character of S. For convenience it will always be assumed for a string S that $S[0] = S[|S| + 1] = m + 1$ are characters not occurring elsewhere in S, so that border effects can be ignored when speaking of the left or right neighbor of a character in S. In our application of comparative genomics, the characters from Σ represent the genes. We will refer to S as a genome or a string interchangeably.

Definition 1 (character set). *Given a string S, the character set of a substring $S[i, j]$ is defined by*

$$CS(S[i, j]) := \{S[k] \mid i \le k \le j\} \subset \Sigma.$$

A character set represents the set of all genes occurring in a given interval of a genome, where the order and the number of occurrences of paralogous copies of a gene is irrelevant.

Definition 2 (CS-location, maximal). *Given a string S over an alphabet Σ and a subset $C \subseteq \Sigma$, the pair (i, j) is a CS-location of C in S if and only if $CS(S[i, j]) = C$. A CS-location (i, j) of C in S is left-maximal if $S[i - 1] \notin C$, it is right-maximal if $S[j + 1] \notin C$, and it is maximal if it is both left- and right-maximal.*

A CS-location of a subset C of Σ represents a contiguous region in a genome that contains exactly the genes contained in C, allowing for possible multiplicities. Note that C has a CS-location in S if and only if C has a *maximal CS-location* in S.

Definition 3 (common CS-factor of k strings). *Given a collection of k strings $S = (S_1, S_2, \ldots, S_k)$ over an alphabet Σ, a subset $C \subseteq \Sigma$ is a common CS-factor of S if and only if C has a CS-location in each S_l, $1 \le l \le k$.*

A common CS-factor of k genomes represents a gene cluster that occurs in each of the k genomes. This concept is similar to a common interval of k permutations, but it allows the presence of paralogous genes in the genomes and particularly within a gene cluster.

These definitions motivate the following two problems:

Problem 1. Given a collection of k strings $S = (S_1, S_2, \ldots, S_k)$, find all its common CS-factors.

Problem 2. For each common CS-factor of S, find all its maximal CS-locations in each of the S_l, $1 \le l \le k$.

Note that the solution of Problem 2 implies a solution of Problems 1. In this paper we present algorithms that solve both of these problems in optimal time and space.

3   A Simple Pairwise Algorithm

For $k = 2$ sequences, the best known algorithm so far solving Problems 1 and 2 requires $O(n^2 \log n)$ time and linear space [4] where n is the length of the longer of the two strings. Here we present an algorithm "Connecting Intervals" (CI) that solves the two problems in $\Theta(n^2)$ time and requires $\Theta(n^2)$ space. Moreover, we will show in the next section how this algorithm can easily be generalized to more than two genomes.

3.1   Basic Algorithm

The input for Algorithm CI are two strings S_1 and S_2, each of length $\leq n$, with characters drawn from the set $\Sigma = \{1, \ldots, m\}$, $m \leq 2n$. Its output are the pairs of CS-locations of all common CS-factors of S_1 and S_2. Pseudocode is given in Algorithm 1.

In a pre-processing step, the algorithm constructs two simple data structures, illustrated in Fig. 1. The first data structure, POS, contains for each character $c \in \Sigma$ a list $POS[c]$ that holds the positions of occurrence of c in sequence S_1 in ascending order, see Fig. 1 (a). The second data structure, NUM, is a $|S_1| \times |S_1|$ table where entry $NUM(i,j)$ contains the number $|CS(S_1[i,j])|$ of *different* characters in the interval $S_1[i,j]$ for each $1 \leq i \leq j \leq |S_1|$, see Fig. 1 (b). Clearly, POS requires linear space and can be computed in linear time by a simple scan over S_1, while NUM requires $\Theta(n^2)$ space and its computation takes $\Theta(n^2)$ time.

(a) $POS[1] = 2, 5$
     $POS[2] = 3, 7$
     $POS[3] = 1, 4$
     $POS[4] = empty$
     $POS[5] = 6$
     $POS[6] = 8$

(b) $NUM(i,j):$

$i \backslash ^j$	1	2	3	4	5	6	7	8
1	1	2	3	3	3	4	4	5
2		1	2	3	3	4	4	5
3			1	2	3	4	4	5
4				1	2	3	4	5
5					1	2	3	4
6						1	2	3
7							1	2
8								1

Fig. 1. Pre-processing of $S_1 = (3, 1, 2, 3, 1, 5, 2, 6)$ with $\Sigma = \{1, \ldots, 6\}$: **(a)** for each character $c \in \Sigma$, $POS[c]$ holds the positions at which c occurs in S_1; **(b)** the table NUM holding the values $|CS(S_1[i,j])|$.

On a high level, Algorithm CI can be described as follows (see Fig. 2): For a fixed position i in S_2, while reading the substring of S_2 starting at that position, the observed characters in S_1 are marked and simultaneously maximal intervals of marked characters are tracked. This is iterated for all start positions i of substrings in S_2.

The maximal intervals of marked characters in S_1 are candidates for common CS-factors with the current interval $[i,j]$ of S_2. It only needs to be tested (i) if

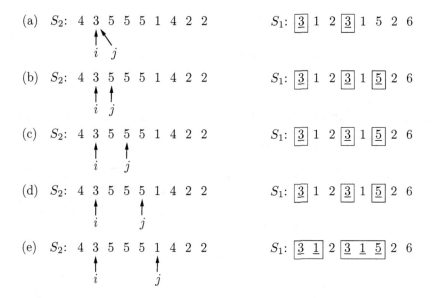

Fig. 2. Algorithm CI at a high level: Position $i = 2$ of S_2 is fixed as the left end of the increasing interval $[i, j]$. While moving j to the right, the observed characters are marked (underlined) in S_1, and maximal intervals of marked characters are tracked (the boxes).

the character set of a candidate interval coincides with that of $S_2[i, j]$, and (ii) if the interval $S_2[i, j]$ is a maximal $\mathcal{CS}$-location of its character set.

In fact, to test (i) it suffices to compare the number of different characters in the two intervals. We know that the maximal marked intervals in S_1 contain a subset of the characters in $S_2[i, j]$, hence if the character sets have equal size, they must be equal. The number of different characters in $S_2[i, j]$ can be tracked while reading the substring of S_2 starting at position i. (In Algorithm 1 we use a binary vector OCC plus a counter $|OCC|$ that counts the number of ones in OCC.) The number of different characters in a maximal marked interval in S_1 can be read from the table NUM that was computed in the preprocessing phase.

Test (ii) is performed implicitly by the way how the value of j is incremented and the while-loop starting in line 5 of Algorithm 1 is terminated. Clearly, during the process of increasing j, once the interval $S_2[i, j]$ is not left-maximal for some $j \geq i$ (i.e. $S_2[i - 1] = S_2[j']$ for some $j' \in \{i, \ldots, j\}$), it will never be left-maximal for any $j'' > j$. Hence it is a valid action to terminate the while loop as soon as $S_2[i, j]$ is not left-maximal, and left-maximality is guaranteed whenever the body of the while loop is entered. Right-maximality is explicitly tested in line 8 of Algorithm 1. This can be done in constant time by testing if $OCC[S_2[j + 1]] = false$.

This establishes the correctness of the algorithm. For the analysis we have to show how the marking and tracking of maximal intervals in S_1 is performed. Obviously, marking the r occurrences of character $c = S_2[j]$ in S_1 is possible in

$O(r)$ time using the list $POS[c]$. Further, if for each maximal interval of marked positions in S_1 the interval boundaries $(start, end)$ are stored at the left and right end of the interval, then it is easy to test, whenever a position p of S_1 is newly marked, if it connects to already existing intervals (ending at position $p-1$ or starting at position $p+1$ or both), and to increase these intervals by index p (if p connects to only one interval) or merge the two intervals (if p connects to two intervals). All this can be done in constant time for each newly marked position p of S_1.

Algorithm 1 Connecting Intervals (CI).

```
 1: pre-processing: build data structures POS and NUM
 2: for i = 1, ..., |S₂| do
 3:     OCC[c] ← 0 for each character c in Σ, |OCC| ← 0
 4:     j ← i
 5:     while j ≤ |S₂| and (i, j) is left-maximal in S₂ do
 6:         c ← S₂[j]
 7:         OCC[c] ← 1
 8:         while (i, j) is not right-maximal in S₂ do
 9:             j ← j + 1
10:         end while
11:         for each position p in POS[c] do
12:             mark position p in S₁
13:             find the maximal interval (start, end) of positions marked so far that con-
                tains position p
14:             if NUM(start, end) = |OCC| and (start, end) is maximal in S₁ then
15:                 output the pair ((i, j), (start, end))
16:             end if
17:         end for
18:         j ← j + 1
19:     end while
20: end for
```

Theorem 1. *Algorithm CI outputs all common $\mathcal{CS}$-factors of S_1 and S_2, in form of pairs of their maximal $\mathcal{CS}$-locations, in $\Theta(n^2)$ time using $\Theta(n^2)$ space.*

Proof. The `for`-loop starting in line 2 of Algorithm 1 is executed $|S_2| \leq n$ times; and in the outer `while`-loop together with the `while`-loop in line 8, j is incremented at most $|S_1| \leq n$ times. More difficult is the analysis of the `for`-loop starting in line 11. Here, observe that due to the test for right-maximality in line 8, this `for`-loop is reached for each character $c = S_2[j]$ only once, and hence for each i the body of the loop is executed at most $\sum_{c \in \Sigma} |POS[c]| = |S_1| \leq n$ times, where $|POS[c]|$ is the number of occurrences of character c in S_1. Together with the pre-processing, this yields the overall $\Theta(n^2)$ time and space complexity. Due to the fact that the number of common $\mathcal{CS}$-factors can be as large as $n(n+1)/2$, e.g. assume $S_1 = S_2 = (1, 2, \ldots, n)$, this algorithm is time-optimal in the sense of worst case analysis. □

This simple quadratic-time algorithm solves the problem of detecting all common $\mathcal{CS}$-factors of two strings. It can also easily be extended to more than two strings (see Section 4), and it provides a good opportunity to address variations of the model (e.g. intervening non-coding regions, coding directions, or pseudogenes) while still staying within feasible computation time. The price for this simplicity is paid in space consumption. The table NUM, which is calculated during the pre-processing, consumes $\Theta(n^2)$ space. In Section 5 we will discuss a quadratic-time solution for Problems 1 and 2 that uses only linear space. Before extending the algorithm, we shortly discuss the form of the output.

3.2   Generating Non-redundant Output

Algorithm CI outputs the common $\mathcal{CS}$-factors by their maximal $\mathcal{CS}$-locations in S_1 and S_2, leading to a redundant output for paralogous gene clusters. For example, given $S_1 = (1, 2, 3, 1, 2)$ and $S_2 = (1, 2, 4, 1, 2, 5, 1, 2)$, the algorithm outputs the $\mathcal{CS}$-locations for the common $\mathcal{CS}$-factor $\{1, 2\}$ in the following way:

$$((1,2),(1,2)),((1,2),(4,5)),((1,2),(7,8)),((4,5),(1,2)),((4,5),(4,5)),((4,5),(7,8)).$$

A non-redundant output of the following form, should be preferred, though:

$$S_1 : (1,2),(4,5) \quad - \quad S_2 : (1,2),(4,5),(7,8).$$

This output can be obtained by a modification of Algorithm CI that we only sketch here. Two additional tables LOC_1 and LOC_2, each of size $|S_1| \times |S_1|$, are used to store lists of intervals.

In a first step, Algorithm CI is applied to S_1 as first *and* second input sequence, yielding the paralogous gene clusters within S_1. These are stored in LOC_1 such that if (i', j') is contained in list $LOC_1(i, j)$, then $\mathcal{CS}(S_1[i', j']) = \mathcal{CS}(S_1[i, j])$, in the following way. Initially, all lists $LOC_1(i, j)$ are empty. Whenever a common $\mathcal{CS}$-factor with maximal $\mathcal{CS}$-locations (i, j) and (i', j'), $i' \neq i$, of a paralogous cluster is detected, then the $\mathcal{CS}$-location (i', j') is appended to the list in $LOC_1(i, j)$, and the interval (i', j') is marked, so that it is not being tested again.

In the second step, Algorithm CI is applied to S_1 and S_2, detecting the orthologous gene clusters between these two genomes. Whenever a common $\mathcal{CS}$-factor with maximal $\mathcal{CS}$-locations (i, j) in S_1 and (k, l) in S_2 is found, the $\mathcal{CS}$-location (k, l) is appended to $LOC_2(i, j)$. Finally, the output for each non-empty entry $LOC_2(i, j)$ is

$$S_1 : (i,j), LOC_1(i,j) \quad - \quad S_2 : LOC_2(i,j).$$

4   Multiple Genomes

To solve Problems 1 and 2 for any given $k \geq 2$, Algorithm CI can easily be extended to more than two strings. The general idea is that a set of characters $C \subseteq \Sigma$ is a common $\mathcal{CS}$-factor of $\mathcal{S} = \{S_1, \ldots, S_k\}$ if and only if it is a (pairwise) common $\mathcal{CS}$-factor of one fixed sequence (w.l.o.g. S_1) and all other sequences in $\mathcal{S}$. Therefore, Algorithm CI is applied to each pair of input strings (S_1, S_r) with $S_r \in \mathcal{S}$ and $1 \leq r \leq k$. Since the first input string is always S_1, the pre-processing step has to be performed only once. The k-fold application of Algorithm CI leads to an overall worst-case time and space complexity of $O(kn^2)$.

Unfortunately, with an increasing number of genomes, the probability to have a conserved gene cluster in all genomes decreases rapidly. For the use on biological data, it is hence even more interesting to find gene clusters which appear in only a subset of at least k' out of k given genomes. Based on the iterated use of Algorithm CI for multiple strings, its improvement to detect such gene clusters can be done in a straightforward manner. This yields a worst-case time complexity of $O(k(1 + k - k')n^2)$. The space complexity is $O(kn^2)$ if non-redundant output is written, and if only Problem 1 is to be solved, it can be reduced to $\Theta(n^2)$.

5   Saving Space

The basic algorithm for two sequences presented in Section 3 uses $\Theta(n^2)$ space, because for each interval $[i, j]$ of S_1 we store the number of different characters in that interval in table NUM. In this section we present an algorithm that runs in quadratic time and uses only linear space.

This algorithm is a modified version of the $O(n^2 \log n)$ time algorithm by Didier [4]. We sketch Didier's algorithm here and in detail discuss only those parts that need to be modified in order to obtain the improved time bound.

Similar to the main structure of Algorithm CI, Didier's algorithm generates, for a fixed left index i and variable right index $j = i, i+1, \ldots$ of intervals of S_2, candidate intervals $(start, end)$ in S_1, and then tests which of these candidates are indeed maximal locations of common $\mathcal{CS}$-factors. Didier uses a stack algorithm for generating the candidates, but the way we generate the candidates in Algorithm CI could be used as well. The main difference then is that Didier stores the intervals in a hierarchical manner according to their overlap relationship. Indeed, in Fig. 2, one can see this hierarchy for the boxes on the right hand side of the figure.

The key idea in the testing phase of Didier's algorithm is the notion of an i-path which is defined in the following way:

Definition 4 (i-rank, left-/right-neighbor, i-distance, successor, i-path).

1. *For a fixed position i in S_2, associate to each character $c \in \Sigma$ its i-rank $\mathbf{r}_i(c)$, i.e. the position of c in the list of different characters as they occur in left-to-right order in the suffix of S_2 starting at position i, and $+\infty$ if c does not occur in this list.*
2. *If $k \leq k'$ are positions of S_1, the i-distance $\mathbf{d}_i(k, k')$ between k and k' is the maximum i-rank of characters occurring in the substring $S_1[k, k']$.*

3. For any position k of S_1 with a finite i-rank $r = \mathbf{r}_i(S_1[k])$, the left-neighbor (resp. the right-neighbor) of k is the greatest position smaller (resp. the smallest position greater) than k with i-rank $r + 1$, if it exists.

4. For a position k in S_1 of finite i-rank, its successor is its (left or right) neighbor with smaller i-distance. If both neighbors have infinite i-distance, k does not have a successor.

5. The i-path of position k of S_1 is the sequence of positions $p = (p_1, p_2, \ldots, p_d)$ of S_1 such that $p_1 = k$ and p_j is the successor of p_{j-1} for all $1 < j \leq d$.

An important observation is then the following.

Theorem 2 (Didier [4]). *An interval candidate* $(start, end)$ *in* S_1 *with i-distance* $\mathbf{d}_i(start, end) = d$ *is a maximal occurrence of a common $\mathcal{CS}$-factor with the interval* $[i, j]$ *in* S_2 *if and only if it contains an i-path* $(p_1, p_2, \ldots, p_d)$ *of length* d.

Based on this theorem, Didier's algorithm traverses for each position k with $S_1[k] = 1$ its i-path $(p_1, p_2, \ldots, p_d)$ and, for each position p_j on this path, it tests if all the positions traversed on the path are contained in the interval $(start, end)$ where $start$ is the leftmost index $k' \leq p_j$ such that $\mathbf{d}_i(k', p_j) = j$ and end is the rightmost index $k' \geq p_j$ such that $\mathbf{d}_i(p_j, k') = j$. In order to avoid that paths are traversed more than once, positions of S_1 are marked whenever the test has been done for the first time, and whenever a path that started at another position k' enters a path that was already traversed before, the procedure is stopped. For a fixed value of i, this part of the algorithm runs in linear time. However, the algorithm suggested in [4] for finding the i-successors and hence the i-paths takes time $O(n \log n)$ since a binary search in the sorted list of i-ranks occurring between p_j and its left- respectively right-successor is performed in order to compute the two i-distances. Repeated for each i, this is the reason for the $O(n^2 \log n)$ overall time complexity.

However, the problem of computing the i-distances is an application of the Range Maximum Query problem for which Bender and Farach [2] have shown how it can be solved in constant time per query after linear time preprocessing.

Hence we can state the following theorem.

Theorem 3. *All common $\mathcal{CS}$-factors of two strings* S_1 *and* S_2 *of maximal length n can be found in* $O(n^2)$ *time using* $\Theta(n)$ *space.*

6   Experimental Results

In order to show the positive effect of our model extension (sequences instead of permutations), we applied our algorithms to five bacterial genomes: *Corynebacterium glutamicum*, *Bacillus subtilis*, *Bacillus halodurans*, *Pseudomonas aeruginosa*, and *Mesorhizobium meliloti*, selected due to their varying pairwise evolutionary distance. All five genomes are included in the COG (Clusters of Orthologous Groups of proteins) database [11], and we assume that two genes are homologous (orthologous or paralogous) if they are in the same COG cluster.

For these five genomes Algorithm CI reported 3428 gene clusters[1], where 197 clusters (6%) contain at least one paralogous gene, 216 clusters (6%) cover at least one region of internal duplication, and 86 clusters (3%) belong to both groups, see Fig. 3 (a). This results in 499 clusters (15%) containing at least one paralogous gene or one region of internal duplication, which an algorithm based on permutations would not be able to find.

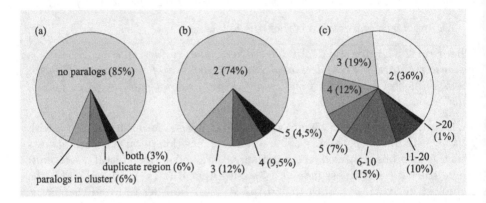

Fig. 3. Results of Algorithm CI applied to five bacterial genomes: (a) distribution of cluster type, (b) number of genomes where a cluster is found, (c) distribution of cluster sizes.

Fig. 3 (b) shows that the majority (74%) of gene clusters was found only between two genomes. However, a large portion of them ($\sim$72%) stems from the comparison of *B. subtilis* and *B. halodurans* as a result of their very low evolutionary distance. Here, the prediction of functional roles from cluster conservation could be expected to fail, because the conservation results less from selection than from the fact that these genomes did not have enough time to diverge. A more detailed analysis of the gene clusters appearing in all five, or at least four sequences revealed that they contain so-called house-keeping genes, whose products are essential for the organism (e.g., ribosomal proteins, ABC transporters, or transcription related proteins). We also found some clusters of essential genes detected only very fragmentary, because the missing of just one gene in one of the genomes is sufficient to 'destroy' this cluster.

The evaluation of the cluster size, see Fig. 3 (c), showed that $\sim$90% of the clusters contain less than 10 genes. The gene cluster with the maximum number of genes was a highly conserved region from *B. subtilis* and *B. halodurans* containing 27 genes. Based on these numbers, it seems to be possible to limit the maximum length of a gene cluster to a fixed value, and thus reducing the time complexity of Algorithm CI to $O(n)$.

---

[1] Here, we call a common $\mathcal{CS}$-factor a *gene cluster* if it contains at least two different genes and has a $\mathcal{CS}$-location in at least two sequences.

7   Discussion and Future Work

In this paper, we have presented a gene cluster model based on common intervals that includes the notion of paralogous genes and regions of internal duplication. We also presented Algorithm CI, which is the first quadratic time algorithm that detects these gene clusters in two genomes, and sketched the extension of this algorithm to be used on any given number of genomes. The evaluation on a set of five bacterial genomes revealed that Algorithm CI finds $\sim$15% more gene clusters than any algorithm working on permutations. To use gene clusters for functional prediction, it is necessary to use genomes with a sufficient evolutionary distance to avoid finding gene clusters that did not have enough time to diverge. The evaluation of the cluster size provides the opportunity to reduce the time complexity to linear by setting a fixed maximum cluster size. In future, this reduction possibly allows to generalize the model to report also gene clusters with a small symmetric set difference.

Acknowledgments

The authors wish to thank Gilles Didier, Mathieu Raffinot, Sven Rahmann, and David Sankoff for helpful discussions on the topic of gene clusters.

References

1. A. Amir, A. Apostolico, G.M. Landau, and G. Satta. Efficient text fingerprinting via parikh mapping. *J. Discr. Alg.*, 26:1–13, 2003.
2. M. A. Bender and M. Farach-Colton. The LCA problem revisited. In *Proceedings of the 4th Latin American Symposium on Theoretical Informatics, LATIN 2000*, volume 1776 of *LNCS*, pages 88–94. Springer Verlag, 2000.
3. P. Bork, B. Snel, G. Lehmann, M. Suyama, T. Dandekar, W. Lathe III, and M. A. Huynen. Comparative genome analysis: exploiting the context of genes to infer evolution and predict function. In D. Sankoff and J. H. Nadeau, editors, *Comparative genomics*, pages 281–294. Kluwer Academic Publishers, 2000.
4. G. Didier. Common intervals of two sequences. In *Proceedings of the Third International Workshop on Algorithms in Bioinformatics, WABI 2003*, pages 17–24.
5. D. Durand and D. Sankoff. Tests for gene clustering. *J. Comput. Biol.*, 10(3/4):453–482, 2002.
6. S. Heber and J. Stoye. Finding all common intervals of k permutations. In *Proceedings of the 12th Annual Symposium on Combinatorial Pattern Matching, CPM 2001*, pages 207–218, 2001.
7. W.C. Lathe III, B. Snel, and P. Bork. Gene context conservation of a higher order than operons. *Trends Biochem. Sci.*, 25:474–479, 2000.
8. R. Overbeek, M. Fonstein, M. D'Souza, G.D. Pusch, and N. Maltsev. The use of gene clusters to infer functional coupling. *Proc. Natl. Acad. Sci. USA*, 96:2896–2901, 1999.
9. I.B. Rogozin, K.S. Makarova, J. Murvai, E. Czabarka, Y.I. Wolf, R.L. Tatusov, L.A. Szekely, and E.V. Koonin. Connected gene neighborhoods in prokaryotic genomes. *Nucleic Acids Res.*, 30:2212–2223, 2002.

10. J. Tamames, G. Casari, C. Ouzounis, and A. Valencia. Conserved clusters of functionally related genes in two bacterial genomes. *J. Mol. Evol.*, 44:66–73, 1997.

11. R.L. Tatusov, D.A. Natale, I.V. Garkavtsev, T.A. Tatusova, U.T. Shankavaram, B.S. Rao, B. Kiryutin, M.Y. Galperin, N.D. Fedorova, and E.V. Koonin. The COG database: new developments in phylogenetic classification of proteins from complete genomes. *Nucleic Acids Res.*, 29:22–28, 2001.

12. T. Uno and M. Yagiura. Fast algorithms to enumerate all common intervals of two permutations. *Algorithmica*, 26:290–309, 2000.

13. I. Yanai and C. DeLisi. The society of genes: networks of functional links between genes from comparative genomics. *Genome Biol.*, 3:0064.1–12, 2002.

Maximal Common Connected Sets
of Interval Graphs[*]

Michel Habib[1], Christophe Paul[1], and Mathieu Raffinot[2]

[1] CNRS – Université de Montpellier 2, LIRMM
161 rue Ada, 34392 Montpellier Cedex 5, France
{habib,paul}@lirmm.fr
[2] CNRS – Laboratoire Génome et Informatique, Evry
and Ecole Normale Supérieure, Paris, France
raffinot@genopole.cnrs.fr

Abstract. Given a pair of graph $G_1 = (V, E_1)$, $G_2 = (V, E_2)$ on the same vertex set, a set $S \subseteq V$ is a maximal common connected set of G_1 and G_2 if the subgraphs of G_1 and G_2 induced by S are both connected and S is maximal the inclusion order. The maximal Common Connected sets Problem (CCP for short) consists in identifying the partition of V into maximal common connected sets of G_1 and G_2. This problem has many practical applications, notably in computational biology.
Let $n = |V|$ and $m = |E_1| + |E_2|$. We present an $\mathcal{O}((n + m) \log n)$ worst case time algorithm solving CCP when G_1 and G_2 are two interval graphs. The algorithm combines maximal clique path decompositions of the two input graphs together with an Hopcroft-like partitioning approach.

1 Introduction

Let $G = (V, E)$ be a loopless undirected graph. The degree of a vertex $x \in V$ in the graph G is denoted by $d_G(x)$. Let X be a subset of vertices of G, we denote $G[X]$ the subgraph induced by X: the set of vertices of $G[X]$ is X and its edge set is $E_X = E \cap \{(u, v) \mid u \in X, v \in X\}$. We denote by $m_X = |E_X|$ the number of edges in $G[X]$ and by $|G[X]| = |X| + m_X$ the size of the induced subgraph. A set X of vertices is *connected in* G if $G[X]$ is a connected graph.

Let $\mathcal{F}$ be a non empty family of graphs on (or restricted to) the same vertex set[1], say $\mathcal{F} = \{G_1 = (V, E_1), \dots G_k = (V, E_k)\}$. A set S of vertices ($S \subseteq V$) is said *connected in* $\mathcal{F}$ if X is connected in all $G_i \in \mathcal{F}$.

Definition 1. *A set $S \subseteq V$ of vertices is a* maximal common connected set *of a family $\mathcal{F} = \{G_1 = (V, E_1), \dots G_k = (V, E_k)\}$ of graphs if S is a connected set in $\mathcal{F}$ and no other set $X \supset S$ is connected in $\mathcal{F}$.*

---

[*] Resarch supported by the CNRS *Action spéficique "Nouveaux modèles et algorithmes de graphes pour la biologie"*.

[1] Or equivalently the vertices could be considered uniquely labeled.

Trivially, the maximal common connected sets of $\mathcal{F}$ form a partition of the vertex set. If $\mathcal{F}$ only contains a single graph, a maximal common connected set reduces to the classical notion of *connected component*. Well-known linear time algorithms that identify the connected components partition of a single graph exist. However, when $|\mathcal{F}| \geqslant 2$, the problem becomes much harder. In [8], the problem of finding maximal common connected sets of two graphs, namely the *CC-Problem* (CCP for short), was addressed. If $|\mathcal{F}| > 2$, the problem is named *gen-CCP* (see Section 2).

CC-Problem:
>   *Input:*   two graphs $G_1 = (V, E_1)$ and $G_2 = (V, E_2)$.
>   *Output:* the partition of V into the maximal common connected sets
>   of $\{G_1, G_2\}$.

Before we continue with the discussion, let us define some notations: the number of vertices will be denoted by $n = |V|$, while m will design the total number of edges in the two graphs G_1 and G_2, i.e. $m = m_1 = m_2$ with $m_1 = |E_1|$, $m_2 = |E_2|$.

A natural approach to solve CCP is to first search the maximal connected components of G_1. Then, in each of these components, search the connected components of G_2. In each such new connected component of G_2, search the maximal connected components of G_1, and repeat this process until the two sets of components on G_1 and G_2 are similar. A simple example on which this approach yields to $\Omega(n)$ steps is given in [8], where the two graphs are in fact interval graphs. Since each step consists in a search on a subgraph whose size may decrease one by one, the complexity of this method is $\Omega(n(n + m))$ worst case time. The algorithm proposed in [8] runs in $\mathcal{O}(n \log n + m \log^2 n)$ for general graphs. Their algorithm mixes dynamical connectivity maintenance with a partitioning approach.

However, obtaining faster algorithms for solving CCP is a real challenge, since the graphs currently considered in many applications, like computational biology, are huge: comparing graphs with more than $250\,000$ vertices becomes frequent (see for instance the TERAPROT project [14]).

This paper improves the practical and theoretical complexity of CCP for a restricted graph family, that of *interval* graphs. A graph is an *interval graph* iff there is a one-to-one mapping between its vertices and a set of intervals on the real line such that two vertices are adjacent iff their corresponding intervals intersect [13]. This family of graphs represents a large part of the graphs involved in applications of CCP in computational biology, because a chromosome is naturally represented by construction as the interval graph of smaller sequences (cDNA, ESTs, etc). For instance, comparing the longest "common" contigs of two chromosomes built on the same cDNA database requires solving CCP on interval graphs.

We present an algorithm for solving CCP on interval graph in $\mathcal{O}((n+m) \log n)$ worst case time. The algorithm is both faster and simpler than the algorithm solving CCP on general graphs. It combines an Hopcroft-like partitioning frame-

work together with a kind of dynamical maintenance of a spanning separator forest. Interval graphs are represented through a forest of clique paths that roughly
captures all the possible separators of the graphs. This forest is "dynamically"
maintained, in the sense that we are able to quickly compute the new clique
representation after extracting a set of vertices. Sets of vertices are extracted
following an Hopcroft-like partitioning framework, inspired by the gene teams
identification algorithm [2] that has later been proved to resemble a simplified
Hopcroft partitioning approach [1].

This article is organized as follows. Section 2 explains the whole framework
of a recursive partitioning algorithm to solve CCP. Section 3 presents data-
structures and algorithms that allow us to improve the time complexity for
interval graphs. Finally the whole algorithm and its complexity are explained
and proved.

2   A Recursive Partitioning Algorithm

Solving CCP on two graphs G_1 and G_2 on the same vertex set V consists in
computing a partition of V whose parts are the maximal common connected
sets. A *partition* $\mathcal{P}$ of a set V is a set of disjoint subsets $\{\mathcal{X}_1, \ldots \mathcal{X}_k\}$, whose
union is exactly V. Our partitioning algorithm is based on the following simple
lemma.

Lemma 1. *Let G_1 and G_2 be graphs on the same vertex set V and let C be a
maximal connected component of G_1 distinct from V. Then*

$$CCP(G_1, G_2) = CCP(G_1[C], G_2[C]) \cup CCP(G_1[V \backslash C], G_2[V \backslash C])$$

Proof. Let S be a maximal common connected set of the pair G_1 and G_2. By
definition S is connected in G_1. Since C is a maximal connected component, S is
either included in C or in $V \backslash C$. It follows that any maximal common connected
set of G_1 and G_2 is either a maximal common connected set of $G_1[C]$ and $G_2[C]$
or of $G_1[V \backslash C]$ and $G_2[V \backslash C]$.   □

A simple paradigm for a recursive algorithm derives from Lemma 1. The
inputs are two graphs G_1 and G_2 on the same vertex set V and a partition $\mathcal{P}$
of V. Initially, $\mathcal{P}$ is set to the trivial partition $\{V\}$. Then, it first searches for
a connected component of G_1 or G_2 distinct from V. If such a component C
exists, according to Lemma 1, two recursive calls are launched on the subgraphs
induced respectively by C and $V \backslash C$. A sketch of this algorithm, named **CCP-
Algorithm**, is given in Figure 1.

Lemma 2. *CCP-Algorithm computes the maximal common connected set
partition of a pair of graphs.*

Proof. The algorithm ends since (a) the recursive calls are launched on strict
subgraphs and (b) it stops the recursive calls when both graphs are connected.
The correctness of the algorithm directly derives from Lemma 1.   □

CCP-Algorithm$(G_1 = (V, E_1), G_2 = (V, E_2))$
1.   **If** G_1 and G_2 are both connected **Then**
2.   **Return** $\mathcal{P} = \{V\}$
3.   **Else**
4.   **If** G_1 is not connected **Then**
5.   Let C be a connected component of G_1
6.   **Else**
7.   Let C be a connected component of G_2
8.   **End of if**
9.   Let $\mathcal{P}' =$**CCP-Algorithm**$(G_1[C], G_2[C])$
10.   Let $\mathcal{P}'' =$**CCP-Algorithm**$(G_1[V \backslash C], G_2[V \backslash C])$
11.   **Return** $\mathcal{P} = \mathcal{P}' \cup \mathcal{P}''$
12.   **End of if**

Fig. 1. Recursive algorithm computing the maximal common connected sets partition of the vertex set V of a pair of graphs G_1 and G_2

Generalization to an Arbitrary Number of Graphs. One can also consider the *gen-CC Problem* of computing the maximal common connected sets of an arbitrary number of graphs (i.e. CCP applied to a family of $k \geq 2$ graphs). Lemma 1 can be generalized and the algorithm modified.

Lemma 3. *Let* $\mathcal{F} = \{G_1, \ldots, G_k\}$ *be a family of graphs on the same vertex set V and let $C \neq V$ be a connected component of G_1. Then* **gen-CCP**$(G_1 \ldots G_k) =$ **gen-CCP**$(G_1[C] \ldots G_k[C]) \cup$ **gen-CCP**$(G_1[V \backslash C] \ldots G_k[V \backslash C])$

Proof. Let S be a maximal common connected set of the k graphs. By definition S is connected in G_1. Since C is a connected component of G_1, S is either included in C or in $V \backslash C$. It follows that any maximal common connected set of $\mathcal{F}$ is either a maximal common connected set of the family $\{G_1[C], \ldots, G_k[C]\}$ or of the family $\{G_1[V \backslash C], \ldots G_k[V \backslash C]\}$. □

It is straightforward to modify the algorithm in order to handle an arbitrary number of graphs. The connected component C has just to be a connected component of an arbitrary graph among $G_1, \ldots G_k$. The generalized algorithm is depicted in Figure 2.

Observations on the Complexity. Notice that the main difficulties of the above algorithms are first to compute a connected component C of one input graph if it exists and then to extract the subgraphs induced by C and $V \backslash C$. However, without *ad-hoc* data-structures, such a recursive approach may yield to a $\Theta(n(n + m))$ worst case time algorithm. In [8], using a sophisticated data-structure to maintain dynamic connectivity [11], an $\mathcal{O}(n \log n + m \log^2 n)$ algorithm for CCP on general graphs was proposed.

For simplicity, in the rest of the paper, we restrict the study to the case of 2 graphs. The main difference between CPP and gen-CCP consists in managing

---

gen-CCP-Algorithm$(G_1 = (V, E_1), \ldots G_k = (V, E_k))$
1.   **If** all G_i $(1 \leqslant k)$ is connected **Then**
2.   **Return** $\mathcal{P} = \{V\}$
3.   **Else**
4.   Let G_i be a non connected graph among $G_1, \ldots G_k$
5.   Let C be a connected component of G_i
6.   Let $\mathcal{P}' =$**gen-CCP-Algorithm**$(G_1[C], \ldots G_k[C])$
7.   Let $\mathcal{P}'' =$**gen-CCP-Algorithm**$(G_1[V \backslash C], \ldots G_k[V \backslash C])$
8.   **Return** $\mathcal{P} = \mathcal{P}' \cup \mathcal{P}''$
9.   **End of if**

Fig. 2. Recursive algorithm to compute the maximal common connected sets partition of the vertex set V of a family $\mathcal{F} = \{G_1, \ldots, G_k\}$ of k graphs

the connected graphs and the non-connected graphs. It will clearly appear below that the data-structures we use, enable us to run the test of line 1 of algorithm **CCP-Algorithm** (Figure 1) in $\mathcal{O}(1)$ time. It follows that the complexity of gen-CCP differs from the complexity of CCP only by a factor k. The remainder of the article focuses on improving the worst case complexity of **CCP-Algorithm** on two interval graphs.

Section 3 presents two algorithms that retrieve the connected components of the subgraphs $G_1[C], G_2[C], G_1[V \backslash C], G_2[V \backslash C]$ after having extracted C. These algorithms strongly rely on interval graph structural properties. Their complexities are both $\mathcal{O}(|C| + d(C))$ where $d(C) = \sum_{x \in C} d(x)$, which is proportional, not exactly to the size of their induced subgraph, but close to.

However, even if an arbitrary connected component C can be extracted in $\mathcal{O}(|C| + d(C))$, it would not be enough to reach the announced $\mathcal{O}((n+m) \log(n))$ complexity. It could still lead to $\Theta(n(n + m))$ operations. To lower the whole complexity, we combine the extraction scheme with an Hopcroft's partitioning approach [12]. Only *small* connected components have to be considered. *Small* means that the size of the connected component considered has to be less than or equal to the half of the size of the original graph. Such a connected component always exists if the graph is not connected, but it is quite complicated to retrieve it efficiently. This is the purpose of the SIS algorithm of subsection 3.3. The whole partionning approach is presented in section 4.

3   Clique Path Representation of Interval Graphs

This section presents the material for managing the interval graphs. We first introduce some well-known properties and the data-structures used in the algorithms. Then two algorithms that update the data-structures for induced subgraphs are developed. These algorithms allow efficient recursive calls. Finally, the last SIS algorithm searches for a small connected component of a given interval graph.

3.1   Preliminaries and Data-Structures

Let $G = (V, E)$ be a graph and $G[X]$ be the subgraph induced by X. We set $d(X) = \sum_{x \in X} d_G(x)$. Notice that the definition of interval graphs (see the introduction) directly implies that this family of graphs is *hereditary*: any induced subgraph of an interval graph is an interval graph. A *clique* is a complete induced subgraph (not necessarily *maximal* for the inclusion).

Definition 2. *Let $G = (V, E)$ be a connected interval graph. A* clique decomposition path *of G is a path $P = (\mathcal{C}, F)$ such that:*

1. *any set $C \in \mathcal{C}$ is a set of vertices and $\bigcup_{C \in \mathcal{C}} C = V$;*
2. *any $(u, v) \in E$ is contained in some $C \in \mathcal{C}$;*
3. *the set $\mathcal{C}_u = \{C \in \mathcal{C} \mid u \in C\}$ induces a subpath P_u of P*
4. *any $C \in \mathcal{C}$ is a clique;*

A clique decomposition path will be denoted hereafter by CDP. Notice that a CDP gives an interval intersection model of the interval graph: the underlying path P and the family of subpaths P_u that contains the vertex u. If the condition 4 is not required, a decomposition path can be defined for arbitrary graphs and this is the basis of the pathwidth theory (see [3]).

Dealing with interval graphs, we usually define the *Maximal Clique decomposition Path* (shorten as MCP) where any clique $C \in \mathcal{C}$ has to be a maximal clique.

A *separator* is a set S of vertices whose removal disconnects the graph in two or more connected components. A separator S is *minimal* if there exists a pair of vertices u, v, which are separated by S and such that no subset of S separates u and v in different connected components. Since interval graphs are chordal (graphs with no induced cycle of length larger than 3), any minimal separator is a clique [7]. The following lemma gives some hints on the structure of the set of minimal separators of an interval graph.

Lemma 4 (e.g. [9]).
Let P be a MCP of an interval graph G. A set of vertices S is a minimal separator iff it is the intersection S of some consecutive cliques C_1 and C_2 in P.

For our purposes, we label the edges of a CDP by the intersection of the corresponding cliques. A non-connected interval graph clearly enjoys a CDP: the edges between two cliques of different paths are labelled by the empty set since these cliques belong to different connected components and are disjoint. The number of cliques in a CDP P is denoted $|P|$. We say that the set of paths defines a *linear forest* denoted CPF for Clique Path Forest. When all the paths are maximal, the forest is denoted MCPF.

Lemma 5 (e.g. [9]). *Let G be an interval graph with n vertices and m edges. Any MCP is of size $\mathcal{O}(n + m)$.*

Many linear-time interval graphs recognition algorithms exist. The first is due to Booth and Lueker in 1976 [4]. Most recent ones [10, 6] are much simpler than

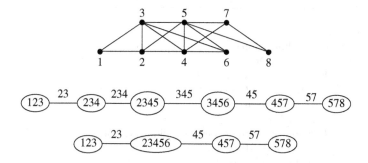

Fig. 3. An interval graph with two CDPs. The second is maximal

the original. All these algorithms can be easily modified to output in $O(n + m)$ a maximal clique path decomposition.

For algorithmic settings, in the case of non-connected interval graphs, the set of paths of a CPF is stored in a list $\mathcal{F}$. The cliques of a CDP are stored in a doubly linked list and the CDP are rooted at one of their extremities. A given clique C stores a pointer to its *father* $f(C)$ and to its *child* $s(C)$; its set of vertices is stored in a doubly linked list and its size is denoted n_C. Moreover, each edge is assigned to a record containing: (a) its two extremities; (b) the label of its minimal separator (see lemma 4) whose vertices are stored in a doubly linked list; (c) the size n_S of this separator. In addition, two lists, namely L_S and L_C, are associated to any vertex x. The list L_S (resp. L_C) contains pointers to the copies of x in each separator (resp. clique) containing x.

3.2 Dynamic Clique Decomposition Path

Lemma 6. *Let $P = (\mathcal{C}, F)$ be a CDP of $G = (V, E)$ and $X \subseteq V$. Then $P' = (\mathcal{C}', F')$ defined as follows is a CDP of $G[V \backslash X]$.*

- *$\mathcal{C}' = \{f(C) \mid C \in \mathcal{C}\}$, where $f(C) = C \backslash X$;*
- *$(f(C_1), f(C_2)) \in F'$ iff $(C_1, C_2) \in F$;*
- *$(f(C_1), f(C_2))$ is labelled by $f(C_1) \cap f(C_2) = (C_1 \cap C_2) \backslash X$.*

Proof. Let us consider two vertices u and v belonging to $V \backslash X$. The proof directly derives from the definition. First any $f(C)$ is a clique and $\bigcup_{C \in \mathcal{C}} f(C) = V \backslash X$. If u and v are adjacent, there exists a clique $C \in \mathcal{C}$ containing both u and v. Clearly $f(C)$ also contains both u and v. Since the set $\mathcal{C}_u \subset \mathcal{C}$ of cliques containing $u \in V \backslash X$ occurs consecutively in P, the set $\mathcal{C}'_u \subset \mathcal{C}'$ also occurs consecutively in P'. □

Notice that some separators may be empty after the extraction of X, in which case the resulting CDP is in fact a CPF.

For complexity issues, the above operation is implemented by two different algorithms. Given a CDP, the first one removes the vertices of a given set from each clique: it is called REMOVE. The second, in contrast, computes the intersection of any clique with a given set: it is called EXTRACT.

REMOVE(P, X)
1.   Let $\mathcal{F}$ be a linear forest containing P
2.   **For** any $x \in X$ **Do**
3.   **For** any clique C st $x \in C$ **Do**
4.   Remove x from C
5.   Decrease n_C by 1
6.   **End of for**
7.   **For** any separator S between cliques C and C' st $x \in S$ **Do**
8.   Remove x from S
9.   Decrease n_S by 1
10.   **If** $n_S = 0$ **Then**
11.   Let (P, C_r) be the CDP containing the edge (C, C')
                    labeled by S
12.   Remove the edge (C, C') from P (C is the father of C')
13.   Create in $\mathcal{F}$ the new CDP (P', C')
14.   **Else**
15.   **If** $n_S = n_C$ **Then**
16.   Remove C from P
17.   Connect $s(C)$ and $f(C)$ with the edge labelled by
                         $s(C) \cap f(C)$
18.   **End of if**
19.   **If** $n_S = n_{C'}$ **Then**
20.   Remove C' from P
21.   Connect $s(C')$ and $f(C')$ with the edge labelled by
                         $s(C') \cap f(C')$
22.   **End of if**
23.   **End of if**
24.   **End of for**
25.   **End of for**
26.   **Return** $\mathcal{F}$

Fig. 4. Maintaining a MCPF of a graph after removing a set of vertices X from an interval graph represented by a MCP P

The pseudo-code of algorithm **REMOVE**(P, X) is given in Figure 4. Lemma 7 states its validity and time complexity.

Lemma 7. *Let P be a MCP of the connected interval graph $G = (V, E)$. The algorithm REMOVE(P, X) (Figure 4) computes in $O(|X| + d(X))$ time a linear forest of $G[V \backslash X]$ where each path is a MCP of the corresponding connected component.*

Proof. First, notice that by lemma 6, when the vertices of X have been removed (lines 4 and 8), $\mathcal{F}$ is a CDP of $G[V \backslash X]$ (but no longer a maximal one). Let S be the intersection between two consecutive cliques C and C' of a given path $P \in \mathcal{F}$ (w.l.o.g. $C = f(C')$).

- $S = \emptyset$ (S is no longer a separator since it is empty): Since P is a CDP C and C' belongs to different connected components, the path P can be split into two CDPs. The first one contains the clique from the root to C while the second one is rooted at C' and contains the clique deriving from C' (lines 10-13).
- $n_S = n_C$ (the case $n_S = n_{C'}$ is similar): C is no longer a maximal clique (it is included in C'). Therefore we can remove C from P (lines 15-22).

It follows that when $\mathcal{F}$ has been cleaned up, any clique is a maximal clique and each new CDP is therefore a MCP. For complexity issue, since the number of copies of elements of X is $\mathcal{O}(|X| + d(X))$ and since each copy is touched once, removing X cost $\mathcal{O}(|X| + d(X))$. The cleaning can be done within the same complexity since (a) removing a separator or a clique costs $\mathcal{O}(1)$; (b) the number of removing operations is bounded by the number of copies of elements of X. □

We now consider the maximal clique path decomposition of the induced subgraph $G[X]$. The pseudo-code of algorithm EXTRACT(P, X) is given in Figure 5. The next lemma 8 states its validity and time complexity.

Lemma 8. *Let F be a MCP of the connected interval graph $G = (V, E)$. The algorithm EXTRACT(F, X) (Figure 5) computes in $\mathcal{O}(|X| + d(X))$ time a MCPF of $G[X]$.*

Proof. A similar proof than that of lemma 7 shows that a linear forest of $G[X]$ can be computed in $\mathcal{O}(|X| + d(X))$. □

3.3 Smaller Induced Subgraph (SIS) Algorithm

The SIS algorithm on two MCPs P_1 and P_2 allows us to find the smallest of the two induced subgraphs in time proportional to the size of this smallest subgraph. The difficulty comes from that the sizes of the two paths are not necessarily representative of the sizes of their induced subgraphs. It may happen that $|P_1| < |P_2|$, but that $|G[V_1]| > |G[V_2]|$, where V_1 (resp. V_2) is the set of vertices contained in the cliques of P_1 (resp. P_2).

To overcome this obstacle, we use a trick. We perform simultaneously a Depth First Search (DFS) on the two paths. We read a new clique (or path node) of each MCP alternatively, until we reach the end of one of the paths. During this search, we compute for each path the sums S_1 and S_2 of the sizes of the cliques we encountered.

At the end of these simultaneous DFS, the smallest MCP, say P_1, has been totally covered, and S_1 is the size of its induced subgraph. If $S_1 \leq S_2$, the simplest case (a), the subgraph induced by P_1 is smaller than that induced by P_2, and SIS returns P_1. Otherwise, if $S_1 > S_2$ we continue the DFS of the second path P_2, computing the new sum S_2' for each new clique encountered. Figure 6 illustrates this search. The process goes on, until, case (b), either the whole MCP P_2 has been visited, in which case $S_2' \leq S_1$ and SIS returns P_2, either, case (c),

EXTRACT(P, X)
1.   Let $\mathcal{F}$ be an empty linear forest
2.   **For any** $x \in X$ **Do**
3.   **For** any clique C containing x **Do**
4.   **If** C has not been already duplicated **Then**
5.   Create a copy $C' = \{x\}$ in $\mathcal{F}$ in a new singleton CDP
6. $n_{C'} \leftarrow 1$
7.   **Else**
8.   Let C' the existing copy of C
9. $C' \leftarrow C' \cup \{x\}$
10. $n_{C'} \leftarrow n_{C'} + 1$
11.   **End of if**
12.   **End of for**
13.   **For** any separator S containing x **Do**
14.   Let (C_1, C_2) be the edge labeled by S in P (wlog $C_1 = f(C_2)$)
15.   **If** (C_1, C_2) has not been duplicated **Then**
16.   Create a new edge (C'_1, C'_2) labelled by x
17. $n_{S'} \leftarrow 1$
18.   **Else**
19.   Let S' the label of the edge (C'_1, C'_2)
20. $S' \leftarrow S' \cup \{x\}$
21. $n_{S'} \leftarrow n_{S'} + 1$
22.   **End of if**
23.   **End of for**
24.   **End of for**
25.   Remove from $\mathcal{F}$ any non maximal clique as in lines 15-22 of REMOVE
        (see Figure 4)
26.   **Return** $\mathcal{F}$

Fig. 5. Maintaining a MCPF of the induced subgraph of $G[X]$ when extracting X from G

S'_2 becomes greater than S_1 and SIS returns P_1. Figure 7 illustrates these two last cases.

It is obvious that SIS returns the MCP which represents the smallest induced graph. We prove in Lemma 9 that its complexity only depends on the size of this smallest induced subgraph.

Lemma 9. *Given two MCP's, Algorithm SIS returns the one that represents the smallest induced subgraph, say $G[X]$ in time $O(|X| + m_X)$.*

Proof. Let c_1 (resp. c_2) be the number of cliques visited in P_1 (resp c_2) at the end of SIS algorithm. The total number of cliques visited is $c_1 + c_2$.

In case (a), P_1 represents the smallest subgraph $G[X]$ of size $S_1 = |X| + m_X$. The number $c_1 + c_2$ is in this case $2c_1$. As $c_1 \leq S_1$ (lemma 5), the complexity of SIS is $O(|X| + m_X)$.

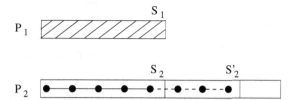

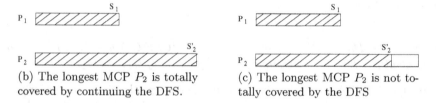

Fig. 6. Continuing the Depth First Search in the longest MCP P_2 if $S_1 > S_2'$ while until either $S_2' \geq S_1$ or P_2 is completely covered

In case (b), $S_2' \leq S_1$. The path P_1 represents the smallest subgraph $G[X]$ of size $S_2' = |X| + m_X$. Therefore SIS returns P_2. As the first DFS stopped first w.l.o.g on P_1, $c_2 > c_1$, and, as $c_2 \leq S_2'$ (lemma 5), $c_1 + c_2 < 2S_2'$ and the complexity of SIS is $O(|X| + m_X)$.

In case (c), the path P_1 represents the smallest subgraph $G[X]$ of size $S_1 = |X| + m_X$. As $c_2 \leq S_1 + 1$, $c_1 + c_2 \leq 2S_1 + 1$ and the complexity of SIS is $O(|X| + m_X)$. ☐

(b) The longest MCP P_2 is totally covered by continuing the DFS.

(c) The longest MCP P_2 is not totally covered by the DFS

Fig. 7. Two ending cases when continuing the DFS on the longest MCP P_2. In the first case (a), the DFS covers all the vertices of P_2. Then as $S_2' \leq S_1$, SIS returns P_2. In the second case, at most S_1 vertices of P_2 have been visited by the DFS without exploring all the tree. Then SIS returns P_1

4   The Whole CCP Algorithm For Interval Graphs

The whole CCP algorithm for interval graphs (CCPI-Algorithm) is shown on Figure 8. The algorithm takes as input two lists L_1 and L_2 that are respectively the clique forest decompositions of the two graphs G_1 and G_2. It outputs a partition of the vertex set. At lines 5 and 7, it searches for a connected component C whose size is at most half of the size of the corresponding graph. By lemma 9, it can be done in time proportional to that connected component. Let P be the MCP of C. W.l.o.g. we assume that C is a connected component of G_1 and P is the first MCP of L_1. Lines 11 and 12 compute the four subgraphs on which the recursive calls are launched. Using EXTRACT(L_2,P) and REMOVE(L_2,P) we compute the subgraphs of G_2 induced respectively by the vertices V_P belonging to the cliques of P and $V \setminus V_P$. As seen in Lemmas 8 and 7, it can be done in $\mathcal{O}(|V_P| + d(V_P))$.

CCPI-Algorithm(L_1, L_2)
1.   **If** $|L_1| = 1$ and $|L_2| = 1$ **Then**
2.   **Return** $\mathcal{P} = \{V\}$ /* G_1 and G_2 are both connected */
3.   **Else**
4.   **If** $|L_1| \geq 2$ **Then**
5. $P \leftarrow \text{SIS}(L_1[1], L_1[2])$ /* G_1 is not connected */
6.   **Else**
7. $P \leftarrow \text{SIS}(L_2[1], L_2[2])$ /* G_2 is not connected */
8.   **End of if**
9.   /* we assume below w.l.o.g that $P = L_1[1]$ */
10. $L_1' \leftarrow P;$ $L_1'' \leftarrow L_1 \backslash P$
11. $L_2' \leftarrow \text{EXTRACT}(L_2, V[P]);$ $L_2'' \leftarrow \text{REMOVE}(L_2, V[P])$
12.   Let $\mathcal{P}' = $**CCPI-Algorithm**($L_1', L_2'$)
13.   Let $\mathcal{P}'' = $**CCPI-Algorithm**($L_1'', L_2''$)
14.   **Return** $\mathcal{P} = \mathcal{P}' \cup \mathcal{P}''$
15.   **End of if**

Fig. 8. Recursive algorithm to compute the partition of the vertex set into maximal common connected sets of two interval graphs G_1 and G_2 represented respectively by the MCPF $L1$ and L_2

Theorem 1. *The CCPI-algorithm applied on MCPF(G_1) and MCPF(G_2) correctly identifies the common connected components of G_1 and G_2.*

Proof. The CCPI-Algorithm fully respects the general algorithm framework described in section 2. Indeed lemmas 8 and 7 ensures that the recursive calls are done on the right subgraphs. The only difference is that we now choose which maximal component we extract first.   □

To analyse its complexity, we use an amortized argument that is common to many Hopcroft-like approaches, but did not appear in the original paper [12]. To our knowledge, it is due to [5].

Theorem 2. *The CCPI-Algorithm is worst case $\mathcal{O}((n + m) \log n)$ time .*

Proof. We first focus on the number of times a vertex x and a transition (y, z) may participate to EXTRACT and REMOVE. W.l.o.g., let S_1 be the size of the subgraph of G_1 at the beginning of a recursive call of CCCIA. If a connected component of G_1 is extracted through its MCP P, then the size of the induced subgraph of P is less than or equal to $S_1/2$. This is straightforward since P has been isolated through SIS as the smallest of the two induced subgraphs. By induction, if x and (y, z) participate to many EXTRACT and REMOVE, they are contained in subgraphs whose sizes are divided by at least two at each recursive call. Therefore, they may only participate to $\log(n + m)$ EXTRACT and REMOVE calls.

Secondly, we amortized the cost of each EXTRACT and REMOVE of a path P on all the vertices and edges of the induced subgraph of P. The complexity of EXTRACT and REMOVE (lemmas 8 and 7)) for extracting a set X out of a graph G is $|X| + d(X)$. We amortize the cost $|X|$ over each vertex of X, and therefore a vertex $x \in X$ participates for a constant amount of time. The term $d(X)$ is amortized over the edges. As an edge (x, y) may be visited when considering x and when considering y, an edge can be visited only twice and therefore participates for a constant amount of time.

In consequence, each vertex or each edge costs at most $\log(n + m)$. This leads to an overall complexity of $\mathcal{O}((n + m)\log(n + m))$. As in the worst case, $m = \mathcal{O}(n^2)$, the final complexity is $\mathcal{O}((n + m)\log n)$ worst case time.   □

The space complexity is $O(n + m)$, since the two MCPFs are space linear in $n + m$ and that the recursive call of CCCIA algorithm can be managed with a list of at least $O(\log(n + m))$ pointers on the MCPFs.

5   Conclusion

We presented an $\mathcal{O}((n+m)\log n)$ worst case time and $\mathcal{O}(n+m)$ space algorithm for solving CCP on interval graphs. For this kind of graph, our algorithm is both faster and simpler than the actual best algorithm for general graphs, running in $\mathcal{O}(n\log n + m\log^2 n)$ [8]. Our algorithm combines an Hopcroft partitioning approach with a maintenance of a spanning clique forest decomposition of the two graphs. Designing faster algorithms or proving a lower bound for CCP remains open, on interval and general graphs. It is also worthwhile to notice that even on chordal graphs the general upper bound can still not be improved.

References

1. M.-P. Béal, A. Bergeron, and M. Raffinot. Gene Teams and Hopcroft's Partionning Framework. 2003. Submitted.
2. A. Bergeron, S. Corteel, and M. Raffinot. The algorithmic of gene teams. In *Workshop on Algorithms in Bioinformatics (WABI)*, number 2452 in Lecture Notes in Computer Science, pages 464–476. Springer-Verlag, Berlin, 2002.
3. H. Bodlaender. A tourist guide through treewidth. *Acta Cybernetica*, 11(1-2), 1993.
4. K.S. Booth and G.S. Lueker. Testing for the consecutive ones properties, interval graphs and graph planarity using pq-tree algorithm. *J. Comput. Syst. Sci.*, 13:335–379, 1976.
5. A. Cardon and M. Crochemore. Partitioning a graph in $O(|A|\log_2 |V|)$. *Theoretical Computer Science*, 19(1):85–98, 1982.
6. D.G. Corneil, S. Olariu, and L. Stewart. The ultimate interval graph recognition algorithm? In *Proceedings of the ninth Annual ACM-SIAM Symposium on Discrete Algorithms (SODA)*, pages 175–180, 1998.
7. G.A. Dirac. On rigid circuit graphs. *Abh. Math. Sem. Uni. Hamburg*, 25, 1961.
8. A.-T. Gai, M. Habib, C. Paul, and M. Raffinot. Identifying Common Connected Components of Graphs. *Technical report*, (RR-LIRMM-03016), 2003. http://www.lirmm.fr/~paul.

9. P. Galinier, M. Habib, and C. Paul. Chordal graphs and their clique graph. In M. Nagl (Ed.), editor, *Graph-Theoretic Concepts in Computer Science, WG'95*, volume 1017 of *Lecture Notes in Computer Science*, pages 358–371, Aachen, Germany, June 1995. 21st Internationnal Workshop WG'95, Springer.

10. M. Habib, R. McConnell, C. Paul, and L. Viennot. Lex-bfs and partition refinement, with applications to transitive orientation, interval graph recognition and consecutive ones testing. *Theoretical Computer Science*, 234:59–84, 2000.

11. J. Holm, K. De Lichtenberg, and M. Thorup. Poly-logarithmic deterministic fully-dynamic algorithms for connectivity, minimum spanning tree, 2-edge and biconnectivity. *Journal of the ACM*, 48(4):723–760, 2001.

12. J. E. Hopcroft. An $n \log n$ algorithm for minimizing the states in a finite automaton. In Z. Kohavi, editor, *The Theory of Machines and Computations*, pages 189–196. Academic Press, 1971.

13. C.G. Lekkerkerker and J.C. Boland. Representation of a finite graph by a set of intervals on the real line. *Fund. Math.*, 51:45–64, 1962.

14. TERAPROT project. http://www.infobiogen.fr/services/Teraprot/.

Performing Local Similarity Searches with Variable Length Seeds*

Miklós Csűrös

Département d'informatique et de recherche opérationnelle, Université de Montréal
C.P. 6128 succ. Centre-Ville, Montréal, Québec, H3C 3J7, Canada
csuros@iro.umontreal.ca

Abstract. This paper describes a general method for controlling the running time of similarity search algorithms. Our method can be used in conjunction with the seed-and-extend paradigm employed by many search algorithms, including BLAST. We introduce the concept of a seed tree, and provide a seed tree-pruning algorithm that affects the specificity in a predictable manner. The algorithm uses a single parameter to control the speed of the similarity search. The parameter enables us to reach arbitrary levels between the exponential increases in running time that are typical of seed-and-extend methods.

1 Introduction

Finding similarities between sequences is one of the major preoccupations of bioinformatics [1]. All similarities, defined as local alignments scoring above a certain threshold, can be found with dynamic programming using the Smith-Waterman algorithm [2]. While there are many inventions that improve the speed of a full sensitivity search (e.g., [3]), a full-scale search that involves sequences with several million letters cannot be carried out in a reasonable time frame. In such cases, sensitivity is most often sacrificed for speed. This paper describes a method to adjust the running time of local similarity searches on a fine scale.

Most successful alignment heuristics rely on suffix trees [4], or hash tables. Hashing-based methods include FASTA [5], BLAST [6], BLAT [7], BLASTZ [8], and PatternHunter [9, 10]. In their simplest form, hashing-based, or *seed-and-extend*, methods identify common short substrings of a fixed length k between the input sequences S and T. Matching substrings, or hits, are the starting points ("seeds") for computing local alignments. The sensitivity depends on the seed length k: weaker similarities can be found by using smaller values. Unfortunately, by decreasing k, we increase the number of hits due to random matches that do not expand into significant alignments. Roughly speaking, the number of such spurious hits is $\frac{|S| \cdot |T|}{4^k}$ for DNA sequences. As a consequence, increased sensitivity levels are achieved at the cost of fourfold increases in the running time.

We propose a general technique to regulate the specificity of hashing-based methods. The key idea is to vary the seed lengths, which leads to a fine resolution

---

* Research supported by NSERC grant 250391-02.

S.C. Sahinalp et al. (Eds.): CPM 2004, LNCS 3109, pp. 373–387, 2004.

in specificity. Shorter seeds mean better sensitivity, so we aim at a maximal increase in sensitivity by maximizing the number of shorter seeds. The procedure is controlled by a parameter that explicitly sets the permitted increase in running time. More precisely, the parameter limits the number of spurious hits caused by shorter seeds.

Seed-and-extend methods find matching substrings between S and T by using a lookup table. For every length-k word, the table stores a list of positions where it is seen in S. Our method creates shorter seeds by merging lists in the lookup table for words with common prefixes. We estimate the increase in spurious hits by using actual word counts in S and the letter frequencies observed in T. The list merging procedure uses the concept of a seed tree, described in Section 2. The technique is not restricted to fixed-length substrings but can be used in conjunction with spaced seeds, as reviewed in Section 3. The procedure modifies the lookup table, and thus can be incorporated in most seed-and-extend algorithms. Our proposed method is fast and requires only a small amount of additional memory. Specifically, it can be implemented for DNA sequences with the help of about $\frac{2}{3}4^k$ integer variables (the lookup table on its own uses $|S|+4^k$ integers), an increase of about 11 megabytes for $k = 11$ over the basic method. Section 4 analyzes the method's space and time requirements in detail. Section 5 describes the results of a few experiments. Section 6 concludes with a discussion of related methods and future research.

1.1   Seed-and-Extend Methods

We use the following notation. We are interested in sequences over a finite alphabet Σ. The length of a sequence S is denoted by $|S|$. The set of all sequences with length m is denoted by Σ^m. The concatenation of two sequences S and T is denoted by $S \cdot T$. The notation $S[i]$ stands for the i-th character of S, and $S[i..j]$ means the substring of S starting with its i-th character and ending with its j-th character.

Seed-and-extend methods rely on a hashing function $h \colon \Sigma^\ell \mapsto \Sigma^k$. Pairs of positions (i, j) are identified in which $h(S[i..i + \ell - 1]) = h(T[j..j + \ell - 1])$. Such pairs are called *hits*. Hits are further extended into local alignments. In the simplest case, the similarity search uses the identity function for hashing (and thus $\ell = k$): hits are identified by identical substrings. Such fixed length substrings are called *k-mers*. The success of the seed-and-extend approach is due to the speed in which hits can be identified. The first step of the search constructs a hash table using a window of length ℓ that slides over S. For every position $i = 1, \ldots, |S| - \ell + 1$, the hash key $h(S[i..i + \ell - 1])$ is calculated: the table contains the $\langle h(S[i..i+\ell-1]), i \rangle$ pairs. In other words, the *key occurrence list* $\mathrm{Occ}_S(u)$ is calculated for every possible hash key $u \in \Sigma^k$, defined as

$$\mathrm{Occ}_S(u) = \Big\{ i \colon h(S[i..i + \ell - 1]) = u \Big\}. \tag{1}$$

In the second step of the search, a window slides over T. In each position $j = 1, \ldots, |T| - \ell + 1$, the hits (i, j) are considered for extension where i is in the occurrence list $\mathrm{Occ}_S\big(h(T[j..j + \ell - 1])\big)$.

It was discovered recently [9] that *spaced seeds* yield better sensitivity than k-mers. A spaced seed is defined by an ordered set $S = \{s_1, s_2, \ldots, s_k\} \subseteq \{1, \ldots, \ell\}$. Such a seed is referred to as an (ℓ, k)-*seed*. The corresponding hashing function is $h(u) = u[s_1] \cdot u[s_2] \cdots u[s_k]$, i.e., it samples the positions specified by S. Parameter k is called the *weight* of the seed. The use of a (k, k)-seed corresponds to the case of using k-mers. From this point on we assume that the hash function h is defined by an (ℓ, k)-seed.

1.2 Spurious Hits

Using a simple model of random sequences, we estimate the number of spurious hits that spaced seeds generate. Assume that S is a random sequence of independent and identically distributed (i.i.d.) characters, specified by the distribution $\mathbb{P}\{S[i] = \sigma\} = p_\sigma$ for all $i = 1, \ldots, |S|$ and $\sigma \in \Sigma$. Similarly, assume that T is a sequence of i.i.d. characters, specified by the distribution $\mathbb{P}\{T[j] = \sigma\} = q_\sigma$ for all $j = 1, \ldots, |T|$ and $\sigma \in \Sigma$. Let $S = \{s_1, s_2, \ldots, s_k\}$ be the spaced seed that defines the hash function h. The number of window positions in which a specific key u is encountered in S equals

$$n_S(u) = \Big| \mathrm{Occ}_S(u) \Big| = \sum_{i=1}^{|S|-\ell+1} \prod_{j=1}^{|u|} \Big\{ S[i + s_j - 1] = u[j] \Big\},$$

where $\{\cdot\}$ stands for the indicator function. Hence the expected number of such positions is

$$\mathbb{E}n_S(u) = (|S| - \ell + 1) \prod_{j=1}^{|u|} p_{u[j]}. \tag{2}$$

The expected number of hits produced by u is

$$H_{S,T}(u) = \mathbb{E}[n_S(u) n_T(u)] = (|S| - \ell + 1)(|T| - \ell + 1)\left(\prod_{j=1}^{|u|} (p_{u[j]} q_{u[j]}) \right), \tag{3}$$

where we used the independence of the two sequences. The expected total number of hits is

$$H_{S,T} = \sum_{u \in \Sigma^k} H_{S,T}(u) = (|S| - \ell + 1)(|T| - \ell + 1)\left(\sum_{\sigma \in \Sigma} p_\sigma q_\sigma \right)^k = NM\beta^k \tag{4}$$

with $N = |S| - \ell + 1$, $M = |T| - \ell + 1$, and $\beta = \sum_{\sigma \in \Sigma} p_\sigma q_\sigma$. If at least one of the sequences is uniform, i.e., if $\forall \sigma \colon p_\sigma = 1/|\Sigma|$ or $\forall \sigma \colon q_\sigma = 1/|\Sigma|$, then $\beta = 1/|\Sigma|$. However, genomic sequences often display biased nucleotide frequencies. For instance, the mouse genome [11] has a a 42% (G+C) content, and (G + C) content varies between 38% and 48% on different human chromosomes. In order to model the effect of a given (G + C) content f_{G+C}, let $p_A = p_T = (1 - f_{G+C})/2$, $p_C = p_G = f_{G+C}/2$, and $q_\sigma = p_\sigma$. Equation (4) then gives $H_{S,T} = NM\left(\frac{1}{2} - f_{G+C}(1 - f_{G+C})\right)^k$. For $f_{G+C} = 40\%$, we obtain $H_{S,T} = NM \cdot 3.85^{-k}$ and for $f_{G+C} = 30\%$, the formula gives $H_{S,T} = NM \cdot 3.45^{-k}$. The base of the exponential may actually be larger than 4, when e.g., S has (G+C) > 50% and T has (G + C) < 50%.

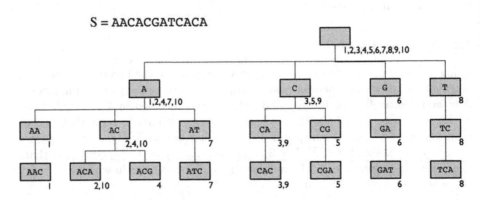

Fig. 1. Example of a seed tree built over 3-mers in the sequence S. Numbers at the nodes are the window positions at which the node's t-mer is seen. Only those nodes are shown that are sampled by at least one sequence position.

2   Variable Length Seeds

A *seed tree* is obtained by placing the hash keys in a trie data structure [12]. Specifically, the seed tree is a rooted tree defined as follows. Every non-leaf node has a child corresponding to every character $\sigma \in \Sigma$. The root is at level 0. Nodes at level $t = 0, \ldots, k$ correspond to t-mers. The children of node $u \in \Sigma^t$ are $\{u \cdot \sigma : \sigma \in \Sigma\}$. The leaves are (initially) at level k. Figure 1 shows an example of a seed tree. We now define the seeds corresponding to higher levels, and extend Equation (1) to inner nodes. Recall the definition of the occurrence list for a key u, $\mathrm{Occ}_S(u) = \{i\colon h(S[i..i+\ell-1]) = u\}$. The list Occ is defined recursively by $\mathrm{Occ}_S(v) = \cup_{\sigma \in \Sigma}\mathrm{Occ}_S(v \cdot \sigma)$ whenever $|v| < k$. It is not hard to see that in the case of hashing with k-mers, this definition corresponds to the idea that the lists Occ_S at level t are produced by the t-mers in S. Accordingly, $\mathrm{Occ}_S(u) = \{i\colon i \le |S| - k + 1, S[i..i+|u|-1] = u\}$ holds for all nodes u in that case.

Equation (3) is particularly interesting when the nucleotide frequencies are not uniform. Figure 2 shows the frequency of different k-mers in a sequence with

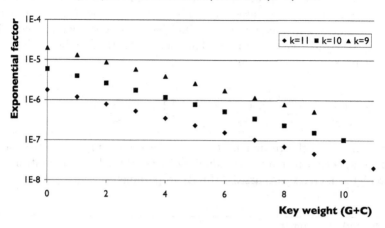

Fig. 2. Frequency of k-mers in the case of non-uniform nucleotide distribution. Under the assumption that S is a sequence of i.i.d. nucleotides with frequencies $p_A = p_T = 0.3$ and $p_C = p_G = f_{G+C}/2 = 0.2$, Equation (2) predicts that a k-mer with t $(G + C)$ content is encountered $(|S| - \ell + 1)\alpha$ times where $\alpha = \frac{f_{G+C}^t(1 - f_{G+C})^{k-t}}{2^k}$. The plot shows α as a function of t for $k = 9, 10, 11$. For example, a 9-mer with eight $(G + C)$ is seen about as often as a 10-mer with five $(G + C)$ or an 11-mer with two $(G + C)$.

40% $(G + C)$ content. The frequencies may vary widely for a given k. Given the fact that some shorter k-mers are rarer than others, we prune the seed tree and use shorter or longer k-mers depending on the number of hits they generate. In every pruning operation, a node is selected whose children are leaves, and the children are removed. Our aim is to balance the number of hits generated by the leaves. After a certain amount of pruning, we stop, and use the modified seed tree for finding hits. Using a sliding window over T, instead of looking for the key $h(T[j..j + \ell - 1])$ in every position j, we find its longest prefix u that is in the pruned seed tree. The corresponding hits are $\text{Occ}_S(u) \times \{j\}$.

The key to the pruning procedure is the criterion by which nodes are selected. Our criterion relies on predicting the increase in the number of hits at each pruning. First, the sequence S is analyzed to build the key occurrence lists at the leaf level. At the same time $n_S(u) = |\text{Occ}_S(u)|$ is calculated for all $u \in \Sigma^k$. For inner nodes we rely on the recursions

$$\text{Occ}_S(v) = \bigcup_{\sigma \in \Sigma} \text{Occ}_S(v \cdot \sigma); \quad \text{and} \quad n_S(v) = \sum_{\sigma \in \Sigma} n_S(v \cdot \sigma). \quad (5)$$

The number of hits a node $v \in \Sigma^t$ generates is predicted as

$$\text{hits}(v) = M n_S(v) \prod_{j=1}^{|v|} \hat{q}_{v[j]},$$

where $M = |T| - \ell + 1$ and $\hat{q}_\sigma$ are the nucleotide frequencies observed in T. Notice that while Equation (3) calculates the number of hits from letter frequencies in S and T, hits(v) is the expected number of hits, given observed key frequencies in S, and observed letter frequencies in T. The increase in the number of hits when pruning at node v is predicted as

$$\text{hits}^+(v) = \text{hits}(v) - \sum_{\sigma \in \Sigma} \text{hits}(v \cdot \sigma) = M\left(\prod_{j=1}^{|v|} \hat{q}_{v[j]}\right) \sum_{\sigma \in \Sigma}(1 - \hat{q}_\sigma)n_S(v \cdot \sigma), \quad (6)$$

where we employed the recursion of (5). We keep on pruning the tree, always selecting a node that increases the number of hits by the least amount, until the total predicted increase surpasses a threshold parameter.

Procedure EstimateHits
Input: hash key size k, occurrence lists $\text{Occ}_S(\cdot)$, character frequencies $\hat{q}$, sequence length M
Output: places every $u \in \Sigma^{<k}$ in the appropriate bin by setting up the linked lists BinList

H1 set total $\leftarrow 0$; **for all** $b \leftarrow 0, \dots, B$ **do** set BinList $\leftarrow \emptyset$
H2 **for all** $u \in \Sigma^{k-1}$ **do**
H3   set $n_\sigma \leftarrow |\text{Occ}_S(u \cdot \sigma)|$ for all $\sigma \in \Sigma$ // $n_\sigma = n_S(u \cdot \sigma)$
H4   set $h \leftarrow M\left(\prod_{i=1}^{|u|} \hat{q}_{u[i]}\right)$
H5   set total $\leftarrow$ total $+ h \sum_{\sigma \in \Sigma} \hat{q}_\sigma n_\sigma$
H6   set $n_S(u) \leftarrow \sum_{\sigma \in \Sigma} n_\sigma$
H7   set $H \leftarrow h \sum_{\sigma \in \Sigma}(1 - \hat{q}_\sigma)n_\sigma$ // $H = \text{hits}^+(u)$
H8   add u to the end of BinList$[b(H)]$
H9 **for** $t \leftarrow k - 2, \dots, 0$ **do**
H10   **for all** $u \in \Sigma^t$ **do**
H11   set $n_S(u) \leftarrow \sum_{\sigma \in \Sigma} n_S(u \cdot \sigma)$
H12   set $H \leftarrow M\left(\prod_{i=1}^{|u|} \hat{q}_{u[i]}\right) \sum_{\sigma \in \Sigma}(1 - \hat{q}_\sigma)n_S(u \cdot \sigma)$
H13   add u to the end of BinList$[b(H)]$
H14 **return** total. // (*returns number of hits without pruning*)

Fig. 3. Procedure EstimateHits. EstimateHits places the inner nodes of the seed tree in the appropriate bins, based on the number of hits they produce. Notice that $n_S(\cdot)$ is not needed outside this procedure.

In order to carry out the pruning, we use a binning procedure. Proceeding from the leaves towards the root, we calculate $n_S(v)$ using Equation (5), and hits$^+(v)$ using Equation (6). Using a monotone binning function $b\colon [0, \infty) \mapsto \{0, 1, \dots, B\}$, we place every tree node v into the bin $b(\text{hits}^+(v))$. The pruning is carried out by selecting nodes first from bin 0, then from bin 1, and so on, until the threshold for the number of hits is reached, or there are no more nodes available. Lemma 1 shows that the increase in the number of hits is larger at parents than it is at the children unless the sequence T consists mostly of the same character. The importance of this lemma is that it suggests how nodes should be stored: using a linked list at each bin, nodes are added at the list tail as we calculate hits^+ from the leaves upwards. Pruning proceeds by traversing the list in each bin. By Lemma 1, whenever a node is considered for pruning, all of its descendants are pruned already.

Lemma 1. *If $\hat{q}_\sigma \leq 1/2$ for all $\sigma \in \Sigma$, then for all v with $|v| < k - 1$ and $x \in \Sigma$, $\mathsf{hits}^+(v) \geq \mathsf{hits}^+(v \cdot x)$.*

Proof. By Equation (6), $\mathsf{hits}^+(v) \geq \mathsf{hits}^+(v \cdot x)$ if and only if

$$\sum_{\sigma \in \Sigma} (1 - \hat{q}_\sigma) n_S(v \cdot \sigma) \geq \hat{q}_x \sum_{\sigma' \in \Sigma} (1 - \hat{q}_{\sigma'}) n_S(v \cdot x \cdot \sigma'). \qquad (*)$$

We prove that the term for $\sigma = x$ on the left-hand side already ensures the inequality, i.e., that

$$(1 - \hat{q}_x) n_S(v \cdot x) \geq \hat{q}_x \sum_{\sigma' \in \Sigma} (1 - \hat{q}_{\sigma'}) n_S(v \cdot x \cdot \sigma'), \qquad (**)$$

which implies (*) and thus the lemma. By Equation (5), $n_S(v \cdot x) = \sum_{\sigma'} n_S(v \cdot x \cdot \sigma')$. Consequently, Equation (**) holds if $(1 - \hat{q}_x) n_S(v \cdot x \cdot \sigma') \geq \hat{q}_x(1 - \hat{q}_{\sigma'}) n_S(v \cdot x \cdot \sigma')$, that is, if $(1 - \hat{q}_x) \geq \hat{q}_x(1 - \hat{q}_{\sigma'})$ for all σ'. Since $\hat{q}_x \leq \frac{1}{2}$, $1 - \hat{q}_x \geq \hat{q}_x$, and Equation (**) follows. $\qquad\square$

Figure 3 sketches the procedure used to estimate the increased number of hits, and to place the nodes in their appropriate bins. Figure 4 sketches the pruning itself. It is important to notice that the seed tree is not used later when hit extension is performed, but rather that Line P5 updates the lookup table entry $\mathsf{Occ}_S(u)$ for every leaf in the subtree rooted at an inner node v when pruning at v. Line P4 calculates the parent's list by merging the lists at the children to preserve the ordering of positions. (It is often exploited during hit extension that positions are listed in decreasing order for each key.) When T is finally processed, we can use the lists $\mathsf{Occ}_S(u)$ directly with every hash key u. This technique allows for interfacing with other lookup table-based algorithms. Procedures EstimateHits and Prune update the hash table after it is constructed, and the search can proceed as usual without any additional data structures.

Figure 5 sketches a possible way to incorporate the seed tree pruning procedures in a local similarity search algorithm. It follows the logic of such tools as BLAST [6] and PatternHunter [9]. The difference is in Lines A4–A6, where the hash table is updated based on the seed tree pruning procedure. Lines A1–A3 build the hash table as usual. Lines A7–A10 find hits and extend just as if the pruning never happened.

3   Spaced Seeds

A seed tree can be used immediately in conjunction with a spaced seed $S = \{s_1, \cdots, s_k\}$, since the pruning works regardless of how the hash keys are obtained. It is advantageous, however, to select a good permutation $s_{i_1}, \ldots, s_{i_k}$ of

---

Procedure Prune
Input: total increase in hits max, list BinList of nodes in every bin
Output: pruned seed tree
P1 set increase $\leftarrow 0$ and bin $\leftarrow 0$
P2 **while** increase $<$ max and bin $\leq B$ **do**
P3   **for all** $u \in$ BinList[bin] starting at the list head, **do**
P4   set $\mathrm{Occ}_S(u) \leftarrow \bigcup_{\sigma \in \Sigma} \mathrm{Occ}_S(u \cdot \sigma)$
P5   set $\mathrm{Occ}_S(v) \leftarrow \mathrm{Occ}_S(u)$ for every leaf v in the subtree of u
P6   set increase $\leftarrow$ increase $+ b^{-1}(\text{bin})$
P7   **if** increase $\geq$ max **then return**
P8   set bin $\leftarrow$ bin $+ 1$

Fig. 4. Procedure Prune. Line P5 implements the pruning: the lists at the leaves are updated to accelerate the ensuing extension process when scanning T. (Accordingly, for every inner node u, $\mathrm{Occ}_S(u)$ is stored physically at the leftmost leaf in its subtree.) Notice that BinList is not needed after the pruning. The expression $b^{-1}(\text{bin})$ in Line P6 denotes the average value of hits$^+$ in bin.

the sampled positions that maximizes sensitivity. For a permutation $i_1, \ldots, i_k$, define the series of seeds S_t for $t = 0, \ldots, k$ by $S_0 = \emptyset$ and $S_t = S_{t-1} \cup \{s_{i_t}\}$ for $t = 1, \ldots, k$. The keys at level t of the seed tree correspond to hashing with the spaced seed S_t. An optimal permutation is obtained by iteratively selecting $i_k, i_{k-1}, \ldots, i_1$. Starting with $S_k = S$, S_{t-1} is computed by selecting the element of S_t that can be removed to obtain the highest increase in sensitivity. Sensitivity is measured as the probability of detecting a region of a given length and similarity. The probability is calculated under the assumption that every position of the region mutates independently with the same probability. Formally, the following model is used. Assume that a region of length $L \geq \ell$ of similarity g is "hidden" in the two sequences: there exist i_0, j_0 such that $S[i_0..i_0 + L - 1]$ and $T[j_0..j_0 + L - 1]$ are identical in about gL positions. More precisely, the corresponding substrings have such a distribution that for all $\sigma, \sigma' \in \Sigma$, and position offsets $j = 1, \ldots, L$,

$$\mathbb{P}\left\{T[j_0 + j - 1] = \sigma' \mid S[i_0 + j - 1] = \sigma\right\} = \begin{cases} g & \text{if } \sigma = \sigma'; \\ \frac{1-g}{|\Sigma|-1} & \text{if } \sigma \neq \sigma', \end{cases}$$

and that different positions are independent. Sensitivity of a hashing function is measured as the probability of producing a hit within this region, i.e., the probability that there exists $i \in \{-s_1+1, -s_1+2, \ldots, L-s_k\}$ for which $h(S[i_0 + i..i_0 + i + \ell - 1]) = h(T[j_0 + i..j_0 + i + \ell - 1])$. The probability of detecting such a homology region by a given seed can be calculated explicitly using an appropriately defined Markov chain [13,14].

We implemented the Markov chain method and calculated the optimal permutation for a number of seeds. Table 1 shows the seeds used in our experiments. At first sight it may seem that it is best to remove the leftmost or rightmost positions from the seed, since that increases the expected number of hits within the similarity region (which is $(L - (\max S - \min S + 1))g^{|S|}$). Interestingly, this

---

Algorithm CompareSequences
Input: sequences S, T, hashing function $h\colon \Sigma^\ell \mapsto \Sigma^k$, relative increase in running time R
A1 initialize $\mathsf{Occ}_S(u) \leftarrow \emptyset$ for all $u \in \Sigma^k$
A2 **for all** $i \leftarrow 1, \ldots, |S| - \ell + 1$ **do**
A3   set $u \leftarrow h(S[i..i + \ell - 1])$ and add i to $\mathsf{Occ}_S(u)$
A4 calculate $M = |T| - \ell + 1$, and $\hat{q}_\sigma = \dfrac{\sum_{j=1}^{|T|} \{T[j] = \sigma\}}{|T|}$ for all $\sigma \in \Sigma$
A5 set $\mathsf{T} \leftarrow$ EstimateHits$(k, \mathsf{Occ}_S, \hat{q}, M)$ // T *is the number of spurious hits*
    without pruning
A6 do Prune$(R\mathsf{T})$
A7 **for** $j \leftarrow 1, \ldots, |T| - \ell + 1$ **do**
A8   set $u \leftarrow h(T[j..j + \ell - 1])$
A9   **for all** $i \in \mathsf{Occ}_S(u)$ **do**
A10   extend the hit at (i, j) and report the alignment if significant

---

Fig. 5. The comparison algorithm.

is not always the case: for instance, the sensitivity of the $(18, 12)$-seed increases the most by removing a sampling position in the middle.

Clearly, the principle of choosing a permutation based on sensitivity is not restricted to the case of independent mutations and constant mutation rates. In particular, it can be applied in conjunction of other, more sophisticated models for assessing sensitivity, such as the Markov models for coding regions employed in [14] and [15].

4 Performance Analysis

Theorem 1. *The procedures EstimateHits and Prune can be implemented using* $2\frac{|\Sigma|^k - 1}{|\Sigma| - 1} + O(1)$ *integer variables.*

Proof. Procedure EstimateHits has to keep track of $n_S(u)$ for every non-leaf node u. The number of non-leaf nodes in the seed tree equals $\sum_{t=0}^{k-1} |\Sigma|^t = \frac{|\Sigma|^k - 1}{|\Sigma| - 1}$. Procedure Prune relies on the BinList array of linked lists. The chaining within the lists can be implemented by using an array that gives the next element after each non-leaf node u in the bin that contains u. Therefore, $\frac{|\Sigma|^k - 1}{|\Sigma| - 1}$ integers can be used for the chaining if every node address is encoded by an integer value. An additional array of $(B + 1)$ elements specifies the first element in each bin. The total number of integer variables is therefore $2\frac{|\Sigma|^k - 1}{|\Sigma| - 1} + (B + 1) + O(1)$, where the $O(1)$ term accounts for auxiliary local variables. $\square$

As discussed in [9], the lookup table can be implemented so that every occurrence list $\mathsf{Occ}_S(u)$ is a linked list of positions in decreasing order, with the aid of $(|S| + |\Sigma|^k)$ integers. (In an array of size $|\Sigma|^k$, the last position is given where each hash key is seen in S. Another array of size $|S|$ gives the previous position for every i where the hash key $h(S[i..i + \ell - 1])$ is seen.) Theorem 1 thus implies that the DNA similarity search algorithm of Figure 5 can be implemented with $(4|S| + \frac{5}{3}4^{k+1} + \epsilon)$ bytes (allowing four bytes per integer) in addition to the storage of the input sequences. The term ϵ accounts for the runtime environment, all local variables, and the data structure tracking recent extensions in Line A10. As an illustration of space efficiency, if S and T are around 150 Mbp in length, if every nucleotide is stored in one byte, and if $k = 11$, then the algorithm uses less than 1 Gbytes of memory. In our experiments with trees over the $(18, 12)$-seed comparing human and rat X chromosomes (152 and 163 million letters), the maximum memory usage was about 990MB.

Theorem 2. *EstimateHits and Prune update the lookup table in* $O\left(k|S|+k|\Sigma|^k\right)$ *time.*

Proof. EstimateHits executes Line H3 for every leaf node. The total time spent counting the size of every list Occ in Line H3 is thus $O(|\Sigma|^k + |S|)$. Lines H11–H13 take $O(|\Sigma| + t)$ time for every node at level $t = 0, \ldots, k - 2$, and thus the total time of executing the loop is of $O(k|\Sigma|^{k-2} + |\Sigma|^{k-1})$. EstimateHits thus runs in $O(k|\Sigma|^{k-2} + |\Sigma|^k + |S|)$ time. In an extreme scenario, procedure Prune is called with $\max = \infty$, and all nodes are pruned. Since list merging in Line P4 takes linear time in the sum of the list lengths, pruning all nodes at level t takes $O(|S|)$ time. Hence the total time spent on merging in Line P4 is $O(|S|k)$. Line P5 takes $O(1)$ time for every leaf v, and thus $O(|\Sigma|^k)$ time is spent in Line P5 for every tree level. Consequently, Prune runs in $O(|\Sigma|^k k + |S|k)$ time. $\square$

It is worth pointing out that in contrast to the worst-case analysis of Theorem 2, pruning should be carried out with a parameter $R = O(\beta^{-1})$ where β is defined in Equation (4). Otherwise, too many nodes of the seed tree are pruned, resulting in more hits than a shorter seed would produce. Since the shorter seed would require less memory and might provide better sensitivity, it is better to avoid large R values. For a reasonable choice of R, procedure Prune takes a time of $O(|\Sigma|^k + |S|)$, rather than $O(k)$ times as much, as pruning takes place only on one or a few tree levels.

5   Experiments

We implemented the algorithm in Figure 5 to test its performance on DNA sequences. We carried out a number of experiments with our prototype implementation with two goals in mind. First, we wanted to assess how valid the estimation of spurious hits is for biological sequences, and whether the fine-tuning of specificity vs. sensitivity is truly achieved by using a single parameter. Secondly, we wanted to measure how much time the tree pruning takes with respect to other steps in the search. We implemented only gapless seed extension since our aim was not to develop another similarity search tool, but to produce a testbed for seed tree pruning. The experiments rely on a simple scoring function

Table 1. Seeds used in our experiments. "Best" refers to highest sensitivity, measured by the probability of detecting a 70%-similarity region of length 64. Sampled positions are ordered by maximizing the sensitivity of shortened seeds as described. The weight-10 seeds were not used in conjunction with tree pruning, and thus their permutation is arbitrary.

Seed	Ordered positions $s_1, s_2, \ldots, s_k$	Remark
$(20, 13)$-seed	$10, 9, 6, 4, 3, 14, 16, 18, 19, 2, 13, 1, 20$	Best weight-13 seed
$(18, 12)$-seed	$2, 3, 8, 10, 13, 14, 5, 1, 16, 17, 18, 7$	Best weight-12 seed
$(18, 11)$-seed	$2, 3, 8, 10, 13, 14, 5, 1, 16, 17, 18$	Best weight-11 seed (PatternHunter)
$(18, 10)$-seed	$1, 2, 3, 6, 9, 12, 14, 16, 17, 18$	Best $(18,10)$-seed
$(16, 10)$-seed	$1, 2, 4, 5, 9, 10, 12, 14, 15, 16$	Best weight-10 seed

with a match score of 1 and mismatch penalty of 1. The extension calculates high-scoring segment pairs (HSPs) for every hit (i, j). The extensions are found by exploring the diagonal of (i, j) in two directions until the score drops below a threshold. The highest scoring segment is selected in both directions to obtain the HSP. The significance of the alignment is assessed using standard techniques [16]. We keep track of the longest extension along each diagonal, and attempt the extension only if the hit does not overlap with a previously explored region. In all experiments we used a cutoff of $E = 0.1$ for defining HSPs, i.e., we considered local alignments that have a score that is expected to occur 0.1 times in the sequence comparison.

The seeds we used in the experiments are shown in Table 1. In a first set of experiments, we compared the genome sequence of *H. influenzae* (1.8 million base pairs) to that of *E. coli* (4.6 Mbp). Figure 6a shows the accuracy of predicting the number of spurious hits using two seed trees, one built over the $(18, 12)$-seed and the other over the $(18, 11)$-seed. The parameter R predicts the increase in number of spurious hits very well[1]. Figure 6b plots the number of HSPs in function of the number of hits. The plot shows that the seed trees attain intermediate levels of sensitivity. In fact, the seed trees perform better than the weight-10 seeds.

We also compared the concatenated genome sequence of the budding yeast *S. cerevisiae* (12 Mbp) to the concatenated genome sequence of the fission yeast *S. pombe* (12.5 Mbp). Figure 7a shows the accuracy of predicting the number of spurious hits. Again, the predicted and measured values of specificity are very close. Figure 7b plots the number of HSPs for different levels of specificity: the seed tree over the $(18, 12)$-seed falls behind the $(18, 11)$-seed, but the seed tree over the latter slightly outperforms a weight-10 seed.

The final example of the method's application is the comparison of human and rat X chromosomes (152 and 163 million base pairs, respectively). We con-

---

[1] Using the seed tree over the $(18, 11)$-seed as an example, Equation (4) predicts $H \approx 1.9 \cdot 10^6$ spurious hits (*H. influenzae* genome has a 38% ($\mathtt{G} + \mathtt{C}$) content, while that of *E. coli* is close to 50%). The $(18, 11)$-seed finds $2.6 \cdot 10^6$ hits without pruning. We expect therefore that the number of hits grows as $H(R) = 2.6 \cdot 10^6 + 1.9 \cdot 10^6 R$.

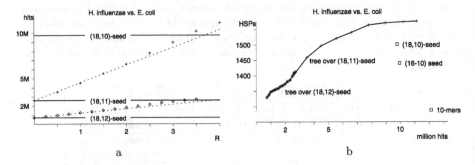

a

b

Fig. 6. Comparison of *H. influenzae* and *E. coli* genomes. The left-hand side plots the number of hits in function of the algorithm's parameter R for two seed trees. Each dotted line shows the expected slope of the function. The right-hand side shows the specificity-sensitivity trade-off for HSPs with E-value 10^{-1}. Both curves start at $R = 0$, i.e., where hits are produced by spaced seeds without tree pruning.

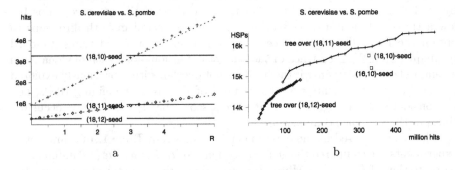

a

b

Fig. 7. Comparison of two yeast genomes. The left-hand side plots the number of hits in function of the algorithm's parameter R for two trees. Dotted lines show the expected slopes. The right-hand side shows the specificity-sensitivity trade-off for HSPs with E-value 10^{-1}. Both curves start at $R = 0$.

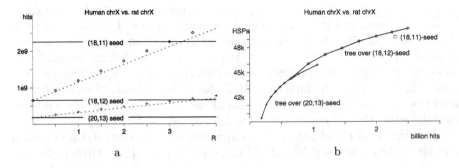

a

b

Fig. 8. Comparison of two mammalian X chromosomes. The figure on the left-hand side plots the number of hits for different R values; dotted lines indicate the expected slopes. The right-hand side illustrates the sensitivity-specificity trade-off. Both curves start at $R = 0$.

catenated the sequence contigs for each chromosome, using sequences that were masked for repeat and low-complexity regions. We ignored masked regions for hit generation but included them for hit extension. The experimental results are summarized in Figure 8. Again, the parameter accurately predicts the increases in running time, and specificity levels between those of spaces seeds are reached.

The time it takes to carry out the pruning is comparable to the time of building the lookup table. Table 2 shows measured running times. (The hit extension in our implementation is not optimized, and only gapless extensions are found, but comparable running times were reported for PatternHunter [9], which does construct gapped extensions.) The time increments for the X chromosomes are particularly instructive, as the chromosomes are of average size in the two genomes. Comparison of the entire human and rat genomes would take about four hundred times more. Using weight-13, 12, and 11 seeds, the chromosome comparison takes $\frac{1}{2}$, 1.25, and 4 hours without pruning. Multiplying by 400, the genome comparisons would take 8, 21, or 67 CPU days. Seed trees provide a flexible way to perform the genome comparisons at various levels of specificity in addition to these three very different choices.

Table 2. Measured running times for different parts of the comparison algorithm, excluding input-output. Column "Table" shows the time it takes to build the initial hash table after reading the first sequence. Column "Pruning" gives the running time for EstimateHits and Prune procedures. Column "Extensions" gives the running time for processing all hits and calculating HSPs. Timing was done on an Apple PowerBook 1.25GHz G4 with 1 GBytes of memory running MacOS X, using Java's System.currentTimeMillis() method.

	Table	Pruning	Extensions
H. influenzae–E. coli, (18,12)-seed $R = 0$	2s	0s	8s
H. influenzae–E. coli, (18,12)-seed $R = 3.5$	2s	9s	15s
H. influenzae–E. coli, (18,11)-seed $R = 3.5$	2s	3s	31s
S. cerevisiae–S. pombe, (18,12)-seed $R = 0$	13s	0s	3m 31s
S. cerevisiae–S. pombe, (18,12)-seed $R = 3.5$	14s	21s	6m 27s
S. cerevisiae–S. pombe, (18,11)-seed $R = 3.5$	12s	13s	32m 24s
Human chrX–rat chrX, (20,13)-seed $R = 0$	82s	0s	31m
Human chrX–rat chrX, (20,13)-seed $R = 3.5$	76s	6m 34s	78m
Human chrX–rat chrX, (18,12)-seed $R = 3.5$	119s	105s	4h 21m

6   Discussion

Seed-and-extend methods, exemplified by the NCBI BLAST suite, have been thoroughly successful in providing a way to perform highly sensitive similarity searches between long molecular sequences in reasonable time. Traditional seed-and-extend methods use fixed length hash keys. Our proposed method creates a hashing function with varying key lengths, based on statistics gathered from the input sequences. Shorter keys are introduced to maximize the sensitivity with a predictable compromise of specificity. We introduced the concept of the seed tree, which guides the selection of shorter seeds. After gathering statistics about the hash keys, and pruning, the seed tree is not needed anymore. The second sequence in the comparison is processed using the updated lookup table, and thus hash keys and their occurrence lists are still found in $O(1)$ time.

There are many variations on the basic seed-and-extend idea that affect the sensitivity of the search. We consider two such variations: multiple spaced seeds [10, 14] and vector seeds [17]. Li *et al.* [10] explore the use of multiple spaced seeds in homology searches. In order to use a set of m seeds, one needs to construct a lookup table for every seed, requiring enough memory for $(m|S|+m|\Sigma|^k)$ integers. As a consequence, long sequences may need to be split into smaller segments. The number of spurious hits grows linearly with m, allowing for no fine-tuning between consecutive values of m. Our method can be used instead of multiple seeds to control the specificity on a fine scale. It can also be used in conjunction with multiple lookup tables in the same way as with a single one: a seed tree is pruned for every table. In order to track the number of hits created, the procedure Prune would have to be modified to consider the seed trees built for the different seeds simultaneously.

Brejová *et al.* [17] consider relaxing the definition of a match between hash keys by introducing the concept of vector seeds. Vector seeds permit weighted mismatches between hash keys for producing hits. Relaxing or tightening the hit criteria allows for reaching different levels of sensitivity and specificity. Our approach of pruning a seed tree creates new hits in a data-dependent manner, and offers a way of setting different levels of specificity for different keys. Brudno *et al.* [18] also use a trie to manipulate hash keys, but do not consider either variable length keys, or pruning.

While our implementation focused on nucleotide sequences, we attempted to present the idea of pruning seed trees independently of the underlying alphabet. In particular, our algorithms in Figures 3 and 4 can be implemented for protein sequences, with consequences for space and time requirements as analyzed in Section 4. In the case of proteins, hits are usually defined by similarity rather than identity, so the ramifications of pruning are more complex. We are currently exploring the practicality of our method on other than DNA sequences.

References

1. Miller, W.: Comparison of genomic DNA sequences: solved and unsolved problems. Bioinformatics **17** (2001) 391–397
2. Smith, T.F., Waterman, M.S.: Identification of common molecular subsequences. J. Mol. Biol. **147** (1981) 195–197
3. Myers, G., Durbin, R.: A table-driven full sensitivity similarity search algorithm. J. Comput. Biol. **10** (2003) 103–117
4. Delcher, A.L., Phillippy, A., Carlton, J., Salzberg, S.L.: Fast algorithms for large-scale genome alignment and comparison. Nucleic Acids Res. **30** (2002) 2478–2483
5. Pearson, W.R., Lipman, D.J.: Improved tools for biological sequence comparison. Proc. Natl. Acad. Sci. USA **85** (1988) 2444–2448

6. Altschul, S.F., Madden, T.L., Schäffer, A.A., Zhang, J., Zhang, Z., Miller, W., Lipman, D.J.: Gapped BLAST and PSI-BLAST: a new generation of protein database search programs. Nucleic Acids Res. **25** (1997) 3389–3402
7. Kent, W.J.: BLAT — the BLAST-like alignment tool. Genome Res. **12** (2002) 656–664
8. Schwartz, S., Kent, W.J., Smit, A., Zhang, Z., Baertsch, R., Hardison, R.C., Haussler, D., Miller, W.: Human-mouse alignments with BLASTZ. Genome Res. **13** (2003) 103–107
9. Ma, B., Tromp, J., Li, M.: PatternHunter: faster and more sensitive homology search. Bioinformatics **18** (2002) 440–445
10. Li, M., Ma, B., Kisman, D., Tromp, J.: PatternHunter II: highly sensitive and fast homology search. Journal of Bioinformatics and Computational Biology (2004) To appear.
11. MGSC: Initial sequencing and comparative analysis of the mouse genome. Nature **420** (2002) 520–562
12. Friedkin, E.: Trie memory. Comm. ACM **3** (1960) 490–500
13. Nicodème, P., Salvy, B., Flajolet, P.: Motif statistics. In Nešetřil, J., ed.: Algorithms — ESA'99: 7th Annual European Symposium. Volume 1643 of LNCS., Heidelberg, Springer-Verlag (1999) 194–211
14. Buhler, J., Keich, U., Sun, Y.: Designing seeds for similarity search in genomic DNA. In Vingron, M., Istrail, S., Pevzner, P., Waterman, M., eds.: Proc. 7th Annual International Conference on Computational Molecular Biology (RECOMB), New York, NY, ACM Press (2003) 67–75
15. Brejová, B., Brown, D., Vinař, T.: Optimal spaced seeds for homologous coding regions. Journal of Bioinformatics and Computational Biology **1** (2004) 595–610
16. Karlin, S., Altschul, S.F.: Methods for assessing the statistical significance of molecular sequence features by using general scoring schemes. Proc. Natl. Acad. Sci. USA **87** (1990) 2264–2268
17. Brejová, B., Brown, D., Vinař, T.: Vector seeds: an extension to spaced seeds allows substantial improvements in sensitivity and specificity. In Benson, G., Page, R., eds.: Algorithms and Bioinformatics: 3rd International Workshop (WABI). Volume 2812 of LNCS., Heidelberg, Springer-Verlag (2003) 39–54
18. Brudno, M., Chapman, M.A., Gottgens, B., Batzoglou, S., Morgenstern, B.: Fast and sensitive multiple alignment of large genomic sequences. BMC Bioinformatics **4** (2003) 66

Reversal Distance
without Hurdles and Fortresses

Anne Bergeron[1], Julia Mixtacki[2], and Jens Stoye[3]

[1] LaCIM, Université du Québec à Montréal, Canada
[2] Fakultät für Mathematik, Universität Bielefeld, Germany
[3] Technische Fakultät, Universität Bielefeld, Germany

Abstract. This paper presents an elementary proof of the Hannenhalli-Pevzner theorem on the reversal distance of two signed permutations. It uses a single PQ-tree to encode the various features of a permutation. The parameters called *hurdles* and *fortress* are replaced by a single one, whose value is computed by a simple and efficient algorithm.

1 Introduction

Computing the reversal distance of two signed permutations is a delicate task since some reversals unexpectedly affect deep structures in permutations. In 1995, Hannenhalli and Pevzner proposed the first polynomial-time algorithm to solve this problem [7], developing along the way a theory of how and why some permutations were particularly resistant to sorting by reversals.

Hannenhalli and Pevzner relied on several intermediate constructions that have been subsequently simplified [8], [10], but grasping all the details remained a challenge. Before [1], all the criteria given for choosing a *safe* reversal involved the construction of an associate permutation on $2n$ points, and the analysis of cycles and/or connected components of graphs associated to this permutation.

Another puzzling aspect of the Hannenhalli-Pevzner theory is the complex, but always colorful, classification of *hurdles*. In this paper, we show that simple results on trees are at the root of all results on hurdles, either maximal or simple, super-hurdles, and fortresses. We give an elementary proof of the Hannenhalli-Pevzner *duality theorem* in terms of a PQ-tree associated to the permutation, yielding efficient and simple algorithms to compute the reversal distance.

The next section presents classical material and results in a simpler form, and describes the tree associated to the permutation. Section 3 gives a new proof and formula for the Hannenhalli-Pevzner theorem, and Section 4 discusses the algorithms.

2 Background

A *signed permutation* is a permutation on the set of integers $\{0, 1, 2, \ldots, n\}$ in which each element has a sign. We also assume that all permutations begin with

S.C. Sahinalp et al. (Eds.): CPM 2004, LNCS 3109, pp. 388–399, 2004.
© Springer-Verlag Berlin Heidelberg 2004

0 and end with n, for example: $P_1 = (0\ -2\ -1\ 4\ 3\ 5\ -8\ 6\ 7\ 9)$. A *point* $p{\cdot}q$ is defined by a pair of consecutive elements in the permutation. For example, $0{\cdot}-2$ and $-2{\cdot}-1$ are the first two points of P_1. When a point is of the form $i\cdot i+1$, or $-(i+1)\cdot -i$, it is called an *adjacency*, otherwise it is called a *breakpoint*. For example, P_1 has two adjacencies, $-2\cdot-1$ and $6\cdot 7$. All other points of P_1 are breakpoints.

We will make an extensive use of intervals of consecutive elements in a permutation. An interval is easily defined by giving its *endpoints*. The *elements* of the interval are the elements between the two endpoints. When the two endpoints are equal, the interval contains no elements. A non-empty interval can also be specified by giving its first and last element, such as $(i..j)$, called the *bounding elements* of the interval.

A *reversal* of an interval of a signed permutation is the operation that consists of reversing the order of the elements of the interval, while changing their signs. The reversal of an interval modifies the points of a signed permutation in various ways. Points $p\cdot q$ that are inside the interval are transformed to $-q\cdot -p$, the endpoints of the interval exchange their flanking elements, and points that are outside the interval are unaffected.

The *reversal distance* $d(P)$ of a permutation P is the minimum number of reversals needed to transform P into the identity permutation.

2.1 Elementary Intervals and Cycles

Let P be a signed permutation on the set $\{0, 1, 2, \ldots, n\}$ that begins with 0 and ends with n. When sign is irrelevant, we will refer to an element as an *unsigned element*. The *right*, or *left*, point of an element of P is the point immediately to its right, or left.

Definition 1. *For each pair of unsigned elements* $(k, k+1)$, $0 \le k < n$, *define the* elementary interval I_k *associated to the pair to be the interval whose endpoints are:*

1) *The right point of k, if k is positive, otherwise its left point.*
2) *The left point of $k+1$, if $k+1$ is positive, otherwise its right point.*

Elements k and $k+1$ are called the extremities *of the elementary interval.*

Note that an elementary interval can contain zero, one, or both of its extremities. For example, in the following permutation, interval I_0 contains one of its extremities, interval I_3 contains both, and interval I_5 contains none.

Note that empty elementary intervals correspond to the adjacencies in the permutation.

When the extremities of an elementary interval have different signs, the interval is said to be *oriented*, otherwise it is *unoriented*. Oriented intervals play a basic role in the problem of sorting by reversals since they can be used to create adjacencies. Namely, we have:

Proposition 1. *Reversing an oriented interval I_k creates, in the resulting permutation, either the adjacency $k \cdot k + 1$ or the adjacency $-(k+1) \cdot -k$.*

When a point is the endpoint of two elementary intervals, these are said to *meet* at that point. Definition 1 implies that a point is used at most twice as an endpoint, and since there are as many non-empty elementary intervals as there are breakpoints, we have:

Proposition 2. *Exactly two elementary intervals meet at each breakpoint of a permutation.*

Therefore, starting from an arbitrary point, one can follow elementary intervals on a unique path that eventually comes back to the original point. More formally:

Definition 2. *A cycle is a sequence $b_1, b_2, \ldots, b_k$ of points such that two successive points are the endpoints of an elementary interval, including b_k and b_1.*

For example, the following permutation has four cycles, two of them are adjacencies, and the other two contain, respectively, 4 and 3 breakpoints.

One of the cornerstones of the sorting by reversals problem is to study the effects of a reversal on elementary intervals and cycles. The following result, due to [9], quantifies the effect of a reversal on the number of cycles. It is a consequence of the fact that, for all points except the endpoints of a reversal, the elementary intervals that meet at those points will meet at the same points after the reversal.

Proposition 3. *A reversal modifies the number of cycles by $+1$, 0, or -1.*

Finally, note that the identity permutation on the set $\{0, 1, 2, \ldots, n\}$ is the only one with n cycles. Thus, Proposition 3 implies that $d(P) \geq n - c$, where c is the number of cycles of P.

2.2   Components

Elementary intervals and cycles are organized in higher structures called *components*. These were first identified in [5] as *subpermutations* since they are intervals that contain a permutation of a set of consecutive integers.

Definition 3. *Let P be a signed permutation on the set $\{0, 1, 2, \ldots, n\}$. A component of P is an interval from i to $(i + j)$ or from $-(i + j)$ to $-i$, for some $j > 0$, whose set of unsigned elements is $\{i, \ldots, i+j\}$, and that is not the union of two such intervals. Components with positive – respectively negative – bounding elements are referred to as direct – respectively reversed – components.*

For example, consider the permutation:

$$P_2 = (0 \ {-3} \ 1 \ 2 \ 4 \ 6 \ 5 \ 7 \ {-15} \ {-13} \ {-14} \ {-12} \ {-10} \ {-11} \ {-9} \ 8 \ 16).$$

It has 6 components, one of them being the interval $(0..4)$, which contains all unsigned elements between 0 and 4; another is the interval $(-15.. - 12)$. Note that a component, such as the interval $(1..2)$, can contain only two elements.

Components of a permutation can be represented by the following diagram, in which the bounding elements of each component have been boxed, and the elements between them are enclosed in a rectangle. Elements which are not bounding elements of any component are also boxed.

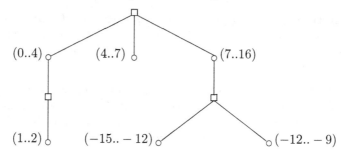

Proposition 4 ([3]). *Two different components of a permutation are either disjoint, nested with different endpoints, or overlapping on one element.*

When two components overlap on one element, we say that they are *linked*. Successive linked components form a *chain*. A chain that cannot be extended to the left or right is called *maximal*. Note that a maximal chain may consist of only a single component. If one component of a chain is nested in a component A, then all other components of the chain are also nested in A.

The nesting and linking relations between components turn out to play a major role in the sorting by reversal problem. Another way of representing these relations is in form of the following tree:

Definition 4. *Given a permutation P on the set $\{0, 1, \ldots, n\}$ and its components, define the tree T_P by the following construction:*

1) *Each component is represented by a round node.*
2) *Each maximal chain is represented by a square node whose (ordered) children are the round nodes that represent the components of this chain.*
3) *A square node is the child of the smallest component that contains this chain.*

For example, the tree associated to permutation P_2 is:

It is easy to see that, if the permutation begins with 0 and ends with n, the resulting graph is a single tree with a square node as root. The tree is similar to the PQ-tree used in different context such as the consecutive ones test [4]. The following properties of paths in T_P are elementary consequences of the definition of T_P.

Proposition 5. *Let C be a component on the (unique) path joining components A and B, then C contains either A or B, or both.*

1) If C contains both A and B, it is unique.
2) If no component on the path contains both A and B, then A and B are included in two components that are in the same chain.

Components organize hierarchically the points of a permutation.

Definition 5. *A point $p \cdot q$ belongs to the smallest component that contains both p and q.*

Note that this does not prevent p and q to be contained, separately, in smaller components, such as 5 and 7 in:

$$\boxed{0}\ \boxed{2}\ \boxed{4}\ \boxed{3}\ \boxed{5}\ \boxed{7}\ \boxed{9}\ \boxed{8}\ \boxed{10}\ \boxed{1}\ \boxed{6}\ \boxed{11}$$

Proposition 6. *The endpoints of an elementary interval belong to the same component, thus all the points of a cycle belong to the same component.*

Proof: Consider an elementary interval I_k and any component C of the form

$$(i..i+j) \quad \text{or} \quad (-(i+j).. - i)$$

such that $i \le k < i+j$. Then both endpoints of I_k are contained in C. This is obvious if k is different from i and $k+1$ is different from $i+j$, since both k and $k+1$ will be in the interior of the component. If $k = i$, then k and i have the same sign, and the first endpoint of I_k belongs to the component. If $k+1 = i+j$, then $k+1$ and $i+j$ have the same sign, and the second endpoint of I_k belongs to the component.

Thus endpoints of I_k are either both contained, or not, in any given component, and the result follows. ∎

Finally, components can be classified according to the nature of the points they contain:

Definition 6. *The sign of a point $p \cdot q$ is positive if both p and q are positive, it is negative if both are negative. A component is unoriented if it has one or more breakpoints and all of them have the same sign. Otherwise the component is oriented.*

All the elementary intervals of an unoriented component are unoriented. Therefore, it is impossible to sort unoriented components using oriented reversals. In the next section, we discuss the type of reversals that can be used to create oriented intervals in unoriented components.

2.3   Effects of a Reversal on Components

The next two propositions describe the effects of a reversal whose endpoints are in the same unoriented component. These are classical results from the Hannenhalli-Pevzner theory.

Proposition 7. *If a component C is unoriented, no reversal with its two endpoints in C can split one of its cycles, or create a new component.*

Proposition 8. *If a component C is unoriented, the reversal of an elementary interval whose endpoints belong to C orients C, and leaves the number of cycles of the permutation unchanged.*

Orienting a component as in Proposition 8 is called *cutting* the component. Cutting an unoriented component is seldom used in optimal sorting of a permutation since it is possible, with a single reversal, to get rid of more than one unoriented component. The following proposition describes how to *merge* several components, and the relations of this operation to paths in T_P.

Proposition 9. *If a reversal has its two endpoints in different components A and B, then only the components on the path from A to B in T_P are affected.*

1) A component C is destroyed if and only if it contains either A or B, but not both.

2) If A or B is unoriented, any component C that contains both A and B, and that is on the path that joins A and B, will be oriented.

3) A new component C is created if and only if A and B are included in two components that are in the same chain. If either A or B is unoriented, C will be oriented.

Sketch of proof: 1) One of the bounding elements of C will change sign, but not the other. 2) By 1), all components between A and C, and all components between B and C will be destroyed. Suppose that A is unoriented, then reversing one bounding element of A will introduce, in C, at least one oriented interval. 3) If A is included in A', and B is included in B', such that $A' = (a..a')$ precedes $B' = (b..b')$ in the same chain, then $C = (a..b')$ will be a new component.   ∎

Proposition 9 thus states that merging two unoriented components destroys or orients all components on the path, without creating new unoriented components.

3   The Hannenhalli-Pevzner Theorem

In this section, we develop a formula for computing the reversal distance of a permutation. There are two basically different problems: the contribution of oriented components to the total distance is treated in Section 3.1, and the general formula is given in Section 3.2.

3.1   Sorting Oriented Components

Sorting oriented components is done by choosing oriented reversals that do not create new unoriented components. Several different criteria for choosing such reversals exist in the literature, and we give here the simplest one.

Definition 7. *The* score *of a reversal is the number of oriented elementary intervals in the resulting permutation.*

Theorem 1 ([1]). *The reversal of an oriented elementary interval of maximal score does not create new unoriented components.*

Corollary 1. *If a permutation P on the set $\{0, \ldots, n\}$ has no unoriented components and c cycles, then $d(P) = n - c$.*

Proof: As stated following Proposition 3, we have that $d(P) \geq n - c$ since any reversal adds at most 1 cycle, and the identity permutation has n cycles. Any oriented reversal adds one cycle, thus Theorem 1 guarantees that there will be always enough oriented reversals to sort the permutation.   ∎

3.2   Computing the Reversal Distance

Definition 8. *A* cover *$\mathcal{C}$ of T_P is a collection of paths joining all the unoriented components of P, such that each terminal node of a path belongs to a unique path.*

By Propositions 8 and 9, each cover of T_P describes a set of reversals that orients all the components of P. A path that contains two or more unoriented components, called a *long* path, corresponds to merging the two components at its terminal nodes. A path that contains only one component, a *short* path, corresponds to cutting the component.

The *cost* of a cover is defined to be the sum of the costs of its paths, given that:
1) The cost of a short path is 1.
2) The cost of a long path is 2.

An *optimal* cover is a cover of minimal cost. Define t as the cost of any optimal cover T_P.

Theorem 2. *If a permutation P on the set $\{0, \ldots, n\}$ has c cycles, and the associated tree T_P has minimal cost t, then*

$$d(P) = n - c + t.$$

Proof: We first show that $d(P) \leq n - c + t$. Let $\mathcal{C}$ be an optimal cover. Apply to P the sequence of m merges and q cuts induced by the cover $\mathcal{C}$. Note that $t = 2m + q$. By Proposition 6, the resulting permutation P' has $c - m$ cycles, since merging two components always merges two cycles, and cutting components does not change the number of cycles. Thus, by Corollary 1, $d(P') = n - c + m$. Since $m + q$ reversals were applied to P, we have:

$$d(P) \leq d(P') + (m + q) = n - c + 2m + q = n - c + t.$$

In order to show that $d(P) \geq n - c + t$, consider any sequence of length d that optimally sorts the permutation. By Proposition 3, d can be written as

$$d = s + m + q,$$

where s is the number of reversals that split cycles, m is the number of reversals that merge cycles, and q is the number of reversals that do not change the number of cycles. Since the m reversals remove m cycles, and the s reversals add s cycles, we must have:

$$c - m + s = n, \text{ implying } d = n - c + 2m + q.$$

The sequence of d reversals induces a cover of T_P. Indeed, any reversal that merges a group of components traces a path in T_P, of which we keep the shortest segment that includes all unoriented components of the group. Of these paths, suppose that m_1 are long paths, and m_2 are short paths. Clearly, $m_1 + m_2 \leq m$. The $q' \leq q$ remaining unoriented components are all cut. Thus

$$2m_1 + m_2 + q' \leq 2m + q.$$

Since $t \leq 2m_1 + m_2 + q'$, we get $d \geq n - c + t$. $\blacksquare$

The last task is to give an explicit formula for t. Let T' be the smallest subtree of T_P that contains all unoriented components of P. Formally, T' is obtained by recursively removing from T_P all dangling oriented components and square nodes. All leaves of T' will thus be unoriented components, while internal round nodes may still represent oriented components. For example, the tree T' obtained from T_{P_2} contains one oriented and three unoriented components.

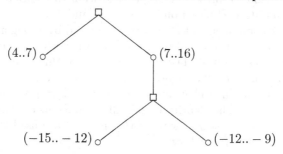

Define a *branch* of a tree as the set of nodes from a leaf up to, but excluding, the next node of degree ≥ 3. A *short* branch of T' contains 1 unoriented component, and a *long* branch contains 2 or more unoriented components. We have:

Theorem 3. *Let T' be the subtree of T_P that contains all the unoriented components.*

(1) If T' has $2k$ leaves, then $t = 2k$.
(2) If T' has $2k + 1$ leaves, one of them on a short branch, then $t = 2k + 1$.
(3) If T' has $2k + 1$ leaves, none of them on a short branch, then $t = 2k + 2$.

Proof: Let $\mathcal{C}$ be an optimal cover of T', with m long paths and q short ones. By joining any pair of short paths into a long one, $\mathcal{C}$ can be transformed into an optimal cover with $q = 0$ or 1.

Any optimal cover has only one path on a given branch, since if there were two, one could merge the two paths and lower the cost. Thus if a tree has only long branches, there always exists an optimal cover with $q = 0$.

Since a long path covers at most two leaves, we have $t = 2m + q \geq l$, where l is the number of leaves of T'. Thus cases (1) and (2) are lower bounds. But if $q = 0$, then t must be even, and case (3) is also a lower bound.

To complete the proof, it is thus sufficient to exhibit a cover achieving these lower bounds. Suppose that $l = 2k$. If $k = 1$, the result is obvious. For $k > 1$, suppose T' has at least two nodes of degree ≥ 3. Consider any path in T' that contains two of these nodes, and that connects two leaves A and B. The branches connecting A and B to the tree T' are incident to different nodes of T'. Thus cutting these two branches yields a tree with $2k - 2$ leaves. If the tree T' has only one node of degree ≥ 3, the degree of this node must be at least 4, since the tree has at least 4 leaves. In this case, cutting any two branches yields a tree with $2k - 2$ leaves.

If $l = 2k + 1$ and one of the leaves is on a short branch, select this branch as a short path, and apply the above argument to the rest of the tree. If there is no short branch, select a long branch as a first (long) path.   ∎

4   Algorithms

In this section we present a simple algorithm to compute the reversal distance of a permutation P based on Theorems 2 and 3. The algorithm consists of two parts. The components of P are first computed by an algorithm presented in [2], then the tree T_P is created by a simple pass over the components of P.

For completeness, we briefly recall the algorithm from [2], called Algorithm 1 here. The input of the algorithm is a signed permutation P, separated into an array of unsigned elements $\pi = (\pi_0, \pi_1, \ldots, \pi_n)$ and an array of signs $\sigma = (\sigma_0, \sigma_1, \ldots, \sigma_n)$. The algorithm finds all components of P in linear time. It makes use of four stacks, two of which (M_1 and M_2) are used to compute two arrays M and m, defined as follows:

$M[i]$ is the nearest element of π that precedes π_i and is greater than π_i,
$m[i]$ is the nearest element of π that precedes π_i and is smaller than π_i.

The algorithm to find the components uses two stacks S_1 and S_2 that store potential start points s of components, which are then tested by the following criterion: $(s..i)$ is a direct component if and only if:

1) both σ_s and σ_i are positive,
2) all elements between π_s and π_i in π are greater than π_s and smaller than π_i, the latter being equivalent to the simple test $M[i] = M[s]$, and
3) no element "between" π_s and π_i is missing, i.e. $i - s = \pi_i - \pi_s$.

Algorithm 1 (Find the components of a signed permutation $P = (\pi, \sigma)$)

1: M_1 and M_2 are stacks of integers; initially M_1 contains n and M_2 contains 0
2: S_1 and S_2 are stacks of integers; initially S_1 contains 0 and S_2 contains 0
3: $M[0] \leftarrow n$, $m[0] \leftarrow 0$
4: **for** $i \leftarrow 1, \ldots, n$ **do**

(* Compute the $M[i]$ *)
5:     **if** $\pi[i-1] > \pi[i]$ **then**
6:         push $\pi[i-1]$ on M_1
7:     **else**
8:         pop from M_1 all entries that are smaller than $\pi[i]$
9:     **end if**
10:    $M[i] \leftarrow$ the top element of M_1

(* Find direct components *)
11:    pop the top element s from S_1 as long as $\pi[s] > \pi[i]$ or $M[s] < \pi[i]$
12:    **if** $\sigma[i] = +$ **and** $M[i] = M[s]$ **and** $i - s = \pi[i] - \pi[s]$ **then**
13:        report the component $(s..i)$
14:    **end if**

(* Compute the $m[i]$ *)
15:    **if** $\pi[i-1] < \pi[i]$ **then**
16:        push $\pi[i-1]$ on M_2
17:    **else**
18:        pop from M_2 all entries that are larger than $\pi[i]$
19:    **end if**
20:    $m[i] \leftarrow$ the top element of M_2

(* Find reversed components *)
21:    pop the top element s from S_2 as long as $(\pi[s] < \pi[i]$ or $m[s] > \pi[i])$ and $s > 0$
22:    **if** $\sigma[i] = -$ **and** $m[i] = m[s]$ **and** $i - s = \pi[s] - \pi[i]$ **then**
23:        report the component $(s..i)$
24:    **end if**

(* Update stacks *)
25:    **if** $\sigma[i] = +$ **then**
26:        push i on S_1
27:    **else**
28:        push i in S_2
29:    **end if**

30: **end for**

A symmetric criterion allows to find reverse components. For details, see Algorithm 1. Without much overhead it is also possible to tell whether each component is oriented or not. Again, details can be found in [2].

Note that Algorithm 1 reports the components in left-to-right order with respect to their right end. For each index i, $0 \leq i \leq n$, at most one component can start at position i and at most one component can end at position i. Hence, it is possible to create a data structure that tells in constant time if there is a component beginning or ending at position i and, if so, reports such components. Given this data structure, it is a simple procedure to construct the tree T_P in one left-to-right scan along the permutation. Initially one square root node and one round node representing the component with left bounding element 0 are created. Then, for each additional component, a new round node p is created as the child of a new or an existing square node q, depending if p is the first component in a chain or not. For details, see Algorithm 2.

Algorithm 2 (Construct T_P from the components $C_1, \ldots, C_k$ of P)

1: create a square node q, the root of T_P and a round node p as the child of q
2: **for** $i \leftarrow 1, \ldots, n-1$ **do**
3:      **if** there is a component C starting at position i **then**
4:          **if** there is no component ending at position i **then**
5:              create a new square node q as a child of p
6:          **end if**
7:          create a new round node p (representing C) as a child of q
8:      **else if** there is a component ending at position i **then**
9:          $p \leftarrow$ parent of q
10:         $q \leftarrow$ parent of p
11:     **end if**
12: **end for**

To generate tree T' from tree T_P, a bottom-up traversal of T_P recursively removes all dangling round leaves that represent oriented components, and square nodes. Given the tree T', it is easy to compute the reversal distance: perform a depth-first traversal of T' and count the number of leaves and the number of long and short branches. Then use the formula from Theorem 3 to obtain t, and the formula from Theorem 2 to obtain d.

Altogether we have:

Theorem 4. *Using Algorithms 1 and 2, the reversal distance $d(P)$ of a permutation P on the set $\{0, \ldots, n\}$ can be computed in linear time $O(n)$.*

5   Conclusion

In this paper, we presented a simpler formula for the Hannenhalli-Pevzner reversal distance equation. It captures the notion of hurdles, super-hurdles and fortresses in a single parameter whose value can be computed with the help of a PQ-tree. Our next goal is to apply this kind of simplification to the harder problem of comparing multi-chromosomal genomes, whose treatment currently involves half a dozen parameters [6].

References

1. A. Bergeron. A very elementary presentation of the Hannenhalli-Pevzner theory. In *CPM 2001 Proceedings*, volume 2089 of *LNCS*, pages 106–117. Springer Verlag, 2001.
2. A. Bergeron, S. Heber, and J. Stoye. Common intervals and sorting by reversals: A marriage of necessity. *Bioinformatics*, 18(Suppl. 2):S54–S63, 2002. (Proceedings of ECCB 2002).
3. A. Bergeron and J. Stoye. On the similarity of sets of permutations and its applications to genome comparison. In *Proceedings of COCOON 03*, volume 2697 of *LNCS*, pages 68–79. Springer Verlag, 2003.
4. K. S. Booth and G. S. Lueker. Testing for the consecutive ones property, interval graphs and graph planarity using *PQ*-tree algorithms. *J. Comput. Syst. Sci.*, 13(3):335–379, 1976.
5. S. Hannenhalli. Polynomical algorithm for computing translocation distance between genomes. *Discrete Appl. Math.*, 71(1-3):137–151, 1996.
6. S. Hannenhalli and P. A. Pevzner. Transforming men into mice (polynomial algorithm for genomic distance problem). In *Proceedings of FOCS 1995*, pages 581–592. IEEE Press, 1995.
7. S. Hannenhalli and P. A. Pevzner. Transforming cabbage into turnip: Polynomial algorithm for sorting signed permutations by reversals. *J. ACM*, 46(1):1–27, 1999.
8. H. Kaplan, R. Shamir, and R. E. Tarjan. A faster and simpler algorithm for sorting signed permutations by reversals. *SIAM J. Computing*, 29(3):880–892, 1999.
9. J. D. Kececioglu and D. Sankoff. Efficient bounds for oriented chromosome inversion distance. In *Proceedings of CPM 94*, volume 807 of *LNCS*, pages 307–325. Springer Verlag, 1994.
10. Berman P. and Hannenhalli S. Fast sorting by reversal. In *CPM 1996 Proceedings*, volume 1075 of *LNCS*, pages 168–185. Springer Verlag, 1996.

A Fast Set Intersection Algorithm
for Sorted Sequences

Ricardo Baeza-Yates

Center for Web Research
Departamento de Ciencias de la Computación,
Universidad de Chile, Casilla 2777, Santiago, Chile
rbaeza@dcc.uchile.cl

Abstract. This paper introduces a simple intersection algorithm for two sorted sequences that is fast on average. It is related to the multiple searching problem and to merging. We present the worst and average case analysis, showing that in the former, the complexity nicely adapts to the smallest list size. In the later case, it performs less comparisons than the total number of elements on both inputs when $n = \alpha m$ ($\alpha > 1$). Finally, we show its application to fast query processing in Web search engines, where large intersections, or differences, must be performed fast.

Keywords: Set operations, merging, multiple search, Web search engines, inverted indices.

1 Introduction

Our problem is a particular case of a generic problem called multiple searching [2] (see also [15], research problem 5, page 156). Given an n-element data multiset, D, drawn from an ordered universe, search D for each element of an m-element query multiset, Q, drawn from the same universe. An algorithm solving the problem must report any elements in both multisets. The metric is the number of three-way comparisons $(<, =, >)$ between any pair of elements, worst case or average case. Throughout this paper $n \geq m$ and logarithms are base two unless explicitly stated otherwise.

Multiply search is directly related to computing the intersection of two sets. In fact, the elements found is the intersection of both sets. Although in the general case, D and Q are arbitrary, an important case is when D and Q are sets (and not multisets) already ordered. In this case, multiply search can be solved by merging both sets. However, this is not optimal for all possible cases. In fact, if m is small (say if $m = o(n/\lg n)$), it is better to do m binary searches obtaining an $O(m \lg n)$ algorithm. Can we have an adaptive algorithm that matches both complexities depending on the value of m? We present an algorithm which on average performs less than $m + n$ comparisons when both sets are ordered under some pessimistic assumptions. Fast average case algorithms are important for large n and/or m.

S.C. Sahinalp et al. (Eds.): CPM 2004, LNCS 3109, pp. 400–408, 2004.

This problem is motivated by Web search engines. Most search engines use inverted indices, where for each different word, we have a list of positions or documents where it appears. In some settings those lists are ordered by position or by a global precomputed ranking, to facilitate set operations between lists (derived from Boolean query operations), which is equivalent to the ordered case. In other settings, the lists of positions are sorted by frequency of occurrence in a document, to facilitate ranking based on the vectorial model [1, 3]. The same happens with word positions in each file (full inversion to allow sentence searching). Therefore, the complexity of this problem is interesting also for practical reasons, as in search engines, partial lists can have hundreds of millions elements for very frequent words.

In section 2 we present related work. Section 3 presents our intersection algorithm and its analysis. Section 4 presents the motivation for our problem and some practical issues. We end with some concluding remarks and on-going work.

2   Related Work

If an algorithm determines whether any elements of a set of $n + m$ elements are equal, then, by the element uniqueness lower bound in algebraic-decision trees (see [11]), the algorithm requires $\Omega((n + m) \lg(n + m))$ comparisons in the worst case. However, this lower bound does not apply to the search problem because a search algorithm need not need to determine the uniqueness of either D or Q; it need only to determine whether $D \cap Q$ is empty. For example, an algorithm for $m = 1$ must find whether some element of D equals the element in Q, not whether any two elements of D are equal. Conversely, however, lower bounds on the search problem (or, equivalently, the set intersection problem) apply to the element uniqueness problem [10]. In fact, this idea was exploited by Demaine et al. to define an adaptive multiple set intersection algorithm [8, 9]. They also defined the difficulty of a problem instance, which was refined later by Barbay and Kenyon [6].

For the ordered case, lower bounds on set intersection are also lower bounds for merging both sets. However, the converse is not true, as in set intersection we do not need to find the actual position of each element in the union of both sets, just if it is in D or not. Although there has been a lot of work on minimum comparison merging in the worst case, almost no research has been done on the average case because it does not make much of a difference. However, this is not true for multiple search, and hence for set intersection [2].

In the case of merging, Fernandez de la Vega et al. [13] analyzed the average case of a simplified version of Hwang-Lin's binary merge [14] finding that if $\alpha = n/m$ with $\alpha > 1$ and not a power of 2, then the expected number of comparisons was

$$\left(r + \frac{1}{1 - \left(\frac{\alpha}{\alpha+1}\right)^{2^r}} \right) \frac{n}{\alpha}$$

with $r = \lfloor \lg_2 \alpha \rfloor$. When α is a power of 2, the result is more complicated, but similar. Fernandez de la Vega *et al.* [12] also designed a probabilistic algorithm that improved upon Hwang-Lin's algorithm on the worst case for $1.618m \leq n \leq 3m$.

In the case of upper bounds, good algorithms for multiple search can be used to compute the intersection of two sets, obtaining the same time complexity. They can be also used to compute the union of two sets, by subtracting the intersection of both sets to the set obtained by merging both sets. Similarly to compute the difference of two sets.

3   A Simple but Good Average Case Algorithm

Suppose that D is sorted. In this case, obviously, if Q is small, will be faster to search every element of Q in D by using binary search. Can we do better if both sets are sorted? In this case set intersection can be solved by merging. In the worst or average case, straight merging requires $m + n - 1$ comparisons. Can we do better for set intersection? The following simple algorithm improves on average under some pessimistic assumptions. We call it double binary search and can be seen as a balanced version of Hwang and Lin's [14] algorithm adapted to our problem.

We first binary search the median (middle element) of Q in D. If found, we add that element to the result. Found or not, we have divided the problem in searching the elements smaller than the median of Q to the left of the position found on D, and the elements bigger than the median to the right of that position. We then solve recursively both parts using the same algorithm. If in any case, the size of the subset of Q to be considered is larger than the subset of D, we exchange the roles of Q and D. Note that set intersection is symmetric in this sense. If any of the subsets is empty, we do nothing.

In the best case, the median element in each iteration always falls outside D (that is, all the elements in Q are smaller or larger than all the elements in D). Hence, the total number of comparisons is $\lceil \lg(m+1) \rceil \lceil \lg(n+1) \rceil$, which for $m = O(n)$ is $O(\lg^2 n)$. This shows that there is room for doing less work. The worst case happens when the median is not found and divides D into two sets of the same size (intuitively seems that the best and worst case are reversed). Hence, if $W(m, n)$ is the cost of the set intersection on the worst case, for m of the form $2^k - 1$, we have

$$W(m, n) = \lceil \lg(n+1) \rceil + W((m-1)/2, \lceil n/2 \rceil) + W((m-1)/2, \lfloor n/2 \rfloor)$$

It is not difficult to show that

$$W(m, n) = 2(m+1) \lg((n+1)/(m+1)) + 2m + O(\lg n)$$

That is, for small m the algorithm has $O(m \lg n)$ worst case, while for $n = \alpha m$ it is $O(n)$. In this case, the ratio between this algorithm and merging is $2(1 + \lg(\alpha))/(1 + \alpha)$ asymptotically, being 1 when $\alpha = 1$. The worst case is worse than merging for $1 < \alpha < 6.3197$ having its maximum at $\alpha = 2.1596$ where it is 1.336 times slower than merging (this is shown in figure 1). Hence the worst case of the algorithm matches the complexity of both, the merging and the multiple binary search, approaches, adapting nicely to the size of m.

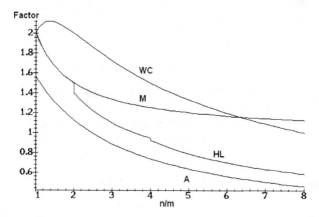

Fig. 1. Constant factor on n depending on the ratio $\alpha = n/m$.

Let us consider now the average case. We use two assumptions: first, that we never find the median of Q and hence we assume that some elements never appear in D; and second, that the median will divide D in sets of size i and $n-i$ with the same probability for all i (this is equivalent to consider every element on D as random, like in the average case analysis of Quicksort). The first assumption is pessimistic, while the second considers that overlaps are uniformly distributed, which is also pessimistic regarding our practical motivation as we do not take in account that word occurrences may and will have locality of reference. Figure 2 shows the actual number of comparisons for $n = 128$ and all powers of 2 for $m \leq n$, for all the cases already mentioned.

If $A(m, n)$ denotes the average case number of comparisons, for m of the form $2^k - 1$ we have

$$A(m, n) = \lceil \lg(n + 1) \rceil + \frac{1}{n + 1} \sum_{i=0}^{n} (A((m - 1)/2, i) + A((m - 1)/2, n - i))$$

We now show that

$$A(m, n) = (m + 1)(\ln((n + 1)/(m + 1)) + 3 - 1/\ln(2)) + O(\lg n)$$

The recurrence equation can be simplified to

$$A(m, n) = \lceil \lg(n + 1) \rceil + \frac{2}{n + 1} \sum_{i=0}^{n} A((m - 1)/2, i)$$

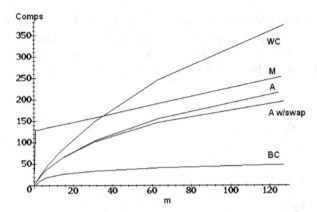

Fig. 2. Number of comparisons in the best, worst and average case (with and without swaps) for $n = 128$, as well as for merging (M).

As the algorithm is adaptive on the size of the lists, we have

$$A(m,n) = \lceil \lg(n+1) \rceil + \frac{2}{n+1} \sum_{i=0}^{(m-1)/2} A(i, (m-1)/2)$$

$$+ \frac{2}{n+1} \sum_{(m+1)/2}^{n} A((m-1)/2, i)$$

by noticing that we switch the sets when $m > n$. However, solving this version of the recurrence is too hard, so we do not use this improvement. Nevertheless, this does not affect the main order term. Notice that our analysis allows any value for n.

Making the change of variable $m = 2^k - 1$ and using k as subindex we get

$$A_k(n) = \lceil \lg(n+1) \rceil + \frac{2}{n+1} \sum_{i=0}^{n} A_{k-1}(i)$$

Eliminating the sum, we obtain

$$(n+1)A_k(n) = nA_k(n-1) + 2A_{k-1}(n) + \lceil \lg(n+1) \rceil + n\delta(n = 2^j)$$

where $\delta(n = 2^j)$ is 1 if n is a power of 2, or 0 otherwise. Let $T_n(z) = \sum_k A_k(n)z^k$ be the generating function of A in the variable k. Hence

$$T_n(z) = \frac{n}{n+1-2z} T_{n-1}(z) + \frac{\lceil \lg(n+1) \rceil + n\delta(n = 2^j)}{(n+1-2z)(1-z)}$$

Unwinding the recurrence in the subindex of the generating function, as $T_0(z) = 0$, we get

$$T_n(z) = \frac{n!}{(1-z)\Gamma(n+2-2z)} \sum_{i=1}^{n} \frac{\Gamma(i+1-2z)}{i!} (\lceil \lg(i+1) \rceil + i\delta(i = 2^j))$$

where $\Gamma(x)$ is the Gamma function (if x is a positive integer, then $\Gamma(x) = (x-1)!$). Then, $A(2^k - 1, n) = [z^k]T_n(z)$ where $[z^k]$ is the coefficient of z^k. Expanding the terms in z using partial fractions and extracting the coefficient, with the help of the Maple symbolic algebra system, we obtain the main order terms sought.

For $n = \alpha m$, the ratio between this algorithm and merging is $(\ln(\alpha) + 3 - 1/\ln(2))/(1 + \alpha)$ which is at most 0.7913 when $\alpha = 1.2637$ and 0.7787 when $\alpha = 1$. This is also shown in figure 1, where we also include the average case analysis of Hwang and Lin's algorithm [13]. Recall that this analysis uses different assumptions, however shows the same behavior, improving over merging when $\alpha \geq 2$.

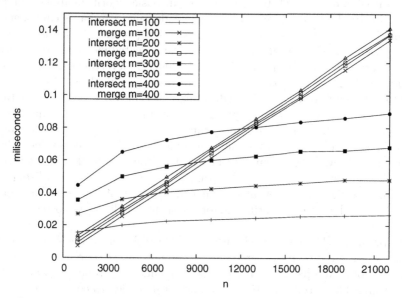

Fig. 3. Experimental results for the new algorithm and merging for several values of m and n.

Figure 3 shows experimental results for four values of m and varying n, for the new algorithm (intersect) compared to merge. We use twenty random instances per case and ten thousand runs to be able to measure the running times. We use uniformly random integer numbers in the range 1 to 10^9 and we implemented the programs using the Gcc 3.3.3 compiler in a Linux platform running an Intel Xeon of 3GHz. We can see that the new algorithm becomes better than merging for larger n.

A simple way to improve this algorithm is to start comparing the smallest elements of both sets with the largest elements in both sets. If both sets do not overlap, we use just $O(1)$ time. Otherwise, we search the smallest and largest element of D in Q, to find the overlap, using just $O(\lg m)$ time. Then we apply the previous algorithm just to the subsets that actually overlaps. This improves both, the worst and the average case. The dual case is also valid, but then finding the overlap is $O(\lg n)$, which is not good for small m.

4   Application to Query Processing in Inverted Indices

Inverted indices are used in most text retrieval systems [3]. Logically, they are a vocabulary (set of unique words found in the text) and a list of references per word to its occurrences (typically a document identifier and a list of word positions in each document). In simple systems (Boolean model), the lists are sorted by document identifier, and there is no ranking (that is, there is no notion of relevance of a document). In that setting, our basic algorithm applies directly to compute Boolean operations on document identifiers: union is equivalent to merging, intersection is the complement operation (we only keep the repeated elements), and subtraction implies deleting the repeated elements. In practice, long lists are not stored sequentially, but in blocks. Nevertheless, these blocks are large, and the set operations can be performed in a block-by-block basis.

In complex systems ranking is used. Ranking is typically based in word statistics (number of word occurrences per document and the inverse of the number of documents having it). Both values can be precomputed and the reference lists are then stored by decreasing intra-document word frequency order to have first the most relevant documents. Lists are then processed by decreasing inverse extra-document word frequency order (that is, we process the shorter lists first), to obtain first the most relevant documents. However, in this case we cannot always have a document identifier mapping such that lists are sorted by that order.

The previous scheme was used initially on the Web, but as the Web grew, the ranking deteriorated because word statistics do not always represent the content and quality of a Web page and also can be "spammed" by repeating and adding (almost) invisible words. In 1998, Page and Brin [7] described a search engine (which was the starting point of Google) that used links to rate the quality of a page. This is called a global ranking based in popularity, and is independent of the query posed. It is out of the scope of this paper to explain Pagerank, but it models a random Web surfer and the ranking of a page is the probability of the Web surfer visiting it. This probability induces a total order that can be used as document identifier. Hence, in a pure link based search engine we can use our intersection algorithm as before. However, nowadays hybrid ranking schemes that combine link and word evidence are used. In spite of this, a link based mapping still gives good results as approximates well the true ranking (which can be corrected while is computed).

Another important type of query is sentence search. In this case we use the word position to know if a word follows or precedes a word. Hence, as usually sentences are small, after we find the Web pages that have all of them, we can process the first two words to find adjacent pairs and then those with the third word and so on. This is like to compute a particular intersection where instead of finding repeated elements we try to find correlative elements (i and $i + 1$), and therefore we can use again our algorithm as word positions are sorted. The same is true for proximity search. In this case, we can have a range k of possible valid positions (that is $i \pm k$) or to use a different ranking weight depending on the proximity.

Finally, in the context of the Web, our algorithm is in practice much faster because the uniform distribution assumption is pessimistic. In the Web, the distribution of word occurrences is quite biased. The same is true with query frequencies. Both distributions follow a power law (a generalized Zipf distribution) [3, 5]. However, the correlation of both distributions is very small [4]. That implies that the average length of the lists involved in the query are not that biased. That means that the average lengths of the lists, n and m, when sampled, will satisfy $n = \Theta(m)$ (uniform), rather than $n = m + O(1)$ (power law). Nevertheless, in both cases our algorithm makes an improvement.

5   Concluding Remarks

We have presented a simple set intersection algorithm that performs quite well in average and does not inspect all the elements involved. It can be seen as a natural hybrid of binary search and merging. We are currently studying how this algorithm behaves with other word occurrence distributions, for example, a Zipf distribution.

In practice, queries are short (on average 2 to 3 words [5]) so there is almost no need to do multiset intersection and if so, they can be easily handled by pairing the smaller sets firsts, which seems to be the most used algorithm [9]. In addition, we do not need to compute the complete result, as most people only look at less than two result pages [5]. Moreover, computing the complete result is too costly if one or more words occur several millions of times as happens in the Web and that is why most search engines use an intersection query as default. Hence, lazy evaluation strategies are used. If we use the straight classical merging algorithm, this naturally obtains first the most relevant Web pages. For our algorithm, it is not so simple, because although we have to process first the left side of the recursive problem, the Web pages obtained do not necessarily appear in the correct order. A simple solution is to process the smaller set from left to right doing binary search in the larger set. However this variant is efficient only for small m, achieving a complexity of $O(m \lg n)$ comparisons. An optimistic variant can use a prediction on the number of pages in the result and use an intermediate adaptive scheme that divides the smaller sets in non-symmetric parts with a bias to the left side. We are currently working on this problem for multiple sets.

Acknowledgements

We thank Phil Bradford, Joe Culberson, and Greg Rawlins for many useful discussions on this and similar problems a long time ago. We also thank Alejandro Salinger for his help in the experimental results.

References

1. R.A. Baeza-Yates. *Efficient Text Searching*. PhD thesis, Dept. of Computer Science, University of Waterloo, May 1989. Also as Research Report CS-89-17.
2. Ricardo Baeza-Yates, Phillip G. Bradford, Joseph C. Culberson, and Gregory J. E. Rawlins. The Complexity of Multiple Searching, unpublished manuscript, 1993.
3. R. Baeza-Yates and B. Ribeiro-Neto, *Modern Information Retrieval*, ACM Press/Addison-Wesley, England, 513 pages, 1999.
4. R. Baeza-Yates, and Felipe Saint-Jean. A Three Level Search Engine Index based in Query Log Distribution. SPIRE 2003, Springer LNCS, Manaus, Brazil, October 2003.
5. Ricardo Baeza-Yates. Query Usage Mining in Search Engines. In Web Mining: Applications and Techniques, Anthony Scime, editor. Idea Group, 2004.
6. Jérémy Barbay and Claire Kenyon. Adaptive Intersection and t-Threshold Problems. In *Proceedings of the 13th Annual ACM-SIAM Symposium on Discrete Algorithms*, pp. 390–399, San Francisco, CA, January 2002.
7. S. Brin and L. Page. The anatomy of a large-scale hypertextual Web search engine. In *7th WWW Conference*, Brisbane, Australia, April 1998.
8. Erik D. Demaine, Alejandro López-Ortiz, and J. Ian Munro, Adaptive set intersections, unions, and differences. In *Proceedings of the 11th Annual ACM-SIAM Symposium on Discrete Algorithms*, pp. 743–752, San Francisco, CA, January 2000.
9. Erik D. Demaine, Alejandro López-Ortiz, and J. Ian Munro, Experiments on Adaptive set intersections for text retrieval systems. In *Proceedings of the 3rd Workshop on Algorithm Engineering and Experiments*, LNCS, Springer, Washington, DC, January 2001.
10. Dietz, Paul, Mehlhorn, Kurt, Raman, Rajeev, and Uhrig, Christian; "Lower Bounds for Set Intersection Queries," *Proceedings of the 4th Annual Symposium on Discrete Algorithms,* 194–201, 1993.
11. Dobkin, David and Lipton, Richard; "On the Complexity of Computations Under Varying Sets of Primitives," *Journal of Computer and Systems Sciences,* **18**, 86–91, 1979.
12. W. Fernandez de la Vega, S. Kannan, and M. Santha. Two probabilistic results on merging, *SIAM J. on Computing* 22(2), pp. 261–271, 1993.
13. W. Fernandez de la Vega, A.M. Frieze, and M. Santha. Average case analysis of the merging algorithm of Hwang and Lin. *Algorithmica* 22 (4), pp. 483–489, 1998.
14. F.K. Hwang and S. Lin. A Simple algorithm for merging two disjoint linearly ordered lists, *SIAM J. on Computing* 1, pp. 31–39, 1972.
15. Rawlins, Gregory J. E.; *Compared to What?: An Introduction the the Analysis of Algorithms,* Computer Science Press/W. H. Freeman, 1992.

Faster Two Dimensional Pattern Matching with Rotations

Amihood Amir[1,*], Oren Kapah[2], and Dekel Tsur[3]

[1] Department of Computer Science,
Bar-Ilan University, 52900 Ramat-Gan, Israel
and Georgia Tech
Tel. (972-3)531-8770
amir@cs.biu.ac.il

[2] Department of Computer Science,
Bar-Ilan University, 52900 Ramat-Gan, Israel
Tel. (972-3)531-8408
kapaho@cs.biu.ac.il

[3] Caesarea Rothschild Institute of Computer Science,
Haifa University, Haifa 31905, Israel
Tel. (972-4)828-8363
dekelts@cs.haifa.ac.il

Abstract. The most efficient currently known algorithms for two dimensional matching with rotation have a worst case time complexity of $O(n^2 m^3)$, where the size of the text is n^2 and the size of the pattern is m^2. In this paper we present two algorithms for the two dimensional rotated matching problem whose running time is $O(n^2 m^2)$. The preprocessing time of the first algorithms is $O(m^5)$ and the preprocessing time of the second algorithm is $O(m^4)$.

Keywords: Design and analysis of algorithms, two dimensional pattern matching, rotation.

1 Introduction

One of the main motivation for research in two dimensional pattern matching is the problem of searching aerial photographs. The problem is a basic one in computer vision, but it was felt that pattern matching can not be of use in its solution. Such feelings were based on the belief that pattern matching algorithms are only good for *exact matching* whereas in reality one seldom expects to find an exact match of the pattern. Rather, it is interesting to find all text locations that "approximately" match the pattern. The types of differences that make up these "approximations" in the aerial photograph case are: (1) *Local Errors* – introduced by differences in the digitization process, noise, and occlusion (the pattern partly obscured by another object). (2) *Scale* – size difference between the image in the pattern and the text, and (3) *Rotation* – angle differences

---
* Partly supported by NSF grant CCR-01-04494 and ISF grant 282/01.

between the images. Progress has been made on local errors and scaling (e.g. [3, 6–8, 17, 18, 4]), but rotation had proven challenging.

The pattern matching with rotation problem is that of finding all occurrences of a two dimensional pattern in a text, in all possible rotations. An efficient solution to the problem proved elusive even though many researchers were thinking about it for over a decade. Part of the problem was lack of a rigorous definition to capture the concept of rotation in a discrete pattern.

The major breakthrough came when Fredriksson and Ukkonen [13] gave an excellent combinatorial definition of rotation. They resorted to a geometric interpretation of text and pattern and provided the following definition.

Let P be a two-dimensional $m \times m$ array and T be a two-dimensional $n \times n$ array over some finite alphabet Σ. The array of *unit pixels* for T consists of n^2 unit squares, called *pixels* in the real plane R^2. The corners of the pixel for $T[i, j]$ are $(i - 1, j - 1)$, $(i, j - 1)$, $(i - 1, j)$, and (i, j). Hence the pixels for T form a regular $n \times n$ array covering the area between $(0, 0)$, $(n, 0)$, $(0, n)$, and (n, n). The *center* of each pixel is the geometric center point of the pixel. Each pixel $T[i, j]$ is identified with the value from Σ that the original text had in that position. We say that the pixel has a *color* from Σ.

The array of pixels for pattern P is defined similarly. A different treatment is necessary for patterns with odd sizes and for patterns with even sizes. For simplicity's sake we assume throughout the rest of this paper that the pattern is of size $m \times m$ and m is even. The *rotation pivot* of the pattern is its exact center, the point $(\frac{m}{2}, \frac{m}{2}) \in R^2$. See Figure 1 for an example of the rotation pivot of a 4×4 pattern P.

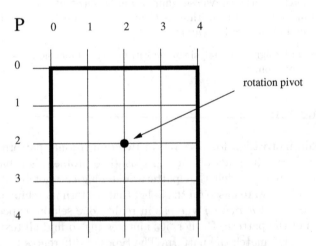

Fig. 1. The rotation pivot of a 4×4 pattern P.

Consider now a rigid motion (translation and rotation) that moves P on top of T. Consider the special case where the translation moves the grid of P precisely on top of the grid of T, such that the grid lines coincide.

Assume that the rotation pivot of P is at location (i, j) on the text grid, and that the pattern lies *under* the text. The pattern is now rotated, centered at (i, j), creating an angle α between the x-axes of T and P. P is said to be at location $((i, j), \alpha)$ *under* T. Pattern P is said to have an *occurrence* at location $((i, j), \alpha)$ if the *center* of each pixel in T has the same color as the pixel of P under it, if there is such a pixel. When the center of a text pixel is exactly over a vertical (horizontal) border between text pixels, the color of the pattern pixel left (below) to the border is chosen. Consider some occurrence of P at location $((i_0, j_0), \alpha)$. This occurrence defines a non-rectangular substring of T that consists of all the pixel of T whose centers are inside pixels of P. We call this string P *rotated by* α, and denote it by P^α. Note that there is an occurrence of P at location $((i, j), \alpha)$ if and only if P^α occurs at (i, j).

Fredriksson, Navarro and Ukkonen [11] give two possible definitions for rotation. One is as described above and the second is, in some way, the opposite. P is placed *over* the text T. More precisely, assume that the rotation pivot of P is on top of location (i, j) on the text grid. The pattern is now rotated, centered at (i, j), creating an angle α between the x-axes of T and P. P is said to be at location $((i, j), \alpha)$ *over* T. Pattern P is said to have an *occurrence* at location $((i, j), \alpha)$ if the center of each pixel in P has the same color as the pixel of T under it.

While the two definitions of rotation, "over" and "under", seem to be quite similar, they are not identical. For example, in the "pattern over text" model there exist angles for which two pattern pixel centers may find themselves in the same text pixel. Alternately, there are angles where there are "holes" in the rotated pattern, namely there is a text pixel that does not have in it a center of a pattern pixel, but all text pixels around it have centers of pattern pixels. See Figure 2 for an example.

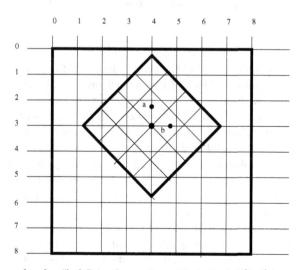

Fig. 2. An example of a "hole" in the pattern. Text pixel $T[3, 5]$ has no pattern pixel over it, but the pixels $T[2, 5]$ and $T[3, 6]$ have pattern pixel centers.

The challenges of "discretizing" a continuous image are not simple. In the Image Processing field, stochastic and probabilistic tools need to be used because the images are "smoothed" to compensate for the fact that the image is presented in a far coarser granularity than in reality. The aim of the the pattern matching community has been to fully discretize the process, thus our different definitions. However, this puts us in a situation where some "gut" decisions need to be made regarding the model that best represents "reality". It is our feeling that in this context the "pattern under text" model is more intuitive since it does not allow anomalies such as having two pattern centers over the same text pixel (a contradiction) nor does it create "holes" in the rotated pattern For examples of the rotated patterns in the two models see Figure 3.

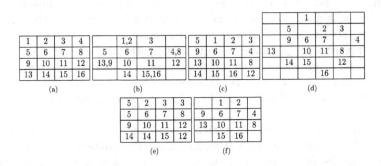

Fig. 3. An example of some possible 2-dimensional arrays that represent one pattern. Fig (a) – the original pattern. Figures (b)–(d) are computed in the "pattern over the text" model. Fig (b) – a representation of the pattern rotated by 19^0. Fig (c) – Pattern rotated by 21^0. Fig (d) – Pattern rotated by 26^0. Figures (e)–(f) are computed in the "pattern under the text" model. Fig (e) – Pattern rotated by 17^0. Fig (f) – Pattern rotated by 26^0.

In this paper we seek an efficient algorithm for rotated matching in the "pattern under text" model. Most of the algorithms for rotated matching are filtering algorithms that behave well on average but that have a bad worst case complexity (e.g. [14, 11, 15]). In three papers ([12, 2, 10]), there is a $O(n^2m^3)$ worst case algorithm for rotated matching. All worst-case algorithms basically work by enumerating all possible rotated patterns and solving a two dimensional dictionary matching problem on the text. In [2] it was proven that there are $\Theta(m^3)$ such rotated patterns. The high complexity results from the fact that the dictionary patterns have "don't care" symbols in them and thus, essentially, every pattern needs to be sought separately.

In this paper we present the first rotated matching algorithms whose time is better than $O(n^2m^3)$. The scanning time of our algorithms is $O(n^2m^2)$. These results are achieved by identifying monotonicity properties on the rotated patterns. These properties allow using *transitivity*-based dictionary matching algorithms, cutting the worst-case time by an m factor. The first algorithm is more simple, and is brought to give a "taste" of monotonicity properties. Its pattern preprocessing time is $O(m^5)$. The second algorithm is more sophisticated and its pattern processing time is $O(m^4)$. Thus its total running time is truly $O(n^2m^2)$.

For simplicity's sake we assume a fixed sized alphabet. The algorithm can be extended in the standard way to deal with arbitrary alphabets with a $O(\log m)$ degradation in the time complexity.

2   The Dictionary Matching Solution

An immediate idea for solving such a problem is *dictionary matching*. Construct a dictionary of all possible rotations of the pattern and then search for them in the text. While there exist efficient two dimensional dictionary matching algorithms (e.g. [5, 16]), none of them works with *don't cares* (a "wildcard" alphabet symbol that matches every character), since they are based on the transitivity of alphabet symbol equality. The "don't care" symbol breaks that transitivity. The only known algorithms for solving string matching with don't cares are based on convolutions. The Fischer-Paterson algorithm [9] finds all occurrences of a string of length m in a string of length n, both possibly having occurrences of the don't care symbol, in time $O(n \log m)$ for a fixed sized alphabet. Unfortunately, this method does not generalize to dictionary matching.

Therefore an immediate suggestion for a rotation algorithm is the following.

Preprocessing: Construct a database of all possible pattern rotations P_i, $i = 1, \ldots, k$.

Text scanning: Let T be the input text (an $n \times n$ array).
For every pattern rotation P_i in the data base do:
  Find all occurrences of P_i in T.

Time: The algorithm's running time is $O(kn^2 \log m)$. In [2] it was proven that the number of different pattern rotations k is $\Theta(m^3)$. Thus the time for the above algorithm is $O(n^2m^3 \log m)$.

Some previous algorithms ([12, 2]) improved the worst case complexity. However, this was only by a logarithmic factor. Their running time is $O(n^2m^3)$.

3   Algorithm 1

3.1   The Algorithm's Idea

If one considers the pattern as a continuous shape, then it will look the same in every rotated angle. The text pixels whose centers are in the area of this square define the borders and contents of the discretized rotated pattern.

The key observation that enables our algorithm's efficiency is the following.

Observation 1: The intersection of all rotated patterns centered at point (x_0, y_0) is a circle of radius m around (x, y). Consider now only quadrant I, i.e. $\{(x, y) | x \geq x_0, y \geq y_0\}$. Assume we start with the pattern at location $((x_0, y_0), 0)$ and we consider only the pattern points under $\{(x, y) | x \geq x_0, y < y_0\}$. Rotating the pattern $90°$ counterclockwise means the triangular "wedge" attached to the circle's circumference counterclockwise along quadrant I.

Discretized Meaning of Observation 1: The pixels whose centers fall in the circle of radius m around (x_0, y_0) form a convex area without "don't cares". In that area transitivity of symbol equality holds. Thus any linearization of that area can be a basis for dictionary matching. In particular, if we split this convex area into four parts by passing a horizontal and vertical line through its center, each "quarter circle" has the *border extensibility property* of [5]. Thus the intersection area in every text pixel can be identified in linear time in a manner even simpler than that of [5].

The problem is the "wedge" movement, since every angle defines a different part with relevant symbols, and a different part with "don't cares". However, consider the $O(m)$ circular strips $C_i = \{(x, y) | i \leq x^2 + y^2 \leq i + 1\}$, $i = m, \ldots, \lceil \sqrt{2}m \rceil$. If we succeed in linearizing each of the C_i without "don't care"s we can employ a linear-time dictionary matching on the linearization.

In Figure 4, the lightly shaded area is the convex region which is the intersection of all rotated patterns. The darker shaded area are the text pixels whose centers are in C_m.

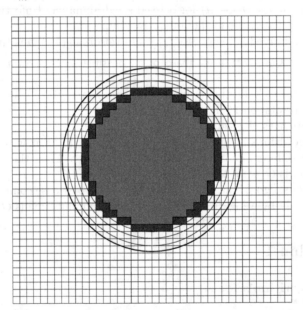

Fig. 4. An example of the convex intersection and of the discretized C_m.

The linearization of C_i: Linearize the discretized C_i in the order that the edge of the rotated pattern wedge covers the pixel centers.

As the pattern is rotated, the non-"don't care" part of the wedge snakes its way along each linearized C_i. Denote each such non-"don't care" part of a wedge by a *strip*.

Easy geometric considerations allow us to conclude the following two claims.

Claim. As the pattern rotates counterclockwise in quadrant I, the strip in the linearization of C_i does not have any internal "don't care".

Claim. A linearized strip that begins at a given pixel of the linearization, may end at one of two pixels in the linearization.

The above two claims allow us to do a linear-time dictionary matching on every one of the $O(m)$ linearized C_i's. Since every linearized C_i has length $O(m)$, this means $O(m^2)$ work at every one of the $O(n^2)$ text locations.

3.2   The Algorithm

The algorithm is now clear.

Preprocessing: In time $O(m^5)$ we construct all $O(m^3)$ different rotated patterns. We class these into sets whose inner circle area is equal. Each of these sets is further split into appropriate strings in each linearization of each C_i. The different cores and the different strings in the linearizations of C_i are all processed for dictionary matching.

Text Scanning: For each text pixel, perform the dictionary matching to decide the core set. Then for the linearization of each C_i perform dictionary matching to determine the possible sets that agree with the text around this pixel. There will be no more than $O(m)$ such possibilities. Take the intersection of the possibilities as we advance on the linearizations of the C_i's.

Total Text Scanning Time: $O(n^2 m^2)$.

4   Algorithm 2

4.1   The Algorithm's Idea

Let $\alpha_1 \le \alpha_2 \le \cdots \le \alpha_k$ be the angles in which some pixel center changes its unit square. We assume that in each α_i, only one pixel center changes its unit square (when several center change their squares simultaneously, we assign them into angles $\alpha_i, \alpha_{i+1}, \ldots$ such that $\alpha_i = \alpha_{i+1} = \cdots$).

Let B_0 be the $m \times m$ square whose bottom left corner is at origin $(0,0)$, and whose other three corners clockwise from the origin are at $(0, m)$, (m, m) and $(m, 0)$. Let B_i denote the square whose corners are (i, i), $(i, m-i)$, $(m-i, m-i)$, and $(m-i, i)$. We denote by B_i^α the square obtained by rotating B_i by an angle of α with rotation center at $(m/2, m/2)$.

The main idea of the algorithm is to partition the pattern into two parts: an *inner part* and an *outer part*. By carefully choosing the partition, we will be able to efficiently search for occurrences of each part in T. The inner and outer parts are defined as follows:

Definition 1. *The* inner part *of* $P^0 = P$ *is the array containing all the pixels whose centers are inside the square* B_3. *The* outline *of the inner part is the set* *of coordinates of the pixels in the inner part. In the case of the inner part of P^0, the outline is $I_1 = \{(i, j) \mid i = 3, \ldots, m-3, \ j = 3, \ldots, m-3\}$.*

Recall the set of angles $\{\alpha_i\}_{i=1}^k$. Let i_1 be the minimum index such that there is a pixel $(i, j) \in I_1$ which is not a pixel of $P^{\alpha_{i_1}}$ (that is, the center of the pixel is outside the square $B_0^{\alpha_{i_1}}$).

For $j = 1, \ldots, i_1 - 1$, define the inner part of P^{α_j} to be the array containing all the pixels of P^{α_j} in I_1. Note that while the outlines of the inner parts of P^{α_j}, $j = 1, \ldots, i_1 - 1$ are identical, the values (colors) of elements of these arrays may differ depending on the internal perturbations caused by the rotations, thus they may have different inner parts.

Let I_2 be the set of coordinates of pixels whose centers are in $B_3^{\alpha_{i_1}}$. I_2 is the outline of the inner part of $P^{\alpha_{i_1}}$. Let i_2 be the minimum index such that there is a pixel $(i, j) \in I_2$ which is not a pixel of $P^{\alpha_{i_2}}$. The inner part of P^{α_j}, $j = i_1, \ldots, i_2 - 1$ is the array of all pixels of P^{α_j} in I_2.

The inner parts of $P^{\alpha_{i_2}}, \ldots, P^{\alpha_k}$ are defined in the same manner.

Let $i_1, \ldots, i_l$ be all the indices in which the outline of the inner part is changed.

For every α, the outer part of P^α is the two-dimensional array containing all the pixels of P^α which are not in the inner part.

4.2   Matching the Inner Part

First, we build a dictionary D containing all the rows of all the inner parts. For each string in D assign a distinct label from $1, \ldots, |D|$. Then, find all the occurrences of the strings in D in the rows of T. In other words, for each location (a, b) in T, we store a table $R_{a,b}$ such that $R_{a,b}[i]$ is the label of the dictionary word of length i that begins in row a of T in position b, if there is such a word, and 0 otherwise.

Claim. $|D| = O(m^3)$. The length of every string in D is $O(m)$.

Proof. P has m rows. For each i, P^{α_i} differs from $P^{\alpha_{i+1}}$ in exactly one row and by [2] there are $\Theta(m^3)$ different angles α_i. Note also that the length of every string in D is at most $\lceil \sqrt{2}m \rceil$.

Therefore, this stage takes $O(m^4 + n^2 m)$ time when using, for example, the dictionary matching algorithm of [1].

We now need to put together rows that create an inner part. We will handle each different outline separately. We will prove that there are $O(m)$ different outlines.

Lemma 1. *For every j, $1 \leq j \leq l$, $i_j - i_{j-1} = \Omega(m^2)$.*

Proof. Fix a value of j. By definition, there is pixel $(a, b) \in I_j$ which is not a pixel of $P^{\alpha_{i_j}}$. In other words, the pixel center $(a - 0.5, b - 0.5)$ is inside the square $B_3^{\alpha_{i_{j-1}}}$ and outside the square $B_0^{\alpha_{i_j}}$.

Denote by $\hat{I}_j$ the set of centers of the pixels in I_j. Consider the process that begins with the points of $\hat{I}_j$ rotated by $-\alpha_{i_{j-1}}$, and then continuously rotating them until reaching an angle of $-\alpha_{i_j}$. We have that the point $(a-0.5, b-0.5) \in \hat{I}_j$ is inside the square B_3 when rotated by $-\alpha_{i_{j-1}}$, and outside the square B_0 when rotated by $-\alpha_{i_j}$. Therefore, the point $(a - 0.5, b - 0.5)$ moves a distance of at least 3 during the rotation process. As the points in $\hat{I}_j - \{(a - 0.5, b - 0.5)\}$ are always inside B_0 during the rotation process, every such point contributes 1 to the total number of α_i's angles when it changes its unit square. There are $\Omega(m^2)$ points in $\hat{I}_j$ whose distance from $(m/2, m/2)$ is at least $m/2$. Denote the set of these points by I_j'.

Since the distance of $(a-0.5, b-0.5)$ from $(m/2, m/2)$ is less than $\sqrt{2} \cdot m/2$, it follows that every point in I_j' moves a distance of at least $3/\sqrt{2}$. For large enough m, the maximum distance a point can move without changing its unit square is $1.01 \cdot \sqrt{2} < 3/\sqrt{2}$, so it follows that every point in I_j' changes its unit square at least once. Therefore the number of different angles for any given outline is $\Omega(m^2)$.

Corollary 1. $l = O(m)$.

Proof. Immediate from Lemma 1 and the fact that there are $\Theta(m^3)$ different angles altogether.

We now proceed to put together the different rows to create inner parts. Fix an outline I_j and consider the set of its inner parts, namely, the inner parts of $P^{\alpha_{i_j}}, \ldots, P^{\alpha_{i_{j+1}-1}}$. Denote this set of inner parts by P_j. Each inner part P' can be encoded by a string of length at most $\lceil \sqrt{2}m \rceil$ by writing the labels of its rows from the top row to the bottom row. We construct a dictionary D_j containing the strings that encode the inner parts in P_j (note that the strings in D_j all have the same length).

We will do the dictionary matching in a brute force fashion, spending time $O(m)$ for every text location. Simply check if the text substring of the appropriate length at the appropriate offset matches an expected pattern row. More precisely, consider some text location (a, b) of T. We wish to check whether a string from P_j occurs at (a, b). This is done by building a string $T_{a,b,j}$ as follows: Let P' be some string in P_j. If the i-th row of P' has length ℓ, and its leftmost pixel is (x, y), then the i-th letter of $T_{a,b,j}$ is $R_{a+x-m/2, b+y-m/2}[\ell]$. Clearly, a string $P'' \in P_j$ appears at location (a, b) in T if and only if $T_{a,b,j}$ is equal to the string in D_j that corresponds to P''. The time complexity for building $T_{a,b,j}$ and searching whether it appears in D_j is $O(m)$ (for fixed a, b, and j). Thus, the time complexity of this stage is $O(n^2 m l)$.

From Corollary 1, the time complexity for matching the inner parts is $O(n^2 m^2)$.

4.3 Matching the Outer Part

Lemma 2. *For every i, the pixel centers of the outer part of P^{α_i} are inside the square $B_9^{\alpha_i}$.*

Proof. Let $(a - 0.5, b - 0.5) \in \hat{I}_j$ be the point which is inside $B_3^{\alpha_{i_{j-1}}}$ and outside $B_0^{\alpha_{i_j}}$. The distance between $(a - 0.5, b - 0.5)$ and $B_0^{\alpha_{i_j}}$ is exactly 0. Moreover, the distance between $(a - 0.5, b - 0.5)$ and $B_3^{\alpha_{i_{j-1}}}$ is at most 1, so the distance between $(a - 0.5, b - 0.5)$ and $B_0^{\alpha_{i_{j-1}}}$ is at most 4. In other words, during the rotation of B_0 from angle $\alpha_{i_{j-1}}$ to angle α_{i_j}, one point of the segments of B_0 moves a distance of at most 4. Therefore, every the point of B_0 moves a distance of at most $4\sqrt{2}$. Hence, for $i_{j-1} \leq r \leq i_j$, the distance between a point in $B_3^{\alpha_{i_{j-1}}}$ and a point on $B_0^{\alpha_r}$ is at most $3 + 4\sqrt{2} < 9$. It follows that for every pixel in the outer part of P^{α_r}, its center is within distance of at most 9 of $B_0^{\alpha_r}$.

From Lemma 2, the outer part depends on the position of $O(m)$ pixel centers. Each pixel center can at $O(m)$ different unit squares, so it follows that the number of angles in which the outer part of the pattern can change (namely a pixel center in the outer part changes its unit square, or the outline of the inner part is changed) is $O(m^2)$. Therefore, we can check for all occurrences of the outer parts at some text location (a, b) in T in $O(m^2)$ time using the algorithm of [2].

5   Conclusion and Open Problems

We have used geometric considerations and the fact that there is a monotonicity in the rotated patterns to provide an $O(n^2 m^2)$ algorithm for pattern matching with rotations. Previous algorithms only used the rotation structure for efficiency. Our algorithms also exploit the alphabet symbols in the rotated patterns. Nevertheless, it seems like there is more knowledge to be exploited. For example, we did not make any use of point (a, b) in the text to say something about its neighboring point, even though they share a vast number of pixels. Thus it is our feeling that the pattern matching with rotation problem can be solved more efficiently than $O(n^2 m^2)$.

References

1. A.V. Aho and M.J. Corasick. Efficient string matching. *Comm. ACM*, 18(6):333–340, 1975.
2. A. Amir, A. Butman, M. Crocehmore, G.M. Landau, and M. Schaps. Two-dimensional pattern matching with rotations. In *Proc. 14th Annual Symposium on Combinatorial Pattern Matching (CPM 2003)*, number 2676 in LNCS, pages 17–31. Springer, 2003.
3. A. Amir, A. Butman, and M. Lewenstein. Real scaled matching. *Information Processing Letters*, 70(4):185–190, 1999.
4. A. Amir, A. Butman, M. Lewenstein, and E. Porat. Real two dimensional scaled matching. In *Proc. 8th Workshop on Algorithms and Data Structures (WADS)*, pages 353–364, 2003.
5. A. Amir and M. Farach. Two dimensional dictionary matching. *Information Processing Letters*, 44:233–239, 1992.

6. A. Amir and G. Landau. Fast parallel and serial multidimensional approximate array matching. *Theoretical Computer Science*, 81:97–115, 1991.

7. A. Apostolico and Z. Galil (editors). *Pattern Matching Algorithms*. Oxford University Press, 1997.

8. M. Crochemore and W. Rytter. *Text Algorithms*. Oxford University Press, 1994.

9. M.J. Fischer and M.S. Paterson. String matching and other products. *Complexity of Computation, R.M. Karp (editor), SIAM-AMS Proceedings*, 7:113–125, 1974.

10. K. Fredriksson, V. Mäkinen, and G. Navarro. Rotation and lighting invariant template matching. In *Proceedings of the 6th Latin American Symposium on Theoretical Informatics (LATIN'04)*, LNCS, 2004. To appear. Available at http://www.dcc.uchile.cl/~gnavarro/ps/latin04.ps.gzx.

11. K. Fredriksson, G. Navarro, and E. Ukkonen. An index for two dimensional string matching allowing rotations. In *Prof. IFIP International Conference on Theoretical Computer Science (IFIP TCS)*, volume 1872 of *LNCS*, pages 59–75. Springer, 2000.

12. K. Fredriksson, G. Navarro, and E. Ukkonen. Optimal exact and fast approximate two dimensional pattern matching allowing rotations. In *Proc. 13th Annual Symposium on Combinatorial Pattern Matching (CPM)*, volume 2373 of *LNCS*, pages 235–248. Springer, 2002.

13. K. Fredriksson and E. Ukkonen. A rotation invariant filter for two-dimensional string matching. In *Proc. 9th Annual Symposium on Combinatorial Pattern Matching (CPM 98)*, pages 118–125. Springer, LNCS 1448, 1998.

14. K. Fredriksson and E. Ukkonen. A rotation invariant filter for two-dimensional string matching. In *Proc. 9th Annual Symposium on Combinatorial Pattern Matching (CPM)*, volume 1448 of *LNCS*, pages 118–125. Springer, 1998.

15. K. Fredriksson and E. Ukkonen. Combinatorial methods for approximate pattern matching under rotations and translations in 3d arrays. In *Proc. 7th Symposium on String Processing and Information Retrieval (SPIRE'2000)*, pages 96–104. IEEE CS Press, 2000.

16. R. Giancarlo and R. Grossi. On the construction of classes of suffix trees for square matrices: Algorithms and applications. *Information and Computation*, 130(2):151–182, 1996.

17. Dan Gusfield. *Algorithms on Strings, Trees, and Sequences: Computer Science and Computational Biology*. Cambridge University Press, 1997.

18. G. M. Landau and U. Vishkin. Pattern matching in a digitized image. *Algorithmica*, 12(3/4):375–408, 1994.

Compressed Compact Suffix Arrays

Veli Mäkinen[1] and Gonzalo Navarro[2,*]

[1] Department of Computer Science, P.O. Box 26 (Teollisuuskatu 23)
FIN-00014 University of Helsinki, Finland
vmakinen@cs.helsinki.fi
[2] Department of Computer Science, University of Chile
Blanco Encalada 2120, Santiago, Chile
gnavarro@dcc.uchile.cl

Abstract. The *compact suffix array* (CSA) is a space-efficient full-text index, which is fast in practice to search for patterns in a static text. Compared to other *compressed suffix arrays* (Grossi and Vitter, Sadakane, Ferragina and Manzini), the CSA is significantly larger (2.7 times the text size, as opposed to 0.6–0.8 of compressed suffix arrays). The space of the CSA includes that of the text, which the CSA needs separately available. Compressed suffix arrays, on the other hand, *include* the text, that is, they are *self-indexes*. Although compressed suffix arrays are very fast to determine the *number* of occurrences of a pattern, they are in practice very slow to *report* even a few occurrence positions or text contexts. In this aspect the CSA is much faster. In this paper we contribute to this space-time trade off by introducing the *Compressed CSA* (CCSA), a self-index that improves the space usage of the CSA in exchange for search speed. We show that the occ occurrence positions of a pattern of length m in a text of length n can be reported in $O((m + occ) \log n)$ time using the CCSA, whose representation needs $O(n(1 + H_k \log n))$ bits for any k, H_k being the k-th order empirical entropy of the text. In practice the CCSA takes 1.6 times the text size (and includes the text). This is still larger than current compressed suffix arrays, and similar in size to the LZ-index of Navarro. Search times are by far better than for self-indexes that take less space than the text, and competitive against the LZ-index and versions of compressed suffix arrays tailored to take 1.6 times the text size.

1 Introduction and Related Work

The classical problem in string matching is to determine the occ occurrences of a short pattern $P = p_1 p_2 \ldots p_m$ in a large text $T = t_1 t_2 \ldots t_n$. Text and pattern are sequences of characters over an alphabet Σ of size σ. In practice one wants to know the text positions of those occ occurrences, and usually also a text context around them. Usually the same text is queried several times with different patterns, and therefore it is worthwhile to preprocess the text in order to speed up the searches. Preprocessing builds an index structure for the text.

* Supported in part by Fondecyt Grant 1-020831.

S.C. Sahinalp et al. (Eds.): CPM 2004, LNCS 3109, pp. 420–433, 2004.
© Springer-Verlag Berlin Heidelberg 2004

To allow fast searches for patterns of any size, the index must allow access to all suffixes of the text. These kind of indexes are called *full-text indexes*. Optimal query time, which is $O(m + occ)$ as every character of P must be examined and the occ occurrences must be reported, can be achieved by using the *suffix tree* [19] as the index. In a suffix tree every suffix of the text is represented by a path from the root to a leaf. The space requirement of a suffix tree is very high. It can be $12n$ bytes in practice, even with a careful implementation [7]. In addition, in any practical implementation there is always an alphabet dependent factor on search times.

The *suffix array* (SA) [11] is a reduced form of the suffix tree. It represents only the leaves of the suffix tree, via pointers to the starting positions of all the suffixes. The array is lexicographically sorted by the pointed suffixes. A suffix array takes $4n$ bytes, and searches in $O(m \log n + occ)$ time via two binary searches. One finds the first cell i pointing to a suffix $\geq P$ (lexicographically), and the other finds the first cell j pointing to a suffix $\geq p_1 p_2 \ldots p_{m-1}(p_m + 1)$. Then all the cell values at suffix array positions $i \ldots j - 1$ are the initial positions of occurrences of P in T.

There is often a significant amount of redundancy in a suffix array, such that some array areas can be represented by links to other areas. Basically, it is rather common that one area contains the same pointers of the other area, all shifted by one text position. This observation has been intensively used recently in different ways to obtain succinct representations of suffix arrays and still provide fast search time [8, 18, 5].

The *compact suffix array* (CSA) [14] makes direct use of that redundancy to reduce the space usage of suffix arrays. Areas similar to others (modulo a shift in text positions) are found and replaced by a direct link to the similar areas. In practice the CSA takes less than $2n$ bytes and can search in $O(m \log n + occ)$ time, which in practice turns out to be about twice as slow as the plain suffix array. Note that, like suffix trees and arrays, the CSA needs the text itself separately available.

A recent trend in compressed data structures is that of *self-indexes*, which include the text. Hence the text can be discarded and the index must provide functions to obtain any desired text substring in reasonable time. Self-indexes open the exciting possibility of the index taking less space than the text, even including it. Existing implemented self-indexes are the compressed suffix array CSArray of Sadakane [18] (built on [8]), the FM-index of Ferragina and Manzini [5, 6], and the LZ-index of Navarro [16]. The first two take 0.6–0.8 times the text size, while the LZ-index takes about 1.5 times the text size on English text.

In this paper we introduce the *compressed CSA* (CCSA), a self-index based on the CSA which is more compact and represents a relevant space-time trade off in practice. We retain the links of the CSA, but encode them in a compact form. We also encode the text inside the CCSA by using small additional structures that permit searching and displaying the text without accessing T. We show that the CCSA needs $O(n(1 + H_k \log n))$ bits for any k, and that it can find all the occurrences of P in T in $O((m + occ) \log n)$ time. In an 80 Mb English text example, the CCSA need $1.6n$ bytes, replacing the text. This is much less than the $2.7n$ bytes needed by the CSA, about the same space of the LZ-index, and 2–3 times larger than other compressed suffix arrays. Searching the CCSA is 50 times slower than the CSA, but 50–75 times faster than any other self-index that takes less space than the text. The CCSA is competitive against the LZ-index, and against compressed suffix arrays versions tailored to use the same $1.6n$ space to boost their search time.

Our space analysis represents indeed a contribution with independent interest, as we relate the space requirement of CCSA and CSA to the number of runs in Burrows-Wheeler transformed text [2]. We show that this quantity is at most $|\Sigma|^k + 2H_k n$.

2   The Compact Suffix Array (CSA)

Let Σ be an ordered alphabet of size $\sigma = |\Sigma|$. Then $T = t_1 t_2 \ldots t_n \in \Sigma^*$ is a (text) *string* of length $n = |T|$. A *suffix* of text T is a substring $T_{i\ldots n} = t_i \ldots t_n$. We assume that the last text character is $t_n = \$$, which does not occur elsewhere in T and is lexicographically smaller than any other character in Σ.

Definition 1 *The* suffix array *of text T of length $n = |T|$ is an array $SA[1\ldots n]$ that contains all starting positions of the suffixes of the text T, such that $T_{SA[1]\ldots n} < T_{SA[2]\ldots n} < \cdots < T_{SA[n]\ldots n}$, that is, array SA gives the lexicographic order of all suffixes of the text T.*

The idea of compacting the suffix array is the following: Let $\ell \geq 0$. Find two areas $j \ldots j + \ell$ and $i \ldots i + \ell$ of SA that are repetitive in the sense that the suffixes represented by $j \ldots j + \ell$ are obtained, in the same order, from the suffixes represented by $i \ldots i + \ell$ by inserting the first symbol. In other words, $SA[j + k] = SA[i + k] - 1$ for $0 \leq k \leq \ell$. Then replace the area $j \ldots j + \ell$ of SA by a link, stored in $SA[j]$, to the area $i \ldots i + \ell$. This is called a *compacting operation*. The areas may be compacted recursively, meaning that area $i \ldots i + \ell$ (or some parts of it) may also be replaced by a link.

Due to the recursive definition, we need three values to represent a link:

- A pointer p to the entry that contains the start of the linked area.
- A value δ such that entry $p + \delta$ denotes the actual starting point after entry p is uncompacted.
- The length of the linked area ℓ.

Definition 2 *A* compact suffix array (CSA) *of text T of length $n = |T|$ is an array $CSA[1\ldots n']$ of length $n' \leq n$, such that for each entry $1 \leq i \leq n'$, $CSA[i]$ is either an explicit suffix or a triple (p, δ, ℓ), where p, δ, and ℓ denote a link to an area obtained by a compacting operation from the suffix array of T. The optimal CSA for T is such that its length n' is the smallest possible.*

The original idea of using CSA as an index [14] is to guarantee that a CSA is *binary searchable*. That is, not all areas of the suffix array are compacted; it is required that each other entry of the CSA contains a suffix. The search for a pattern then consists of three phases: (i) A binary search is executed over the entries of the CSA that contain suffixes, (ii) the entries in the range found by the initial binary search are uncompacted, and (iii) the start and end of occurrences is found by binary searches over the uncompacted area.

3   The Compressed CSA

The Compressed CSA (CCSA) is conceptually built on top of the CSA. It involves some slight changes in the structure itself, and radical changes in its representation. A complete example is given in Fig. 1

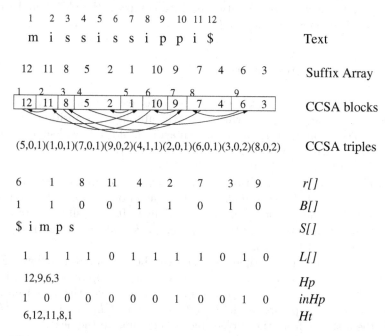

Fig. 1. Example of our CCSA structure for the text "mississippi$".

3.1   Conceptual Structure

The CCSA data structure is conceptually composed of an array of entries (p_i, δ_i, ℓ_i), $1 \le i \le n'$, just like the CSA. This array, however, differs slightly from that of the CSA. It corresponds to the *optimal* CSA defined in previous section, without any explicit suffix. The CCSA represents the original suffix array SA as follows. Entry i in the CCSA represents a block of ℓ_i entries of SA, namely entries $\left(\sum_{1 \le j < i} \ell_j \right) + 1$ to $\left(\sum_{1 \le j < i} \ell_j \right) + \ell_i$. The actual content is obtained by copying ℓ_i positions of SA from another area and subtracting 1 from their cell values. The pair (p_i, δ_i) is a reference to the SA position where the area to copy begins. The reference indicates position inside the CCSA array and offset inside the p_i-th block (so it should hold $0 \leq \delta_i < \ell_{p_i}$). The corresponding absolute SA position is $sapos(p_i, \delta_i)$, where $sapos(p, \delta) = 1 + \delta + \sum_{1 \leq j < p} \ell_j$.

The only case where no proper reference exists is for SA entry with value n. In this case we state that the entry should reference position 1.

Furthermore, the CCSA array has to be of minimum size. That is, it cannot happen that $sapos(p_i, \delta_i) = sapos(p_{i-1}, \delta_{i-1}) + \ell_{i-1}$, as in this case the CCSA entry i could be merged with entry $i - 1$. However, we limit areas that can be extended so that the first characters of all the suffixes pointed by the SA area represented by a single CCSA entry are equal.

Conceptually, the CCSA structure needs the text separately available. However, we propose now a representation both to compress the CCSA and to get rid of the explicit representation of T.

3.2   A Compact Representation

The CCSA array will be represented as follows. For each block (p_i, δ_i, ℓ_i) we will store number $r_i = sapos(p_i, \delta_i)$, which gives the absolute SA position where the i-th CCSA block points to. Additionally, an array L of n bits will signal the SA positions that start a block in the CCSA. That is, $L[j] = 1$ iff there is a value $1 \leq i \leq n'$ such that $sapos(i, 0) = j$ in the CCSA.

We will be interested in performing $rank$ and $select$ queries over array L. These are defined as follows: $rank(L, j)$ is the number of 1's in L up to position j, and $select(L, i)$ is the position j of the i-th "1" in L. It is possible to preprocess L so that, using only $o(n)$ additional space, $rank$ and $select$ queries can be answered in constant time [13, 3].

Now, the components of triple (p_i, δ_i, ℓ_i) can be computed as follows. First, $p_i = rank(L, r_i)$, that is, the number of blocks beginnings up to position r_i in the SA. Second, $\delta_i = r_i - select(L, p_i)$, since $select(L, p_i)$ gives the initial position of the block where r_i points inside. Finally, $\ell_i = select(L, i + 1) - select(L, i)$, which is the distance from the current block beginning to the next.

In order to discard the text, we need to supply a structure to replace it. It turns out that we will never need to access T_j directly but, rather, given suffix array entry $SA[i]$, we will access $T_{SA[i]}$. This is much easier, because the characters of T are sorted by index i, that is, given two text characters $a < b$, all the text occurrences of a appear before those of b in the SA. Moreover, since the first characters of each CCSA block are the same, we will only require the characters of the form $T_{SA[sapos(i,0)]}$.

We store an array B of n' bits, so that $B[i] = 1$ iff $T_{SA[sapos(i,0)]} \neq T_{SA[sapos(i-1,0)]}$ or $i = 1$, that is, if the first character of suffixes in CCSA block i differ from that in the previous block. We also store an array of characters S, of size at most σ, where all the distinct characters appearing in T are stored in lexicographic order. Hence, $T_{SA[sapos(i,0)]} = S[rank(B, i)]$, since $rank(B, i)$ tells how many times the first suffix character has changed since the beginning of the

CCSA array, and S maps this number to the corresponding character. Therefore, bit array B will be also preprocessed for $rank$ queries.

The above structures require $n' \log n + n + n' + \sigma \log \sigma + o(n)$ bits[1]. With them we have enough information to determine the SA range that contains the occurrences of a pattern P. In the following we will depict the search algorithms. Later, we will consider the problem of showing the text positions and contexts for the occurrences, and introduce a few more structures for that.

3.3   Search Algorithm

Our aim is to binary search the CCSA just like the SA. Even if the SA is not explicitly represented, we can perform such a binary search provided we are able to extract the first m characters of a given entry $SA[i]$, so as to compare it against our search pattern P. Therefore, our problem is to extract $T_{SA[i]...SA[i]+m-1}$ without having T nor SA.

Let us first concentrate in obtaining character $T_{SA[i]}$. Let $j = rank(L, i)$ be the CCSA block that contains SA entry i. The offset corresponding to entry i inside CCSA block j is $\delta = i - select(L, j)$, so $i = sapos(j, \delta)$. Since all the first letters of blocks inside CCSA block j are the same, we can rather fetch character $T_{SA[sapos(j,0)]}$. As explained above, this is precisely $S[rank(B, j)]$. Hence we can obtain the first character $T_{SA[i]} = S[rank(B, j)]$.

We need now to move to the next character $T_{SA[i]+1}$. But this is easy to to obtain from the CCSA. Since $SA[i]$ corresponds to reference (j, δ) in the CCSA, then position $SA[i] + 1$ corresponds to CCSA reference $(p_j, \delta_j + \delta)$. The corresponding SA entry is thus $sapos(p_j, \delta_j + \delta) = r_j + \delta$.

Hence, the algorithm obtains the m characters by repeatedly computing $j \leftarrow rank(L, i)$, getting character $S[rank(B, j)]$, and then moving to $i \leftarrow r_j + i - select(L, j)$. This clearly takes $O(m)$ time, and the whole binary search takes $O(m \log n)$.

3.4   Reporting Occurrence Positions

Once we determine the SA range where the occurrences of P lie, we wish to show those text positions where P occurs. With the current structures we do not have enough information to do that.

We sample text positions at regular intervals of length I, that is, text positions $h + I$, $h + 2I$, ..., so that text position n is sampled, $h = n \bmod I$. For each sampled text position pos, pointed to by SA entry i, we store (i, pos) in an array H_p, in increasing i order. At reporting time, given a position i of SA to report, we search for i in H_p. If present, we immediately know its text position pos. Otherwise, we switch to $i' \leftarrow r_j + i - select(L, j)$, where $j = rank(L, i)$, which is the SA position pointing to text position $pos + 1$ (we do not yet know pos), and repeat the process. If we find (i', pos') in H_p, then the original text position is $pos' - 1$. We repeat the process until we find a reference in array H_p.

---

[1] Our logarithms are all in base 2.

Fast searching of array H_p is possible by storing a bit array $inH_p[1 \ldots n]$, such that $inH_p[i] = 1$ iff entry (i, pos) is present in H_p. If present, it is at H_p entry number $rank(inH_p, i)$, since H_p entries are stored in increasing order of i. Hence inH_p is precomputed to answer $rank$ queries in constant time. We note that only pos has to be stored in H_p, since i is actually the search key.

If we sample one text position out of $I = \log n$, then we can execute at most $\log n$ steps in our quest for the text position, since some text position must be sampled in the range $pos \ldots pos + \log n - 1$. Hence the total cost of the process is $O(\log n)$. The extra space needed is $2n + o(n)$ bits, since each of the $n/\log n$ text positions needs $\log n$ bits for pos and inH_p needs $n + o(n)$ bits.

3.5   Showing Text Contexts

Since the CCSA is a self-index, we must be able to show not only the text context around an occurrence, but any text substring we are asked to. Say that, in general, we wish to show a text string of length ℓ starting at text position pos, that is, retrieve $T_{pos \ldots pos+\ell-1}$.

When we considered the binary search, we saw that we can retrieve as many characters as we wish from the suffix pointed to by $SA[i]$, given i. This time, however, we are given $pos = SA[i]$ instead of i, so the first step is to find some suitable i.

We store in array H_t the same entries (i, pos) implicitly stored in H_p, this time in increasing order of pos. Actually, pos does not need to be stored since at array position j we have $pos = h + jI$. Hence, at position $H_t[\lfloor (pos - h)/I \rfloor]$ we find entry (i, pos'), where pos' is the largest sampled text position $pos' \leq pos$. (For this to work properly we must add an entry $H_t[0]$ corresponding to text position 1.) Then, we can extract $\ell + pos - pos'$ text characters from $SA[i]$ with the same method used in the binary search. This will give us $T_{pos \ldots pos+\ell-1}$ as desired. The overall time is $O(\ell + \log n)$ and we need other n bits to store the entries of H_t.

3.6   The Whole Picture

Our final CCSA structure is composed of the following elements:

- Array r of n' entries r_i.
- Array L of n bits with structures for $rank$ and $select$ operations.
- Array B of n' bits with structures for $rank$ operations.
- Array S of at most σ characters.
- Array H_p storing $1 + \lfloor n/\log n \rfloor$ values i, plus bit vector inH_p of n bits with structures for $rank$ operation.
- Array H_t, storing $1 + \lfloor n/\log n \rfloor$ values pos.

Together, these structures add $n' \log n + 4n + n' + \sigma \log \sigma + o(n)$ bits. We remark that the text needs not be stored separately. It is clear that the CCSA can be built in $O(n)$ time from the suffix array, since the most complex part is similar to the CSA construction, which can be done in linear time [14].

We can do better in terms of space, at least in theory. A bit array of size n where only k bits are set can be preprocessed for constant-time *rank* and *select* queries and stored in $\log \binom{n}{k} + o(n)$ bits [1]. In particular, our array B requires only $O(\sigma \log n')$ space, while array inH_p requires $O(n \log \log n / \log n) = o(n)$ space.

The final result, taking σ as a small constant to simplify, is that we need $n' \log n + 3n + o(n)$ bits. With this CCSA structure, we can search for the occ occurrences of a pattern of length m and show a text context of length ℓ around each occurrence in worst-case time $O((m \log n + occ(\ell + \log n)))$. If we only want to show the text positions, the complexity is $O((m + occ) \log n)$. If we only want to know how many occurrences there are, the complexity is $O(m \log n)$.

We can attain $n' \log n + n + o(n)$ space by sampling one out of $\log n \log \log n$ entries in arrays H_p and H_t. In this case the time to report the occurrences raises to $O(occ \log n \log \log n)$, and a text string can be displayed in $O(\ell + \log n \log \log n)$ time.

All our space analysis is given in terms of n'. In the next section we show that $n' = O(H_k n)$, and therefore the CCSA structure needs $O(n(1 + H_k \log n))$ bits of space.

4   An Entropy Bound on the Length of CSA and CCSA

We will now prove that the length n' of the optimal CSA and the CCSA is at most $|\Sigma|^k + 2H_k n$, where H_k is the *k-th order empirical entropy* of T [12]. To be precise, we obtain the bound when the indexes are built on the *inverse string* $T^{-1} = t_1^{-1} t_2^{-1} \cdots t_n^{-1} = t_n t_{n-1} \cdots t_1$ of T.

Let us first recall some basic facts and definitions from [12]. Let n_i denote the number of occurrences in T of the i-th symbol of Σ. The zero-order empirical entropy of the string T is

$$H_0(T) = -\sum_{i=1}^{\sigma} \frac{n_i}{n} \log \frac{n_i}{n}, \qquad (1)$$

where $0 \log 0 = 0$. If we use a fixed codeword for each symbol in the alphabet, then $H_0 n$ bits is the smallest encoding one can achieve for T ($H_0 = H_0(T)$). If the codeword is not fixed, but it depends on the k previous symbols that may precede it in T, then $H_k n$ bits is the smallest encoding one can achieve for T, where $H_k = H_k(T)$ is the k-th order empirical entropy of T. It is defined as

$$H_k(T) = \frac{1}{n} \sum_{W \in \Sigma^k} |W_T| H_0(W_T), \qquad (2)$$

where W_T is a concatenation of all symbols t_j (in arbitrary order) such that $W t_j$ is a substring of T. String W is the *k-context* of each such t_j. Note that the order in which the symbols t_j are permuted in W_T does not affect $H_0(W_T)$, and hence we have not fixed any particular order for W_T.

The *Burrows-Wheeler transform* [2], denoted by $bwt(T)$, is a permutation of the text. Run-length encoding of $bwt(T)$ is closely related to the compression achieved by the CSA. The *runs* in $bwt(T)$ (maximal repeats of one symbol) correspond to links in the CCSA; if we construct the optimal CCSA for string T with the restriction that the suffixes inside each linked area must start with the same symbol, then the length of the CCSA is equal to the number of runs in $bwt(T)$. To state this connection formally, recall from [12] that $bwt(T) = t^{-1}_{SA[1]-1} t^{-1}_{SA[2]-1} \cdots t^{-1}_{SA[n]-1}$, where $t^{-1}_0 = t^{-1}_n = \#$ and SA is the suffix array of T^{-1}. Symbol $\# \notin \Sigma$ precedes all symbols of Σ in the lexicographic order[2]. Now, if suffixes $SA[j], SA[j+1], \ldots, SA[j+\ell]$ are replaced by a link to suffixes $SA[i], SA[i+1], \ldots, SA[i+\ell]$ in CCSA, then $SA[j+r] = SA[i+r] - 1$ and $t^{-1}_{SA[i+r]-1} = t^{-1}_{SA[i+r']-1}$ for all $0 \le r, r' \le \ell$. Since the linked areas are maximal in CCSA, each run in $bwt(T)$ corresponds to exactly one link in CCSA (omitting the degenerate case of t_n). Thus, the length n' of the optimal CCSA equals the number of runs in $bwt(T)$.

We will now prove that the number of runs in $bwt(T)$ is at most $|\Sigma|^k + 2H_k n$.

Let $rle(S)$ be the *run-length encoding* of string S, that is, a sequence of pairs (s_i, ℓ_i) such that $s_i s_{i+1} \cdots s_{i+\ell-1}$ is a *maximal run* of symbol s_i (i.e., $s_{i-1} \ne s_i$ and $s_{i+\ell} \ne s_i$), and all such maximal runs are listed in $rle(S)$ in the order they appear in S. The length $|rle(S)|$ of $rle(S)$ is the number of pairs in it. Notice that $|rle(S)| \le |rle(S_1)| + |rle(S_2)| + \cdots + |rle(S_p)|$, where $S_1 S_2 \cdots S_p = S$ is any partition of S.

Recall string W_T as defined in Eq. (2) for a k-context W of a string T. Note that we can apply any permutation to W_T so that (2) still holds. Now, $bwt(T)$ can be given as a concatenation of strings W_T for $W \in \Sigma^k$, if we fix the permutation of each W_T and the relative order of all strings W_T appropriately [12]. As a consequence, we have that

$$|rle(bwt(T))| \le \sum_{W \in \Sigma^k} |rle(W_T)|, \tag{3}$$

where the permutation of each W_T is now fixed by $bwt(T)$. In fact, Eq. (3) holds also if we fix the permutation of each W_T so that $|rle(W_T)|$ is maximized. This observation gives a tool to upper bound $|rle(bwt(T))|$ by the sum of code lengths when zero-order entropy encoding is applied to each W_T separately. We next show that $|rle(W_T)| \le 1 + 2|W_T|H_0(W_T)$.

First notice that if $|\Sigma_{W_T}| = 1$ then $|rle(W_T)| = 1$ and $|W_T|H_0(W_T) = 0$, so our claim holds. Let us then assume that $|\Sigma_{W_T}| = 2$. Let x and y ($x \le y$) be the number of occurrences of the two letters, say a and b, in W_T, respectively. We have that

$$H_0(W_T) = -(x/(x+y)) \log(x/(x+y)) - (y/(x+y)) \log(y/(x+y)) \ge x/(x+y), \tag{4}$$

---

[2] We follow the convention of Manzini [12]; the original transformation [2] uses T instead of T^{-1}.

since $- \log(x/(x+y)) \geq 1$ (because $x/(x+y) \leq 1/2$) and $-(y/(x+y)) \log(y/(x+y)) > 0$. The permutation of W_T that maximizes $|rle(W_T)|$ is such that there is no run of symbol a longer than 1. This makes the number of runs in $rle(W_T)$ to be $2x + 1$. By using Eq. (4) we have that

$$|rle(W_T)| \leq 2x + 1 = 1 + 2|W_T|x/(x + y) \leq 1 + 2|W_T|H_0(W_T). \qquad (5)$$

We are left with the case $|\Sigma_{W_T}| > 2$. This case splits into two sub-cases: (i) the most frequent symbol occurs at least $|W_T|/2$ times in W_T; (ii) all symbols occur less than $|W_T|/2$ times in W_T. Case (i) becomes analogous to case $|\Sigma_{W_T}| = 2$ once x is redefined as the sum of occurrences of symbols other than the most frequent. In case (ii) $|rle(W_T)|$ can be $|W_T|$. On the other hand, $|W_T|H_0(W_T)$ must also be at least $|W_T|$, since it holds that $- \log(x/|W_T|) \geq 1$ for $x \leq |W_T|/2$, where x is the number of occurrences of any symbol in W_T. Therefore we can conclude that Eq. (5) holds for any W_T.

Combining Eqs. (2) and (5) we get the following result:

Theorem 3 *The length of the run-length encoded Burrows-Wheeler transformed text of length n is at most $|\Sigma|^k + 2H_k n$, for any fixed $k \geq 1$.*

As a direct consequence of Theorem 3

$$n' \leq |rle(bwt(T))| \leq |\Sigma|^k + 2H_k n, \qquad (6)$$

where n' is the length of the optimal CCSA (or CSA) for text T^{-1}.

5   Implementation and Experiments

We implemented our CCSA structure almost exactly as described. The main difference is that we changed the constant time *select* implementation described in [13,3], as it has a huge constant factor (an asymptotic constant that is usually > 300). Instead, we implemented a tailored algorithm to compute $i - select(L, rank(L, i))$, which is the way we use *select*. In this case we know position i and simply want the last bit set before position i in array L. We implemented a word-wise followed by bit-wise upward scan from position i until the first bit set appears. Currently we have only implemented the counting of occurrences and reporting of text positions, but not yet displaying the context around the occurrences. The implementation is available at
http://www.cs.helsinki.fi/u/vmakinen/software/.

We tried out several alternative implementations for reporting the occurrences. The main idea in these alternative implementations is to exploit the common search paths for consecutive suffixes. This property is used in the original recursive reporting algorithm for compact suffix arrays [14]. We implemented an analogous recursive reporting algorithm for CCSA, but it was only slightly faster than the direct method described in Sect. 3.4. However, an algorithm that only exploits the common search paths for minimizing the (costly) computation of $i - select(L, rank(L, i))$ turned out to be practical; it is about 25% faster than the direct computation.

Our experiments were run over 83.37 Mb of text obtained from the "ZIFF-2" disk of the TREC-3 collection [9]. The tests ran on a Pentium IV processor at 2 GHz, 512 Mb of RAM and 512 Kb cache, running Linux SuSE 7.3. We compiled the code with gcc 2.95.3 using optimization option -O3. Times were averaged over 10,000 search patterns. As we work only in main memory, we only consider CPU times. The search patterns were obtained by pruning text lines to their first m characters. We avoided lines containing tags and non-visible characters such as '&'.

The CCSA index takes 1.6 times the text size. Some quick tests showed that the CCSA is about 50 times slower than the CSA (2.7 times the text size) and 50 to 75 times faster than the standard implementations of the FM-index [5, 6] and the CSArray [18] using default parameters (around 0.7 times the text size). This shows that the CCSA is a valid trade off alternative.

A much more interesting experiment is to determine *how well* does the CCSA use the space it takes. Both the FM-index and the CSArray can be tuned to use more space, so the natural question is how would the CCSA compare against them if we let them use $1.6n$ bytes. Similarly, the LZ-index takes $1.5n$ bytes over our text, so a direct comparison is fair.

The original FM-index implementation (http://butirro.di.unipi.it/~ferrax/fmindex/) does not permit using as much as $1.6n$ bytes. Instead, we used the implementation from G. Navarro (http://www.dcc.uchile.cl/~gnavarro/software), which takes more space than the text and makes good use of it (see details in [17]), and tuned it to use $1.5n$ bytes. On the other hand, the CSArray original implementation by K. Sadakane (also available at http://www.dcc.uchile.cl/~gnavarro/software), let us tuning it to use near $1.6n$ bytes.

Figure 2 shows the result for counting queries (just telling the number of occurrences) and for reporting queries (telling also all the text positions where they appear). For counting, the CCSA is much faster than the LZ-index, albeit slower than the FM-index and the CSArray. It is interesting that the search cost of the CCSA seems to grow slower with m: For $m = 5$ it is 5–15 times slower, but for $m = 60$ it is only 1.5–4 times slower. The reason is evidently in the expected running time; for larger m, only small portion of the pattern is compared against each suffix in the binary search.

For reporting, the CCSA is about 3.5 faster than the FM-index to process each occurrence. This is clear for $m = 5$, where the number of occurrences is high and reporting them dominates overall time. For $m > 20$ their number is low enough to make the counting superiority of the FM-index to show up and dominate the CCSA. The situation is reversed with the LZ-index, which is 10 times faster than the CCSA at reporting occurrences, but its inferiority to find them shows up for $m > 10$, where it loses against the CCSA. Finally, the CSArray is consistently nearly twice as fast as the CCSA.

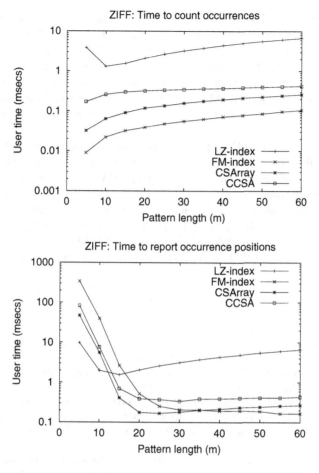

Fig. 2. Query times for our CCSA versus alternative succinct indexes tuned to use about the same space.

6   Conclusions

Compact suffix array represents an analogous improvement to suffix arrays as compact DAWG [4] for suffix trees; both are examples of *concrete optimization* (using the terminology of Jacobson [10]). The research on compressed index structures has recently concentrated on compressing suffix arrays and trees. Such compression is called *abstract optimization* ([10]), as an analogy to the goal to represent a data structure in as small space as possible while supporting the functionality of the abstract definition of the structure.

In this paper, we have presented the first data structure, compressed compact suffix array, that simultaneously exploits *both* concrete optimization and abstract optimization. The resulting structure is competitive against the counterparts that only use abstract optimization.

Our experiments, however, reveal that the structure does not in practice dominate the best current implementations on any domain. Namely, the compressed suffix array implementation of Sadakane [18] is always slightly better. We note that the situation might easily change: Our structure uses heavily the select-function. A more efficient implementation of this function would make our structure a good alternative. Also, if the link structure could be compressed to $O(H_k n)$ bits instead of the $O(H_k n \log n)$ bits, our structure would become very appealing.

The entropy bound on the size of compact suffix array is itself interesting. It could be possible to obtain similar bound also for the size of compact DAWGs, to explain the well-known fact that compact DAWGs have usually much less nodes than suffix trees.

In our subsequent work [15], we have developed an index that is a cross between CCSA and FM-index [5, 6]. From the same entropy analysis as used here follows that this index occupies $O(n + H_k n \log |\Sigma|)$ bits. It supports counting queries in time $O(m \log |\Sigma|)$, and reports occ occurrences in time $O(occ \log |\Sigma| \log n)$.

References

1. A. Brodnik and I. Munro. Membership in constant time and almost-minimum space. *SIAM J. on Comp.* 5:1627–1640, 1999.
2. M. Burrows and D. J. Wheeler. A block-sorting lossless data compression algorithm. *DEC SRC Research Report 124*, 1994.
3. D. Clark. *Compact Pat Trees*. PhD thesis, University of Waterloo, 1996.
4. M. Crochemore and Renaud Vérin. Direct Construction of Compact Directed Acyclic Word Graphs. In *Proc. CPM'97*, Springer-Verlag LNCS 1264, pp. 116-129, 1997.
5. P. Ferragina and G. Manzini. Opportunistic Data Structures with Applications. In *Proc. IEEE Symp. on Foundations of Computer Science (FOCS'00)*, pp. 390–398, 2000.
6. P. Ferragina and G. Manzini. An Experimental Study of an Opportunistic Index. In *Proc. 12th Symposium on Discrete Algorithms (SODA'01)*, pp. 269–278, 2001.
7. R. Giegerich, S. Kurtz, and J. Stoye. Efficient Implementation of Lazy Suffix Trees. In *Proc. 3rd Workshop on Algorithmic Engineering (WAE'99)*, LNCS 1668, pp. 30–42, 1999.
8. R. Grossi and J. Vitter. Compressed suffix arrays and suffix trees with applications to text indexing and string matching. In *Proc. 32nd Symposium on Theory of Computing (STOC'00)*, pp. 397–406, 2000.
9. D. Harman. Overview of the Third Text REtrieval Conference. In *Proc. TREC-3*, pages 1–19, 1995. NIST Special Publication 500-207.
10. G. Jacobson. *Succinct Static Data Structures*. PhD thesis, CMU-CS-89-112, Carnegie Mellon University, 1989.
11. U. Manber and G. Myers. Suffix arrays: A new method for on-line string searches. *SIAM J. Comput.* 22, pp. 935–948, 1993.
12. G. Manzini. An Analysis of the Burrows-Wheeler Transform. *J. of the ACM* 48(3):407–430, 2001.
13. I. Munro. Tables. In *Proc. FSTTCS'96*, pp. 37–42, 1996.

14. V. Mäkinen. Compact Suffix Array — A Space-efficient Full-text Index. *Fundamenta Informaticae* 56(1-2), pp. 191–210, 2003.
15. V. Mäkinen and G. Navarro. New search algorithms and time/space trade offs for succinct suffix arrays. *Technical report C-2004-20*, Dept. of Computer Science, Univ. of Helsinki, April 2004.
16. G. Navarro. Indexing Text using the Ziv-Lempel Trie. In *Proc. 9th String Processing and Information Retrieval (SPIRE'02)*, LNCS 2476, pp. 325–336, 2002. Extended version to appear in *J. of Discrete Algorithms*.
17. G. Navarro. The LZ-index: A Text Index Based on the Ziv-Lempel Trie. *Technical Report TR/DCC-2003-1*, Dept. of Computer Science, Univ. of Chile, January 2003.
18. K. Sadakane. Compressed text databases with efficient query algorithms based on the compressed suffix array. In *Proc. 11th Algorithms and Computation (ISAAC'00)*, LNCS 1969, pp. 410–421, 2000.
19. P. Weiner. Linear pattern matching algorithms. In *Proc. IEEE 14th Annual Symposium on Switching and Automata Theory*, pp. 1–11, 1973.

Approximate String Matching
Using Compressed Suffix Arrays

Trinh N.D. Huynh[1,*], Wing-Kai Hon[2], Tak-Wah Lam[2], and Wing-Kin Sung[1,**]

[1] School of Computing, National University of Singapore, Singapore
{huynhngo,ksung}@comp.nus.edu.sg
[2] Department of Computer Science and Information Systems, The University of Hong Kong,
Hong Kong
{wkhon,twlam}@csis.hku.hk

Abstract. Let T be a text of length n and P be a pattern of length m, both strings over a fixed finite alphabet A. The k-difference (k-mismatch, respectively) problem is to find all occurrences of P in T that have edit distance (Hamming distance, respectively) at most k from P. In this paper we investigate a well-studied case in which $k = 1$ and T is fixed and preprocessed into an indexing data structure so that any pattern query can be answered faster [16–19]. This paper gives a solution using $O(n)$ bits indexing data structure with $O(m \log^2 n)$ query time. To the best of our knowledge, this is the first result which requires linear indexing space. The results can be extended for the k-difference problem with $k \geq 1$.

1 Introduction

Let T be a text of length n and P be a pattern of length m, both strings over a fixed finite alphabet A. The string matching problem is to find all occurrences of P in T which satisfy some criteria. Depending on the criteria, we have three different problems. (1) The exact string matching problem requires us to find all exact occurrences of P in T; (2) The k-difference problem is to find all occurrences of P in T that have edit distance at most k from P; and (3) the k-mismatch problem is to find all occurrences of P in T that have Hamming distance at most k from P. Edit distance between two strings is defined to be the number of character insertions, deletions and replacements to convert one string to another. When only character replacements are allowed, we have Hamming distance. These problems are well-studied. They find applications in many areas including computational biology, signal processing, text retrieval, handwritting recognition, pattern recognition, etc.

In the past, most of the research are on the online version of the string matching problem. This version of the problem assumes both the text T and the pattern P are not known in advance. For exact online matching, well-known algorithms include Boyer-Moore [1] and Knuth-Morris-Pratt [2] algorithms. For approximate online matching, standard dynamic programming solves this in $O(mn)$ time. Landau and Vishkin[3] improved this to $O(kn)$. Recently, there are results by Baeza-Yates and Navarro [5] and Amir et al [4]. We refer to Navarro [6] for a survey.

---

* (Undergraduate) student coauthor
** This work was supported in part by the NUS Academic Research Grant R-252-000-119-112.

S.C. Sahinalp et al. (Eds.): CPM 2004, LNCS 3109, pp. 434–444, 2004.
© Springer-Verlag Berlin Heidelberg 2004

Recently, people start to consider the offline version of the problem, which assumes the text T is given in advance and we can preprocess it to build an indexing data structures so that any pattern query can be answered faster. One of the motivating applications of the offline version of the problem is the DNA sequence searching. This application requires us to find DNA subsequences (like genes, promoter concsensus sequences) over some known DNA genome sequences like the human genome. Since the genome sequence is very long, people would like to preprocess it to accelerate pattern queries.

In the literature, there are a number of indexing data structures for the exact offline pattern matching problem. Suffix trees [7] and suffix arrays [8] are some well-known solutions. For a text T of length n, building a suffix tree takes $O(n)$ time. After that, exact occurrences of a pattern P can be located in $O(|P| + occ)$ time where occ is the number of occurrences. For suffix arrays, construction and searching takes $O(n \log n)$ time and $O(|P| + \log n + occ)$ time, respectively. Both data structure require $O(n \log n)$ bits space, though suffix arrays are associated with a smaller constant. Recently, two compressed versions of suffix arrays have been devised, which are compressed suffix arrays (CSA) [9] and FM-index [10]. Both of them occupy only $O(n)$ bits space, yet still supporting exact pattern searching as efficient as suffix arrays.

Besides exact matching, the k-difference and the k-mismatch problems are also important since the text and the pattern may contain "errors" (for example, in gene hunting, there may be some mutations in a gene). Jokinen and Ukkonen [24] were the first to treat the approximate offline matching problem in which the text T can be preprocessed. Since then, there are many different approaches proposed [11–20]. We refer to Navarro et al [25] for a brief survey. Note that many existing approaches actually incur a query time complexity depending on n, i.e., they are inefficient even if the pattern is very short and k is as small as one. The first solution which has query time complexity depends only on m and k (and the logarithm of n) is proposed by Ukkonen [19]. The problem is indeed not trivial even for $k = 1$, and there are several interesting results focusing on one single error [16–18]. Cobbs [16] gave an indexing data structure using $O(n \log n)$ bits space and having $O(m^2 + occ)$ query time for $k = 1$, where occ is the number of occurrences. More recently, Amir et al [17] proposed a result with $O(n \log^2 n)$ preprocessing time, $O(n \log^2 n)$ bits indexing space and $O(m \log n \log \log n + occ)$ query time and Buchsbaum et al [18] propsed another with $O(n \log n)$ preprocessing time, $O(n \log n)$ bits indexing space and $O(m \log \log n + occ)$ query time.

The contribution of this paper is the first solution in literature for 1-differrence and 1-mismatch problems using linear space indexing data structure. We show in this paper that a suffix array plus an inverse suffix array can give a simple solution that uses $O(n \log n)$-bit space and $O(m \log n + occ)$ query time. We combine the techniques of forward searching and backward searching on suffix arrays data structure to achieve this time bound. Furthermore, this solution allows us to exploit compressed suffix arrays to reduce the space to $O(n)$ bits, while increasing the query time by an $O(\log n)$ factor only. Though this solution is a bit slower, it uses optimal space[1]. Moreover, our index-

---

[1] It is non-trival to change the data-structures of [17, 18] to linear space data structure. For Cobbs[16], its space can be reduced by replacing suffix trees by compressed suffix trees [9]. However, the query time becomes $\omega(m^2 + occ)$.

ing data structure can be constructed in $O(n)$ time, and hence our solution requires only linear preprocessing time.

We also extend our algorithm for k-differrence and k-mismatch problems for $k > 1$. Previous result proposed by Cobbs [16] uses $O(n \log n)$ bits indexing data structure and takes $O(m^{k+2}|A|^k + occ)$ query time. Our solution takes only $O(m^k |A|^k \log n + occ)$ or $O(m^k |A|^k \log^2 n + occ \log n)$ query time, when using $O(n \log n)$ bits or $O(n)$ bits indexing data structure, respectively.

The structure of this paper is as follows: In the next section we formally define the approximate pattern matching problems and give an introduction about suffix arrays and compressed suffix arrays data structure. Then, Section 3 shows our algorithm for approximate pattern matching problems with $k = 1$. Next, Section 4 extends our solution for $k \geq 1$. Finally, Section 5 concludes the paper.

2 Preliminaries

2.1 Definition of the Problems

The k-mismatch problem and the k-difference problem are formally defined as follow. Consider a text T of length n and a pattern P of length m, both strings over a fixed finite alphabet A.

1. **k-difference problem:** We are to find all approximate occurrences of P in T that have edit distances at most k from P. The edit distance between two strings is defined to be the minimum number of character insertions, deletions, and replacements to convert one string to the other.
2. **k-mismatch problem:** We are to find all approximate occurrences of P in T that have Hamming distances at most k from P. The Hamming distance between two strings is defined to be the minimum number of character replacements to convert one string to the other.

In this paper we focus on the case $k = 1$ and the offline version of the two problems, in which T is fixed and can be preprocessed to accelerate pattern queries.

2.2 Suffix Arrays and Inverse Suffix Arrays

Let $T[0..n] = T[0]T[1] \ldots T[n]$ be a text of length n over an alphabet A where $T[n] = \$$ is a special symbol that is not in A and smaller than any other symbol in A. The j-th suffix of T is defined as $T[j..n] = T[j]T[j + 1] \ldots T[n]$ and is denoted by T_j.

The *suffix array* $SA[0..n]$ of T is an array of integers j that represent suffixes T_j and the integers are sorted in lexicographic order of the corresponding suffixes. We have that $SA[0] = n$.

Together with suffix arrays, we also use *inverse suffix arrays* to support searching in our algorithm. The inverse suffix array of T is denoted as $SA^{-1}[0..n]$, that is, $SA^{-1}[i]$ equals the number of suffixes which are lexicographically smaller than T_i.

The sizes of SA, and SA^{-1} are $O(n \log n)$ bits. There are many algorithms to construct SA in linear time [22, 26, 27].

In this paper, we let $[st..ed]$ be the range of the suffix array of T corresponding to a string P if $[st..ed]$ is the largest range such that P is a prefix of every suffix T_j for $j = SA[st], SA[st+1], \ldots, SA[ed]$.

We have the following lemma based on the forward searching technique on suffix arrays.

Lemma 1. [23] *Given a text T of length n together with its suffix array, assume $[s..e]$ is the range of the suffix array of T corresponding to a string P. For any character c, we can find the range $[s'..e']$ of the suffix array of T corresponding to the string Pc in $O(\log n)$ time.*

Hence a pattern P can be found forwardly in T in $O(|P| \log n)$ time by applying the above lemma $|P|$ times.

We have another lemma.

Lemma 2. *Given the range $[st_1..ed_1]$ of the suffix array of T corresponding to a string P_1 and the range $[st_2..ed_2]$ of the suffix array of T corresponding to a string P_2, we can find the range $[st..ed]$ of the suffix array of T corresponding to the string $P_1 P_2$ in $O(\log n)$ time using the suffix array and the inverse suffix array of T.*

Proof. To find the range $[st..ed]$ for $P_1 P_2$, we have to find the smallest st and the largest ed such that both $T_{SA[st]}$ and $T_{SA[ed]}$ have $P_1 P_2$ as their prefixes. So $[st..ed]$ is a subrange of $[st_1..ed_1]$.

Let the length of P_1 be m_1. By the definition of suffix arrays, the lexicographic orders of $T_{SA[st_1]}, T_{SA[st_1+1]}, \ldots, T_{SA[ed_1]}$ are increasing. Since they share the same prefix P_1, the lexicographic orders of $T_{SA[st_1]+m_1}, T_{SA[st_1+1]+m_1}, \ldots, T_{SA[ed_1]+m_1}$ are also increasing. Thus $SA^{-1}[SA[st_1] + m_1] < SA^{-1}[SA[st_1 + 1] + m_1] < \ldots < SA^{-1}[SA[ed_1] + m_1]$.

To find st and ed, we find the smallest st such that $st_2 \leq SA^{-1}[SA[st]+m_1] \leq ed_2$ and the largest ed such that $st_2 \leq SA^{-1}[SA[ed] + m_1] \leq ed_2$. This can be done by binary search on the range $[st_1..ed_1]$ and make $O(\log n)$ calls to the suffix array and inverse suffix array. □

Assuming we have an array C such that for any c in A, $C[c]$ stores the total number of occurrences of all characters c' in T, where $c' \leq c$. This gives us the following backward searching technique.

Lemma 3. *Given the suffix array and the inverse suffix array of T, assume $[s..e]$ is the range of the suffix array of T corresponding to P. For any character c, assume we have in advance the array C, we can find $[s'..e']$ which is the range of the suffix array of T corresponding to cP in $O(\log n)$ time.*

Proof. The lemma follows directly from Lemma 2 where $P_1 = c$ and $P_2 = P$. Note that the range corresponding to P_1 is $[C[c'] + 1..C[c]]$ where c' is a character immediately before c in A. □

Hence a pattern P can be found backwardly in T in $O(|P| \log n)$ time by applying the above lemma $|P|$ times.

2.3   Compressed Suffix Array

Compressed suffix arrays data structure (CSA) $\Psi[1..n]$ [9] is a compressed version of suffix arrays and has the size of the same order as the text itself. It is defined as

$$\Psi[i] = SA^{-1}[SA[i] + 1]$$

Compressed suffix array can be constructed in $O(n)$ time and can be stored in $O(n)$ bit space while every $\Psi[i]$ can be accessed in constant time [22]. We have the following result from Sadakane and Shibuya [21].

Lemma 4. [21] *We can store compressed suffix array together with a supporting data structure in $O(n)$ bits space so that $SA[i]$ and $SA^{-1}[i]$ can be captured in $O(\log n)$ time.*

So instead of using suffix arrays and inverse suffix arrays, we can use compressed suffix arrays which will slow down by a factor of $O(\log n)$ every time we access suffix array and inverse suffix array from CSA. Henceforth, when we refer to compressed suffix arrays data structure we refer to itself together with the additional supporting data structure.

By Lemma 4, we can simulate operations done on suffix arrays and inverse suffix arrays on Ψ with the time complexity slowed down by a factor of $O(\log n)$. Thus for compressed suffix arrays, we also have the same results for Lemmas 1, 2 and 3, but in time $O(\log^2 n)$ [2].

3   Approximate Matching with One Error Using Linear Indexing Space

First, for the explanation of the algorithm, we use suffix array and inverse suffix array as indexing data structures of the text. Then, we replace the suffix array and the inverse suffix array by compressed suffix array data structure to achieve the linear space complexity.

We use suffix array and inverse suffix array as the indexing data structures of the text. Lemma 2 gives us an idea to solve the k-difference problem for $k = 1$. For $k = 1$, there is at most 1 "error" between the pattern $P[1..m]$ and any of its occurrences in $T[0..n]$. An error may be a character insertion, replacement or deletion. We can try to put the error at each position in the pattern to form an edited pattern P' which has edit distance 1 from P and check if P' is in T. Normally, checking if P' is in T requires $O(|P'|)$ time. Based on Lemmas 2 and 3, such checking can be done in $O(\log n)$ time as follows. We let P'_L and P'_R be the portions of the pattern P to the left and to the right of the error, respectively. Assume we know in advance the range $[s..e]$ of the suffix array of T corresponding to P'_L and the range $[s'..e']$ of the suffix array of T corresponding to P'_R. By using Lemma 3, we can find in $O(\log n)$ time the range $[s''..e'']$ corresponding to P'_R prefixed with the error. Then, given $[s..e]$ and $[s''..e'']$, by Lemma 2, we can find the range $[st..ed]$ corresponding to the whole edited pattern P' using $O(\log n)$ time.

---

[2] Backward search (Lemma 3) actually can be done in $O(1)$ time using compressed suffix arrays [22].

The algorithm is shown in Fig. 1. We first construct $F_{st}[0..m]$ and $F_{ed}[0..m]$ which are arrays such that $[F_{st}[i]..F_{ed}[i]]$ is the range of the suffix array of T corresponding to $P[1..i]$. Here we define $P[1..0]$ empty and $F_{st}[0] = 0$, $F_{ed}[0] = n$. F_{st} and F_{ed} are used later as the range corresponding to P'_L.

Then, in Step 3, the loop will iterate $m + 1$ times for $i = m$ downto 0. In iteration i, the algorithm first forms an edited pattern P' by considering an error at position i and check if P' exists in the text T. The algorithm maintains an invariant that $[s'..e']$ is the range corresponding to P'_R, the portion of the pattern just right to the error. Using the array F_{st} and F_{ed}, we can get the range $[s..e]$ corresponding to P'_L. By the above idea and together with Lemmas 2 and 3, we can get the range $[st..ed]$ corresponding to P'.

Then we have the following theorem.

Theorem 1. *After an $O(n)$ time preprocessing of the text T of length n, an $O(n \log n)$-bit data-structure can be constructed such that the k-different (k-matching) problem with $k = 1$ can be solved in $O(|A|m \log n + occ)$ time where m is the length of the pattern P, A is the alphabet and occ is the number of approximate occurrences of P in T.*

Proof. We use the suffix array and inverse suffix array as the index of T. Its size is of $O(n \log n)$ bits.

We will do a time analysis of the algorithm in Fig. 1 using suffix array and inverse suffix array as an index to T.

In the first step we build two arrays F_{st} and F_{ed}. These arrays can be found using forward searching, $F_{st}[i]$ and $F_{ed}[i]$ are updated from $F_{st}[i-1]$ and $F_{ed}[i-1]$ (using Lemma 1) where $F_{st}[0] = 0$ and $F_{ed}[0] = n$. This can be done in $O(m \log n)$ time.

The algorithm iterates i from m to 0. At each iteration, it assumes the error is at position i. $[s'..e']$ is the range of the suffix array corresponding to the second half of the pattern just right to the error, that is, $P[i+1..m]$. In the begining, $[s'..e']$ is set to $[0..n]$. At each step i, there are 3 cases:

1. The character at position i is deleted from the pattern $(1 \le i \le m)$. The pattern becomes $P' = P[1..i-1]P[i+1..m]$. We already have $[F_{st}[i-1]..F_{ed}[i-1]]$ corresponding to $P[1..i-1]$ and $[s'..e']$ corresponding to $P[i+1..m]$, we find $[st..ed]$ for P'. By Lemma 2, this takes $O(\log n)$ time.
2. The character at position i is replaced by each character c in A $(1 \le i \le m)$. The pattern becomes $P' = P[1..i-1]cP[i+1..m]$. Having the range $[s'..e']$ corresponding to $P[i+1..m]$, we find the range $[s''..e'']$ corresponding to $cP[i+1..m]$ in $O(\log n)$ time using Lemma 3. Then we find $[st..ed]$ corresponding to P' from $[s''..e'']$ and $[F_{st}[i-1]..F_{ed}[i-1]]$ using Lemma 2. This step takes $O(|A| \log n)$ time.
3. Each symbol c in A is inserted to the pattern at position i. The pattern becomes $P' = P[1..i]cP[i+1..m]$ $(0 \le i \le m)$. Having the range $[s'..e']$ corresponding to $P[i+1..m]$, we find the range $[s''..e'']$ corresponding to $cP[i+1..m]$ in $O(\log n)$ time using Lemma 3. Then we find $[st..ed]$ for P' from $[s''..e'']$ and $[F_{st}[i]..F_{ed}[i]]$ using Lemma 2. This step takes $O(|A| \log n)$ time.

Reporting occurrences takes $O(occ)$ time using suffix array.

1. Construct $F_{st}[0..m]$ and $F_{ed}[0..m]$, that is, for every $P[1..i]$, find the range $[F_{st}[i]..F_{ed}[i]]$ of the suffix array of T corresponding to $P[1..i]$. If the range $[F_{st}[m]..F_{ed}[m]]$ exists, then P has exact occurrences in T, we report occurrences in this range.

2. $s' := 0, e' := n$

3. For $i = m$ downto 0
   (a) (Case 1:deletion at $i > 0$, ignored for 1-mismatch problem)
       i. $P' = P[1..i-1]P[i+1..m]$
       ii. Given the range $[s'..e']$ corresponding to $P[i+1..m]$ and the range $[F_{st}[i-1]..F_{ed}[i-1]]$ corresponding to $P[1..i-1]$, find $[st..ed]$ corresponding to P'.
       iii. Report $[st..ed]$ if exist.
   (b) (Case 2:replacement at $i > 0$) for each $c \neq P[i]$ in A
       i. $P' = P[1..i-1]cP[i+1..m]$
       ii. Given the range $[s'..e']$ corresponding to $P[i+1..m]$, find $[s''..e'']$ corresponding to $cP[i+1..m]$.
       iii. Given the range $[s''..e'']$ corresponding to $cP[i+1..m]$ and the range $[F_{st}[i-1]..F_{ed}[i-1]]$ corresponding to $P[1..i-1]$, find $[st..ed]$ corresponding to P'.
       iv. Report $[st..ed]$ if exist.
   (c) (Case 3:insertion at i, ignored for 1-mismatch problem) for each c in A
       i. $P' = P[1..i]cP[i+1..m]$
       ii. Given the range $[s'..e']$ corresponding to $P[i+1..m]$, find $[s''..e'']$ corresponding to $cP[i+1..m]$.
       iii. Given the range $[s''..e'']$ corresponding to $cP[i+1..m]$ and the range $[F_{st}[i]..F_{ed}[i]]$ corresponding to $P[1..i]$, find $[st..ed]$ corresponding to P'.
       iv. Report $[st..ed]$ if exist.
   (d) Given $[s'..e']$ corresponding to $P[i+1..m]$, find $[s..e]$ corresponding to $P[i..m]$. If not exist, exit out of the loop.
       $s' := s, e' := e$

Fig. 1. Algorithm for 1-difference and 1-mismatch problem.

Finally $[s'..e']$ is updated to be the range corresponding to $P[i..m]$ using backward searching (Lemma 3) in $O(\log n)$ time. For an error to appear at i, $P[i+1..m]$ must exist somewhere in T. If $[s'..e']$ is not exist, we exit out of the loop and stop the algorithm.

So the time complexity of the algorithm is $O(m \log n + m(\log n + |A| \log n + |A| \log n + \log n) + occ)$ which is $O(|A| m \log n + occ)$. For A is fixed, the time is reduced to $O(m \log n + occ)$. □

Now we have the following corollary which is the main result in this paper.

Corollary 1. *After an $O(n)$ time preprocessing of the text T of length n, an $O(n)$-bit data-structure can be constructed such that the k-different (k-matching) problem with $k = 1$ can be solved in $O(|A| m \log^2 n + occ \log n)$ time where m is the length of the pattern P, A is the alphabet and occ is the number of approximate occurrences of P in T.*

---

1. Construct $F_{st}[0..m]$ and $F_{ed}[0..m]$, that is, for every $P[1..i]$, find the range $[F_{st}[i]..F_{ed}[i]]$ of the suffix array of T corresponding to $P[1..i]$.
2. Call $kapproximate([0..n], m, k, P)$.

$kapproximate([s'..e'], i, k', P')$
- invariant 1: $[s'..e']$ is the range corresponding to $P'[i+1..|P'|]$
- invariant 2: $P'[1..i] = P[1..i]$
begin
  (a) Given $[F_{st}[i]..F_{ed}[i]]$ corresponding to $P[1..i] = P'[1..i]$ and $[s'..e']$ corresponding to $P'[i+1..|P'|]$, by Lemma 2 find $[st..ed]$ corresponding to P'.
  (b) Report $[st..ed]$ if exist.
  (c) If $(k' = 0)$ return.
  (d) For $j = i$ downto 0
    i. (Case 1:deletion at $j > 0$, ignored for k-mismatch)
      Call $kapproximate([s'..e'], j-1, k'-1, P'[1..j-1]P'[j+1..|P'|])$.
    ii. (Case 2:replacement at $j > 0$) for each c in A
      A. Given $[s'..e']$ corresponding to $P'[j+1..|P'|]$, by Lemma 3 find $[s''..e'']$ corresponding to $cP'[j+1..|P'|]$.
      B. Call $kapproximate([s''..e''], j-1, k'-1, P'[1..j-1]cP'[j+1..|P'|])$.
    iii. (Case 3:insertion at j, ignored for k-mismatch) for each c in A
      A. Given $[s'..e']$ corresponding to $P'[j+1..|P'|]$, by Lemma 3 find $[s''..e'']$ corresponding to $cP'[j+1..|P'|]$.
      B. Call $kapproximate([s''..e''], j, k'-1, P'[1..j]cP'[j+1..|P'|])$.
end

Fig. 2. Algorithm for k-difference and k-mismatch problem.

Proof. We use compressed suffix array data structure to index the text T, then we achieve the $O(n)$ bits space. We use the same algorithm as in Theorem 1. As compressed suffix array incurs the penalty of $O(\log n)$ for every access to suffix array and inverse suffix array (Lemma 4) every time we use Lemma 1, 2 or 3 and report occurrences, thus from the time analysis in theorem 1, the time complexity becomes $O(m \log^2 n + m(\log^2 n + |A| \log^2 n + |A| \log^2 n + \log^2 n) + occ \log n)$ which is $O(|A|m \log^2 n + occ \log n)$ and reduced to $O(m \log^2 n + occ \log n)$ when $|A|$ is constant. □

4   Approximate Matching Problem with $k \geq 1$

This section generalizes Theorem 1 for $k \geq 1$. The idea is the same as in Theorem 1.

We also use the suffix array and the inverse suffix array as the indexing data structures. Instead of just finding 1 error, now we try to locate k errors in the pattern P to get an edited pattern P' which has edit distance at most k from P and check if P' is in T. By recursion we locate k errors from right to left one by one, with the first error at the rightmost and the kth error at the leftmost.

The algorithm is shown in Fig. 2. It has two main phases. The first phase is to construct $F_{st}[0..m]$ and $F_{ed}[0..m]$, these arrays are the same as in Theorem 1, that is, for every $P[1..i]$, find the range $[F_{st}[i]..F_{ed}[i]]$ of the suffix array of T corresponding to $P[1..i]$. These arrays are used to speedup the search in the second phase.

The second phase executes recursive function $kapproximate$, which takes 4 parameters: a range $[s'..e']$, i, k', and a pattern P'. When called, the function will recursively put k' errors to $P'[1..i]$ to get a new pattern P'' having edit distance at most k' from P' and check if P'' is in T. By other words, it is used to locate all the approximate occurrences of P' in T such that there are at most k' errors all in $P'[1..i]$. The parameters to $kapproximate$ satisfy: (a) P' is an approximate match of the queried pattern P with $k - k'$ errors where all $k - k'$ errors are located after position i and (b) $[s'..e']$ is the range of the suffix array of T corresponding to $P'[i + 1..|P'|]$. Note that constraint (a) implies that $P'[1..i] = P[1..i]$. To find all approximate occurrences of P with at most k errors, the second phase executes $kapproximate$ with $[s'..e'] = [0..n]$, $i = m$, $k' = k$, and $P' = P$.

Here the parameter P' is for the sake of explanation only. In an actual implemetation, we do not need P'.

$kapproximate([s'..e'], i, k', P')$ can be subdivided into two steps. The first step (Steps (a)-(c) in Fig. 2) finds and reports the range $[st..ed]$ of the suffix array of T corresponding to P'. Given $[s'..e']$ and $[F_{st}[i]..F_{ed}[i]]$, by Lemma 2, $[st..ed]$ can be found in $O(\log n)$ time.

The second step (Step (d) in Fig. 2) tries to introduce one more error into P' and call recursively $kapproximate$ to locate all approximate occurrences of P' with at least one and at most k' errors in $P'[1..i]$. This step is a loop which iterates $i + 1$ times for $j = i$ downto 0. For each j, the algorithm introduces an error at position j and generates a new pattern from P' with one more error. Then, recurrsively call $kapproximate$ to report more approximate matches.

From the discussion, we have the following theorem.

Theorem 2. *After an $O(n)$ time preprocessing of the text T of length n, an $O(n \log n)$-bit data-structure can be constructed such that the k-different (k-matching) problem can be solved in $O(|A|^k m^k \log n + occ)$ time where m is the length of the pattern P, A is the alphabet and occ is the number of approximate occurrences of P in T.*

Corollary 2. *After an $O(n)$ time preprocessing of the text T of length n, an $O(n)$-bit data-structure can be constructed such that the k-different (k-matching) problem can be solved in $O(|A|^k m^k \log^2 n + occ \log n)$ time where m is the length of the pattern P, A is the alphabet and occ is the number of approximate occurrences of P in T.*

5   Conclusion

We have described an algorithm for approximate string matching problem with 1 error which uses a linear space indexing data structure and takes $O(m \log^2 n + occ \log n)$ query time. To the best of our knowledge, this is the first solution which approaches linear space complexity and searches efficiently for small patterns. Another advantage of our solution over other approaches is that our data structures require only linear preprocessing time. In addition, we also extend the result to solve the approximate string matching problem with more than 1 error. As a future work, we would like to improve the solution to achieve linear query time.

References

1. R. Boyer and S. Moore. A fast string matching algorithm. CACM, 20(1977), 762–772
2. D. E. Knuth, J. Morris, V. Pratt. Fast pattern matching in strings. SAIM Journal on Computing 6 (1977), 323–350.
3. G. M. Landau and U. Vishkin. Fast parallel and serial approximate string matching. J. Algorithms,10:157–169, 1989.
4. A. Amir, M. Lewenstein, Ely. Porat. Faster algorithms for string matching with k mismatches. In *Proc. 11th Annual ACM-SIAM Symposium on Discrete Algorithms*, pages 794–803, 2000
5. R. A. Baeza-Yates, G. Navarro. A Faster Algorithm for Approximate String Matching. In *Proc. 7th Ann. Symp. on Combinatorial Pattern Matching (CPM'96)*, pages 1–23.
6. G. Navarro. A guided tour to approximate string matching. *ACM Computing Surveys*, 33(1): 31-88, 2001.
7. E. M. MCreight. A space economical suffix tree construction algorithm. *Journal of the ACM*, 23(2):262–272, 1976.
8. U. Manber and G. Myers. Suffix arrays: a new method for on-line string searches. *SIAM Journal on Computing*, 22(5):935–948, 1993.
9. R. Grossi and J. S. Vitter. Compressed suffix arrays and suffix trees with applications to text indexing and string matching. In *Proceedings of the 32nd ACM Symposium on Theory of Computing*, pages 397–406, 2000.
10. P. Ferragina, G. Manzini. Opportunistic data structures with applications In *Proc. 41st IEEE Symp. Foundations of Computer Science (FOCS'00)*, pages 390-398, 2000.
11. F. Shi. Fast Approximate String Matching with q-Blocks Sequences. In *Proceedings of the 3rd South American Workshop on String Processing (WSP'96)*, Carleton University Press, 1996.
12. R. A. Baeza-Yates, G. Navarro. A practical index for text retrieval allowing errors. In *CLEI*, volume 1, pages 273–282, November 1997.
13. G. Navarro, E. Sutinen, J. Tanninen, J. Tarhio. Indexing Text with Approximate q-Grams. In *Proceedings of the 11th Annual Symposium on Combinatorial Pattern Matching*, number 1848 in LNCS, Springer, 2000.
14. G. Navarro, R. A. Baeza-Yates. A hybrid indexing method for approximate string matching. J. of Discrete Algorithms, 1(1):205–239, 2000, 18.
15. E. Sutinen, J. Tarhio Filtration with q-Samples in Approximate String Matching. In *Proc. 7th Ann. Symp. on Combinatorial Pattern Matching (CPM'96)*, pages 50–63.
16. A. Cobbs. Fast Approximate Matching using Suffix Trees. In *Proc. 6th Ann. Symp. on Combinatorial Pattern Matching (CPM'95)*, LNCS 807, pages 41–54, 1995
17. A. Amir, D. Keselman, G. M. Landau, M. Lewenstein, N. Lewenstein, and M. Rodeh. Indexing and dictionary matching with one error. In *Proc. 6th WADS*, volume 1663 of LNCS, pages 181-92. Springer-Verlag, 1999.
18. Adam L. Buchsbaum, Michael T. Goodrich, Jeffery Westbrook. Range Searching Over Tree Cross Products. In *ESA 2000*: pages 120-131
19. E. Ukkonen. Approximate matching over suffix trees. In *Proc. Combinatorial Pattern Matching 1993*, volume 4, pages 228–242. Springer-Verlag, June 1993.
20. G. Navarro, R. A. Baeza-Yates. A New Indexing Method for Approximate String Matching. In *Proc. 10th Ann. Symp. on Combinatorial Pattern Matching (CPM'99)*, pages 163–185.
21. Kunihiko Sadakane and Tetsuo Shibuya. Indexing Huge Genome Sequences for Solving Various Problems. *Genome Informatics* 12:175–183, 2001
22. W. K. Hon, K. Sadakane and W. K. Sung. Breaking a Time-and-Space Barrier in Constructing Full-Text Indices. In *Proc. of IEEE Symposium on Foundations of Computer Science*, 2003.

23. D. Gusfield. *Algorithms on strings, trees, and sequences: computer science and computational biology*, Cambridge University Press, Cambridge, 1997.
24. P. Jokinen and E. Ukkonen. Two algorithms for approximate string matching in static texts. In *Proc. MFCS'91*, Lect. Notes in Computer Science 520 (Springer-Verlag 1991), pages 240–248.
25. G. Navarro, R. Baeza-Yates, E. Sutinen, and J. Tarhio. Indexing Methods for Approximate String Matching. *IEEE Data Engineering Bulletin*, 24(4):19-27, 2001
26. D. Kyue. Kim, J. S. Sim, Heejin Park, Kunsoo Park. Linear-Time Construction of Suffix Arrays. In *CPM 2003*: 186-199
27. Pang Ko, Srinivas Aluru. Space Efficient Linear Time Construction of Suffix Arrays. In *CPM 2003*: 200-210

Compressed Index for a Dynamic Collection of Texts

Ho-Leung Chan, Wing-Kai Hon, and Tak-Wah Lam*

Department of Computer Science
The University of Hong Kong, Hong Kong
{hlchan,wkhon,twlam}@csis.hku.hk

Abstract. Let T be a string with n characters over an alphabet of bounded size. The recent breakthrough on compressed indexing allows us to build an index for T in optimal space (i.e., $O(n)$ bits), while supporting very efficient pattern matching [2, 4]. This paper extends the work on optimal-space indexing to a dynamic collection of texts. Precisely, we give a compressed index using $O(n)$ bits where n is the total length of texts, such that searching for a pattern P takes $O(|P| \log n + occ \log^2 n)$ time where occ is the number of occurrences, and inserting or deleting a text T takes $O(|T| \log n)$ time.

1 Introduction

Indexing a text to support efficient pattern matching has been studied extensively. The recent breakthrough allows us to build an index in optimal space (i.e., $O(n)$ bits), without sacrificing the speed of pattern matching. This paper extends the work of optimal-space indexing to a dynamic setting, i.e., allowing efficient updating and searching a set of texts. The problem we consider is also known as the library management problem [9] in the literature, which is defined as follows: We need to maintain a collection $\mathcal{L}$ of texts; from time to time, a text may be inserted or deleted from $\mathcal{L}$, and a pattern P may be given and its occurrences in $\mathcal{L}$ are to be reported. This problem occurs naturally in the management of homepages [3], DNA/protein sequences [10], and many other real-life applications.

In the static version of the problem, the texts in $\mathcal{L}$ never change. A simple solution is to concatenate all the texts and then build a suffix tree [9, 13] or a suffix array[8]; the searching time for a pattern P with length p is $O(p + occ)$ and $O(p + \log n + occ)$, respectively, where occ is the total number of occurrences of P in $\mathcal{L}$. Note that a suffix tree of a string is a compact trie containing all suffixes of the string, while a suffix array is an array of all suffixes of the string arranged in lexicographical order. Both indexes occupy $O(n \log n)$ bits storage. For indexing the huge amount of web pages or DNA/protein sequences, this space requirement may be too demanding. For example, the suffix tree for the human genome (totally 3G characters) takes 40G bytes of memory, while the suffix array takes 13G bytes [6].

---

* This work was supported in part by the HKU RGC Grant HKU-7042/02E.

S.C. Sahinalp et al. (Eds.): CPM 2004, LNCS 3109, pp. 445–456, 2004.

Recently, two exciting results have been made on providing indexes occupying only $O(n)$ bits, yet supporting efficient pattern searching. Essentially, both of them can be considered as another form of the suffix array, storing in a compact manner. The first one is the compressed suffix arrays (CSA) [4] proposed by Grossi and Vitter, which supports pattern searching in $O(p \log n + occ \log^\epsilon n)$ time, for any fixed $\epsilon > 0$. The second one is the FM-index [2] by Ferragina and Manzini, with which pattern searching can be done in $O(p + occ \log^\epsilon n)$ time. These data structures are also sound in practice. Using CSA or FM-index, one can index the human genome with 1.5G bytes of memory [5].

For a dynamic collection $\mathcal{L}$ of texts, texts can be inserted into or deleted from $\mathcal{L}$. If $O(n \log n)$ bits of space is allowed, one can build a generalized suffix tree (i.e., a single compact trie containing the suffixes of each text in $\mathcal{L}$). Then, to insert or delete a text of length t in $\mathcal{L}$, we update the generalized suffix tree by adding or removing all suffixes of the text, which can be done in $O(t)$ time. For searching a pattern P, the time remains $O(p + occ)$.

To reduce space, one may attempt to 'dynamize' a compressed index such as CSA or FM-index. Indeed, Ferragina and Manzini have demonstrated in [2] how to maintain multiple FM-indexes so as to support a dynamic collection of texts. Their solution requires $O(n + m \log n)$ bits, where m is the number of texts in the collection. Pattern matching is slowed down slightly, using $O(p \log^3 n + occ \log n)$ time. But insertion and deletion has only an amortized performance guarantee; precisely, insertion and deletion of a text of length t take $O(t \log n)$ and $O(t \log^2 n)$ amortized time, respectively. Note that in the worst case, a single insertion or deletion may require re-constructing many of the FM-indexes, using $\Theta(n / \log^2 n)$ time even if t is very small.

Our Results. In this paper, we introduce a compressed index for the dynamic library management problem, which requires only $O(n)$ bits. Inserting or deleting a text of length t takes $O(t \log n)$ time, while searching for a pattern takes $O(p \log n + occ \log^2 n)$ time. Note that the time complexities of all operations are measured in the worst case (instead of the amortized case). To our knowledge, this is the first result that requires only $O(n)$ bits, yet supporting both update and searching efficiently, i.e., in $O(t \log^{O(1)} n)$ and $O((p + occ) \log^{O(1)} n)$ time, respectively.

Technically speaking, our compressed index is based on CSA and FM-index. Yet a few more techniques are needed in order to achieve the optimal space requirement and efficient updating. Firstly, recall that the index proposed in [2] requires $O(n + m \log n)$ bits. We have a simple but useful trick in organizing the texts to avoid using a lot of space when the collection involves a lot of very short strings, thus eliminating the $m \log n$ term. Secondly, the original representations of CSA and FM-index do not support updates efficiently. For instance, the index proposed in [2], essentially requires re-building one or more FM-index whenever a text is inserted. Inspired by a dynamic representation of CSA in [7], we manage to dynamize the CSA and the FM-index to support efficient updates to a collection of texts. With either of them, we can immediately obtain an $O(n)$-bit index that supports updates in $O(t \log^2 n)$ time. For pattern matching, using FM-index alone can achieve $O(p \log n + occ \log^2 n)$ time, and using CSA alone takes $O(p \log^2 n + occ \log^2 n)$ time.

Last but not the least, we find that FM-index and CSA can complement each other nicely to further improve the update time. Roughly speaking, in the process of updating such suffix-array based compressed indexes, we need two pieces of crucial information; we observe that one of them can be provided quickly by FM-index, and the other can be provided quickly by CSA. Thus, by maintaining both CSA and FM-index together, we can perform the update in a straightforward manner, improving the update time to $O(t \log n)$.

Remarks. (1) In the above discussion, we assume that the alphabet Σ has a constant size. For a variable size alphabet, our compressed index occupies $O(n|\Sigma|)$ bits, which may become a problem if $|\Sigma|$ is huge. Nevertheless, our compressed index based on CSA alone achieves a space complexity of $O(n \log |\Sigma|)$ bits. As noted before, update and pattern searching suffer a slowdown by a logarithmic factor. (2) In practice, for indexing a collection of DNA texts, the data structures suggested in this paper occupy $5n$ or $9n$ bits (corresponding to using CSA alone, or using CSA plus FM-index, respectively).

2   Review on CSA and FM-Index

Let $T[1..n] = T[1]T[2]\cdots T[n]$ be a string of length n over an alphabet Σ. For any $i = 1, \ldots, n$, $T[i..n]$ is a suffix of T. The suffix array $SA[1..n]$ of T is an array of integers such that $T[SA[i]..n]$ is lexicographically the i-th smallest suffix of T.

The main component of CSA is the function $\Psi[1..n]$ where $\Psi[i] = SA^{-1}[SA[i]+1]$. Let i be the lexicographical order of the suffix $T[SA[i]..n]$. $\Psi[i]$ gives the lexicographical order of the suffix $T[SA[i]+1..n]$. Unlike SA, Ψ admits an $O(n)$-bit representation and $O(1)$-time retrieval. We can count the number of occurrences of a pattern P in T using $O(|P| \log n)$ queries to Ψ [4].

The main component of FM-index is the function $count$, which is defined based on the BWT array [1]. For $i = 1, \ldots, n$, $BWT[i]$ is the character $T[SA[i] - 1]$. For each character $c \in \Sigma$ and $i = 1, \ldots, n$, the function $count(c, i)$ is the number of character c appearing in $BWT[1..i]$. Similar to CSA, $count(c, i)$ admits an $O(n)$-bit representation and $O(1)$-time retrieval, and we can count the number of occurrences of a pattern P in T using $O(|P|)$ queries to $count$ [2]. See the figure below for an example of the Ψ, BWT and $count$ functions.

$$T = \texttt{\$ a b b a a a b a}$$

i	suffixes in sorted order	$SA[i]$	$\Psi[i]$	$BWT[i]$	$count(\texttt{a}, i)$	$count(\texttt{b}, i)$
1	\$ a b b a a a b a	1	6	a	1	0
2	a	9	1	b	1	1
3	a a a b a	5	4	b	1	2
4	a a b a	6	5	a	2	2
5	a b a	7	7	a	3	2
6	a b b a a a b a	2	9	\$	3	2
7	b a	8	2	a	4	2
8	b a a a b a	4	3	b	4	3
9	b b a a a b a	3	8	a	5	3

The two lemmas below give the relationship between the Ψ function of CSA and the *count* function of FM-index.

Lemma 1. *Consider a text $T[1..n]$ over the alphabet Σ. For any $c \in \Sigma$ and $i = 1, \ldots, n$, we can compute $count(c, i)$ using $O(\log n)$ queries to Ψ.*

Proof. Observe that $T[SA[j]] = c$ if and only if $\text{BWT}[\Psi[j]] = c$. In other words, $count(c, i)$ is the number of j such that $T[SA[j]] = c$ and $\Psi[j] \leq i$.

Note that $T[SA[j]] = c$ means that j is in the consecutive region of SA whose suffixes have the first character c. Thus, $count(c, i)$ is the number of j in the region with $\Psi[j] \leq i$.

As shown in [12], the Ψ values are increasing in the above region for j. This implies that $count(c, i)$ can be found by a binary search on Ψ in the above region, using $O(\log n)$ queries.

Lemma 2. *Consider a text $T[1..n]$ over the alphabet Σ. For any $i = 1, \ldots, n$, we can compute $\Psi[i]$ using $O(\log n)$ queries to the count function.*

Proof. Consider the suffix $T[SA[i]..n]$. Let c be its first character and y be the lexicographical order of which among all suffixes that begins with c. This implies that $\text{BWT}[\Psi[i]] = c$, and $\Psi[i]$ is the y-th entry in BWT which is c. This means that $\Psi[i]$ is the first entry in $count(c, \cdot)$ array whose value is y. Thus, $\Psi[i]$ can be found based on binary search on $count(c, \cdot)$, using $O(\log n)$ queries.

Next, we state a lemma to demonstrate the searching ability provided by the *count* function.

Lemma 3. *([2]) Consider a text $T[1..n]$ over an alphabet Σ. Let P be any pattern and let c be any character over Σ. Denote the lexicographical order of P among all suffixes of T (i.e., $1 +$ the number of such suffixes less than P) as i. Then, $count(c, i - 1) + 1 + offset_c$ is the lexicographical order of cP among all suffixes of T, where $offset_c$ denotes the number of suffixes of T starting with a character less than c.*

We refer to an execution of the above lemma a *backward search step*. Applying backward search steps repeatedly, we can find the number of occurrences of any pattern P in T using $O(|P|)$ queries to the *count* function. Such a searching method is also known as the *backward search* algorithm in the literature.

3   High Level Organization

This section gives the high level description of a compressed index for a dynamic collection of texts. In particular, we show how to exploit three dynamic data structures, namely, *COUNT*, *MARK*, and *PSI*, to support efficient pattern matching and text updating. The implementation details will be shown in the next section.

Below we first introduce *COUNT*, which is the core data structure that already supports counting the occurrences of a pattern P efficiently, and quick insertion or deletion of texts. Afterwards, we discuss how to exploit $MARK$ and PSI to support efficient enumeration of the positions where a pattern P occurs, and further speed up the updating process.

Consider a set of texts $\mathcal{L} = \{T_1, T_2, \ldots, T_m\}$ over a finite alphabet Σ. We assume that the texts are distinct, and each text T starts with a special character '$' in Σ, which is alphabetically smaller than all other characters in Σ and '$' does not appear in any other part of a text. Denote the total length of texts as n. We always label (relabel) the existing texts in $\mathcal{L}$ in such a way that T_j refers to the lexicographically j-th texts currently in $\mathcal{L}$.

Conceptually, we want to construct a suffix array SA for the texts by listing out all suffixes of all texts in lexicographical order. For $i = 1, 2, \ldots, n$,

$$SA[i] = (j, \ell)$$

if the suffix $T_j[\ell..]$ is the lexicographically i-th suffix. To insert a text T to $\mathcal{L}$, we insert all suffixes of T into the SA. Similarly, to delete a text from $\mathcal{L}$, we delete all suffixes of T from SA. Searching for a pattern P is done by determining the interval $[x, y]$ such that each suffix from $SA[x]$ up to $SA[y]$ has P as a prefix. $SA[x], SA[x+1], \ldots, SA[y]$ gives all locations where P occurs in $\mathcal{L}$. However, the SA table needs $O(n \log n)$ bits.

3.1 Basic Data Structure

Due to the space restriction, we cannot directly store the SA table. Instead, we use the FM-index, which requires only $O(n)$ bits, to represent the SA table implicitly. Recall that FM-index requires the function $count(c, i)$ which returns the total number of occurrences of character c in $BWT[1..i]$. We implement the $count(c, i)$ function with a dynamic data structure $COUNT$ and its performance is summarized in the lemma below. The proof (i.e., the detailed construction of $COUNT$) will be given in Section 4.

Lemma 4. *We can maintain the $COUNT$ data structure using $O(n)$ bits space such that each of the following operations is supported in $O(\log n)$ time.*

- *Report(c, i): Returns the value of $count(c, i)$.*
- *Insert(c, i): Updates $count(\cdot, \cdot)$ due to a character c inserted to position i of BWT.*
- *Delete(i): Updates $count(\cdot, \cdot)$ due to a character deleted from position i of BWT.*

Note that $COUNT$ allows efficient updates which is needed when the BWT array changes due to insertion or deletion of texts. Our implementation is different from the original one in [2], where the $count(c, i)$ function is stored in a data structure which is difficult to update (but allows constant time query).

Pattern Matching. We maintain an array FC such that $FC[c]$ stores the frequency count of the character c in $\mathcal{L}$, which equals the number of suffixes whose first character is c. Together with the $COUNT$ data structure, we can support counting the occurrences of $P[1..p]$ in $\mathcal{L}$ using $O(p \log n)$ time as follows. Firstly, the lexicographical order of $P[p]$ can be computed by $\sum_{c < P[p]} FC[c] + 1$. Then, by Lemma 3, we can find the lexicographical order of $P[p-1..p]$ using one query to the *count* function. The process is repeated, so that eventually, we can find the lexicographical order (say, x) of $P[1..p]$. Similarly, we can find the lexicographical order (say, y) of $P[1..p-1]c$, where c is the character in Σ just greater than $P[p]$. We can easily see that the number of occurrences of P is $y - x$. On the other hand, the whole process requires $O(p)$ queries to the *count* function, and each query takes $O(\log n)$ time. Thus, the total time follows.

Text Insertion. To insert a text $T[1..t]$, we insert all the suffixes of T to SA implicitly, starting from the shortest one. The lexicographical order of $T[t]$, denoted as i, is $1 +$ number of suffixes in SA that starts with a character less than $T[t]$. Thus, we want to insert $T[t]$ at the i-th row SA implicitly. To do so, we insert the character $T[t-1]$ to the i-th position of the BWT array, using the function $Insert(T[t-1], i)$ provided by $COUNT$. Then, let i' be the lexicographical order of $T[t-1..t]$, which can be found easily by one backward search step, and we insert $T[t-2]$ to the i'-th position of BWT. The process continues until the longest suffix $T[1..t]$ is inserted implicitly to SA, which is simulated by inserting $T[1]$ to BWT. The whole process takes $O(t \log n)$ time.

Text Deletion. Deleting a text $T[1..t]$ from the collection of texts is more troublesome because among all those suffixes that is a single character $T[t]$, we do not know which one belongs to T. To handle the problem, we perform a backward search for $T[1..t]$ and let $[x, y]$ be the interval such that for any $i \in [x, y]$, $T[1..t]$ is a prefix of $SA[i]$. Recall that all texts in the collection are distinct and each of them starts with a special character '\$' which is alphabetically smaller than all other characters. Thus, we can conclude that $SA[x]$ corresponds to the text $T[1..t]$ to be deleted because no other text can have $T[1..t]$ as a prefix and have lexicographical order less than $T[1..t]$. Then, we delete all suffixes of $T[1..t]$, starting from the longest one. Note that if the Ψ function of CSA is given, we can determine the lexicographical order of $T[2..t]$ easily from the lexicographical order $T[1..t]$. However, as the Ψ function is not available, we need to simulate each query to Ψ by $O(\log n)$ queries to the *count* function. Since a query to *count* takes $O(\log n)$ time, deleting each suffix of $T[1..t]$ takes $O(\log^2 n)$ time, and the whole process takes $O(t \log^2 n)$ time.

Summarizing the discussion, we have the following theorem.

Theorem 1. *Let $\mathcal{L} = \{T_1, T_2, \cdots, T_k\}$ be a set of k distinct strings over a finite alphabet Σ. Let n be the total length of all strings in $\mathcal{L}$. We can maintain $\mathcal{L}$ in $O(n)$-bit space such that counting the occurrences of a pattern $P[1..p]$ takes $O(p \log n)$ time, inserting a text $T[1..t]$ takes $O(t \log n)$ time, and deleting a text $T[1..t]$ takes $O(t \log^2 n)$ time.*

3.2   Additional Data Structures

The basic data structure in the previous discussion does not support retrieving $SA[x]$ efficiently and thus cannot report the positions where a pattern occurs. In

the following, we give an additional data structure for such a purpose. Afterwards, we introduce another data structure that is based on Ψ to speed up the
deletion process.

Enumerating Pattern Occurrences. To compute $SA[x]$ for any given x efficiently, we use the data structure $MARK$ to selectively store some of the entries
of SA as follows: $MARK$ stores the entries $SA[i] = (j, \ell)$ whenever $\ell > 0$ is an
integral multiple of $\log n$.

Recall that we require all texts in $\mathcal{L}$ to start with the '$\$$' character which
is lexicographically smaller than any other character in Σ. As a result, for a
set of m texts, the first m entries of SA corresponds to the m texts sorted in
lexicographical order. Now, suppose that given a certain x, we want to find the
value of $SA[x]$ (which is actually (j, ℓ)). We first check whether $SA[x]$ is stored
in $MARK$. If so, we obtain the value of $SA[x]$ immediately. Otherwise, we check
whether $x \leq m$, which would imply that $SA[x] = (x, 1)$. If both cases are false,
we can determine the lexicographical order of the suffix $T_j[\ell - 1..]$, denoted as x',
easily using backward search with the $COUNT$ data structure. We check whether
the entry $SA[x']$ is stored in $MARK$ or $x' \leq m$. The process continues and after
$k \leq \log n$ steps, we will either meet a suffix $T_j[\ell - k..]$ such that $\ell - k$ is a multiple
of $\log n$, or $\ell - k = 1$. In both cases, the value of (j, ℓ) can be found accordingly.
As shown in Lemma 5, for any value x, testing whether $MARK$ stores the tuple
$SA[x]$ takes $O(\log n)$ time. Thus, it takes $O(\log^2 n)$ time to find the value of $SA[x]$
for any value x.

Note that $MARK$ should allow updates easily when texts are inserted to
or deleted from the dynamic collection of texts. In particular, when a suffix
is inserted to or deleted from SA, originally stored tuples (j, ℓ) corresponds to
different SA entries. For example, when a suffix is inserted at position u of SA, a
tuple (j, ℓ), corresponding to $SA[v]$ originally, becomes the a tuple corresponding
to $SA[v + 1]$ if $v > u$. That is, we need to update the correspondence for tuples
and SA values whenever a suffix is inserted or deleted from SA. We summarize
the performance of $MARK$ in the lemma below. Note that $MARK$ stores at most
$\frac{n}{\log n}$ entries of SA. We give the actual construction of $MARK$ in Section 4.

Lemma 5. *We can maintain a data structure $MARK$ in $O(n)$ bits such that it
stores a tuple $(i, (j, \ell) = SA[i])$ for all entries $SA[i]$ whenever ℓ is a multiple of
$\log n$. $MARK$ supports each of the following operations in $O(\log n)$ time.*

- *Report(i): Returns the $(i, (j, \ell))$ if this tuple is stored. Else, return false.*
- *Insert(i, j, \ell): Inserts the tuple $(i, (j, \ell))$ to $MARK$.*
- *Delete(x): Deletes the tuple $(i, (j, \ell))$ from $MARK$.*
- *Increment_lexico(k): For each tuple stored, the j value is incremented by one
  if the original j value is at least k. This function allows us to update the
  lexicographical order of the texts after a new text with lexicographical order
  k is inserted.*
- *Decrement_lexico(k): For each tuple stored, the j value is decrement by one
  if the original j value is greater than k.*

- $Shift\_up(k)$: For each tuple stored, the i value is incremented by one if the original i value is at least k. This function allows us to update the correspondence between tuples and SA after a suffix is insert to position k of SA.
- $Shift\_down(k)$: For each tuple stored, the i value is decremented by one if the original i value is greater than k.

With $COUNT$ and $MARK$, we can find the positions where a pattern $P[1..p]$ occurs in the collection of texts in $O(p \log n + occ \log^2 n)$ time.

Speeding up the Deletion. Recall that to delete a text $T[1..t]$, we first determine the location of $T[1..t]$ in SA. Then, we delete all the suffixes of T starting from the longest one. The bottleneck for the deletion operation is determining the lexicographical order of $T[r+1..t]$ after the deletion of the suffix $T[r..t]$. We observe that CSA provides a good solution for it. In fact, the Ψ function of CSA stores exactly the information we needed.

However, we cannot use the original implementation of Ψ as we need to update Ψ efficiently. We dynamize Ψ with the data structure PSI, whose performance is summarized in the lemma below. Recall that Ψ is a list of n integers defined as follows. Given an integer i, let $S[r..s]$ is the lexicographically i-th suffix in SA. $\Psi[i]$ is the lexicographical order of the suffix $S[r+1..s]$ in SA.

Lemma 6. *We can maintain the PSI data structure in $O(n)$ bits such that each of the following operations can be done in $O(\log n)$ time.*

- $Report(i)$: Return $\Psi(i)$.
- $Insert(i, x)$: Insert the integer x to position i of the list of integers. This function is needed when we insert a suffix to SA.
- $Delete(i)$: Delete the integer from position i of the list.
- $Shift\_up(k)$: Increment by one for each integer in the list with value at least k. This function is needed when we insert a suffix to position i of SA.
- $Shift\_down(k)$: Decrement by one fore each integer in the list with value greater than k.

With the PSI data structure, insertion and deletion of a text of length t can both be done in $O(t \log n)$ time.

3.3   Summary

We summarize how the search, insert and delete operations are performed with $COUNT$, $MARK$, and PSI.

Searching for a Pattern $P[1..p]$. We perform backward search to determine the interval $[x, y]$ such that for each $i \in [x, y]$, SA$[i]$ corresponds to an occurrence of P. This can be done in $O(p \log n)$ time using the $COUNT$ data structure. Then, for each $i \in [x, y]$, the value of SA$[i]$ is obtained by at most $\log n$ backward search steps, with one query to $MARK$ in each step. Thus, the time is $O(p \log n + occ \log^2 n)$.

Inserting a Text $T[1..t]$. Intuitively, we insert each suffix of T to SA starting from the shortest one. For $r = t, t-1, \ldots, 1$, we first determine the lexicographical order of $T[r..t]$ in SA and let that value be k. To simulate the effect of inserting $T[r..t]$ into position k of SA, we need to update $COUNT$ by inserting $T[r-1]$ to position k of BWT. Then, we need to update PSI by incrementing all integers in PSI whose value at least k, followed by inserting the lexicographical order of $T[r+1..t]$ to the i-th position of PSI. We also need to increment the i value for any tuples in $MARK$ whose i value is at least k. Note that the longest suffix will be inserted to some position x of SA, where x is the lexicographical order of T in the collection of texts. Therefore, for each tuple in $MARK$ with j value at least x, we increment its j value by one. Finally, we insert tuples corresponding to T to $MARK$. The total time required is $O(t \log n)$.

Deleting a Text $T[1..t]$. Intuitively, we delete each suffix of T starting from the longest one. We first determine the lexicographical order of T in the collection of texts. Afterwards, the lexicographical order of the other suffixes of T can be found using the PSI. Updating of $COUNT$, $MARK$, and PSI are done similarly to that of inserting a text, except that we are decrementing the values this time. The total time required is $O(t \log n)$ as well.

Adjustment Due to Huge Updates. Note that in the above discussion, our data structures require the value of $\log n$ as a parameter, and we have assumed that this value is fixed over the time. This is not true in general as texts can be inserted or deleted in the collection. Thus, when the value of $\log n$ changes, our data structures become different. A simple way to handle this is to reconstruct everything when necessary, but this would imply huge update time, say, $O(n)$ time, on the single update operation that induces the change. Nevertheless, using a standard technique for global rebuilding [11], we can distribute the reconstruction process over each update operation, so that we can bound the update time to be $O(t \log n)$, while having a new data structure ready when $\log n$ is changed. Details will be shown in the full paper.

Summarizing the results, we have the following theorem.

Theorem 2. *Let $\mathcal{L} = \{T_1, T_2, \cdots, T_k\}$ be a set of k distinct strings over a finite alphabet Σ. Let n be the total length of all strings in $\mathcal{L}$. We can maintain $\mathcal{L}$ in $O(n)$-bit space such that inserting or deleting a text $T[1..t]$ takes $O(t \log n)$ time and searching for a pattern $P[1..p]$ takes $O(p \log n + occ \log^2 n)$ time, where occ is the total of occurrences.*

4   Implementing the Data Structures

In this section, we explain how each of data structure $COUNT$, $MARK$, and PSI are implemented.

4.1   *COUNT*

Recall that the $COUNT$ data structure maintains the function $count(c, i)$, which returns the total number of occurrences of the character c in BWT$[1..i]$. To implement $COUNT$, we store $|\Sigma|$ lists of bits, denoted as $COUNT_c$ for each $c \in \Sigma$. Each list is n bits long and $COUNT_c[i] = 1$ if $\text{BWT}[i] = c$ and $COUNT_c[i] = 0$ otherwise.

To support updates easily, for each list $COUNT_c$, we partition it into segments of $5 \log n$ to $10 \log n$ bits long. The segments are stored in a red-black tree, so that a left to right traversal of the tree gives the list $COUNT_c$. Precisely, each node u in the tree contains the following fields.

- A color bit (red or black), a pointer to parent, a pointer to the left child and a pointer to the right child.
- A segment of bits, with length $5 \log n$ to $10 \log n$.
- An integer $size$ indicating the total number of bits contained in the subtree rooted at u.
- An integer sum indicating the total number of 1 contained in the subtree rooted at u.

To support the function $count(c, i)$, we search the tree of $COUNT_c$ for the node u that contains the i-th bit. We record the number of 1's in the segment of u preceding the i-th bit, and also the sum of the left child of u. Then, we traverse from u to the root. For every left parent v on the path, we record the number of 1's in the segment of v and also the sum of the left child of v. Summing up all these recorded values gives the number of 1's in the list of $COUNT_c$ up to the i-th bit, which equals $count(c, i)$. The whole process takes $O(\log n)$ time.

To update the $COUNT$ data structure when a character c is inserted to position i of BWT column, we insert a bit 1 to position i of $COUNT_c$ and insert a bit 0 to position i of $COUNT_{c'}$ for each $c' \neq c$. The time required is $O(\log n)$. Deletion of a character from BWT can be done in the opposite way in $O(\log n)$ time.

For the space requirement, we note that each node takes $O(\log n)$ bits and there are $O(\frac{n}{\log n})$ nodes. Thus, the space requirement is $O(n)$ bits.

4.2   MARK

Recall that $MARK$ is a set of at most $\frac{n}{\log n}$ tuples, each in the format of (i, j, ℓ). Note that no two tuples have the same i value, but there may be more than one tuple having the same j value.

To support efficient update, we maintain two red-black trees, one for the i values and the other for the j and ℓ values as follows.

For all the i values of the tuples, they are stored in a red-black tree R_i, such that the left to right traversal of the tree gives the i values of the tuples stored in sorted order.

For all the j values of the tuples (allowing duplication), they are stored in a red-black tree, denoted as R_j, such that the left to right traversal of the tree gives the j values of all the tuples stored in sorted order.

Let $i(u)$ be the i value stored in the node u in R_i. Let $j(v)$ be the j value stored in the node v in R_j. To represent the tuples, we store a pointer at each node u in R_i, pointing to a node v in R_j if $i(u)$ and $j(v)$ belongs to the same tuple. Furthermore, the ℓ value of the tuple $(i(u), j(v), \ell)$ is stored in the node v in R_j.

More precisely, each node in R_i has the following fields.

- A color bit (red or black), a pointer to the left child and a pointer to the right child.
- An integer $diff(u) = i(u) - i(lp(u))$, where $lp(u)$ denotes the left parent of u. $i(lp(u)) = 0$ if $lp(u)$ does not exist.
- A pointer to a node v in R_j.

Each node in R_j has the following fields.

- A color bit (red or black), a pointer to the left child and a pointer to the right child.
- An integer $diff(v) = j(v) - j(lp(v))$, where $lp(v)$ denotes the left parent of u. $j(lp(v)) = 0$ if $lp(v)$ does not exist.
- A pointer to a node u in R_i.
- An integer ℓ.

Although we do not store the value $i(u)$ explicitly for every node $u \in R_i$, its value can be recovered when we traverse down the tree R_i starting from the root. More precisely, when we traverse down the tree, for every node x we meet on the path, we can compute the values $lp(x)$ and $i(x)$ in constant time, as follows. Let x' be the parent of x and assume that $lp(x')$ and $i(lp(x'))$ are known. If x is the left child of x', $lp(x) = lp(x')$. Else, $lp(x) = x'$. In both cases, $i(x) = i(lp(x)) + diff(x)$.

Note that R_i and R_j are very similar to a red-black tree and they inherit the advantages of a balanced binary search tree. Searching, inserting and deleting a tuple can be done easily in $O(\log n)$ time.

For any integer k, let $X_k = \{u | u \in R_i \text{ and } i(u) \geq k\}$. Recall that we need to support the function that given k, we increment $i(u)$ by 1 for all $u \in X_k$. We observe that the actual value of $i(u)$ is not stored for every $u \in R_i$. Instead, we store $diff(u) = i(u) - i(lp(u))$. Thus, if the value $i(lp(u))$ is increased by 1, the value $i(u)$ is also increased by 1 automatically. To increment $i(u)$ by 1 for all $u \in X_k$, we search R_i for the node u with smallest $i(u)$ such that $i(u) \geq k$. Then, for any node $w \neq u$ on the path from root to u, we increment the value $diff(w)$ by 1 if w is a right parent of some other node on the path. We increment $diff(u)$ by 1 if u is the left child of some other node on the path. It can be verified that $i(u)$ is incremented by 1 for all $u \in X_k$. The process takes $O(\log n)$ time.

4.3   PSI

The *PSI* data structure maintains the Ψ function where for $i = 1, 2, \ldots, n$, $\Psi[i] = \text{SA}^{-1}[\text{SA}[i] + 1]$. As shown in [12], $\Psi[1..n]$ can be partitioned into $|\Sigma|$ increasing sequences. We store each sequence in a separate data structure PSI_c for each $c \in \Sigma$. To support easy insertion or deletion of integers, each PSI_c is based on a red-black tree. The exact construction is similar to that in [7]. We omit the proof here.

References

1. M. Burrows and D. J. Wheeler. A Block-sorting Lossless Data Compression Algorithm. Technical Report 124, Digital Equipment Corporation, Paolo Alto, California, 1994.
2. P. Ferragina and G. Manzini. Opportunistic Data Structures with Applications. In *Proceedings of Symposium on Foundations of Computer Science*, pages 390–398, 2000.
3. The Google Homepage Search Engine. http://www.google.com/.
4. R. Grossi and J. S. Vitter. Compressed Suffix Arrays and Suffix Tree with Applications to Text Indexing and String Matching. In *Proceedings of Symposium on Theory of Computing*, pages 397–406, 2000.
5. W. K. Hon, T. W. Lam, W. K. Sung, W. L. Tse, C. K. Wong and S. M. Yiu. Practical Aspects of Compressed Suffix Arrays and FM-index in Searching DNA Sequences. To appear in *Proceedings of Workshop on Algorithm Engineering and Experiments*, 2004.
6. S. Kurtz. Reducing the Space Requirement of Suffix Trees. *Software Practice and Experience*, 29(13):1149–1171, 1999.
7. T. W. Lam, K. Sadakane, W. K. Sung, and S. M. Yiu. A Space and Time Efficient Algorithm for Constructing Compressed Suffix Array. In *Proceedings of International Conference on Computing and Combinatorics*, pages 401–410, 2002.
8. U. Manber and G. Myers. Suffix Arrays: A New Method for On-Line String Searches. *SIAM Journal on Computing*, 22(5):935–948, 1993.
9. E. M. McCreight. A Space-economical Suffix Tree Construction Algorithm. *Journal of the ACM*, 23(2):262–272, 1976.
10. H. W. Mewes and K. Heumann. Genome Analysis: Pattern Search in Biological Macromolecules. In *Proceedings of Symposium on Combinatorial Pattern Matching*, pages 261–285, 1995.
11. M. H. Overmars. The Design of Dynamic Data Structures. Lecture Notes in Computer Science 156, pages 34–35, 1983.
12. K. Sadakane. Compressed Text Databases with Efficient Query Algorithms based on Compressed Suffix Array. In *Proceedings of International Symposium on Algorithms and Computation*, pages 410–421, 2000.
13. P. Weiner. Linear Pattern Matching Algorithm. In *Proceedings of Symposium on Switching and Automata Theory*, pages 1–11, 1973.

Improved Single
and Multiple Approximate String Matching

Kimmo Fredriksson[1,*] and Gonzalo Navarro[2,**]

[1] Department of Computer Science, University of Joensuu
kfredrik@cs.joensuu.fi
[2] Department of Computer Science, University of Chile
gnavarro@dcc.uchile.cl

Abstract. We present a new algorithm for multiple approximate string matching. It is based on reading backwards enough ℓ-grams from text windows so as to prove that no occurrence can contain the part of the window read, and then shifting the window. Three variants of the algorithm are presented, which give different tradeoffs between how much they work in the window and how much they shift it. We show analytically that two of our algorithms are optimal on average. Compared to the first average-optimal multipattern approximate string matching algorithm [Fredriksson and Navarro, CPM 2003], the new algorithms are much faster and are optimal up to difference ratios of $1/2$, contrary to the maximum of $1/3$ that could be reached in previous work. This is also a contribution to the area of single-pattern approximate string matching, as the only average-optimal algorithm [Chang and Marr, CPM 1994] also reached a difference ratio of $1/3$. We show experimentally that our algorithms are very competitive, displacing the long-standing best algorithms for this problem. On real life texts, our algorithms are especially interesting for computational biology applications.

1 Introduction and Related Work

Approximate string matching is one of the main problems in classical string algorithms, with applications to text searching, computational biology, pattern recognition, etc. Given a text $T_{1...n}$, a pattern $P_{1...m}$, and a maximal number of differences permitted, k, we want to find all the text positions where the pattern matches the text up to k differences. The differences can be substituting, deleting or inserting a character. We call $\alpha = k/m$ the *difference ratio*, and σ the size of the alphabet Σ. For the average case analyses it is customary to assume a random text over a uniformly distributed alphabet.

A natural extension to the basic problem consists of *multipattern searching*, that is, searching for r patterns $P^1 \ldots P^r$ simultaneously in order to report all their occurrences with at most k differences. This has also several applications such as virus and intrusion detection, spelling, speech recognition, optical

---

* Supported by the Academy of Finland.
** Partially supported by Fondecyt grant 1-020831.

S.C. Sahinalp et al. (Eds.): CPM 2004, LNCS 3109, pp. 457–471, 2004.

character recognition, handwriting recognition, text retrieval under synonym or thesaurus expansion, computational biology, multidimensional approximate matching, batch processing of single-pattern approximate searching, etc. Moreover, some single-pattern approximate search algorithms resort to multipattern searching of pattern pieces. Depending on the application, r may vary from a few to thousands of patterns. The naive approach is to perform r separate searches, so the goal is to do better.

The single-pattern problem has received a lot of attention since the sixties [10]. For low difference ratios (the most interesting case) the so-called *filtration algorithms* are the most efficient ones. These algorithms discard most of the text by checking for a necessary condition, and use another algorithm to verify the text areas that cannot be discarded. For filtration algorithms, the two important parameters are the filtration speed and the maximum difference ratio α up to which they work.

In 1994, Chang & Marr [2] showed that the average complexity of the problem is $O((k + \log_\sigma m)n/m)$, and gave the first (filtration) algorithm that achieved that average-optimal cost for $\alpha < 1/3 - O(1/\sqrt{\sigma})$.

The multipattern problem has received much less attention, not because of lack of interest but because of its difficulty. There exist algorithms that search permitting only $k = 1$ difference [8], that handle too few patterns [1], and that handle only low difference ratios [1]. No effective algorithm exists to search for many patterns with intermediate difference ratio. Moreover, as the number of patterns grows, the difference ratios that can be handled get reduced, as the most effective algorithm [1] works for $\alpha < 1/\log_\sigma(rm)$.

Very recently [4], we have proposed the first average-optimal multiple approximate string matching algorithm, with average complexity $O((k + \log_\sigma(rm))n/m)$. It consists of a direct extension of Chang & Marr algorithm to multiple patterns. The algorithm performs well in practice. However, just as Chang & Marr on single patterns, it can only cope with difference ratios $\alpha < 1/3$, and even lower in practice due to memory limitations. Both algorithms are based on reading enough ℓ-grams (substrings of length ℓ) from text windows so as to prove that no occurrence can contain the part of the window read, and then shifting the window.

In this paper we present a new algorithm in this trend. Compared to our previous algorithm [4], the present one is algorithmically more novel, as it is not a direct extension of an existing single pattern algorithm but a distinct approach. We use a sliding window rather than fixed windows. This permits us having a window length of $m - k$ instead of $(m - k)/2$, which is the crux of our ability to handle higher difference ratios. In order to maximize shifts on a sliding window, we have to read it backwards, which was not an issue in [2, 4]. In this sense, our algorithm inherits from ABNDM algorithm [11]. We present also two variants of our algorithm, which trade amount of work done inside a window by amount of window shifting obtained.

The analysis of the new algorithms is more difficult than previous ones [2, 4]. We manage to prove that two of our new algorithms are also average-optimal.

However, they are superior to previous ones because they are optimal for difference ratios up to $\alpha < 1/2$, while previous work permitted up to $\alpha < 1/3$. Hence, the new algorithms not only improve our previous algorithm but also, in the area of single pattern matching, are better than what was achieved by any previous average-optimal algorithm.

In practice, our new algorithms are much faster than our previous algorithm [4], and they are also faster than all other multipattern search algorithms for a wide range of parameters that include many cases of interest in computational biology (DNA and protein searching). Moreover, our new algorithms are also relevant in the much more competitive area of single pattern matching algorithms, where they turn out to be the fastest for low difference ratios.

2 Our Algorithm

2.1 Basic Version

Given r search patterns $P^1 \ldots P^r$, the preprocessing fixes value ℓ (to be discussed later) and builds a table $D : \Sigma^\ell \to \mathbb{N}$ telling, for each possible ℓ-gram, the minimum number of differences necessary to match the ℓ-gram *anywhere* inside *any* of the patterns.

The scanning phase proceeds by sliding a window of length $m - k$ over the text. The invariant is that any occurrence starting before the window has already been reported. For each window position $i + 1 \ldots i + m - k$, we read successive ℓ-grams backwards: $S^1 = T_{i+m-k-\ell+1\ldots i+m-k}$, $S^2 = T_{i+m-k-2\ell+1\ldots i+m-k-\ell}$, $\ldots$, $S^t = T_{i+m-k-t\ell+1\ldots i+m-k-(t-1)\ell}$, and so on. Any occurrence starting at the beginning of the window must fully contain those ℓ-grams. We accumulate the D values for the successive ℓ-grams read, $M_u = \sum_{1 \le t \le u} D[S^t]$. If, at some point, the sum M_u exceeds k, then it is not possible to have an occurrence containing the sequence of ℓ-grams read $S^u \ldots S^1$, as merely matching those ℓ-grams inside *any* pattern in *any* order needs more than k differences.

Therefore, if at some point we obtain $M_u > k$, then we can safely shift the window to start at position $i + m - k - u\ell + 2$, which is the first position not containing the ℓ-grams read.

On the other hand, it might be that we read all the $U = \lfloor (m - k)/\ell \rfloor$ ℓ-grams fully contained in the window without surpassing threshold k. In this case we must check the text area of the window with a non-filtration algorithm, as it might contain an occurrence. We scan the text area $T_{i+1\ldots i+m+k}$ for each of the r patterns, so as to cover any occurrence starting at the beginning of the window and report any match found. Then, we shift the window by one position and resume the scanning.

This scheme may report the same ending position of occurrence several times, if there are several starting positions for it. A way to avoid this is to remember the last position scanned when verifying the text, so as to prevent retraversing the same text areas but just restarting from the point we left the last verification. We use Myers' algorithm [9] for the verification of single patterns, which makes the cost $O(m^2/w)$ per pattern, w being the number of bits in the computer word. Otherwise, the standard $O(m^2)$ dynamic programming could be used [12].

Several optimizations are proposed over our previous algorithm [4]: optimal choice of ℓ-grams, hierarchical verification, reduction of preprocessing costs, and bit-parallel packing of counters. All these can be equally applied to our new algorithm, but we have omitted this discussion for lack of space. In the experiments, however, we have applied all these improvements except the optimal choice of ℓ-grams, that did not work well in practice. The preprocessing cost with hierarchical verification is $O(r\sigma^\ell m/w)$.

2.2   A Stricter Matching Condition

In the basic algorithm we permit that the ℓ-grams match anywhere inside the patterns. This has the disadvantage of being excessively permissive. However, it has an advantage: When the ℓ-grams read accumulate more than k differences, we know that no pattern occurrence can contain them *in any position*, and hence can shift the window next to the first position of the leftmost ℓ-gram read. We show now how the matching condition can be made stricter without losing this property.

First consider S^1. In order to shift the window by $m-k-\ell+1$, we must ensure that S^1 cannot be contained in any pattern occurrence, so the condition $D[S^1] > k$ is appropriate. If we consider $S^2 : S^1$, to shift the window by $m - k - 2\ell + 1$, we must ensure that $S^2 : S^1$ cannot be contained inside any pattern occurrence. The basic algorithm uses the sufficient condition $D[S^1] + D[S^2] > k$.

However, a stricter condition can be enforced. In an approximate occurrence of $S^2 : S^1$ inside the pattern, where pattern and window are aligned at their initial positions, S^2 cannot be closer than ℓ positions from the end of the pattern. Therefore, for S^2 we precompute a table D_2, which considers its best match in the area $P_{1...m-\ell}$ rather than $P_{1...m}$. In general, S^t is input to a table D_t, which is preprocessed so as to contain the number of differences of the best occurrence of its argument inside $P_{1...m-(t-1)\ell}$, for any of the patterns P. Hence, we will shift the window to position $i + m - k - u\ell + 2$ as soon as we find the first (that is, smallest) u such that $M_u = \sum_{t=1}^{u} D_t[S^t] > k$.

The number of tables is $U = \lfloor (m - k)/\ell \rfloor$ and the length of the area for D_U is at most $2\ell - 1$. Fig. 1(a) illustrates.

Since $D_t[S] \geq D[S]$ for any t and S, the smallest u that permits shifting the window is never smaller than for the basic method. This means that, compared to the basic method, this variant never examines more ℓ-grams, verifies more windows, nor shifts less. So this variant can never work more than the basic algorithm, and usually works less. In practice, however, it has to be shown whether the added complexity, preprocessing cost and memory usage of having several D tables instead of just one, pays off. We consider this issue in Section 4. The preprocessing cost is increased to $O(r(\sigma^\ell(m/w + m) + U)) = O(r\sigma^\ell m)$.

2.3   Shifting Sooner

We now aim at abandoning the window as soon as possible. Still the more powerful variant developed above is too permissive in this sense. Actually, the ℓ-grams

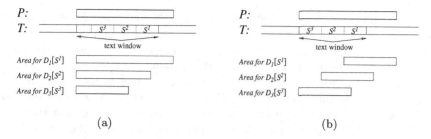

Fig. 1. Pattern P aligned over a text window. (a): The text window and ℓ-grams for the algorithms of Secs. 2.1 and 2.2. The window is of length $m - k$. The areas correspond to the Sec. 2.2. All the areas for S^t for the basic algorithm are the same as for D_1. (b): The areas for the algorithm of Sec. 2.3. The areas overlap by $\ell + k - 1$ characters. The window length is $U\ell$.

should match the pattern more or less at the same position they have in the window.

This fact has been previously used in algorithm LAQ [13]. The idea is that we are interested in occurrences of P that start in the range $i - \ell + 1 \ldots i + 1$. Those that start before have already been reported and those that start after will be dealt with by the next windows. If the current window contains an approximate occurrence of P beginning in that range, then the ℓ-gram at window position $(t - 1)\ell + 1 \ldots t\ell$ can only match inside $P_{(t-1)\ell+1 \ldots t\ell+k}$.

We have $U = \lfloor (m - k - \ell + 1)/\ell \rfloor$ tables D_t, one per ℓ-gram position in the window. Table D_{U-t+1} gives distances to match the t-th window ℓ-gram, so it gives minimum distance in the area $P_{(t-1)\ell+1 \ldots t\ell+k}$ instead of in the whole P, see Fig. 1(b).

At a given window, we compute $M_u = \sum_{0 \le t < u} D_t[S^t]$ until we get $M_u > k$ and then shift the window. It is clear that D_t is computed over a narrower area that before, and therefore we detect sooner that the window can be shifted. The window is shifted sooner, working less per window.

The problem this time is that it is not immediate how much can we shift the window. The information we have is only enough to establish that we can shift by ℓ. Hence, although we shift the window sooner, we shift less. The price for shifting sooner has been too high.

A way to obtain better shifting performance resorts to bit-parallelism. Values $D_t[S]$ are in the range $0 \ldots \ell$. Let us define l, as the number bits necessary to store one value from each table D_t. If our computer word contains at least Ul bits, then the following scheme can be applied.

Let us define table $D[S] = D_1[S] : D_2[S] : \ldots : D_U[S]$, where the U values have been concatenated, giving l bits to each, so as to form a larger number (D_1 is in the area of the most significant bits of D). Assume now that we accumulate $M_u = \sum_{t=1}^{u} (D[S^t] << (t-1)l)$, where "$<< (t-1)l$" shifts all the bits to the left by $(t - 1)l$ positions, enters zero bits from the right, and discards the bits that fall to the left. The leftmost field of M_u will hold the value $\sum_{t=1}^{u} D_t[S^t]$, that is, precisely the value that tells us that we can shift the window when it exceeds k. Similar idea was briefly proposed in [13], but their (single pattern) proposal was based on direct extension of Chang & Marr algorithm [2].

In general, the s-th field of M_u, counting from the left, contains the value $\sum_{t=s}^{u} D_t[S^{t-s+1}]$. In order to shift by ℓ after having read u ℓ-grams, we need to ensure that the window $S^U \ldots S^1$ contains more than k differences. A sufficient condition is the familiar $\sum_{t=1}^{u} D_t[S^t] > k$, that is, when the leftmost field exceeds k. In order to shift by 2ℓ, we need also to ensure that the window $S^{U-1} \ldots S^1 S^0$ contains more than k differences. Here S^0 is the next ℓ-gram we have not yet examined. A lower bound to the number of differences in that window is $\sum_{t=2}^{u} D_t[S^{t-1}] > k$, where the summation is precisely the content of the second field of M_u, counting from the left. In general, we can shift by $s\ell$ whenever all the s leftmost fields of M_u exceed k.

Let us consider the value for l. The values in D_t are in the range $0 \ldots \ell$. In our M_u counters we are only interested in the range $0 \ldots k + 1$, as knowing that the value is larger than $k + 1$ does not give us any further advantage. We will manage to keep our counters in the range $0 \ldots k + 1$, so after computing $M_{u+1} = M_u + (D[S^{u+1}] << ul)$, the counters may have reached $k + 1 + \ell$. So we define $l = 1 + \lceil \log_2(k + \ell + 2) \rceil$. We actually store value $2^{l-1} - (k+1) + c$, where c is the value of the counter. Therefore, the counters activate their highest bits whenever they exceed k, and they still can reach $k + \ell + 1$ without overflowing. After the sum occurs, we have to reset them to 2^{l-1} if they are not smaller than 2^{l-1}. A simple way to do this is $X \leftarrow M_u \;\&\; (10^{l-1})^U$, $M_u \leftarrow M_u \;\&\; \sim (X - (X >> (l-1)))$. The "&" is the bitwise "and", and $(10^{l-1})^U$ is a constant with bits sets at the highest counter positions, so X has bits set at the highest bits of the exceeded counters. Later, $X - (X >> (l-1))$ has all (but the highest) bits set in the fields of the exceeded counters. The "$\sim$" operation reverses the bits, so at the end we clean the trailing bits of the counters that exceed 2^{l-1}, leaving them exactly at 2^{l-1}.

The same mask X serves for the purpose of determining the shift. We want to shift by $s\ell$ whenever the s leftmost highest counter bits in X are set. Let us rather consider $Y = X^{\wedge}(10^{l-1})^U$, where "$\wedge$" is the bitwise "xor" operation, so we now we are interested in how many leftmost highest counter bits in Y are *not* set. This means that we want to know which is the leftmost set bit in Y. If this corresponds to the s-th counter, then we can shift by $(s-1)\ell$. But the leftmost bit set in Y is at position $y = \lfloor \log_2 Y \rfloor$, so we shift by $(U-1-y/l)\ell$. The logarithm can be computed fast by casting the number to float (which is fast in modern computers) and then extracting the exponent from the standardized real number representation.

If there are less than Ul bits in the computer word, there are several alternatives: space the ℓ-grams in the window by more than ℓ, prune the patterns to reduce m, or resort to using more computer words for the counters. In our implementation we have used a combination of the first and last alternatives: for long patterns we use simulated 64 bit words (in our 32 bit machine), directly supported by the compiler. If this is not enough, we use less counters, i.e. use spacing h, where $h > \ell$, for reading the ℓ-grams. This requires that the areas are $\ell + h - 1 + k$ characters now, instead of $2\ell - 1 + k$, and there are only $\lfloor (m - k - h + 1)/h \rfloor$ tables D_t. On the other hand, this makes the shifts to be multiples of h.

The same idea can be applied for multiple patterns as well, by storing the minimum distance over all the patterns in D_t. The preprocessing cost is this time $O(r(\sigma^\ell(m/w + m))) = O(r\sigma^\ell m)$, computing the minimum distances bit-parallelly.

This variant is able of shifting sooner than previous ones. In particular, it never works more than the others in a window. However, even with the bit-parallel improvement, it can shift less. The reason is that it may shift "too soon", when it has not yet gathered enough information to make a longer shift. For example, consider the basic algorithm with $D[S^1] = 1$ and $D[S^2] = k$. It will examine S^1 and S^2 and then will shift by $m - k - 2\ell + 1$. If the current variant finds $D_1[S^1] = k + 1$ and $D_t[S^t] = k$ for $t \geq 2$, it will shift right after reading S^1, but will shift only by ℓ.

This phenomenon is well known in exact string matching. For example, the non-optimal Horspool algorithm [5] shifts as soon as the window suffix mismatches the pattern suffix, while the optimal BDM [3] shifts when the window suffix does not match *anywhere* inside the pattern. Hence BDM works more inside the window, but its shifts are longer and at the end it has better average complexity than Horspool algorithm.

3 Analysis

We analyze our basic algorithm and prove its average-optimality. It will follow that the variant of Section 2.2, being never worse than it, is also optimal.

We analyze an algorithm that is necessarily worse than our basic algorithm for every possible text window, but simpler to analyze. In every text window, the simplified algorithm *always* reads $1 + \lfloor k/(c\ell) \rfloor$ consecutive ℓ-grams, for some constant $0 < c < 1$ that will be considered shortly. After having read them, it checks whether *any* of the ℓ-grams produces less than $c\ell$ differences in the D table. If there is at least one such ℓ-gram, the window is verified and shifted by 1. Otherwise, we have at least $1 + \lfloor k/(c\ell) \rfloor > k/(c\ell)$ ℓ-grams with at least $c\ell$ differences each, so the sum of the differences exceeds k and we can shift the window to one position past the last character read.

Note that, for this algorithm to work, we need to read $1 + \lfloor k/(c\ell) \rfloor$ ℓ-grams from a text window. This is at most $\ell + k/c$ characters, a number that must not exceed the window length. This means that c must observe the limit $\ell + k/c \leq m - k$, that is, $c \geq k/(m - k - \ell)$.

It should be clear that the real algorithm can never read more ℓ-grams from any window than the simplified algorithm, can never verify a window that the simplified algorithm does not verify, and can never shift a window by less positions than the simplified algorithm. Hence an average-case analysis of this simplified algorithm is a pessimistic average-case analysis of the real algorithm. We later show that this pessimistic analysis is tight.

Let us divide the windows we consider in the text into *good* and *bad* windows. A window is good if it does not trigger verifications, otherwise it is bad. We will consider separately the amount of work done over either type of window.

In good windows we read at most $\ell + k/c$ characters. After this, the window is shifted by at least $m - k - (\ell + k/c) + 1$ characters. Therefore, it is not possible to work over more than $\lfloor n/(m - k - (\ell + k/c) + 1) \rfloor$ good windows. Multiplying the maximum number of good windows we can process by the amount of work done inside a good window, we get an upper bound for the total work over good windows:

$$\frac{\ell + k/c}{m - k - (\ell + k/c) + 1} \, n \;=\; O\left(\frac{\ell + k}{m} \, n\right), \tag{1}$$

where we have assumed $k + k/c < x(m - \ell)$ for some constant $0 < x < 1$, that is, $c > k/(x(m - \ell) - k)$. This is slightly stricter than our previous condition on c.

Let us now focus on bad windows. Each bad window requires $O(rm^2)$ verification work, using plain dynamic programming over each pattern. We need to show that bad windows are unlikely enough. We start by restating two useful lemmas proved in [2], rewritten in a way more convenient for us.

Lemma 1 [2] The probability that two random ℓ-grams have a common subsequence of length $(1-c)\ell$ is at most $a\sigma^{-d\ell}/\ell$, for constants $a = (1+o(1))/(2\pi c(1 - c))$ and $d = 1 - c + 2c \log_\sigma c + 2(1 - c) \log_\sigma (1 - c)$. The probability decreases exponentially for $d > 0$, which surely holds if $c < 1 - e/\sqrt{\sigma}$.

Lemma 2 [2] If S is an ℓ-gram that matches inside a given string P (larger than ℓ) with less than $c\ell$ differences, then S has a common subsequence of length $\ell - c\ell$ with some ℓ-gram of P.

Given Lemmas 1 and 2, the probability that a given ℓ-gram matches with less than $c\ell$ differences inside some P^i is at most that of having a common subsequence of length $\ell - c\ell$ with some ℓ-gram of some P^i. The probability of this is at most $mra\sigma^{-d\ell}/\ell$. Consequently, the probability that any of the considered ℓ-grams in the current window matches is at most $(1 + k/(c\ell))mra\sigma^{-d\ell}/\ell$.

Hence, with probability $(1 + k/(c\ell))mra\sigma^{-d\ell}/\ell$ the window is bad and costs us $O(m^2 r)$. Being pessimistic, we can assume that *all* the $n - (m - k) + 1$ text windows have their chance to trigger verifications (in fact only some text windows are given such a chance as we traverse the text). Therefore, the average total work on bad windows is upper bounded by

$$(1 + k/(c\ell))mra\sigma^{-d\ell}/\ell \; O(rm^2) \, n \;=\; O(r^2 m^3 n(\ell + k/c)a\sigma^{-d\ell}/\ell^2). \tag{2}$$

As we see later, the complexity of good windows is optimal provided $\ell = O(\log_\sigma(rm))$. To obtain overall optimality it is sufficient that the complexity of bad windows does not exceed that of good windows. Relating Eqs. (1) and (2) we obtain the following condition on ℓ:

$$\ell \geq \frac{4 \log_\sigma m + 2 \log_\sigma r + \log_\sigma a - 2 \log_\sigma \ell}{d} = \frac{4 \log_\sigma m + 2 \log_\sigma r - O(\log \log(mr))}{d},$$

and therefore a sufficient condition on ℓ that retains the optimality of good windows is

$$\ell = \frac{4\log_\sigma m + 2\log_\sigma r}{1 - c + 2c\log_\sigma c + 2(1-c)\log_\sigma(1-c)}. \tag{3}$$

It is time to define the value for constant c. We are free to choose any constant $k/(x(m-\ell)-k) < c < 1 - e/\sqrt{\sigma}$, for any $0 < x < 1$. Since this implies $k/(m-k) = \alpha/(1-\alpha) < 1 - e/\sqrt{\sigma}$, the method can only work for $\alpha < (1 - e/\sqrt{\sigma})/(2 - e/\sqrt{\sigma}) = 1/2 - O(1/\sqrt{\sigma})$. On the other hand, for any α below that limit we can find a suitable constant x such that, asymptotically on m, there is space for constant c between $k/(x(m-\ell)-k)$ and $1 - e/\sqrt{\sigma}$. For this to be true we need that $r = O(\sigma^{o(m)})$ so that $\ell = o(m)$. (For example, r polynomial in m meets the requirement.)

If we let c approach $1 - e/\sqrt{\sigma}$, the value of ℓ goes to infinity. If we let c approach $k/(m-\ell-k)$, then ℓ gets as small as possible but our search cost becomes $O(n)$. Any fixed constant c will let us use the method up to some $\alpha < c/(c+1)$, for example $c = 3/4$ works well for $\alpha < 3/7$. Having properly chosen c and ℓ, our algorithm is on average

$$O\left(\frac{n(k + \log_\sigma(rm))}{m}\right) \tag{4}$$

character inspections. We remark that this is true as long as $\alpha < 1/2 - O(1/\sqrt{\sigma})$, as otherwise the whole algorithm reduces to dynamic programming. (Let us remind that c is just a tool for a pessimistic analysis, not a value to be tuned in the real algorithm.) Our preprocessing cost is $O(mr\sigma^\ell)$, thanks to the smarter preprocessing of [4]. Given the value of ℓ, this is $O(m^5 r^3 \sigma^{O(1)})$. The space with plain verification is $\sigma^\ell = m^4 r^2 \sigma^{O(1)}$ integers.

It has been shown that $O(n(k + \log_\sigma(rm))/m)$ is optimal [4]. Moreover, used on a single pattern we obtain the same optimal complexity of Chang & Marr [2], but our filter works up to $\alpha < 1/2 - O(1/\sqrt{\sigma})$. The filter of Chang & Marr works only up to $\alpha < 1/3 - O(1/\sqrt{\sigma})$. Hence we have improved the only average-optimal simple approximate search algorithm with respect to its area of applicability.

3.1 Worst-Case Complexity

In the worst case the shift is always 1, and we work $O(m)$ for each each of the $O(n)$ windows, and verify each window for each pattern. The verifications are done incrementally, by saving the state of dynamic programming algorithm, to avoid scanning the same text characters more than once. Hence in the worst case our algorithm takes $O(nm + nr\lceil m/w\rceil)$ time. The worst case verification cost can be improved to $O(rnk)$ by replacing the dynamic programming algorithm with a $O(nk)$ worst case algorithm [7]. The backward scanning filter can be improved to $O(n)$ worst case by combining it with the $O(n)$ time forward scanning filter [4]. The idea is that we stop the backward scanning if $(m-k)/2$ characters have been examined in some window, stopping at character p, and then switch to forward scanning and scan the next $m - k$ characters from p, and then switch again to backward scanning. This guarantees that each text character is examined at most twice. Variations of this trick has been used before, see e.g. [3].

4   Experimental Results

We have implemented the algorithms in C, compiled using `icc 7.1` with full optimizations. The experiments were run in a 2GHz Pentium 4, with 512MB RAM, running Linux 2.4.18. The computer word length is $w = 32$ bits. We measured user times, averaged over five runs.

We ran experiments for alphabet sizes $\sigma = 4$ (DNA) and $\sigma = 20$ (proteins). The test data were randomly generated 64MB files. The randomly generated patterns were 64 characters long. We show also experiments using real texts. These are: the E.coli DNA sequence from Canterbury Corpus[1], and real protein data from TIGR Database (TDB)[2]. In this case the patterns were randomly extracted from the texts. In order to better compare with the experiments with random data, we replicated the texts up to 64MB.

For lack of space we omit the experiments on preprocessing times. The most important result is that the maximum values we can use in practice are $\ell \leq 8$ for DNA, and $\ell \leq 3$ for proteins, otherwise the preprocessing time exceeds by far any reasonable search time. The search times that follow were measured for these maximum ℓ values unless otherwise stated, and we show the total times, preprocessing included.

Observe that the ℓ values required by our analysis (Section 3) in order to have optimal complexity are, depending on r, $\ell = 12 \ldots 20$ for DNA, and $\ell = 6 \ldots 10$ for proteins. These are well above the values we can handle in practice. The result is that, although our algorithms are very fast as expected, they can cope with difference ratios much smaller than those predicted by the analysis. This is the crux of the difference between our theoretical and practical results.

We compare our algorithm against previous work, both for searching single and multiple patterns. Following [10] we have included only the relevant algorithms in the comparison. These are:

Ours: Our basic algorithm (Sec. 2.1).
Ours, strict: Our algorithm, stricter matching (Sec. 2.2).
Ours, strictest: Our algorithm, strictest matching (Sec. 2.3).
CM: Our previous algorithm [4] based on [2].
LT: Our previous linear time filter [4], a variant of CM that cannot skip characters but works up to $\alpha = 1/2$.
EXP: Partitioning into exact search [1], an algorithm for single and multiple approximate pattern matching, implemented by its authors.
MM: Muth & Manber algorithm [8], the first multipattern approximate search algorithm we know of, able of searching only with $k = 1$ differences and until now unbeatable in its niche, when more than 50–100 patterns are searched for. The implementation is also from its authors.

---

[1] http://corpus.canterbury.ac.nz/descriptions/
[2] http://www.tigr.org/tdb

ABNDM: Approximate BNDM algorithm [11, 6], a single pattern approximate
search algorithm extending classical BDM. The implementation is by its authors. We used the version of [11] which gave better results in our architecture, although it is theoretically worse than [6].

BPM: Bit-parallel Myers [9], currently the best non-filtering algorithm for single patterns, using the implementation of its author. We do not expect this
algorithm to be competitive against filtering approaches, but it should give
a useful control value.

For our algorithms we have used hierarchical verification, and for the algorithms of Secs. 2.1 and 2.2 we also applied bit-parallel counters. Both optimizations were applied for our previous algorithms CM and LT too.

We have modified ABNDM and BPM to use the superimposition technique
[1] to handle multiple patterns, combined with hierarchical verification. The
superimposition is useful only if r/σ is reasonably small. In particular, it does
not work well on DNA, and in that case we simply run the algorithms r times.
For proteins we superimposed a maximum of 16 patterns at a time. Note that
the optimal group size depends on k, decreasing as k increases. However, we
used the above group sizes for all the cases.

Since MM is limited to $k = 1$, we compare this case separately in Table 1.
As it can be seen, our algorithm is better by far, 4–40 times faster depending on
the alphabet size.

Table 1. Comparison against Muth & Manber, for $k = 1$.

	$r = 1$	$r = 16$	$r = 64$	$r = 256$	$r = 1$	$r = 16$	$r = 64$	$r = 256$
Alg.	DNA (random)				DNA (Ecoli)			
MM	1.30	3.97	12.86	42.52	1.27	3.85	13.96	45.29
Ours	0.08	0.12	0.21	0.54	0.10	0.15	0.26	0.91
Alg.	proteins (random)				proteins (real)			
MM	1.17	1.19	1.26	2.33	1.15	1.17	1.23	2.21
Ours	0.08	0.11	0.18	0.59	0.08	0.13	0.22	0.75

Figures 2, and 3 show results for the case $r = 1$ (single patterns), as well as
larger r values. Our algorithm is the fastest in the majority of cases. EXP beats
our algorithm for large k, and this happens sooner for larger r.

Our main algorithm is in most cases the clear winner, as expected from the
analysis. The filtering capability of Chang & Marr extension (CM) collapses
already with quite small difference ratios, due to the use of small text blocks,
and this collapse is usually very sharp. Our new algorithms use larger search
windows, so they trigger less verifications and permits larger difference rations.
In addition, even with small difference ratios they skip more characters, and
hence it is faster for all values of k/m. Our algorithm is usually faster than the
linear time filter (LT), because the latter cannot skip text characters. This fact
stays true even for large difference ratios.

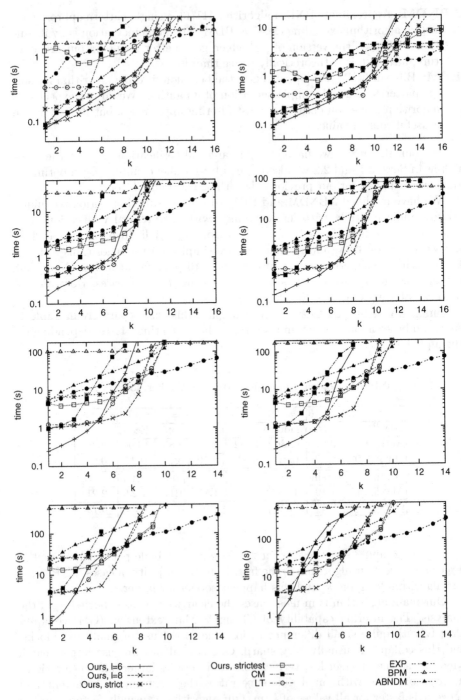

Fig. 2. Comparing different algorithms in DNA, considering both preprocessing and searching time. On the left, random data; on the right E.coli. From top to bottom: $r = 1, 16, 64, 256$. ℓ looks as "l". Note the logarithmic scale.

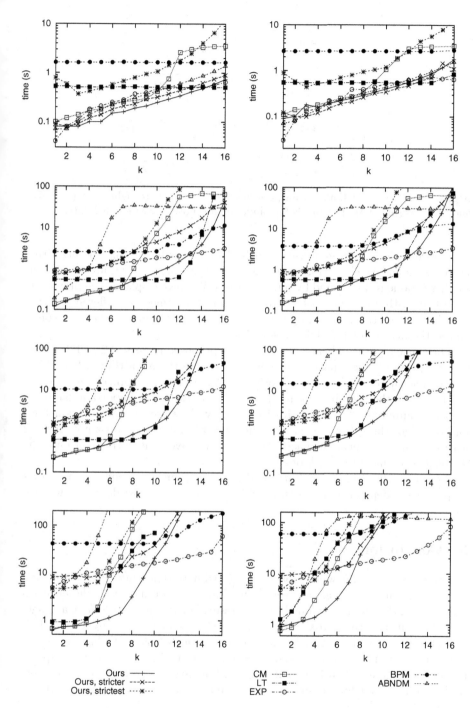

Fig. 3. Comparing different algorithms in proteins, considering both preprocessing and searching time. On the left, random data; on the right real proteins. From top to bottom: $r = 1, 16, 64, 256$. We used $\ell = 3$ for our algorithm. Note the logarithmic scale.

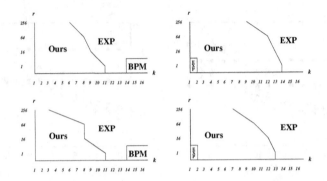

Fig. 4. Areas where each algorithm performs best. From left to right, DNA ($m = 64$), and proteins ($m = 64$). Top row is for random data, bottom row for real data.

Figure 4 shows the areas where each algorithm is best. Our new algorithm becomes the fastest choice for low k, although, the more patterns are sought, the smaller k value is dominated by our algorithm. We have displaced the previously fastest algorithm for this case [1] to the area of intermediate difference ratios. Finally, we note that, when applied to just one pattern, our algorithm becomes indeed the fastest for low difference ratios.

Our basic algorithm beats the extensions proposed in Secs. 2.2 and 2.3, with the exception of single pattern real protein, where the algorithm of Sec. 2.2 is better. The success of our basic algorithm is due to lower preprocessing cost and the fact that the D tables better fit into cache, which is important factor in modern computers. Note that this is true only if we use the same parameter ℓ for both algorithms. If we are short of memory we can use the variant of Sec. 2.2 with smaller ℓ, and beat the basic algorithm. We have verified this, but omit the detailed experiments for lack of space.

Note also that our algorithm would be favored on even longer texts, as its preprocessing depends only on rm, and hence its relative cost decreases when n increases.

5   Conclusions

We have presented a new multipattern approximate string matching algorithm. Our algorithm is optimal on average for low and intermediate difference ratios (up to $1/2$), improving our last-year result [4], which was the first average-optimal multipattern approximate string matching and worked only for $\alpha < 1/3$. Even for $\alpha < 1/3$, our algorithm is in practice much faster. We have also presented two variants, the first variant being also optimal. This algorithm inspects less text characters than the basic version, but becomes interesting only when we have strict limit for the maximum amount of memory we can use for the D tables.

We have experimentally shown that our algorithms perform well in handling from one to a very large number of patterns, and low difference ratios. Our algorithms become indeed the best alternatives in practice for these cases. In particular, this includes being the fastest single-pattern search algorithms for low difference ratios, which is a highly competitive area. On real life texts, we show that our algorithms are especially interesting for computational biology applications (DNA and proteins).

References

1. R. Baeza-Yates and G. Navarro. New and faster filters for multiple approximate string matching. *Random Structures and Algorithms (RSA)*, 20:23–49, 2002.
2. W. Chang and T. Marr. Approximate string matching and local similarity. In *Proc. 5th Combinatorial Pattern Matching (CPM'94)*, LNCS 807, pages 259–273, 1994.
3. M. Crochemore, A. Czumaj, L. Gąsieniec, S. Jarominek, T. Lecroq, W. Plandowski, and W. Rytter. Speeding up two string matching algorithms. *Algorithmica*, 12(4/5):247–267, 1994.
4. K. Fredriksson and G. Navarro. Average-optimal multiple approximate string matching. In *Proc. 14th Combinatorial Pattern Matching (CPM 2003)*, LNCS 2676, pages 109–128, 2003.
5. R. Horspool. Practical fast searching in strings. *Software Practice and Experience*, 10:501–506, 1980.
6. H. Hyyrö and G. Navarro. Faster bit-parallel approximate string matching. In *Proc. 13th Combinatorial Pattern Matching (CPM 2002)*, LNCS 2373, pages 203–224, 2002.
7. G. M. Landau and U. Vishkin. Fast parallel and serial approximate string matching. *J. Algorithms*, 10(2):157–169, 1989.
8. R. Muth and U. Manber. Approximate multiple string search. In *Proc. 7th Combinatorial Pattern Matching (CPM'96)*, LNCS 1075, pages 75–86, 1996.
9. E. W. Myers. A fast bit-vector algorithm for approximate string matching based on dynamic programming. *Journal of the ACM*, 46(3):395–415, 1999.
10. G. Navarro. A guided tour to approximate string matching. *ACM Computing Surveys*, 33(1):31–88, 2001.
11. G. Navarro and M. Raffinot. Fast and flexible string matching by combining bit-parallelism and suffix automata. *ACM Journal of Experimental Algorithmics (JEA)*, 5(4), 2000.
12. P. Sellers. The theory and computation of evolutionary distances: pattern recognition. *Journal of Algorithms*, 1:359–373, 1980.
13. E. Sutinen and J. Tarhio. Filtration with q-samples in approximate string matching. In *Proc. 7th Combinatorial Pattern Matching*, LNCS 1075, pages 50–63, 1996.

Average-Case Analysis
of Approximate Trie Search
(Extended Abstract)

Moritz G. Maaß*

Institut für Informatik, TU München
Boltzmannstr. 3, D-85748 Garching, Germany
maass@informatik.tu-muenchen.de

Abstract. For the exact search of a pattern of length m in a database of n strings of (arbitrary) length the trie data structure allows an optimal lookup time of $O(m)$. If errors are allowed between the pattern and the database strings, no such structure with reasonable size is known. Using a trie some work can be saved and running times superior to the comparison with every string in the database can be achieved. We investigate a comparison-based model where "errors" and "matches" are defined between pairs of characters. When comparing two characters, let p be the probability of an error. Between any two strings we bound the number of errors by D, which we consider a function of n. We study the average-case complexity of the number of comparisons for searching in a trie in dependence of the parameters p and D. Our analysis yields the asymptotic behavior for memoryless sources with uniform probabilities. It turns out that there is a jump in the average-case complexity at certain thresholds for p and D. Our results can be applied for any comparison-based error model, for instance, mismatches (Hamming distance), don't cares, or geometric character distances.

1 Introduction

We study the average-case behavior of the simple problem of finding a given pattern in a set of patterns subject to two conditions. The set of patterns is given in advance and may have been preprocessed with linear space, and we are also interested in occurrences where the pattern is found with a given number of errors.

This research was triggered by a project for the control of animal breeding via SNPs [28]. Without going into details, each SNP can be encoded as a binary string that is an identifier for an individual. Because of errors the search in the dataset of all individuals needs to be able to deal with mismatches and don't cares. The nature of the data allows a very efficient binary encoding, which yields a reasonable fast algorithm that just compares a pattern to each string in the

---

* Research supported in part by DFG, grant Ma 870/5-1 (Leibnizpreis Ernst W. Mayr).

S.C. Sahinalp et al. (Eds.): CPM 2004, LNCS 3109, pp. 472–483, 2004.

set. On the other hand, a trie can be used as an index for the data. It has the same worst-case lookup time, but we expect to save some work because fewer comparisons are necessary. As a drawback, the constants in the algorithm are a bit higher due to the tree structure involved. We call the first variant "Linear Search" (LS) and the second variant "Trie Search" (TS). See Figure 1 for a description in pseudo-code.

An abstract formulation of the problem is the following. Given n strings $X_1, \ldots, X_n$ of the same length m, a pattern P of length m, an error probability p, and a bound D for the maximal number of errors allowed, let L_n^D be the number of comparisons made by the LS algorithm and T_n^D be the number of comparisons made by the TS algorithm using a trie as an index. What is the threshold $D = D(n)$, up to where the average $\mathbf{E}\left[T_n^D\right]$ is asymptotically better than the average $\mathbf{E}\left[L_n^D\right]$? What is the effect of different error probabilities? The answers to these questions give hints at choosing the faster method for our original problem. Note that the parameter p depends on the definition of errors, it gives the relative number of character pairs inducing an error (or mismatch). The parameter D gives an upper bound on the number of errors (mismatched character pairs) allowed for strings considered similar, i.e., if more than D errors occur between two strings, they are considered different.

Let Σ be an arbitrary finite alphabet of size $\sigma := |\Sigma|$. Let $t = t_1 t_2 t_3 \ldots t_n$ be a string with characters $t_i \in \Sigma$, we define $|t| = n$ to be its length. For the average-case analysis we assume that we deal with strings of infinite length, each string $X = \{x_k\}_{k=1}^{\infty}$ is generated independently at random by a memoryless source with uniform probabilities $\Pr\{x_j = s_i\} = 1/\sigma$ for all $s_i \in \Sigma$. We assume that all strings $X_1, \ldots, X_n$ used in the search are different (i.e., $X_i \neq X_j$ for $i \neq j$). Since the strings are generated randomly the probability that two identical strings of infinite length occur is indeed 0. We further assume that a search pattern $P = \{p_k\}_{k=1}^{\infty}$ is generated by a source with the same probabilities and similarly of infinite length.

When two arbitrary, randomly generated characters are compared, let p denote the probability of an error and $q := 1 - p$ the probability of a match. For example, we have a mismatch probability of $p = 1 - 1/\sigma$ and a match probability of $q = 1/\sigma$ for Hamming distance.

It is easy to prove that the average number of comparisons made by the LS algorithm is $\mathbf{E}\left[L_n^D\right] = (D+1)n/p$. Indeed, one can prove almost sure convergence to this value. The LS algorithm has a linear or quasi-linear average-case behavior for small D.

The interesting part is the analysis of the TS algorithm. In the trie each edge represents the same character in a number of strings (the number of leaves in the subtree). Let there be k leaves below an edge. If the TS algorithm compares a character from the pattern to the character at the edge, the LS algorithm needs to make k comparisons. In essence, this can be seen as the trie "compressing" the set of strings. It can be proven that the average number of characters thus "compressed" is asymptotically $n \log_\sigma n + O(n)$. Hence, for $D \geq (1 + \epsilon) \log_\sigma n$ there can be no asymptotic gain when using a trie.

We will show that for $D < p \log_\sigma n$ the TS algorithm performs sublinear, i.e., in time $o(n)$ for n strings in the index, and for $D > p \log_\sigma n$ it performs superlinear, i.e., in time $\omega(n)$ for n strings in the index. When D is a constant the asymptotic running time can be calculated very exactly and is $O\left((\log n)^{D+1}\right)$, for $q = 1/\sigma$, $O\left((\log n)^D n^{\log_\sigma q + 1}\right)$, for $q > 1/\sigma$, and $O(1)$, otherwise.

2   Related Work

Digital tries have a wide range of applications and belong to the most studied data structures in computer science. They have been around for years; for their usefulness and beauty of analysis they have received a lot of attention. Tries were introduced by Brandais [4] and Fredkin [10]. A useful extension are Patricia trees introduced by Morrison [15].

Using tree (especially trie or suffix tree) traversals for indexing problems is a common technique. For instance, in computer linguistics one often needs to correct misspelled input. Schulz and Mihov [22] pursue the idea of correcting misspelled words by finding correctly spelled candidates from a dictionary implemented as a trie or automaton. They build an automaton for the input word and traverse the trie with it in a depth first search. The search automaton is linear in the size of the pattern if only a small number of errors is allowed. A similar approach has been investigated by Oflazer [19], except that he directly calculates edit distance instead of using an automaton.

Flajolet and Puech [9] analyze the average-case behavior of partial matching in k-d-tries. A pattern in k domains with $k - s$ don't cares and s specified values is searched. Each entry in a k-dimensional data set is represented by the binary string constructed by concatenating the first bits of the k domain values, the second bits, the third bits, and so forth. Using the Mellin transform it is proven that the average search time is $O\left(n^{1-s/k}\right)$ under the assumption of an independent uniform distribution of the bits. In terms of ordinary strings this corresponds to matching with a fixed mask of don't cares that is iterated through the pattern.

Baeza-Yates and Gonnet [2] study the problem of searching regular expressions in a trie. The deterministic finite state automaton for the regular expression is built, its size depending only upon the query size (although possibly exponential in the size of the query). The automaton is simulated on the trie and a hit is reported every time a final state is reached. Extending the average-case analysis of Flajolet and Puech [9], the authors are able to show that the average search time depends upon the largest eigenvalue (and its multiplicity) of the incidence matrix of the automaton. As a result, they find that a sublinear number of nodes of the trie is visited. Apostolico and Szpankowski [1] note that suffix trees and tries for independent strings asymptotically do not differ too much, which is an argument for transferring the results on tries to suffix trees.

In another article Baeza-Yates and Gonnet [3] study the average cost of calculating an all-against-all sequence matching. Here, for all strings, the substrings that match each other with a certain (fixed) number of errors are sought. With the use of tries the average time is shown to be subquadratic.

For approximate indexing (with edit distance) Navarro and Baeza-Yates [17] have proposed a method that flexibly partitions the pattern in pieces that can be searched in sublinear time in the suffix tree for a text. For an error rate $\alpha = k/m$, where m is the pattern length and k the allowed number of errors, they show that a sublinear search is possible if $\alpha < 1 - e/\sqrt{\sigma}$, thereby partitioning the pattern into $j = (m + k)/\log_\sigma n$ pieces. The threshold plays two roles, it gives a bound on the search depth in a suffix tree and it gives a bound on the number of verifications needed. In Navarro [16] the bound is investigated more closely. It is conjectured that the real threshold, where the number of matches of a pattern in a text grows subexponentially in the pattern length, is $\alpha = 1 - c/\sqrt{\sigma}$ with $c \approx 1.09$. Higher error rates make a filtration algorithm useless because of too many verifications.

More careful tree traversal techniques can lower the number of nodes that need to be visited. This idea is pursued by Jokinen and Ukkonen [11] (on a DAWG), Ukkonen [27], and Cobbs [7]. No exact average-case analysis is available for these algorithms.

The start of precise analysis of algorithms is contributed to Knuth (i.e., [21]). Especially the analysis of digital trees has yielded a vast amount of results. The Mellin transform and Rice's integrals have been the methods of choice for many results dating back as early as the analysis of radix exchange sort in Knuth's famous books [13]. See [25] for a recent book with a rich bibliography.

Our analysis of the average search time in the trie leads to an alternating sum of the type

$$\sum_{k=m}^{n} \binom{n}{k} (-1)^k f(n, k) \ . \tag{1}$$

The above sum is intimately connected to tries and appears very often in their analysis (see, e.g., [13]). Similar sums, where $f(n, k)$ only depends on k, have also been considered by Szpankowski [23] and Kirschenhofer [12]. The asymptotic analysis can be done through Rice's integrals (a technique that already appears in Nörlund [18], chap. 8, §1). It transfers the sum to a complex integral, which is evaluated by the Cauchy residue theorem.

Our contribution is the general analysis of error models that depend only one the comparison of two characters and limit the number of errors allowed before regarding two strings as different. Unless the pattern length is very short, the asymptotic order of the running time depends on an error-threshold relative to the number of strings in the database and independent of the pattern length. The methods applied here can be used to determine exact asymptotics for each concrete error bound. It also allows to estimate the effect of certain error models in limiting the search time. Furthermore, for constant error bounds we find thresholds with respect to the error probability which reveal an interesting behavior hidden for the most used model, the Hamming distance.

In the following, we will present the main results, the basic analysis, and try to roughly sketch the asymptotic parts. For details and proofs see [14].

3 Main Results

The trie T for a set of strings $X_1, \ldots, X_n$ is a rooted, directed tree where each edge is labeled with a character from Σ, all outgoing edges of any node are labeled with different characters, and the strings spelled out by the edges leading from the root to the leaves are exactly $X_1, \ldots, X_n$. We store $\mathsf{value}(v) = i$ at leaf v, if the path to v spells out the string X_i. The paths from the last branching nodes to the leaves are often compressed to a single edge.

When searching for a pattern P we want to know all strings X_i such that P is a prefix of X_i or vice versa (with the special case of all strings having the same length). The assumption that all strings have infinite length is not severe. Indeed, this reflects the situation that the pattern is not found. Otherwise, the search would be ended earlier, so our analysis gives an upper bound. Pseudo code for the analyzed algorithms is given in Fig. 1.

LS Algorithm	TS Algorithm : rfind(v, P, pos, D)
Input: Strings $X_1, \ldots, X_n$ and pattern P, bound D.	if $D \geq 0$ then
for i **from** 1 **to** n **do**	if v is a leaf **then**
$j := 1$	report match for $X_{\mathsf{value}(v)}$
$c := 0$	**else if** $pos > \mathsf{length}(P)$ **then**
$l := \min\{\mathsf{length}(P), \mathsf{length}(X_i)\}$	**for all** leaves u in the subtree of
while $c \leq D$ **do**	v **do**
while $j \leq l$ **and** $\mathsf{match}(P[j], X_i[j])$	report match for $X_{\mathsf{value}(u)}$
do	**else**
$j := j + 1$	**for each** child u of v **do**
$c := c + 1$	let c be the edge label of (u, v)
$j := j + 1$	**if** $\mathsf{match}(P[pos], c)$ **then**
if $j - 2 = l$ **then**	rfind$(u, P, pos + 1, D)$
report match for X_i	**else**
	rfind$(u, P, pos + 1, D - 1)$

Fig. 1. Pseudo code of the LS and the TS algorithm. The recursive TS algorithm is started with rfind$(r, P, 0, D)$, where r is the root of the trie for the Strings $X_1, \ldots, X_n$.

We focus mainly on the TS algorithm, the LS algorithm is used as a benchmark. Our main result is the following theorem. For fixed D the constants in the Landau symbols and further terms of the asymptotic can also be computed (or at least bounded).

Theorem 1 (Average Complexity of the TS Algorithm).

$$\mathbf{E}\left[T_n^D\right] = \begin{cases} O\left((\log n)^{D+1}\right), & \text{for } D = O(1) \text{ and } q = \sigma^{-1} \\ O\left((\log_\sigma n)^D n^{\log_\sigma q+1}\right), & \text{for } D = O(1) \text{ and } q > \sigma^{-1} \\ O(1), & \text{for } D = O(1) \text{ and } q < \sigma^{-1} \quad (2) \\ o(n), & \text{for } D+1 < p\log_\sigma n \\ \Omega\left(n \log_\sigma n\right), & \text{for } D+1 > p\log_\sigma n. \end{cases}$$

For the cases $q < \sigma^{-1}$, $q = \sigma^{-1}$, and $q > \sigma^{-1}$ exacter bounds are possible through (22), (23), and (24). For instance, (22) tells us that the number of nodes visited grows by $\frac{p\sigma}{1-q\sigma}$ for each additional error allowed. These results can be applied to different models. For instance, for the Hamming distance model with alphabet size 4 we get the exact first order term $\frac{4 \cdot 3^D}{(D+1)!} (\log_4 n)^{D+1}$.

It is well known that the average depth of a trie is asymptotically equal to $\log_\sigma n$ (see, e.g., [20, 24]). When no more branching takes place the TS and LS algorithm behave the same; both algorithms perform a constant number of comparisons on average. If we allow enough errors to go beyond the depth of the trie, they should perform similar. With an error probability of p we expect to make pm errors on m characters. Thus, it comes as no surprise that the threshold is $p \log_\sigma n$.

With respect to the matching probability q we have a different behavior for the three cases $q < \sigma^{-1}$, $q = \sigma^{-1}$, and $q > \sigma^{-1}$. To explain this phenomena we take a look at the conditional probability of a match for an already chosen character. If $q < \sigma^{-1}$, then the conditional probability must be smaller than 1, i.e., with some probability independent of the pattern, we have a mismatch and thus restrict the search independently of the pattern. If $q > \sigma^{-1}$, the conditional probability must be greater than 1. Hence, with some probability independent of the pattern, we have a match and thereby extend our search. This restriction or extension is independent of the number of errors allowed and, hence, the additional factor in the complexity.

For the model where we bound the number of don't cares we have $p = 2/\sigma - 1/\sigma^2$ and $q = 1 - 2/\sigma + 1/\sigma^2$. In the SNP database problem mentioned above, the alphabet size is $\sigma = 4$, including the don't care character. We find that the average-case behavior, bounding only the don't cares, is approximately $O\left((\log n)^D n^{0.585}\right)$ when allowing D don't cares. For the number of mismatches we could resort to the Hamming distance case mentioned above, but in this application a don't care cannot induce a mismatch. Therefore, the average-case complexity is approximately $O\left((\log n)^D n^{0.292}\right)$ when allowing D mismatches. This is significantly worse than Hamming distance only, which is $O\left((\log n)^{D+1}\right)$. It also dominates the bound on the number of don't cares. When deciding whether the LS or the TS algorithm should be used in this problem, we find that for $D > (5/8) \log_4 n$ the LS algorithm will outperform the TS algorithm.

As another application we apply our results to the model used by Buchner et al. [5, 6] for searching protein structures. Here the angles of a protein folding are used for approximate search of protein substructures. The full range of $360°$ degrees is discretized into an alphabet $\Sigma = \{[0°, 15°), \ldots, [345°, 360°)\}$. The algorithm then searches a protein substructure by considering all angles within a number of intervals to the left and right, i.e., for $i = 2$ intervals to both sides, the interval $[0°, 15°)$ matches $[330°, 345°)$, $[345°, 360°)$, $[0°, 15°)$, $[15°, 30°)$, and $[30°, 45°)$. If i intervals to the left or right are allowed, then the probability of a match is $(2i+1)/\sigma$. In their application Buchner et al. [5, 6] allow no mismatch, i.e., the search is stopped if the angle is not within the specified range. The asymptotic running time is thus $O\left(n^{\log_\sigma (2i+1)}\right)$ if i intervals to the left or right are considered. Although a suffix tree is used and the underlying distribution of angles is probably not uniform and memoryless, this result can be used as a (rough) estimate, especially of the effect of different choices of i.

4   Basic Analysis

For completeness we give a quick derivation of the expected value of L_n^D, the number of comparisons made by the LS algorithm. The probability of k comparisons is

$$\Pr\left\{L_n^D = k\right\} = \sum_{i_1+\ldots+i_n=k} \prod_{j=1}^{n} \binom{i_j-1}{D} p^{D+1} q^{i_j-D-1}. \tag{3}$$

From it we can derive the probability generating function

$$g_{L_n^D}(z) = \mathbf{E}\left[z^{L_n^D}\right] = \sum_{k=0}^{\infty} \Pr\left\{L_n^D = k\right\} z^k = \left(\frac{zp}{1-zq}\right)^{n(D+1)}, \tag{4}$$

which yields the expected value $\mathbf{E}\left[L_n^D\right] = \frac{D+1}{p}n$.

For the TS algorithm it is easier to examine the number of nodes visited. Observe that the number of nodes visited is by one larger than the number of character comparisons. Each time a node is visited the remaining leaves split up by a random choice of the next character. Depending on the next character of the pattern the number of allowed mismatches in the subtree may stay the same (with probability q) or may decrease (with probability p). For the average number we can set up the following equation.

$$\mathbf{E}\left[T_n^D\right] = 1 + \sum_{i_1+\cdots+i_\sigma=n} \binom{n}{i_1,\ldots,i_\sigma} \sigma^{-n}\left(\sum_{j=1}^{\sigma} p\mathbf{E}\left[T_{i_j}^{D-1}\right] + \sum_{j=1}^{\sigma} q\mathbf{E}\left[T_{i_j}^{D}\right]\right). \tag{5}$$

The boundary conditions are $\mathbf{E}\left[T_n^{-1}\right] = 1$, counting the character comparison that induced the last mismatch, and $\mathbf{E}\left[T_0^D\right] = 0$. For $n = 1$ we have $\mathbf{E}\left[T_1^D\right] = 1 + \frac{D+1}{p}$, which is the same as $\mathbf{E}\left[L_1^D\right]$, except that additionally the root is counted.

From (5) we can derive the exponential generating function of $\mathbf{E}\left[T_n^D\right]$.

$$t^D(z) = e^z + \sum_{j=1}^{\sigma} pt^{D-1}\left(\frac{z}{\sigma}\right) e^{\left(1-\frac{1}{\sigma}\right)z} + \sum_{j=1}^{\sigma} qt^D\left(\frac{z}{\sigma}\right) e^{\left(1-\frac{1}{\sigma}\right)z} - 1. \tag{6}$$

We multiply with $\exp(-z)$ (which corresponds to applying some kind of binomial inversion) and define $\tilde{t}^D(z) = t^D(z)e^{-z}$. We have

$$\tilde{t}^D(z) = 1 - \exp(-z) + \sigma p\tilde{t}^{D-1}\left(\frac{z}{\sigma}\right) + \sigma q\tilde{t}^D\left(\frac{z}{\sigma}\right). \tag{7}$$

Let y_n^D be the coefficients of $\tilde{t}^D(z)$, then we get the boundary conditions $y_1^D = 1 + (D+1)/p$, $y_0^D = 0$, $y_n^{-1} = (-1)^{n-1}$ for $n > 0$, and $y_0^{-1} = 0$. Comparing coefficients in (7) we find that for $n > 1$

$$y_n^D = \frac{(-1)^{n-1} + y_n^{D-1}\sigma^{1-n}p}{1 - \sigma^{1-n}q} , \tag{8}$$

which by iteration leads to

$$y_n^D = \frac{(-1)^n}{1 - \sigma^{1-n}}\left(\sigma^{1-n}\left(\frac{\sigma^{1-n}p}{1 - \sigma^{1-n}q}\right)^{D+1} - 1\right) . \tag{9}$$

Finally, we translate this back to

$$\mathbf{E}\left[T_n^D\right] = n\left(1 + \frac{D+1}{p}\right)$$
$$+ \sum_{k=2}^{n}\binom{n}{k}\frac{(-1)^k}{\sigma^{k-1} - 1}\left(\frac{p\sigma^{1-k}}{1 - q\sigma^{1-k}}\right)^{D+1} - \sum_{k=2}^{n}\binom{n}{k}\frac{(-1)^k}{1 - \sigma^{1-k}} . \tag{10}$$

Let $A_n := \sum_{k=2}^{n}\binom{n}{k}\frac{(-1)^k}{1-\sigma^{1-k}}$. A similar derivation to the above shows that the sum is the solution to

$$A_n = n - 1 + \sum_{i_1 + \cdots + i_\sigma = n}\binom{n}{i_1, \ldots, i_\sigma}\sigma^{-n}\sum_{j=1}^{\sigma}A_{i_j} , \tag{11}$$

which we call the average "compression number". It gives the average sum of the number of characters "hidden" by all edges, i.e., an edge with n leaves in its subtree "hides" $n - 1$ characters (which would be examined by the LS but not by the TS algorithm). Hence, $n(D + 1)/p - A_n$ is an upper bound for the average performance of the TS algorithm. The following can be proven.

Lemma 2 (Asymptotic Behavior of the Compression Number). *The asymptotic behavior of A_n is*

$$A_n = n\log_\sigma n + n\left(\frac{1}{2} - \frac{1-\gamma}{\ln\sigma} + \frac{\sum_{k\in\mathbb{Z}\setminus\{0\}}n^{-\frac{2\pi ik}{\ln\sigma}}\Gamma\left(-1 + \frac{2\pi ik}{\ln\sigma}\right)}{\ln\sigma}\right) + O(1) . \tag{12}$$

One can show that $\sum_{k=1}^{\infty}\left|\Gamma\left(-1 + \frac{2\pi ki}{\ln\sigma}\right)\right|$ is very small (below 1 for $\sigma \geq 10^6$), but growing in σ. We now turn to the evaluation of the sum

$$S_n^{(D)} := \sum_{k=2}^{n}\binom{n}{k}\frac{(-1)^k}{\sigma^{k-1} - 1}\left(\frac{p}{\sigma^{k-1} - q}\right)^{D+1} . \tag{13}$$

Note that if $S_n^{(D)}$ is sublinear the main term of the asymptotic growth of $\mathbf{E}\left[T_n^D\right]$ is determined by (10) and Lemma 2 to be $n\left(\frac{D+1}{p} - \log_\sigma n\right)$.

5 Asymptotic Analysis

We can prove two theorems regarding the growth of $S_n^{(D)}$ for different bounds D. For constant D we can give a very precise answer.

Theorem 3 (Searching with a Constant Bound). *Let $D = O(1)$, then*

$$S_n^{(D)} = -n\left(1 + \frac{D+1}{p}\right) + A_n + \begin{cases} O\left((\log n)^{D+1}\right), & \text{for } q = \sigma^{-1} \\ O\left((\log_\sigma n)^D n^{\log_\sigma q+1}\right), & \text{for } q > \sigma^{-1} \\ O(1), & \text{otherwise.} \end{cases} \quad (14)$$

For logarithmic D we give a less exact answer, which yields a threshold where the complexity jumps from sublinear to linear-logarithmic.

Theorem 4 (Searching with a Logarithmic Bound). *If $D + 1 = c\log_\sigma n$, then we have*

$$S_n^{(D)} = \begin{cases} -n\left(1 + \frac{D+1}{p}\right) + A_n + o(n), & \text{for } c < p \\ o(n), & \text{for } c > p. \end{cases} \quad (15)$$

The two theorems immediately yield Theorem 1. Both proofs rely on transferring the sum to a complex integral by Rice's Theorem [13, 25, 23, 12]:

$$S_n^{(D)} = \frac{1}{2\pi i} \int_{-\xi-i\infty}^{-\xi+i\infty} \frac{1}{\sigma^{-1-z} - 1} \left(\frac{p}{\sigma^{-1-z} - q}\right)^{D+1} B(n+1, z)dz + O(1) \ . \quad (16)$$

The integral in (16) can be extended to a half-circle to the right because the contribution of the bounding path is very small. Hence, the integral is equal to the sum of the negative of the residues right to the line $\Re(z) = -\xi$. These residues are located at $z = 1$, $z = 0$, $z = 1 \pm 2\pi i k/\ln\sigma$, and $z = -\log_\sigma q - 1 \pm 2\pi i k/\ln\sigma$, $k \in \mathbb{Z}$. The real part of the last ranges from 1 to $1 - \log_\sigma(\sigma^2 - 1) > -1$ under the assumption that $\sigma^{-2} \leq q \leq 1 - \sigma^{-2}$.

The evaluation of the residues proves tricky for the Beta function. We approximate the Beta function using an asymptotic expansion by Tricomi and Erdélyi [26] with help of a result of Fields [8]. This approach was already used by Szpankowski [23]. We concentrate on the first term of the expansion since each further term is by an order of magnitude smaller then the previous. This leads to

$$\mathcal{I}_{\xi,n}^{(D)} := \frac{1}{2\pi i} \int_{-\xi-i\infty}^{-\xi+i\infty} \frac{1}{\sigma^{-1-z} - 1} \left(\frac{p}{\sigma^{-z-1} - q}\right)^{D+1} \Gamma(z)n^{-z}dz \ , \quad (17)$$

with $S_n^{(D)} = \mathcal{I}_{\xi,n}^{(D)} + O(1)$ for $\xi \in (1, 2)$. The most important residues are those where the singularity is at a point with real value -1. We find that

$$-\left(\operatorname{res}\left[g(z), z=-1\right]+\sum_{k\in\mathbb{Z}\setminus\{0\}}\operatorname{res}\left[g(z), z=-1+\frac{2\pi\imath k}{\ln\sigma}\right]\right)$$

$$+n\left(1+\frac{D+1}{p}\right)-A_n=O\left(1\right)\ .\quad(18)$$

Moving the line of integration to $-1+\epsilon$ we get

$$S_n^{(D)}=A_n-n\left(1+\frac{D+1}{p}\right)+\mathcal{I}_{1-\epsilon,n}^{(D)}+O\left(1\right)\ .\quad(19)$$

If we keep the line right of -1 we get

$$S_n^{(D)}=\mathcal{I}_{1+\epsilon,n}^{(D)}+O\left(1\right)\ .\quad(20)$$

One can show that $\int_{-\infty}^{+\infty}\left|B(n,x+\imath y)\right|dy=O\left(n^{-x}\right)$ for large n and constant $x\notin\{0,-1,\dots\}$. Therefore, we can bound the integral for some constants c,C and for $D+1=c\log_\sigma n$ as

$$\mathcal{I}_{\xi,n}^{(D)}\leq\frac{C}{\sigma^{\xi-1}-1}n^{c\log_\sigma\left(\frac{p}{\sigma^{\xi-1}-q}\right)+\xi}\ .\quad(21)$$

Let $\mathcal{E}_{c,q,\xi}:=c\log_\sigma\left(\frac{1-q}{\sigma^{\xi-1}-q}\right)+\xi$ be the exponent. We can bound the exponent as follows.

Lemma 5. *For $0<q<1$, $c\geq 0$, and $c\neq 1-q$ there exists a $\xi<2$ such that $\mathcal{E}_{c,q,\xi}<1$. If $c<1$, then $\mathcal{E}_{c,q,\xi}$ has a minimum at $\xi^*=-\log_\sigma\left(1-c\right)+\log_\sigma q+1$. If $c\geq 1$ or $\xi^*\geq 2$, then some value $\xi\in(1,2)$ satisfies $\mathcal{E}_{c,q,\xi}<1$.*

We can now prove Theorem 4. If $\xi^*>1$ or $c\geq 1$ we find a $\xi\in(1,2)$ such that $\mathcal{E}_{c,q,\xi}<1$, thus $\mathcal{I}_{\xi,n}^{(D)}=o(n)$ and $S_n^{(D)}=o(n)$ by (20).

If $\xi^*<1$, we move the line of integration to $-\xi^*=-1+\epsilon$ and find that $\mathcal{I}_{1-\epsilon,n}^{(D)}=o(n)$, and by (19) we have $S_n^{(D)}=A_n-n\left(1+\frac{D+1}{p}\right)+o(n)$. A special situation occurs only for $\xi^*<0$ because we have to take the singularity of the Gamma function at $z=0$ into account (the singularities at $z=-\log_\sigma q-1\pm 2\pi\imath k/\ln\sigma$ are always to the right of ξ^*). One easily proves that for $q>0$ the singularity contributes only

$$\frac{\sigma}{\sigma-1}\left(\frac{p\sigma}{1-q\sigma}\right)^{D+1}\quad\left(=o(n),\text{ for }D+1=c\log_\sigma n\right)\ .\quad(22)$$

Thus, the complexity has two cases depending on whether ξ^* is left or right of 1. This translates to $\xi^*<1$ if and only if $c<1-q=p$, which proves Theorem 4.

To prove Theorem 3 we calculate the remaining residues at $z=0$ and $z=-\log_\sigma q-1\pm 2\pi\imath k/\ln\sigma$. We consider D,q,p,σ constant, so we look for largest term in n. If $q=\sigma^{-1}$, this term is

$$-\frac{\sigma(\sigma-1)^D}{(D+1)!}\left(\log_\sigma n\right)^{D+1}\ ,\quad(23)$$

otherwise, this term is

$$-\frac{(1-q)^D}{D!q^{D+1}}\left(\log_\sigma n\right)^D n^{\log_\sigma q+1}n^{-\frac{2\pi\imath k}{\ln\sigma}}\Gamma\left(-\log_\sigma q-1+\frac{2\pi\imath k}{\ln\sigma}\right) \quad . \tag{24}$$

For $\log_\sigma q + 1 < 0$ this is $o(1)$. In this case the residue at $z = 0$ yields $O(1)$, see (22). Note also that for real value x $\Gamma(x)$ has different signs left and right of 0. For the calculation of $\mathcal{I}_{\xi,n}^{(D)}$ we sum up the negative of the residues. There are infinitely many residues, but due to the behavior of the Gamma function for large imaginary values we have

$$\left|\sum_{k\in\mathbb{Z}} n^{-\frac{2\pi\imath k}{\ln\sigma}}\gamma_{l-i}^{\left(-\log_\sigma q-1+\frac{2\pi\imath k}{\ln\sigma}\right)}\right| = O(1) \quad . \tag{25}$$

Hence, the growth for constant D is

$$\mathcal{S}_n^{(D)} = A_n - n\left(1+\frac{D+1}{p}\right) + \begin{cases} O\left((\log n)^{D+1}\right), & \text{for } q = \sigma^{-1} \\ O\left((\log_\sigma n)^D n^{\log_\sigma q+1}\right), & \text{for } q > \sigma^{-1} \\ O(1), & \text{otherwise.} \end{cases} \tag{26}$$

Thus, we have proven Theorem 3.

References

1. A. Apostolico and W. Szpankowski. Self-alignments in words and their applications. *Journal of Algorithms*, 13:446–467, 1992.
2. R. A. Baeza-Yates and G. H. Gonnet. Fast text searching for regular expressions or automaton searching on tries. *Journal of the ACM*, 43(6):915–936, 1996.
3. R. A. Baeza-Yates and G. H. Gonnet. A fast algorithm on average for all-against-all sequence matching. In *String Processing and Information Retrieval Symp. SPIRE*, pages 16–23. IEEE, 1999.
4. R. D. L. Briandais. File searching using variable length keys. In *Proc. of the Western Joint Computer Conference*, pages 295–298, March 1959.
5. A. Buchner and H. Täubig. A fast method for motif detection and searching in a protein structure database. Technical Report TUM-I0314, Fakultät für Informatik, TU München, September 2003.
6. A. Buchner, H. Täubig, and J. Griebsch. A fast method for motif detection and searching in a protein structure database. In *Proceedings of the German Conference on Bioinformatics (GCB'03)*, volume 2, pages 186–188, October 2003.
7. A. L. Cobbs. Fast approximate matching using suffix trees. In *Proc. of the 6th Sym. on Combinatorial Pattern Matching (CPM)*, volume 937 of *LNCS*, pages 41–54. Springer, 1995.
8. J. L. Fields. The uniform asymptotic expansion of a ratio of Gamma functions. In *Proc. of the Int. Conf. on Constructive Function Theory*, pages 171–176, Varna, May 1970.
9. P. Flajolet and C. Puech. Partial match retrieval of multidimensional data. *J. ACM*, 33(2):371–407, 1986.

10. E. Fredkin. Trie memory. *Communications of the ACM*, 3(9):490–499, 1960.
11. P. Jokinen and E. Ukkonen. Two algorithms for approximate string matching in static texts. In *Proc. of the 16th Int'l Symp. on Mathematical Foundations of Computer Science (MFCS)*, volume 520 of *LNCS*, pages 240–248. Springer, 1991.
12. P. Kirschenhofer. A note on alternating sums. *Electronic Journal of Combinatorics*, 3(2), 1996.
13. D. E. Knuth. *The Art of Computer Programming – Sorting and Searching*, volume 3. Addison Wesley, 2nd edition, February 1998.
14. M. G. Maaß. Average-case analysis of approximate trie search. Technical Report TUM-I0405, Fakultät für Informatik, TU München, March 2004.
15. D. R. Morrison. PATRICIA – practical algorithm to retrieve information coded in alphanumeric. *J. of the ACM*, 15(4):514–534, October 1968.
16. G. Navarro. *Approximate Text Searching*. PhD thesis, University of Chile, Dept. of Computer Science, University of Chile, Santiago, Chile, 1998.
17. G. Navarro and R. Baeza-Yates. A hybrid indexing method for approximate string matching. *Journal of Discrete Algorithms (JDA)*, 1(1):205–209, 2000. Special Issue on Matching Patterns.
18. N. E. Nörlund. *Vorlesungen über Differenzenrechnung*. Springer, Berlin, 1924.
19. K. Oflazer. Error-tolerant finite-state recognition with applications to morphological analysis and spelling correction. *Computer Linguist*, 22(1):73–89, 1996.
20. B. Pittel. Paths in a random digital tree: Limiting distributions. *Adv. Appl. Prob.*, 18:139–155, 1986.
21. H. Prodinger and W. Szpankowski (Guest Editors). *Theoretical Computer Science*. Elsevier, 144(1–2) (Special Issue), 1995.
22. K. U. Schulz and S. Mihov. Fast string correction with Levenshtein automata. *Int. J. on Document Analysis and Recognition (IJDAR)*, 5:67–85, 2002.
23. W. Szpankowski. The evaluation of an alternative sum with applications to the analysis of some data structures. *Information Processing Letters*, 28:13–19, 1988.
24. W. Szpankowski. Some results on v-ary asymmetric tries. *J. of Algorithms*, 9:224–244, 1988.
25. W. Szpankowski. *Average Case Analysis of Algorithms on Sequences*. Wiley-Interscience, 1st edition, 2000.
26. F. G. Tricomi and A. Erdélyi. The asymptotic expansion of a ratio of Gamma functions. *Pacific J. of Mathematics*, 1:133–142, 1951.
27. E. Ukkonen. Approximate string-matching over suffix trees. In *Proc. of the 4th Sym. on Combinatorial Pattern Matching (CPM)*, volume 684 of *LNCS*, pages 228–242. Springer, 1993.
28. F. A. O. Werner, G. Durstewitz, F. A. Habermann, G. Thaller, W. Krämer, S. Kollers, J. Buitkamp, M. Georges, G. Brem, J. Mosner, and R. Fries. Detection and characterization of SNPs useful for identity control and parentage testing in major European dairy breeds. *Animal Genetics*, to appear, 2003.

Author Index

Lecture Notes in Computer Science

For information about Vols. 1–3005

please contact your bookseller or Springer-Verlag

Vol. 3052: W. Zimmermann, B. Thalheim (Eds.), Abstract State Machines 2004. Advances in Theory and Practice. XII, 235 pages. 2004.

Vol. 3051: R. Berghammer, B. Möller, G. Struth (Eds.), Relational and Kleene-Algebraic Methods in Computer Science. X, 279 pages. 2004.

Vol. 3050: J. Domingo-Ferrer, V. Torra (Eds.), Privacy in Statistical Databases. IX, 367 pages. 2004.

Vol. 3049: M. Bruynooghe, K.-K. Lau (Eds.), Program Development in Computational Logic. VIII, 539 pages. 2004.

Vol. 3047: F. Oquendo, B. Warboys, R. Morrison (Eds.), Software Architecture. X, 279 pages. 2004.

Vol. 3046: A. Laganà, M.L. Gavrilova, V. Kumar, Y. Mun, C.K. Tan, O. Gervasi (Eds.), Computational Science and Its Applications – ICCSA 2004. LIII, 1016 pages. 2004.

Vol. 3045: A. Laganà, M.L. Gavrilova, V. Kumar, Y. Mun, C.K. Tan, O. Gervasi (Eds.), Computational Science and Its Applications – ICCSA 2004. LIII, 1040 pages. 2004.

Vol. 3044: A. Laganà, M.L. Gavrilova, V. Kumar, Y. Mun, C.K. Tan, O. Gervasi (Eds.), Computational Science and Its Applications – ICCSA 2004. LIII, 1140 pages. 2004.

Vol. 3043: A. Laganà, M.L. Gavrilova, V. Kumar, Y. Mun, C.K. Tan, O. Gervasi (Eds.), Computational Science and Its Applications – ICCSA 2004. LIII, 1180 pages. 2004.

Vol. 3042: N. Mitrou, K. Kontovasilis, G.N. Rouskas, I. Iliadis, L. Merakos (Eds.), NETWORKING 2004, Networking Technologies, Services, and Protocols; Performance of Computer and Communication Networks; Mobile and Wireless Communications. XXXIII, 1519 pages. 2004.

Vol. 3040: R. Conejo, M. Urretavizcaya, J.-L. Pérez-de-la-Cruz (Eds.), Current Topics in Artificial Intelligence. XIV, 689 pages. 2004. (Subseries LNAI).

Vol. 3039: M. Bubak, G.D.v. Albada, P.M. Sloot, J.J. Dongarra (Eds.), Computational Science - ICCS 2004. LXVI, 1271 pages. 2004.

Vol. 3038: M. Bubak, G.D.v. Albada, P.M. Sloot, J.J. Dongarra (Eds.), Computational Science - ICCS 2004. LXVI, 1311 pages. 2004.

Vol. 3037: M. Bubak, G.D.v. Albada, P.M. Sloot, J.J. Dongarra (Eds.), Computational Science - ICCS 2004. LXVI, 745 pages. 2004.

Vol. 3036: M. Bubak, G.D.v. Albada, P.M. Sloot, J.J. Dongarra (Eds.), Computational Science - ICCS 2004. LXVI, 713 pages. 2004.

Vol. 3035: M.A. Wimmer (Ed.), Knowledge Management in Electronic Government. XII, 326 pages. 2004. (Subseries LNAI).

Vol. 3034: J. Favela, E. Menasalvas, E. Chávez (Eds.), Advances in Web Intelligence. XIII, 227 pages. 2004. (Subseries LNAI).

Vol. 3033: M. Li, X.-H. Sun, Q. Deng, J. Ni (Eds.), Grid and Cooperative Computing. XXXVIII, 1076 pages. 2004.

Vol. 3032: M. Li, X.-H. Sun, Q. Deng, J. Ni (Eds.), Grid and Cooperative Computing. XXXVII, 1112 pages. 2004.

Vol. 3031: A. Butz, A. Krüger, P. Olivier (Eds.), Smart Graphics. X, 165 pages. 2004.

Vol. 3030: P. Giorgini, B. Henderson-Sellers, M. Winikoff (Eds.), Agent-Oriented Information Systems. XIV, 207 pages. 2004. (Subseries LNAI).

Vol. 3029: B. Orchard, C. Yang, M. Ali (Eds.), Innovations in Applied Artificial Intelligence. XXI, 1272 pages. 2004. (Subseries LNAI).

Vol. 3028: D. Neuenschwander, Probabilistic and Statistical Methods in Cryptology. X, 158 pages. 2004.

Vol. 3027: C. Cachin, J. Camenisch (Eds.), Advances in Cryptology - EUROCRYPT 2004. XI, 628 pages. 2004.

Vol. 3026: C. Ramamoorthy, R. Lee, K.W. Lee (Eds.), Software Engineering Research and Applications. XV, 377 pages. 2004.

Vol. 3025: G.A. Vouros, T. Panayiotopoulos (Eds.), Methods and Applications of Artificial Intelligence. XV, 546 pages. 2004. (Subseries LNAI).

Vol. 3024: T. Pajdla, J. Matas (Eds.), Computer Vision - ECCV 2004. XXVIII, 621 pages. 2004.

Vol. 3023: T. Pajdla, J. Matas (Eds.), Computer Vision - ECCV 2004. XXVIII, 611 pages. 2004.

Vol. 3022: T. Pajdla, J. Matas (Eds.), Computer Vision - ECCV 2004. XXVIII, 621 pages. 2004.

Vol. 3021: T. Pajdla, J. Matas (Eds.), Computer Vision - ECCV 2004. XXVIII, 633 pages. 2004.

Vol. 3019: R. Wyrzykowski, J.J. Dongarra, M. Paprzycki, J. Wasniewski (Eds.), Parallel Processing and Applied Mathematics. XIX, 1174 pages. 2004.

Vol. 3018: M. Bruynooghe (Ed.), Logic Based Program Synthesis and Transformation. X, 233 pages. 2004.

Vol. 3017: B. Roy, W. Meier (Eds.), Fast Software Encryption. XI, 485 pages. 2004.

Vol. 3016: C. Lengauer, D. Batory, C. Consel, M. Odersky (Eds.), Domain-Specific Program Generation. XII, 325 pages. 2004.

Vol. 3015: C. Barakat, I. Pratt (Eds.), Passive and Active Network Measurement. XI, 300 pages. 2004.

Vol. 3014: F. van der Linden (Ed.), Software Product-Family Engineering. IX, 486 pages. 2004.

Vol. 3012: K. Kurumatani, S.-H. Chen, A. Ohuchi (Eds.), Multi-Agnets for Mass User Support. X, 217 pages. 2004. (Subseries LNAI).

Vol. 3011: J.-C. Régin, M. Rueher (Eds.), Integration of AI and OR Techniques in Constraint Programming for Combinatorial Optimization Problems. XI, 415 pages. 2004.

Vol. 3010: K.R. Apt, F. Fages, F. Rossi, P. Szeredi, J. Váncza (Eds.), Recent Advances in Constraints. VIII, 285 pages. 2004. (Subseries LNAI).

Vol. 3009: F. Bomarius, H. Iida (Eds.), Product Focused Software Process Improvement. XIV, 584 pages. 2004.

Vol. 3008: S. Heuel, Uncertain Projective Geometry. XVII, 205 pages. 2004.

Vol. 3007: J.X. Yu, X. Lin, H. Lu, Y. Zhang (Eds.), Advanced Web Technologies and Applications. XXII, 936 pages. 2004.

Vol. 3006: M. Matsui, R. Zuccherato (Eds.), Selected Areas in Cryptography. XI, 361 pages. 2004.